SECOND EDITION

Near-Infrared Technology

in the Agricultural and Food Industries

Edited by

Phil Williams and Karl Norris

Published by the
American Association of Cereal Chemists, Inc.
St. Paul, Minnesota, USA

Cover images: Top row, reproduced with permission from the Canadian International Grains Institute and the Canadian Grain Commission graphics unit. Bottom row (left to right), reproduced with permission from the Canadian International Grains Institute and the Canadian Grain Commission graphics unit; Copyright Photospin, Inc., 2000/2001; and Rajtar Productions, Phil Aarrestad, photographer.

Library of Congress Card Number: 00-111692
International Standard Book Number: 1-891127-24-1

©1987, 2001 by the American Association of Cereal Chemists, Inc.
First edition published 1987. Second edition 2001.

Reference in this publication to a trademark, proprietary product, or company name by personnel of the U.S. Department of Agriculture is intended for explicit description only and does not imply approval or recommendation of the product by the U.S. Department of Agriculture to the exclusion of others that may be suitable.

Printed in the United States of America on acid-free paper

American Association of Cereal Chemists
3340 Pilot Knob Road
St. Paul, Minnesota 55121-2097 USA

This book is dedicated
to Maxine and Diane.

CONTRIBUTORS

F. E. BARTON, II, U.S. Department of Agriculture, Agricultural Research Service, Richard B. Russell Research Center, Athens, GA, U.S.A.

C. BORGGAARD, Danish Meat Research Institute, Roskilde, Denmark

P. J. BRIMMER, Foss NIRSystems, Inc., Asia-Pacific Operations, Oatley, NSW, Australia

D. J. DAHM, Department of Chemistry and Physics, Rowan University, Glassboro, NJ, U.S.A.

K. D. DAHM, Department of Chemical Engineering, Rowan University, Glassboro, NJ, U.S.A.

F. A. DeTHOMAS, ABB Automation, Analytical Division, Woodstock, MD, U.S.A.

J. W. HALL, Foss NIRSystems, Inc., Silver Spring, MD, U.S.A.

W. R. HRUSCHKA, Instrumentation Research Laboratory, Agricultural Research Service, U.S. Department of Agriculture, Beltsville, MD, U.S.A.

S. E. KAYS, U.S. Department of Agriculture, Agricultural Research Service, Richard B. Russell Research Center, Athens, GA, U.S.A.

H. MARK, Mark Electronics, Suffern, NY, U.S.A.

H. MARTENS, Norwegian Computing Center, Oslo, Norway

W. F. McCLURE, North Carolina State University, Raleigh, NC, U.S.A.

C. E. MILLER, DuPont Engineering Technology, Houston, TX, U.S.A.

T. NAES, Norwegian Food Research Institute, Aas, Norway

K. NORRIS, Private Consultant, Beltsville, MD, U.S.A.

D. L. WETZEL, Kansas State University, Shellenberger Hall, Manhattan, KS, U.S.A.

P. C. WILLIAMS, Grain Research Laboratory, Canadian Grain Commission, Winnipeg, Manitoba, Canada

PREFACE TO THE FIRST EDITION

Near-infrared reflectance technology is the most practicable and exciting analytical technique to hit the agricultural and food industries since Johann Kjeldahl introduced the Kjeldahl test. Furthermore, it's quicker and cheaper than other tests and doesn't use sulfuric acid or sodium hydroxide. Some things are easier to measure with near-infrared than others. Water is easy and methionine is harder.

Imagine yourself walking through a forest. You see lots of trees, big ones and small ones, lots of shrubs and flowers and some grass. You've seen them all before and know what they are. By and by, you see a flower you've never seen before. So you take a book from the library and find out just what that flower is. Then you go back to the forest and, after walking a long time, you see another flower that is the same. You think you know what it is, but you are not quite sure. So you go back to the library and check the book out, then back to the forest. After a while, you get to know that rare flower. Meanwhile, you have seen lots of the trees and shrubs and other flowers that you recognized the first time. It has taken more energy and reading (self-programming) to get to know that rare flower, but then you know it. The observations you have made can be regarded as optical data used for calibration and the reference book as the analytical (wet chemistry) data. Despite the large number of optical data collected during your search, your recognition of the flower will only be as reliable as your reference book.

Near-infrared reflectance spectroscopy also involves optical data, the amount of radiation reflected from a sample relative to the radiation striking it. If you try to look into a room from the outside on a sunny day, it is not easy to see the objects inside the room. Most of the light is reflected back into your eyes, so all that your brain records is the shape of the window. This reflected light is *specular reflectance* or *Fresnel reflectance*. If you shade your eyes so that the sunlight is not reflected directly back at you, you will be able to see into the room and your eyes (the detectors) will receive light from the same light source (the sun) reflected back from objects inside the room. This is analogous to *diffuse reflectance*, since the light reaching your eyes carries information reflected back from the objects inside the room. Your brain (the computer) records what is present in the room because years of observation (calibration) have programmed your brain to recognize familiar signals in terms of colors, shades, furniture, pictures, appliances, people, pets, and other things—even brand names of appliances, names of artists, breeds of dogs, etc. Now suppose the room contains a unique object, such as a new model of a radio or VCR that you haven't seen before; your brain forms an opinion as to what it is but is not really sure, and you are not able to relate (display) exactly what the object is. These objects are analogous to *outliers*, and you will not be able to describe what you are seeing accurately until you have been told or have read about the object and this new knowledge has become part of the enormous bank of data already stored in your memory. The recall mechanism cannot recall knowledge that has not been added to the memory bank. Like the flower in the forest, the object belongs in the room, and you shouldn't remove it.

Near-infrared technology is like a "supermicroscope"—it sees past the surface of the sample and penetrates into the molecules themselves, so that theoretically we can measure everything that's there. All we have to do is teach the instrument to recognize what it is seeing (calibration) and make sure it is telling us the truth about what it sees (verification). There are billions of molecules of protein, water, oil, starch, etc., in the sample that the instrument views, and many of these absorb radiation at the same or closely adjacent wavelengths. We have to enable our instruments to differentiate between them. The best way to achieve this is to ensure that we have told our instrument all it needs to know about the material we expect it to analyze for us. We thus have to know all the features about the material before we can teach the instrument, which amounts to comprehensive sample accumulation and reference analysis by reliable standard methods in equally reliable laboratories. This process is the heart and soul of success in near-infrared technology. The instrument companies have done their part—all modern near-infrared instruments are good, generally more precise in their specifications than the laboratories they can and will replace. So it's up to us to enable these instruments to realize their tremendous potential for saving time, saving money, improving accuracy and precision, saving space, preventing pollution, simplifying operation, demonstrating versatility, and all their other features. The field and its potential have been concisely and comprehensively summarized by D. L. Wetzel (Near-infrared reflectance analysis: Sleeper in analytical techniques, Anal. Chem. 55:1165A-1176A, 1983), who also points out the potential of near-infrared technology in the agricultural and food industries.

We have assembled this book to help in the use of near-infrared instruments and technology. The first five chapters deal with fundamentals: the physics and chemistry without which the biology and its appreciation would not be possible; the mathematics and statistics to reduce the physical and chemical information to its simplest terms (believe me); and the building stones of the computerized spectrophotometer, the "supermicroscope." Chapters 6 and 7 attempt to explain how the system works, what can foul it up, and what we can do to prevent foul-ups and get the best out of the instruments. Chapter 8 deals with the complex material loosely referred to as fiber. It is a transition chapter, leading from the chemistry of the fibrous substances through the instrumental methods available for the study of these and other constituents of biological materials, into the contribution of near-infrared technology.

Chapters 9 through 13 summarize what has been done in terms of published or talked-about work in the near-infrared field. We have not included every published item in these chapters, and those of you who don't see your work referred to, please forgive us: the objective of these reviews is to illustrate the diversity of applications. We have no knowledge of what are probably some of the most ingenious applications, which are specific applications in industry. Also, we have no information from Latin America or from the continent of Africa beyond knowing that a number of instruments have been located there. Chapter 14, on whole grain applications, represents what is in many ways the ultimate application—testing with no sample preparation. We do not refer in detail to applications in industries such as plastics, textiles, pharmaceuticals, or petrochemicals, which are legion. This book deals with applications in the agricultural and food regimes. Principles described apply to any use of near-infrared technology, but industrial requirements involve more on-line installations and different techniques for sampling and sample preparation. Chapter 15 deals with qualitative applications, which are likely to become much more widely exploited in the future, for example, for the identification of materials and physical effects such as weather damage and staling.

Following the chapters are several types of supplementary information. We have included a set of spectra to illustrate what many materials and their constituents look like. The Appendix presents AACC Approved Method 39-10 for near-infrared testing. This method, which is concisely written, contains the basic principles for application to any commodity or constituent. The bibliography contains nearly 1,000 references, including supplementary references that deal mainly with subjects that are fairly closely related to near-infrared technology. The spectra and the bibliography are indexed, to aid the reader in using them. Finally, the index to the text will serve partly as a dictionary because it leads the reader to many definitions.

Some repetition will be apparent, but repetition is the foundation stone of remembrance, and we believe that this book will serve as a starting point for future, more comprehensive works on the subject. We have waited until now to assemble it because a great deal of knowledge has been realized and clarified in the last 12 years. The success of near-infrared technology has been attributed to the development of low-cost amplification, detection, and computing components. In terms of instrumentation, this explanation is true, but the success of the method to us, the users, results from the simplicity of sample preparation and instrument operation and the avoidance of chemicals, polluting or otherwise. Most chemical and instrumental techniques require specialized sample preparation such as drying (moisture reduction), digestion, extractions, and other procedures. Not so with near-infrared technology, since moisture does not prevent accurate testing—it only changes the conditions required, such as the wavelength of measurement. We have attempted to explain all of this in the text. It's been fun writing it and we hope you will enjoy reading it.

Our deepest thanks are accorded to our secretaries, Susan Wiebe and Kathi Suderman, who have prepared the entire manuscript. Debbie Sobering and Tricia Starkey have spent many hours in proofreading and general faultfinding. John Antoniszyn has been involved with both proofreading and discussions on the text. Finally, Sue Stevenson, in addition to her authorship, assembled the bibliography.

Phil Williams
Karl Norris

PREFACE TO THE SECOND EDITION

The original driving force behind compiling the first edition of this monograph was that of providing education in all aspects of near-infrared (NIR) spectroscopy for NIR spectroscopy users from novice to graduate student and beyond. This *raison d'être* has persisted, and we believe that the second edition will contain information of interest and value to the same array of readers.

Since the appearance of the first edition in 1987, NIR spectroscopy has become firmly established as the method of choice for rapid, accurate analysis in the agriculture and food industries, and indeed in a host of other areas, many of which are quite unrelated to agriculture. Because of its speed in testing, its flexibility in sample size and presentation methods (to the instruments), and its relative freedom from the need for sample preparation, NIR spectroscopy has survived the potential threats of related techniques, such as Fourier transform infrared (FTIR) spectroscopy and Raman spectroscopy. Industries, such as the booming pharmaceutical industry, have capitalized on the speed, freedom from chemicals, and flexibility of the technique. Other disciplines have turned to NIR spectroscopy as an analytical tool. For example, NIR spectroscopy is receiving significant attention from environmentalists in monitoring changes in the quality of waterways and lakes and in the disposal of materials such as manures and sewage. The NIR spectroscopy technique is beginning to find applications in medical diagnostics. These and other nonagricultural applications are not discussed in the present monograph—they merit a monograph of their own.

From its modest beginnings about 30 years ago NIR spectroscopy has passed through three "plateaux" of application. The first of these saw NIR spectroscopy in use for prediction of simple constituents, such as protein and moisture in ground materials. The addition of more filters and the introduction of the grating monochromator around 1978 extended the scope of the technique to constituents such as fiber components and in vitro digestibility. Advances in both software and hardware from 1980 onward lead to the third plateau, wherein NIR spectroscopy found applications in the analysis of intact grains and seeds and in the prediction of functionality parameters, such as wheat "strength" and seed texture. The dying years of the past century have seen the technique begin an upward trend toward the fourth plateau, whereupon NIR spectroscopy will be used in grading and classifying materials and in the organoleptic-type categorization of materials and foods.

To the best of our endeavors, we have responded to suggestions received from many readers of the first edition on areas in which improvements could be made. All of the chapters that have been retained have been revised, and some of them completely rewritten. The chapters that described regional NIR spectroscopy research in the original work have been omitted, mainly due to the volume of research that has been described since 1987, the number of publications having ballooned from less than 1,000 to over 11,000 in that time. New chapters have been added. These include chapters on implementation, industrial applications, neural networks, and a new approach to qualitative NIR analysis. The original chapter on whole-grain applications has been omitted, and the salient features concerning whole-grain application have been described in the revised chapter on variables. The bibliography has been passed over to the Council for Near-infrared Spectroscopy. It is now maintained at North Carolina State University, from whence it is available on a compact disk. The annotated collection of spectra is retained, with a few additions.

What does the future portend? The next generation of NIR incumbents may well enter an era wherein the demon calibration will no longer raise its ugly head! Calibration methods will continue to improve as software becomes even more comprehensive than it is at present. Instrument companies will develop or contract the development of calibrations for large-scale applications. Instruments purchased for operating in the field will not require a calibration, but will serve as a medium for scanning samples and relaying the spectral data to the calibration center by e-mail. The results will quickly be returned by e-mail. Instrument performance will need monitoring, but this can also be carried out by arrangement with calibration centers.

Near-infrared spectroscopy is here to stay!

Phil Williams
Karl Norris

CONTENTS

CHAPTER 1

The Physics of Near-Infrared Scattering

DONALD J. DAHM
Department of Chemistry and Physics
Rowan University
Glassboro, NJ
U.S.A.

KEVIN D. DAHM
Department of Chemical Engineering
Rowan University
Glassboro, NJ
U.S.A.

I. INTRODUCTION

This chapter will address the physics involved with light scattering and reflection. Physics, the science of matter and of interactions between matter and energy, is divided into several fields, including electricity, optics, dynamics, and particle science. Near-infrared (NIR) reflectance technology includes the phenomena associated with NIR reflectance itself, the instrumentation, sample preparation laboratory analysis, and data-processing. Of these, physics is involved mainly with the instrumentation and the surface phenomena, including diffuse reflectance. The line between applied physics and engineering is a fine one in which electro-optical instrumentation is concerned, and in this book, instrument design and the manipulation of light and wavelength are accorded separate treatments.

The range of shapes and textures of samples that can be examined using NIR reflectance spectroscopy is virtually endless, and a wide variety of experimental arrangements are possible. In this chapter, we will limit our discussion to samples that are plane parallel: i.e., they have two parallel flat surfaces that are large compared with an incident beam. The direction of illumination will be perpendicular to the flat surfaces, with the surface closest to the light source being called the front surface and the other called the rear surface. There will be references to directed illumination, which means a beam of light with all rays moving parallel to each other and perpendicular to the front surface. On the other hand, reference to diffuse radiation will mean that there is a light flux with individual rays uniformly distributed in all possible directions within 90° of a line normal to the sample surface. As depicted in Figure 1, we assume that the radiation that passes through the sample may be divided into two fluxes traveling at 180° to each other: one in which the radiation is passed *forward* in the direction (*I*) of the incident beam, and another in which the radiation is scattered *backward* in the opposite direction (*J*) toward the illuminated surface. We will refer to light that leaves the sample in the forward direction as being transmitted and to light that leaves the sample in the backward direction as being remitted. The samples that we consider will be either a *slab*

of material with no discontinuities within or *particulate* with many interfaces within the sample.

There will be many references to absorption, remission, and scattering coefficients. In all cases, we will be referring to a natural or linear coefficient that is defined in terms of a small slice of the sample, with the slice being perpendicular to the direction of illumination. An effective linear absorption coefficient, K, will be defined as the fraction of light absorbed, A, by a very small thickness, dx, of a sample divided by the magnitude of that small thickness. Similarly, an effective linear remission coefficient, B, will be defined as the fraction of light remitted, R, by a small thickness of the sample divided by the magnitude of that small thickness. Usually, this will be obtained by extrapolation, as discussed in Section V C below. This is expressed mathematically as

$$K = \lim_{dx \to 0} (A_{dx}/dx) \qquad B = \lim_{dx \to 0} (R_{dx}/dx) \qquad (1)$$

As white light passes through a material it may be selectively absorbed, rendering wavelength dependence to the emergent radiation that is perceived as color. This phenomenon is well known, providing a basis for structural study and quantitative analysis using transmission spectroscopic techniques. The selective scatter of radiation may also impart color. A familiar example is the blue color of the sky, which is a result of the fact that the scatter of sunlight by atmospheric particles is more efficient at shorter wavelengths.

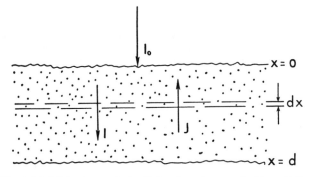

Fig. 1. Model for the travel of radiation through a sample. The radiation field is divided into two components: one traveling forward in the direction of the incident beam (*I*) and one traveling backward toward the illuminated surface (*J*).

Based on THE PHYSICS OF NEAR-INFRARED REFLECTANCE (from the first edition) by Gerald S. Birth and Harry G. Hecht

Both of these phenomena occur within the dense scattering media. If the concentration of scattering centers is sufficiently large, a significant portion of the incident light will be returned to the surface and *remitted*, and the material may be described in terms of its remission properties. The remission of light may range from *specular*, being that of an ideal mirror surface reflecting light only at the single angle dictated by geometrical optics, to *diffuse*, being that of an ideal matte surface having uniform remission at all angles.

Perhaps the first serious attempt to explain the nature of diffuse reflectance was made by Bouguer in 1729. He assumed that a matte surface (e.g., a powder) could be viewed as an assembly of microcrystalline faces statistically distributed over all possible angles. Diffuse reflectance was thought to be a result of the mirror-type reflections from the various crystallites. In other words, Bouguer (1761) viewed diffuse reflectance as strictly a surface phenomenon.

Seeliger (1888) perceived that remission occurred not only at the surface, but from the interior as well. Seeliger considered radiation to penetrate the surface and be partially absorbed, with a portion being returned as a result of reflection, refraction, and diffraction at the various interfaces. An attempt was made by Lommel (1887) to generalize the Seeliger (1888) theory by including the effect of mutual irradiation of each volume element by scatter from every other volume element. The result was too complicated for practical use, but it represented an important conceptual advance in the understanding of the diffuse reflectance process.

It was shown by Pokrowski (1924, 1925, 1926) that both the Bouguer and the Lommel-Seeliger processes are operative in many cases, giving rise to radiation that may be viewed as a superposition of two components. In recent treatments (e.g., Dahm and Dahm, 2001), the origins of these two components are taken as

1. *reflection* (either specular or diffuse) from the external surface of a particle, a process in which NIR radiation is essentially immune from the absorption, and
2. *scatter*, which describes the radiation that enters a particle, is partially absorbed as it transverses the particle, is diffracted by the electron cloud of the molecules, and may be reflected from the internal surfaces of the particle before reemerging from the particle.

There are several theories that satisfactorily describe scattering from individual particles (Bohren and Huffman, 1983). The scatter by very small particles (much smaller than the wavelength of light) is described by Rayleigh's theory and is called Rayleigh scattering. The scatter from very large particles (whose surface is "many square wavelengths of light") may be described by diffraction theory and the theory of geometrical optics (e.g., ray tracing). The scatter from spherical particles is described by Mie's theory. This is most often applied to particles having a size on the same order as the wavelength, since the other size ranges may be well described by simpler treatments.

In reflectance spectroscopy, we are generally concerned about scatter from a collection of particles. In this regard, natural products involve some special considerations. The interaction between light and plant leaves has been a subject of great interest to scientists for many years. To gain an improved understanding of this interaction, Willstätter and Stoll (1918) applied ray tracing to a drawing of a leaf cross section. They proposed a theory to explain reflectance from a leaf on the basis of critical reflection of light at the spongy mesophyll cell wall–air interfaces. Gates et al (1965) measured the reflectance, transmittance, and absorbance of foliage for several plant species in the 400- to 1,000-nm spectral region. This was a major advance in understanding the interaction between light and plant material, because previously published spectra were limited primarily to the visible. A ray tracing included in their report was based largely on critical reflection at cell walls within the leaf, similar to the Willstätter and Stoll approach (1918). Gates et al (1965) postulated that the smaller bodies in the plant cells were of a suitable size to produce considerable light scattering in the leaves. Later, when remote-sensing technology was developed, additional investigations were carried out to gain a more complete understanding of the reflection process from leaves. Kumar and Silva (1973) developed a ray-tracing procedure for a soybean leaf cross section. They ignored the smaller bodies in the cells and considered only air, cell wall, cell sap, and chloroplasts. Using the best estimates of index of refraction for those components of the leaf, they computed the reflectance and transmittance of a leaf to be 45.6 and 54.4% for the nonabsorbing NIR spectral range, i.e., 700–1,300 nm. They compared this to a measured value of 47% reflectance. On the basis of that work (and experimental results by others), they concluded that reflection and refraction at the larger interfaces in the leaf were more significant factors in creating diffuse reflectance than the redistribution of radiation due to scattering by the smaller bodies in the leaves.

Gausman et al (1969) used physical and mathematical models to analyze the reflectance of cotton leaves. In their analysis of leaf reflectance, they concluded that variations in internal structure of upper leaves of cotton plants caused by different salinity stresses resulted in variations of reflectance within the 750–1,300 nm range. The differences in reflectance were attributed to differences in light scattering in the leaves rather than to differences in absorption.

II. PHYSICAL PRINCIPLES

A. Absorption, Remission, and Transmission

Diffuse reflectance spectroscopy is an extension, via instrumentation, of human vision. Instrumentation provides output in numerical form so that mathematical treatments can be used, precise wavelength and radiometric measurements, and the potential for using monochromatic radiation. A primary interest in diffuse reflectance spectroscopy is for the rapid and nondestructive determination of the concentration of certain constituents in a material. The instrumentation and procedures are similar to those used in absorption spectroscopy.

Ideally, the absorption spectrum should be the output of the instrumentation, but most instruments measure the radiation that is not absorbed rather than the radiation that is absorbed. Only if these measurements are made properly can valid absorption spectra be obtained. Fortunately, in many cases, an absolute measure of absorption is not required, but the effect of variables that are dependent on time and wavelength must be eliminated. The objective is to record accurately the wavelength-dependent nature of absorption relative to a stable standard.

According to the Beer-Lambert Law, the concentration of an absorber is directly proportional to the sample absorbance

$$\text{Absorbance} = \log_{10}(I_0/I_t) \quad (2)$$

in which I_0 is the intensity of the incident radiation and I_t is the intensity of the transmitted radiation. This relationship is fundamental to spectroscopy but is strictly applicable only to transmission measurements on samples that have no scatter. However, it is also frequently applied to the diffuse reflectance of light-scattering materials, replacing I_t with I_r, the intensity of the remitted radiation.

The relationship in Equation 2 depends on the Bouguer-Lambert Law, which says that fraction of a light beam absorbed, A, and transmitted, T, while moving through a distance, d, within a continuous (homogeneous) material will be given by

$$\text{Transmission Fraction}: \quad T = \exp(-kd)$$

$$\text{Absorption Fraction}: \quad A = 1 - \exp(-kd) \quad (3)$$

The symbol k denotes the linear absorption coefficient of the *material* in the absence of scatter, which we will refer to as the *absorbing power* of the material. The units of k are the reciprocal of the units used to measure the thickness. Notice that, while base 10 logarithms are usually used to express absorbance, the fundamental relationship is between the *natural* log of the absorption or transmission fraction and the absorption coefficient. (The logarithms are, of course, linearly related through the equality $\ln x = 2.3 \log_{10} x$.)

The situation is more complicated for samples in which scattering also occurs. For a plane-parallel sample (uniform thickness) and a direction of illumination perpendicular to the surface of the sample, there will be a certain fraction of incident intensity absorbed (A), remitted (R), and transmitted (T) by the sample. Assume a sample is divided into two layers, x and y, (in which x is closest to the incident beam) and the A, R, and T fractions for each individual layer are known. The properties for the sample as a whole (indicated by the label $x+y$) are given by the following equations (Benford, 1946):

$$T_{x+y} = T_x T_y / (1 - R_x R_y)$$

$$R_{x+y} = R_x + T_x^2 R_y / (1 - R_x R_y) \qquad (4)$$

$$A_{x+y} = 1 - T_{x+y} - R_{x+y}$$

Note that Equations 4 *do not* require that the properties of the material in layer x are the *same* as those in layer y. For materials composed of multiple *identical layers*, the properties of a sample consisting of $n+1$ such layers can be derived from the properties of a sample consisting of n layers, using Equations 4, and are given by

$$T_{n+1} = T_n T_1 / (1 - R_n R_1)$$

$$R_{n+1} = R_n + T_n^2 R_1 / (1 - R_n R_1) \qquad (5)$$

$$A_{n+1} = 1 - T_{n+1} - R_{n+1}$$

For a sample of a *homogeneous* material twice as thick as a sample of thickness d, use

$$T_{2d} = T_d^2 / (1 - R_d^2)$$

$$R_{2d} = R_d (1 + T_{2d}) \qquad (6)$$

$$A_{2d} = 1 - T_{2d} - R_{2d}$$

For a sample of a *homogeneous* material one half as thick as a sample of thickness d, use

$$R_{d/2} = R_d / (1 + T_d)$$

$$T_{d/2} = \left[T_d \left(1 - R_{d/2}^2 \right) \right]^{0.5} \qquad (7)$$

$$A_{d/2} = 1 - T_{d/2} - R_{d/2}$$

These quantities may be expressed in a form known as the absorption/remission function, $A(R,T)$, which is constant for any thickness of a sample. These relationships may be summarized in the Dahm (Dahm and Dahm, 1999a, b) equation as:

$$A(R,T) = \left[(1 - R)^2 - T^2 \right] / R = (1 + T - R) A / R = (2 - A - 2R) A / R \qquad (8)$$

The absorption/remission function is also proportional to the ratio of both the linear absorption, K, and remission, B, coefficients for the sample and the fraction of light absorbed, A_0, and remitted, R_0, by a sample of infinitesimal thickness, and

$$A(R,T) = 2K/B = 2A_0/R_0 \qquad (9)$$

The absorption coefficient, K, for a sample is different from the absorbing power, k, for the material of which the sample is comprised for several reasons, including

1. *voids*: A scattering sample is likely to contain voids, thus reducing the absorption fraction of a given thickness of material.
2. *surface reflection*: Each particle reflects some of the light incident upon it; thus absorption of this fraction of the light by this particle is prevented.
3. *distance traveled*: The linear absorption coefficient of the sample is dependent on the actual distance that the light travels though a sample (per unit sample thickness). This is different from the distance traveled

through a homogeneous material (in which the distance traveled is equal to the sample thickness), because
   a. scattered light is diffuse and moves through a sample in all directions, and
   b. light is reflected back and forth between the internal surfaces of the particle.

Since scattered light will be diffuse, a simplifying assumption may be made that the incident light will also be diffuse. This eliminates the need to take into account any change in the coherence of the incident beam as it travels through the sample. When this assumption is made, the relationship between the linear absorption coefficient, K, and the absorbing power, k, is given by

$$\exp(-Kd) = {}_0\!\int^{\pi/2} \exp(-kd/\cos\theta) d\theta \qquad (10)$$

For particulate samples, to a good approximation, a $\log(1/T)$ spectrum of a *thin* sample is proportional to k, the absorbing power of the material of which the sample is comprised; and a $\log(1/R)$ spectrum obtained on a *thin* sample in transflection will be proportional to K, the absorption coefficient of the sample. This is illustrated in Figure 2. Note that the two spectra differ by roughly a factor of 2 at all wavelengths.

B. Reflection from a Surface

The reflection and scattering properties of a particle are dependent on the relation between the wavelength of the radiation and the dimensions and orientation of the particle (Wendlandt and Hecht, 1966). Assuming that all particle dimensions are large compared with the wavelength, the reflection from a surface may be calculated by the application of geometrical optics (Bohren and Huffman, 1983). For a given orientation of a particle, two important factors in this calculation are the refractive index and the absorption coefficient of the material that comprises the particle. For a nonabsorbing medium, the refractive index, n, is given simply as the ratio of the velocity of light in a vacuum, c, to that in the sample, υ, so

$$n = c/\upsilon \qquad (11)$$

For specific values of n, a handbook (e.g., Weast, 1984) should be consulted.

An optical interface is the boundary between two media having different indices of refraction. When a radiant beam is incident on an interface, the beam is divided into two parts as shown in Figure 3: the transmitted ray, T, and the reflected ray, R. The direction of the transmitted ray is defined by Snell's Law as

$$(\sin\theta_i) n_1 = (\sin\theta_r) n_2 \qquad (12)$$

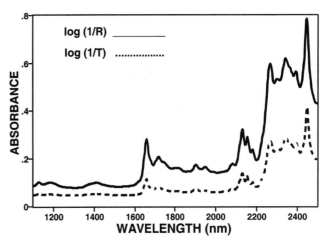

Fig. 2. Absorbance spectra of a plastic sheet collected with transmission and transflection geometry.

The angle of the reflected ray, θ_r, is equal to the angle of incidence θ_i. The fraction of the incident ray that is reflected is a function of the two indices of refraction and can be computed with the Fresnel equations:

$$R_\perp = \left(\left\{ \cos\theta - \left[(n_2/n_1)^2 - \sin^2\theta \right]^{1/2} \right\} \middle/ \left\{ \cos\theta + \left[(n_2/n_1)^2 - \sin^2\theta \right]^{1/2} \right\} \right)^2$$
$$R_\parallel = \left(\left\{ [n_2/n_1]^2\cos\theta - \left[(n_2/n_1)^2 - \sin^2\theta \right]^{1/2} \right\} \middle/ \left\{ [n_2/n_1]^2\cos\theta + \left[(n_2/n_1)^2 - \sin^2\theta \right]^{1/2} \right\} \right)^2 \quad (13)$$

in which $R_\parallel$ is radiation polarized parallel to the plane of incidence, and $R_\perp$ is radiation polarized perpendicular to the plane of incidence.

The plane of incidence is defined by the incident ray and the normal to the interface. The angle of incidence is the angle between the incident ray and the normal to the interface. The reflectance as computed with these equations is illustrated in Figure 4 for angles of incidence from 0–90° for a plane having a refractive index of about 1.5 times its surroundings. The significant points are identified by arrows in the figure.

1. At normal incidence there is no polarization effect and the total reflectance, $R_t = (R_\parallel + R_\perp)/2$, is a minimum.
2. The radiation polarized parallel to the plane of incidence is totally transmitted for an angle of incidence of 56°. Although this angle, Brewster's angle, is a function of the refractive index, it is between 50 and 60° for many common materials.
3. The reflection increases to 100% for high angles of incidence. Except for normal incidence, no radiation propagates away from the interface in a direction coinciding with the direction of the incident ray when n_1 and n_2 differ. The Fresnel equations (Equation 13) yield $R_t = 0$ when $n_1 = n_2$. There is no change in the direction of the transmitted ray in that case. Thus, when a transparent solid is immersed in a fluid having a refractive index identical to the solid, there is no change in the direction of a radiant beam propagating from the fluid into the solid or vice versa. This analysis of change in direction of a radiant beam is an important factor in light scattering associated with diffusely reflecting materials. The regular reflectance (R_r) described here produces the shine or gloss associated with polished surfaces. In literature on color and appearance this is called the *specular component* of reflectance. Other terms used are *Fresnel reflectance* and *surface reflectance*.

C. Absorption, Remission, and Transmission of a Particle

The processes that give rise to scatter are viewed as essentially the same as those (described in Section II B) that give rise to reflection. However, because the processes are taking place within the particle, they are affected by the absorption of the material making up the particle. Even so, scatter is a phenomenon that takes place at an interface. While, as we shall see in Section III, specific orientations and shapes of particles may be exceptions, in general, to a first approximation, the scatter from a particle is dependent on two things: its surface area, and its refractive index (relative to its surroundings).

In this section, we will present a model of absorption, remission, and transmission developed by Dahm and Dahm (2001), based on the work of Simmons (1972, 1975a). Since there is not a definitive model for scatter from a collection of particles, this one should be viewed as illustrative. Other models are described in the references. The specific collection of particles with which we shall be concerned is a layer of spherical particles, so we will begin by considering scattering from a sphere.

The reflectance, r_e, whether diffuse or specular, from the front surface of the particle may be estimated from the refractive index and the cross-sectional surface area that the particle presents to the beam. Figure 5 shows that the reflection from the particle where the angle of incidence is less than 45° to the beam will reverse direction and leave the particle as remitted light. The reflection from any edges of the particle where the angle of incidence is greater than 45° to the beam will pass by the particle as transmitted light. The cross-sectional surface area of a sphere that is at an angle greater than 45° is $1/2$, assuming that the illumination is direct.

Simmons (1975b) estimated that the total reflection from the surface of a particle contributed $3/4 \, r_e$ to remission and $1/4 \, r_e$ to transmission. The plot in Figure 4 shows that the fraction of light reflected is very large for angles of incidence approaching 90°. Consequently, the fraction of the incident intensity that is reflected will be larger for the light rays that contribute to transmission than for those that contribute to remission. Thus, based

I_o = Incident Ray

R = Reflected Ray

T = Transmitted Ray

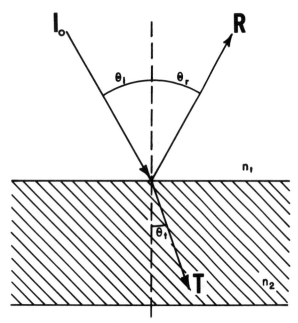

Fig. 3. Reflection and transmission for a ray of optical radiation incident on a plane interface. The media have no absorption but differ in the refractive indexes in which $n_2 > n_1$, I_0 = incident ray, R = reflected ray, and T = transmitted ray.

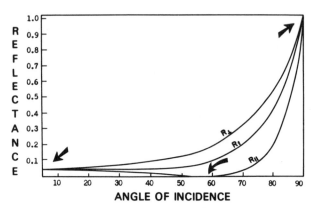

Fig. 4. Intensity of the reflected ray (regular reflection) shown in Figure 3 as related to the angle of incidence. Refractive indices are $n_1 = 1.0$ and $n_2 = 1.5$ for these curves. $R_\perp$ indicates perpendicular to the plane of incidence, $R_\parallel$ indicates parallel to the plane of incidence, and $R_t = (R_\parallel + R_\perp)/2$. Plane of incidence is defined by the ancient ray and the normal to the interface. Radiation not reflected is transmitted through the interface.

on geometrical considerations, even an estimate of $1/2$ r_e to remission and $1/2$ r_e to transmission may be too low. The Melamed (1963) particle theory assigns all the reflection to remission, with the consequence that it fails catastrophically for a material of low relative refractive index.

Within a particle, the incident light is subject to absorption processes. The absorption fraction of a layer of particles is determined by the absorption coefficient of the layer and the thickness of the layer (Equation 3). The average distance through a sphere is $2/3$ that of the diameter, so an estimate for the fraction of light transmitted through a particle, t, may be estimated as

$$t = \exp(-2K_p d/3) \qquad (14)$$

in which K_p is the linear absorption coefficient of the particle. Notice that this does not account for the fact that some distances through a particle are longer than others.

We can apply the mathematics of plane-parallel layers by assuming that a fraction, r_i, of the radiation reflects from an internal surface as the light bounces back and forth between the internal surfaces at the rear and front of the particle. Of the light incident on the particle, the overall remission, R_p, from the particle and transmission, T_p, through and around the particle will be given by

$$R_p = \frac{1}{2}r_e + (1-r_e)(1-r_i)r_i t^2 / (1-r_i^2 t^2)$$
$$T_p = \frac{1}{2}r_e + (1-r_e)(1-r_i)t / 1-r_i^2 t^2 \qquad (15)$$

D. Formation of a Representative Layer

In order to apply meaningfully the mathematics of plane-parallel layers to samples of particulate solids, we need to determine the properties of a layer of the sample from the properties of the individual particles (Dahm and Dahm, 1999b). Each particle may be characterized by its composition, its volume, and the average cross-sectional surface area that it presents to the incident beam. Particles that have the same characteristics belong to a *particle type*.

A layer formed from a single layer of particles may have varying thickness, depending on the diameter of the particles. If the particles are randomly distributed within the sample, each layer of particles will be *representative* of the sample and have the following characteristics.

1. The volume fraction of each particle type is the same in the layer as in the sample as a whole.
2. For all particle types, the cross-sectional surface areas in the layer are in the same proportion as the surface area of the particle type in the sample as a whole.
3. The fraction of the cross-sectional surface area and the volume fraction of the representative layer that is made up of voids are the same as the void fractions of the sample as a whole.

The scatter from a layer of particles will be dependent primarily on the surface area of the particles and the relative refractive index of the material of which they are comprised. The surface area of a particle is proportional to the particle diameter. In a region of low absorption, for a given absorbing power of a material, the absorption fraction of a particle is proportional to its volume. In regions of higher absorption, the absorption fraction of a particle is no longer proportional to the particle volume. As the absorption increases, the absorption becomes less and less dependent on the particle's thickness. In regions of extremely high absorption, the absorption by a particle is proportional to the surface area of the particle and essentially independent of its thickness, since even a very small thickness absorbs essentially all the light that enters a particle.

We can calculate the properties of a representative layer using the following formulas, in which the subscript i denotes a specific particle type, and j denotes a summation over all particle types. The following are definitions.

d_i the thickness of a particle of type i in the direction of the incident beam

ρ_i density of a particle of type i
w_i weight fraction of a particle of type i
v_0 void fraction
v_i fraction of *occupied* volume composed of particle type i
V_i fraction of *total* volume composed of particle type i
s_i fraction of particle surface area that belongs to particle type i
S_i fraction of a cross-sectional surface comprised of particles of type i
k_i the absorption coefficient of particle type i
b_i the remission coefficient of particle type i
$(bd)_i$ the remission power of the material comprising particle type i

Surface area and volume fractions for a given particle type can be computed from weight fractions and particle density as follows.

$$v_i = (w_i/\rho_i)/\sum(w_j/\rho_j) \qquad s_i = (w_i/\rho_i d_i)/\sum(w_j/\rho_j d_j)$$
$$V_i = (1-v_0)v_i \qquad\qquad S_i = (1-v_0)s_i \qquad (16)$$

For a single particle, the fraction of light absorbed is given by the cross-sectional area and the Bouguer-Lambert Law, and the remission fraction is given by the cross-sectional area and remitting power of the material in the particle. Thus, for a representative layer, the absorption (A_1), remission (R_1), and transmission (T_1) fractions are given by

$$A_1 = \sum S_j[1 - \exp(-k_j d_j)]$$
$$R_1 = \sum S_j (bd)_j \qquad (17)$$
$$T_1 = 1 - A_1 + R_1$$

In a representative layer that is one particle thick, for equal weight fraction of two particle size types, the smaller particles make up a larger fraction of the surface area than the large particles. This means that, on a weight basis, smaller particles will be *overrepresented* in the absorption/remission function compared with larger ones. This effect was systematically examined with a combination of image analysis and NIR spectroscopy on mixtures of "fine and course fractions of wheat and rape seed meal" (DeVaux et al, 1995).

In order to allow characterization of the surfaces of the mixtures by image analysis, performed with radiation in the visible range, they were made from "raw materials of contrasting colors.

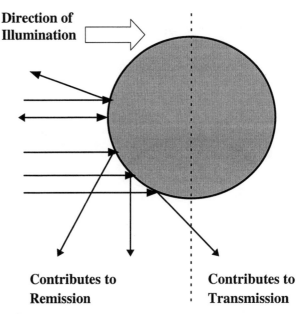

Direction of Illumination

Contributes to Remission

Contributes to Transmission

Fig. 5. Illustration of the fact that regular reflection from the external surface of a particle contributes to transmission as well as remission.

A white product and a black product were chosen: wheat and rape seed meal" (DeVaux et al, 1995). Four series totaling 40 samples were built from "mixtures from 0 to 100% of wheat by steps of 10%: fine wheat with fine rape seed meal, fine wheat with coarse rape seed meal, coarse wheat with fine rape seed meal, and coarse wheat with coarse rape seed meal" (DeVaux et al, 1995). The reflectance of the samples that are mixtures increased with the white wheat proportion and are "not placed on a straight line between the raw fractions" (DeVaux et al, 1995). Figure 6 shows data for the mixtures of fine wheat with fine rape seed meal.

Let us consider the sample surface in a mixture of two particle types, wheat (w) and rape seed meal (r). If the surface is representative of the sample as a whole, the surface layer will be made up of the particle types in proportion to their surface area fraction. Some of the surface will consist of voids, the fraction of which will depend on the sample void fraction (related to bulk density). According to Equations 16, the fraction of surface area of particles that would be observed at a surface which was representative of the sample as a whole would be given by

$$S_w = (1-v_0)(w_w/\rho_w d_w)/(w_w/\rho_w d_w + w_r/\rho_r d_r)$$
$$S_r = (1-v_0)(w_r/\rho_r d_r)/(w_w/\rho_w d_w + w_r/\rho_r d_r) \tag{18}$$

in which the symbols are defined above and the subscripts w and r refer to wheat and rape seed, respectively. The values used for size and density were those reported by Devaux et al (1995).

Equations 17 may be rewritten for this situation as

$$A_1 = S_w a_w + S_r a_r$$
$$R_1 = S_w r_w + S_r r_r \tag{19}$$
$$T_1 = S_w t_w + S_r t_r$$

in which a, r, and t refer to the absorption, remission, and transmission fractions of the individual particles. This neglects the void fraction (or more precisely, in this case, assigns the voids to the wheat fraction).

The fractions A_1, R_1, and T_1 for a hypothetical layer were calculated using the Equations 19, with input values of $a_w = 0.089$, $r_w = 0.43$, and $t_w = 0.48$; with $a_r = 0.83$, $r_r = 0.17$, and $t_r = 0.00$. The R_∞ values for an infinitely thick sample making up the line in Figure 6 were calculated by applying an inverse form of the Dahm Equation 8.

$$R_\infty = \left\{1 + R_1^2 - T_1^2 - \left[\left(1 + R_1^2 - T_1^2\right)^2 - R_1^2\right]^{0.5}\right\} \Big/ 2R_1 \tag{20}$$

Notice that the hypothetical layer of the black rape seed particles would be opaque and the hypothetical layer of wheat particles of

approximately equal size would transmit approximately half of the light. The fact that the calculated line follows the experimental points in Figure 6 is evidence that the weighting by surface area fraction that arises from the representative layer theory is correct.

The preceding situation was one in which the absorption levels were high: a region in which experimenters have encountered a failure of the Kubelka-Munk theory. The situation was described successfully using representative layer theory and the Dahm Equation 8. The next example involves a region of low absorption in the NIR region. In this case, $\log(1/R)$ is a reasonable surrogate for the absorption/remission function, and it is reasonable to assume that the absorption/remission function for an infinitely thick sample is proportional to the ratio of absorption to remission fractions of the representative layer. In regions of low absorption, the fractions may be approximated by

$$A_1 = \sum S_j k_j d_j \qquad R_1 = \sum S_j (bd)_j \qquad T_1 = 1 - A_1 + R_1 \tag{21}$$

and the following conditions will be observed.

1. The contribution of a particle type to absorption is proportional to the volume fraction (including voids) of the particle type and to the absorption coefficient of the material making up the particle.
2. The contribution of a particle type to remission is proportional to the cross-sectional surface area of the particle type in the representative layer and to the remission power of the material making up the particle.
3. In a mixture of two or more particle types of similar remitting power, the absorption/remission function of each particle type is represented in the absorption/remission function of the sample weighted in proportion to surface area-to-volume ratio of the particle type.

The $\log(1/R)$ data at 1,476 nm for various mixtures is shown in Figure 7. This wavelength was chosen because there is an absorption maxima there for rape seed, but the absorption there is low. The series denoted by triangles has wheat and rape seed with the same particle size. Notice that the experimental points generally follow a straight line. The series denoted by diamonds has wheat particles that are twice as large as the rape seed. Here, the points show a departure from linearity, in which the absorbance of the components is weighted in proportion to its total surface area in the sample. (Surface area is inversely proportional to particle size,

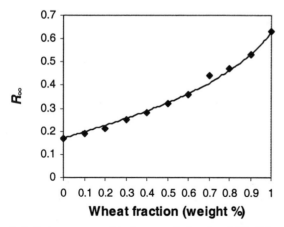

Reflectance at infinite thickness

Fig. 6. Reflectance measured by image analysis using visible light compared with reflectance calculated from the Dahm Equation 8 for mixtures of wheat and rape seed meal.

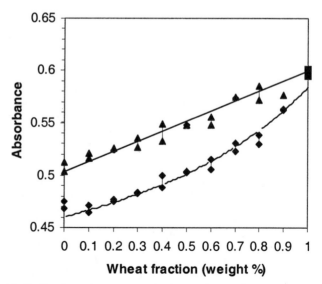

Fig. 7. Absorbance (at a wavelength with low absorption) versus composition of wheat and rape seed mixtures. The lines have been given the shape predicted by the representative layer theory.

so the smaller particles are *overrepresented* compared with their weight fraction.) The squares represent the points in common for the two series.

In addition to showing support for the representative layer theory, this data set illustrates the problems in making reproducible measurements on particle mixtures. The largest source of variation is probably in the void fraction. Notice how the end point of the series with two particle sizes departs from the trend for the central points. In a mixture composed of particles with varied sizes, the small particles fill in the holes between the large ones. This means that samples with only one particle size do not pack as densely as those with two sizes. The effect of this may be explained by the Dahm Equation 8, by including the effects of the voids on each of the fractions. Then, for a sample with the void fraction, v_0,

$$
\begin{aligned}
A(R,T) &= [2-(1-v_0)A-2(1-v_0)R](1-v_0)A/(1-v_0)R \\
&= [2-(1-v_0)A-2(1-v_0)R]A/R \qquad (22) \\
&= [2-A-2R+v_0(A+2R)]A/R
\end{aligned}
$$

This predicts an increase in the absorption/remission function with increasing the void fraction, which is what is observed in the data at the end points.

E. Reflection in Regions of Higher Absorption

A primary concern of reflectance spectroscopy is the relation of reflectance of a material's surface with the absorption of the material. The effect of absorption can be introduced through use of a complex refractive index: $n_2(1-ik')$ in the Fresnel equations, in which k' is known as the absorption index. As absorption increases, the regular reflection remains essentially constant until the absorption index becomes very high. When the absorption index approaches a value large enough to influence the regular reflection, then the absorption fraction is nearly equal to 1.0 for radiation propagating a distance equal to one wavelength into a material. The corresponding absorption coefficient would be of the order 10,000 cm^{-1}. This level of absorption is much greater than that normally encountered with organic materials in the visible and NIR spectral regions.

This can be illustrated by comparing the transmission spectra of a thin layer of water and the reflection spectra (R_r) of a volume of water as shown in Figure 8. Consider the absorption maxima (transmission minima) that occur at 3.0, 4.7, and 6.0 μm. For a 20-μm sample thickness the transmittance is near zero at 3.0 and 6.0 μm, but at 4.7 μm, the transmittance is more than an order of magnitude greater. In these three spectral regions, we see that when the transmittance is essentially zero (and is almost unmeasurable even for a very short pathlength), then reflection is a function of absorption, and information about absorption can be obtained from reflectance measurements. However, in the 4.7-μm wavelength region, the transmittance is relatively high compared with that at 3.0 or 6.0 μm. In such a case, the reflectance does not appear to be a function of absorption, so transmission measurements must be used.

The light remitted by scattering media generally consists of both scattering and regular reflection components. When both are present, the absorption features observed in the diffuse scattering tend to become obscured. The reason for this can easily be seen by a simple qualitative argument. The diffuse component arises through multiple scattering interactions by particles near the surface but inside the material. If the radiation travels a mean pathlength, d, in the sample before it returns to the surface, it would be expected that the diffusely remitted radiation would be attenuated by a factor of $R_d \propto 10^{-k'd}$. Here, k'' is the volume-averaged absorption constant, which is related to the absorption index through $k'' = 4\pi k'/\lambda$. On the other hand, the regular reflectance for normal incidence, corresponding to Equations 4 and 5 for $\theta_i = 0°$ and, taking absorption in medium 2 into account, may be written as

$$
R_r = \left[(n_2-n_1)^2+(n_2k')^2\right]/\left[(n_2+n_1)^2+(n_2k')^2\right] \qquad (23)
$$

We see that $R_d \to 0$ and $R_r \to 1$ as k' becomes large. Hence, these two components are complementary, and as a result, the reflected radiation is less distinct than the true absorption spectrum if both are present. The diffusion of light is an inherent characteristic of the materials concerned, and so methods are considered that may be used to eliminate the regular component.

The scatter of radiation is a result of reflection, refraction, and random diffraction at the surfaces of various particles. The surfaces are assumed to be oriented randomly so that no coherence is maintained in the process and any polarization present in the incident beam is also lost. As the size of the particles is reduced, the number of scattering interfaces is increased, and a greater proportion of the light is returned to the surface without significant penetration of the scattering medium. (This effect is illustrated in Figure 11 in the following section.) If the particle size is reduced sufficiently, the pathlength through the sample is reduced, the absorption is reduced, and the diffuse remission (R_d) becomes sufficiently large so that the regular component (R_r) is negligible compared with it.

In the process of diffusion, the radiation becomes entirely depolarized, but any radiation regularly reflected maintains its state of polarization. Thus, by crossing an analyzer to the polarizer of the incident beam, the regular component can be entirely eliminated. A comparison of reflectance spectra with and without polarizer and analyzer can be used to assess the effectiveness of various efforts to eliminate the regular component. Using such techniques, Kortüm and Vogel (1958) showed that for weakly absorbing materials, such as $CuSO_4 \cdot 5H_2O$, the remitted light is purely diffuse if the material is ground to a fine powder.

For strong absorbers, a large proportion of regular reflection remains even when the particles are reduced to the smallest sizes attainable through grinding (Kortüm and Vogel, 1958). In such cases, it may be necessary to resort to other means. For example, the absorber can be dispersed in a nonabsorbing medium or on the surface of nonabsorbing particles. In extreme cases, it may even be necessary to use polarized irradiation and a crossed analyzer, although this is not usually necessary with natural products for which absorption tends to be rather low.

III. ILLUSTRATIONS OF DIFFUSE REFLECTION

Suppose we have a room with a small window through which direct sunlight is coming. Suppose we have a red table with a relatively smooth surface in the room, and we see a bright area where the direct sunlight is striking the table. We see the details of the surface very clearly in the bright area, more clearly than in the darker area of the surface. The color of the light we see is red. As we move around, the intensity of the light coming off of the spot does not seem to change. The light seems to be scattered by the table equally in all directions. But if we stand in the correct place,

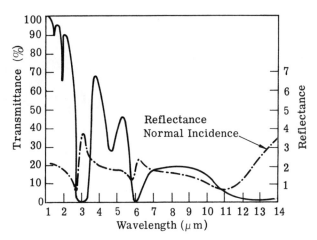

Fig. 8. Transmittance of sea water with a 20-μm pathlength and reflectance of a free sea water surface (Wolfe and Zissis, 1978).

we see a bright glare where light is reflecting off the surface. This light is white, the color of the light coming in through the window, not red, the color of the table. Even if the table were black, the light reflecting from it would still be white.

This simple example leads to the following observations. Light is reflected from a surface at only one angle, with the angle of incidence equal to the angle of reflection. The specular character (color) of reflected light is that of the incident light and does not depend on the absorption characteristics of the object from which the light reflects.

Light is scattered by a surface equally in all directions. The spectral character of the scattered light is that of the incident light, modified by the absorption of the object from which the light is scattered.

We frequently associate the word reflection with the formation of a mirror image. Of course, the rougher the surface of the mirror, the more distortions there are in the image. If the surface is made up of small particles, the distortions are so severe that we see no reflected image at all. But this does not mean that reflection has not occurred. The surface is not one mirror, but many; each placed at a different angle to the light. For each tiny mirror surface, the angle of incidence equals the angle of reflection. Since there are many such angles, the reflections send light back in many different directions. The resulting pattern of reflection is somewhat diffuse. Consequently, it is not always as easy to tell the difference between reflected and scattered light as in the example of the table. However, they still are two distinct manifestations of the effects of interaction of matter and light.

Overall, scientific literature is not uniform in use of the term *diffuse reflection*. For this discussion, we will use the term *regular reflection* to refer to the light that maintains the spectral characteristics of the incident light. The term *scatter* will refer to that light for which the spectral character has been modified by the object, i.e., that radiation which has become diffuse due to the internal structure of the sample.

Effects encountered in measuring absorption when both diffuse and regular components are included in the radiometric measurements are illustrated in Figure 9a and b. A test pattern consisting of a series of straight lines that intersect at a point was photographed using geometrical conditions such that part of the regular reflection was recorded on the film. Since the regular reflection is very directional, only radiation that is directed toward the camera lens would be recorded on film as regular reflection.

Some radiation entered the paper and became diffuse due to light scattering. For areas having no ink, the radiation left the paper without attenuation but was distributed over the entire hemisphere surrounding the illuminated area. The result of this spatial distribution is that less than 0.5% of the radiation entering into the paper actually reaches the camera lens, even though the radiation encountered little or no absorption in the paper. Since

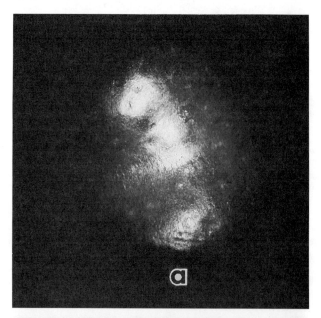

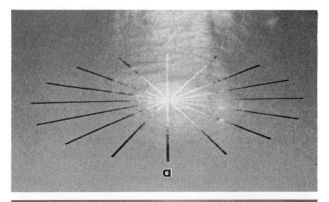

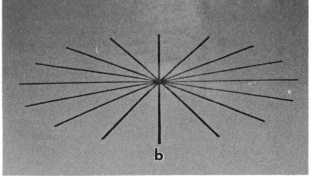

Fig. 9. **a,** Photograph of a test pattern consisting of black lines on white paper and using geometrical conditions required to record the regular reflection (center of the pattern). With regular reflection, more light is reflected by the black lines than by the white paper. **b,** With an angle of incidence near 56°, the regular reflection was blocked with a light-polarizing material. The result is that essentially the only radiation reaching the camera is due to diffuse reflection, so the lines appear completely black.

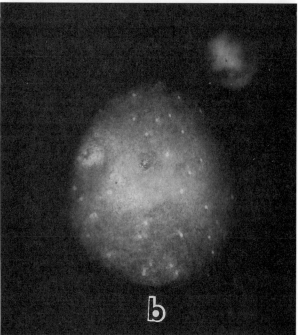

Fig. 10. **a,** Gloss or shine due to regular reflection gives the highlights to a polished apple. **b,** Regular reflection is reduced considerably by photographing the apple when it is in water. Shape of the image is a combination of the diameter of the light beam illuminating the apple and the shape of the fruit. Geometry for the source, apple, and camera is identical for both photographs.

Plate 1. Pulverized green bottle glass. Particle sizes in the respective containers are **A**, less than 0.125 mm; **B**, 0.25–0.50 mm; and **C**, 1.0–2.0 mm.

Plate 2. All three containers have pulverized glass with particles less than 0.125 mm. The media surrounding the particles are **A**, air, $n_1 = 1.0$; **B**, water, $n_1 = 1.33$; and **C**, mineral oil, $n_1 = 1.48$.

the regular reflection was not diffused upon reflection from a smooth surface with high absorption (the black ink), a large fraction of that radiation reached the camera lens and the lines appeared bright. The combination of regular reflection and scatter can result in a condition in which the total remission (over all angles) is independent of absorption. Where the regular reflection was not recorded on film, the lines appeared black because the light was almost totally absorbed by the black ink. If the angle of incidence is equal to Brewster's angle, polarizing the radiation in the vertical plane can eliminate the regular reflectance. This is illustrated in Figure 9b.

The influence of the index of refraction of the surrounding medium is illustrated in Figure 10a in which a polished apple has the shine expected to be associated with a polished surface. Figure 10b is the same apple but photographed under water, where the regular reflectance is reduced by a factor of about 10. Since the regular reflection is directional for smooth surfaces, it can frequently be avoided by proper design of the source-sample-detector geometry. In the absence of regular reflection, the information obtained visually is primarily a response to scatter. That is the process by which the colors of materials such as plant foliage, flowers, textiles, painted surfaces, and the printed word are seen.

Figure 11, which is a series of photographs with water in a glass tank, illustrates how radiation becomes diffuse. A laser is used, with the beam passing vertically through the water.

For Figure 11a, there are no particles in the water and the only place where any radiation was directed toward the camera lens was at the top (air–water) interface and at the bottom of the tank at the water–glass interface.

In Figure 11b, milk was added to the water in the proportion 1:10,000. This created single-particle scattering, in which the radiation was propagated directly from the interaction with particles in the path of the laser beam to the camera lens without any interaction with other particles in the water.

Figure 11c (milk 1:1,000) illustrates the effect of multiple scattering. The radiation is distributed over a larger volume of the water than was occupied by the laser beam, because some radiation reflected or scattered from other particles after the initial interaction with a particle in the path of the laser beam.

In Figure 11d (milk 1:100), multiple scattering predominates and all directional characteristics of the laser beam in the water have been lost. This level of scattering involved a particle density that was much less than encountered in most diffuse reflectance measurements. However, the general semicircular distribution of the radiation is characteristic of the distribution for many materials when photographed in this manner.

In Figure 11e (milk 1:10), the distribution of radiation was similar to Figure 11d, but the radius of the distribution was reduced.

Figure 11f (whole milk, no dilution) showed that the radius of the distribution continued to decrease as the particle density increased. It appeared that all of the radiation left the milk close to the point of incidence. Conversely, the radiation transmitted entirely through the volume of milk was very low.

The procedures for measuring light scattering with diffuse reflectance measurements give a quantity that is proportional to particle density in the sample. The scattering increased from a value of zero in Figure 11a to the highest value for the condition shown in Figure 11f. The scattering was inversely proportional to the radius of the distribution of radiation shown in these illustrations.

Figures 12 and 13 illustrate similar distributions in solid materials. The laser beam was conveyed through a metal tube to avoid an overexposure at the point of incidence.

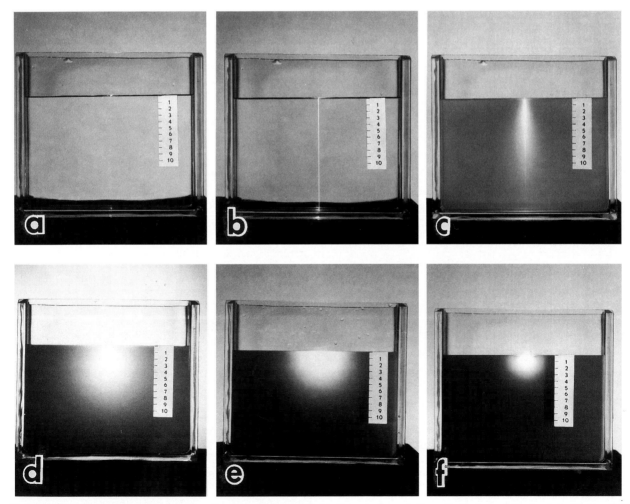

Fig. 11. Laser beam transmitted through water in a glass tank. Milk was added to the water to create the scattering in the following proportions: **a**, 0.0%; **b**, 0.01%; **c**, 0.1%; **d**, 1.0%; **e**, 10.0%; and **f**, 100%. Scales are in centimeters.

Figure 12 shows a potato with the laser beam incident on a cut surface at the top. The distribution of radiation in the flesh was similar to that shown in Figure 11d, e, and f. However, the radius of the distribution should not be compared with the milk, because some room light was used to show the outline of the potato, while the milk was photographed in a darker environment. Actually, the milk (Fig. 11f) had higher light scattering properties and the radius of the distribution was less for the milk than the potato.

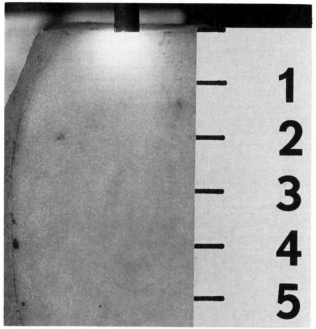

Fig. 12. Semicircular distribution around the point of incidence that is evident in Figure 11 also takes place in solids, as shown here for a potato. Scale is in centimeters.

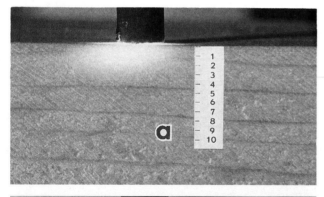

Fig. 13. **a,** Distribution of radiation in wood, a relatively dry material, is elliptical rather than circular, with the long axis parallel to the wood fibers. Scale is in millimeters, so the radius of the distribution is small compared with that shown in Figures 11 and 12. **b,** Illumination on the end grain. Elliptical distribution is still present and parallel to the wood fibers.

Figure 13 is a similar illustration of wood.

Figures 12 and 13 can be compared to observe the distribution of radiation in a high-moisture material (potato) as compared with a dry material (wood). In general, dry materials have higher light scattering properties than materials with high water content.

The cellular structure can influence how the radiation is redistributed in a material. This is illustrated by comparing Figure 13a with b. It can be seen that the radiation propagated along the wood fibers more readily than across the grain of the wood fibers.

In contrast to the change in the distribution of radiation as particle density increases, many materials (Figs. 12 and 13) do not consist of particles but have a cellular structure. The radiation becomes diffuse after reflection at a series of optical interfaces; however, a certain amount of randomness in the position of orientation of the interfaces is essential to bring about complete diffusion. A light ray in an optical glass fiber may undergo millions of reflections and not become diffuse. Figure 14 is a photograph of a laser beam transmitted through an onion skin. The dome-shaped cells act like miniature lenses to refract some of the radiation in a direction away from the centerline of the laser beam. There is a general pattern to the cellular structure, but there are small deviations from that pattern. With additional layers of this cellular structure there would be additional randomization and the net result is complete diffusion of the radiation.

The same distribution of radiation illustrated in Figure 11 can be obtained with solid materials by varying the thickness of the material rather than changing the particle density. The light scattering properties of a solid are considered to be constant, but the transmitted ray must pass through a certain amount of the material to become diffuse. This is illustrated in Figure 15. Potato flesh was used for the scattering medium, and thin slices of tissue were supported on fine wires. The laser beam was conveyed through a metal tube to the tissue as was done for Figures 12 and 13. These photographs were taken with the sample in water with sufficient particles added to create single-particle scattering so that the distribution of the reflected and transmitted radiation could be observed.

In Figure 15a, the slice was 0.025 cm thick. For that thickness there was almost no diffuse reflectance, and the transmitted radiation had predominantly the same directional characteristics as the laser beam.

With a thicker slice, Figure 15b (0.075 cm), the transmitted radiation was partially diffuse but the distribution had a bias in the direction of the laser beam.

Figure 15c (0.16 cm) shows that the transmitted radiation was nearly diffuse or spread out in all directions, and in addition, the reflectance was becoming significant.

In Figure 15d (0.25 cm), the transmitted radiation was essentially diffuse.

Figure 15e and f illustrate that, as thickness increased, the remission increased and transmission decreased. Except for a small amount of radiation that was absorbed, the radiation that was not remitted was transmitted.

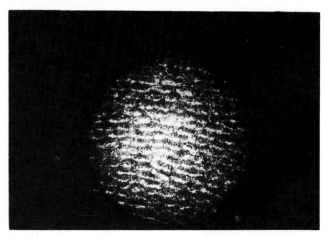

Fig. 14. Laser beam transmitted through an onion skin 0.05 mm thick.

The material thickness required for complete diffusion is a measurable quantity (Birth et al, 1988) and has been termed *diffuse thickness* (Birth, 1982). For potato flesh, the diffuse thickness is 0.37 cm (Birth, 1978).

The combination of light scattering and absorption is of primary interest (Plates 1 and 2). Plate 1 shows the relation between particle size and absorption. Green bottle glass was pulverized and sieved to obtain a range of particle sizes. The effective pathlength increased as particle size increased. As the pathlength of the radiation into the material increased, the absorption due to the green pigment increased, so for large particles, the material appeared dark as a result of the relatively large interaction between the green pigment and the radiation. For smaller particles, the mean pathlength is less and the material appears light, becoming almost white for the smallest particles. The scattering, which is inversely proportional to the mean pathlength, is a result of reflection and refraction of radiation (Fig. 3) at each optical interface. The irregular surface of the particles combined with their random orientation results in the radiation becoming diffuse.

Both the intensity of the reflected ray and the change in direction of the transmitted ray are functions of the change in the index of refraction as the incident ray propagates into the second medium as shown in Figure 3. If there is no change in the refractive index as the radiation propagates from one medium into another, there will be no reflected ray and no change in the direction for the transmitted ray, so the light scattering in these samples of pulverized glass is a function of the refractive index for the medium surrounding the particles. If this medium were changed to a medium having a refractive index closer to the refractive index of the glass particles, scattering would decrease, the mean pathlength

would increase, and the sample would appear darker. Plate 2 shows three samples all having the same particle size but with different media surrounding the particles: (a) air, (b) water, and (c) mineral oil. In each case, the surrounding medium is colorless and free of any particles that could influence the scattering. If the refractive index of the surrounding fluid exactly matched that of the glass particles, the samples would be transparent. When these samples are viewed with transmitted light, samples a and b appear dark, while sample c appears translucent. The physical constants for the glass and the surrounding media are given in Table I. The reflectance of the glass was computed with Equation 23. The green dye in the glass does not absorb sufficiently to influence the regular reflection. This illustration shows why high-moisture materials have lower scattering than the low-moisture materials that are compared in Figures 12 and 13.

IV. THEORETICAL CONSIDERATIONS IN MAKING MEASUREMENTS

The measurement of diffuse reflectance is discussed in Chapter 7 of this book. Here, we will discuss a few points that are critical if the principles described here are to be applied to experimental data.

A. Transmission

In the case of transmission measurements, the intensity of the incident beam is almost always measured directly, and thus the data is on an absolute scale. The are two kinds of detection geometry that are reasonable for transmission data.

A small area detector is used when one desires to approximate the absorption coefficient of the material making up a sample,

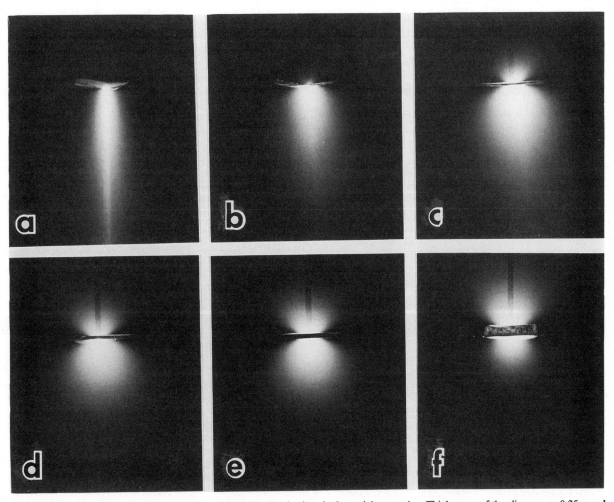

Fig. 15. Potato slices illuminated with a laser and photographed in water having single-particle scattering. Thicknesses of the slices are **a,** 0.25 mm; **b,** 0.75 mm; **c,** 1.60 mm; **d,** 2.50 mm; **e,** 4.20 mm; and **f,** 9.00 mm.

which is the normal case in compositional analysis. It is assumed that a small area detector will detect only the directly transmitted light, and not detect light scattered in a forward direction. If the sample under consideration scatters appreciably, then ~~the~~ should be kept as thin as possible ~~...~~
~~me~~

For ~~reflectance data, hemisp...~~ able. If the data is to be analyzed with any model other than an empirical one, the reflectance must be measured on an absolute scale. For reflectance measurements, a variety of techniques have arisen to do this. When an integrating sphere is used, the incident light beam can be directed toward the sphere wall to determine the intensity of the incident beam at each wavelength. The intensities of the reflectance measurements on a set of samples using that spectrum as a reference are then on an absolute scale.

With other geometries, the remission spectrum of a white ceramic material can be recorded and used as the reference. For empirical uses, the level of reflectance of the standard at every wavelength is frequently arbitrarily set to 100%. Increasingly, reflectance measurements are made on an absolute scale by calibrating the ceramic reference to a reference standard such as Spectralon (Labsphere, Inc., North Sutton, NH). All reflectance calibration tests on Spectralon materials supplied by Labsphere were performed in the Reflectance Spectroscopy Laboratory at Labsphere. The standards used in these tests are directly traceable to The National Institute of Standards and Technology.

V. FUNCTIONAL REPRESENTATION OF ABSORPTION AND SCATTER IN A DIFFUSING MEDIUM

A. The Kubelka-Munk (K-M) Equation

The Kubelka-Munk (K-M) equation (Kubelka and Munk, 1931; Dahm and Dahm, 1999a) provides a model of diffuse reflectance that has been widely used and much misunderstood. While it was derived using differential equations, it can be obtained rather simply by combining the Dahm Equation 8 with Equation 9, which, using the nomenclature of this chapter, yields

$$A(R,T) = \left[(1-R)^2 - T^2\right]/R = 2K/B \qquad (24)$$

For the case of an infinitely thick sample, $T = 0$, so that

$$A(R_\infty, 0) = (1 - R_\infty)^2 / R_\infty = 2K/B$$

A more common form of the K-M equation is obtained when this is divided by two and given the symbol $F(R_\infty)$.

$$F(R_\infty) = A(R_\infty, 0)/2 = (1 - R_\infty)^2 / 2R_\infty = 2K/2B = \mathbf{K/S} \qquad (25)$$

Table I
Data for Pulverized Glass Immersed in Three Different Fluids

Material	Refractive Index	Reflectance of Glass[a]	Sample Reflectance[b]	Scatter Coefficient[c]
Glass	1.52	…	…	…
Air	1.00	0.043	0.74	46
Water	1.33	0.0044	0.44	6
Mineral oil	1.48	0.00018	0.20	1

[a] Computed values for glass in corresponding medium.
[b] The sample is pulverized glass with particles less than 0.125 mm. Reflectance was measured with 1.0-cm sample thickness at 5,500 A.
[c] Computed from Kubelka-Munk theory (Birth, 1986).

The symbols **K** ~~...~~ ing coefficic ~~...~~ cient ~~...~~ ~~use~~ of $2K$ ~~...~~ less complexity and ~~...~~ derivation (Loyalka and Riggs, ~~...~~ generally believed (Wendlandt and ~~...~~ that the K-M absorption coefficient ~~...~~ that of the absorption coefficient of the ~~...~~ sample. The K-M equation was originally ~~...~~ ~~continu~~ous functions, which makes the implicit as~~sumption that the~~ particles in a sample are of infinitesimal size. Kubelka and Munk (1931) also assumed diffuse illumination (for reasons discussed above relative to Equation 6). For diffuse radiation penetrating a continuous, plane-parallel, scattering sample, the average pathlength is twice that of directed radiation. Consequently, in total, the K-M absorption coefficient would be four times the linear absorption coefficient (absorbing power) for the material making up the sample, or $\mathbf{K} = 4k$.

Careful measurements by various workers have shown that the proportionality factor is usually 2 or greater, values of 2.5–3.0 being typical. Since the workers were looking for a factor of 2, they perceived the values as too large and rationalized that the larger values are probably due to the effects of total internal reflection, which tend to increase the pathlength even more for large angles of incidence, but noted that the problem is a complex one involving many factors such as the relative refractive index, particle size and shape, etc. (Hecht, 1976).

For noncontinuous samples, the average path for diffuse illumination is dependent on absorption and is less than twice as long for wavelengths in which there is significant absorption. This was not accounted for in the K-M treatment. As discussed above for the K-M coefficient, the factor would be 4 absent this effect. So we conclude that the results obtained by earlier workers were not surprising.

Since many workers took the K-M absorption coefficient to be directly proportional to the absorbing power of the material and believed that the scattering coefficient would not be dependent on absorption, they were disappointed when plots of the K-M function failed to be linear with concentration at higher absorption levels. This was most often attributed to the effect of regular reflection. Consequently, many believed that if regular reflection were removed from the spectra using techniques described in Section II E, that the function would become linear (Wendlandt, 1966). While this does increase the range of absorption levels over which the function maintains linearity, and certainly makes the data at high absorption levels easier to reliably measure, the function remains nonlinear. The nonlinearity is explained by the Dahm Equation 8. The fact is that both $\log(1/R)$ and the absorption/remission function (along with its special case, the K-M function) are inherently nonlinear as a function of absorption levels in a sample.

B. Using Inherently Nonlinear Functions

If the purpose of an analysis is to measure the concentration of a compound in a sample, the absorption coefficient (absorbing power) of the *material* of which the sample is comprised is the quantity of interest, as it is fundamentally related to the concentration. As noted above, the absorbance obtained in transmission through a nonscattering sample, is linearly related to the concentration of the absorber by the Beer-Lambert Law (Equation 1). If the sample both scatters and absorbs radiation, unless the sample is very thin, the absorbance determined from a transmission measurement is not a linear function of concentration. Further, in some cases, it may not be possible to measure the transmitted radiation in a reproducible manner. In dense scattering media, the pathlength may be difficult to control, and in fact, many systems of practical interest scatter so intensely that very little light is transmitted. For these systems, the reflected radiation must be relied on for an optical analysis.

There is no known way to extract the absorbing power from the reflectance data without resorting to models that make assumptions (or use knowledge) about particle geometry and orientation. Rather sophisticated models have been developed by various workers (Hecht, 1976), but two relationships, log(1/R) and the $F(R_\infty)$ (Fig. 16), both of which are inherently nonlinear with concentration of absorber, are used almost exclusively for NIR reflectance analysis because of their simplicity and broad applicability.

Applying the Beer-Lambert Law to reflectance measurements makes several implicit assumptions, some of which are not particularly good approximations.

1. **The pathlength of the radiation through the medium containing the absorber is constant.** In transmission absorption spectroscopy, the pathlength is maintained constant by the design of the cuvette that contains the absorber. On the other hand, the pathlength of the radiation for diffuse reflectance spectroscopy is a function of the absorption and remission of the particulate material that comprises the sample. This difference in *effective pathlength* constitutes a major difference between diffuse reflectance spectroscopy and absorption spectroscopy.

2. **The absorption coefficient of the sample is proportional to that of the material of which the sample is comprised.** This is true only for situations in which there is a very small absorption loss through the particle.

3. **The scattering coefficient for the sample is independent of absorption.**

In practice, each of the above assumptions is a good approximation over small ranges of sample absorbance. Further, multiple linear regression techniques can correct for some of the effects of the poor approximations. The choice of the function to be a surrogate for the absorbance in diffuse reflection measurements is not necessarily critical to the success of an analytical procedure, but the more nearly linear the function is with respect to the quantity being measured, the better, it is presumed, the linear regression techniques will work.

Despite the fact that an inherently nonlinear relationship is being used, many correlations with the concentration of an absorber show good linearity with computations based on log(1/R). In addition to the fact that the actual change in R for the range of measurement is, in many cases, relatively small, the computations are frequently based on the shape of the reflectance spectrum rather than the specific value of R, by using the first or second derivative of log(1/R). Further, a set of data may be *linearized* by the application of corrections (e.g., the multiplicative scatter correction). The art of applying such techniques is the subject of other chapters in this book (e.g., Chapters 4, 10, and 11).

C. Obtaining Absorption and Remission Coefficients from Reflectance Data

The mathematics of plane-parallel layers have been well worked out (Stokes, 1860; Benford, 1946; Wendlandt and Hecht, 1966; Dahm and Dahm, 1999b). The following formulas derived by Stokes are cumbersome but are an analytical solution to the equation of plane-parallel mathematics described above in Equations 4–7. From the fractions of a layer of a given thickness (given by A_1, R_1, and T_1), we can calculate the fractions for a layer x times as thick (with $A_x + R_x + T_x = 1$) by

$$T_x = (\Omega - 1/\Omega)/(\Omega\Psi^x - 1/\Omega\Psi^x)$$
$$R_x = (\Psi^x - 1/\Psi^x)/(\Omega\Psi^x - 1/\Omega\Psi^x) \qquad (26)$$

Ω and Ψ are defined by

$$\Omega = (1 + R_1^2 - T_1^2 + \Delta)/2R_1$$
$$\Psi = (1 - R_1^2 + T_1^2 + \Delta)/2T_1$$

in which

$$\Delta = [(1 + R_1 + T_1)(1 + R_1 - T_1)(1 - R_1 + T_1)(1 - R_1 - T_1)]^{1/2}$$

R and T may both be determined for a single sample of a finite thickness, and the coefficients will be given by

$$\lim_{K = x \to 0} (A_x/x) \qquad \lim_{B = x \to 0} (R_x/x) \qquad (27)$$

Alternatively, the linear absorption and remission coefficients may be calculated from two samples of different thickness using equations of discontinuous mathematics. Here, we illustrate the procedure for a simple case with one sample twice as thick as the other. If there is a sample of thickness d that has an absorption fraction of A, a remission fraction of R, and a transmission fraction of T, then the fractions absorbed, remitted, and transmitted by a sample of thickness $2d$ will be given by

$$A_{2d} = A_d[1 + T_d/(1 - R_d)]$$
$$R_{2d} = R_d[1 + T_d^2/(1 - R_d^2)] \qquad (28)$$
$$T_{2d} = T_d^2/(1 - R_d^2)$$

If we measure R_d and R_{2d}, the other parameters can be calculated by

$$T_d = [(R_{2d}/R_d - 1)(1 - R_d^2)]^{1/2} \qquad T_{2d} = T_d^2/(1 - R_d^2)$$
$$A_d = 1 - R_d - T_d \qquad A_{2d} = 1 - R_{2d} - T_{2d} \qquad (29)$$

The effective linear absorption and remission coefficients may be calculated by repetitively applying the inverse of the equations

$$A_{d/2} = 1 - T_{d/2} - R_{d/2}$$
$$R_{d/2} = R_d/(1 + T_d)$$
$$T_{d/2} = [T_d(1 - R_{d/2}^2)]^{1/2} \qquad (30)$$

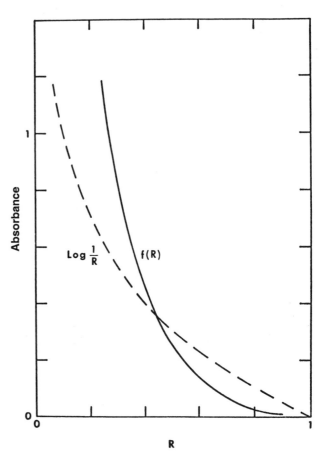

Fig. 16. Comparison of functions log(1/R) and F(R) commonly used to relate near-infrared reflectance to concentration of absorbing material.

until a very small thickness, d_0, is reached; then

$$K = A_{d0}/d_0 \qquad \text{and} \qquad B = R_{d0}/d_0 \qquad (31)$$

In many cases, transflection measurements are more reliable than reflectance measurements from samples of finite thickness. If the remission from the diffuser is measured on an absolute scale and recorded as R_y, and the remission from the diffuser plus a sample is recorded as R_{x+y}, then the two are related by

$$R_{x+y} = R_x + T_x^2 R_y / (1 - R_x R_y) \qquad (32)$$

In Figure 17, we see the absorption and remission coefficients for polyethylene sheets that were determined using the above techniques (Dahm et al, 2000). Notice that the range of absorption coefficient values is approximately 20 times as great as that of the remission coefficients. In general, the remission coefficient shows a shape that correlates inversely with the absorption, but, in the region of highest absorption, the remission coefficient shows a direct correlation. The regular reflection from the front surface causes the direct correlation, and the diffuse scatter from the rear surface causes the inverse correlation. The amount of regular reflection is essentially independent of absorption, while the scatter is greatly affected by it. At high absorption levels there is little scatter, and regular reflection dominates the shape.

In Figure 18, we see absorbance (log[1/R]) data from polyethylene particle samples of 2 mm and 4 mm thicknesses. Similar data was collected on samples of various sizes in the ranges of 0.23–0.28 mm, 0.28–0.38 mm, 0.38–0.52 mm, and greater than 0.52 mm. The particular data shown here is for the largest size fraction. Notice that above the wavelength of 1,700 nm, the 2-mm sample is infinitely thick, so that the spectra of the 2- and 4-mm samples are identical. Consequently, the coefficients can only be calculated for the lower wavelengths. The scattering power of the

sample, taken as the remission coefficient in a region of low absorption, is ~0.6 mm^{-1}. The small ripples in the remission coefficient curve occur because the void fraction and particle size distribution in the two samples are not perfectly matched.

In Figure 19, we see absorption coefficient data calculated for all four particle size ranges. They are displayed against the data calculated for the largest particle size to allow comparison. The data at a particular value of the abscissa all come from the same wavelength of the different samples. At low absorption values, the absorption coefficients determined for each of the sizes are the same. However, at higher absorption levels, the absorption coefficients for the samples of smaller particle size are greater than those calculated of samples having particle sizes that are larger. This trend is predicted by Equation 10 for the smaller particles, the samples reach infinite thickness at lower absorption levels than for the larger particles.

VI. ILLUSTRATIONS OF K-M SCATTERING

A. Scattering from Plastic Particles

Figure 20 displays a series of $F(R_\infty)$ spectra for infinitely thick samples of varying particle size. Also displayed is the absorption coefficient from Figure 17. The magnitude of the function is approximately proportional to the particle size and the absorption coefficient, but, at higher levels of absorption, the linearity is lost as the function levels off. By the K-M equation, $F(R_\infty)$ is equal to K/B. If we make the assumption that K is constant for all particle sizes, we obtain the remission coefficient curve for each particle size shown in Figure 21.

We have seen in Figure 19 that the absorption coefficient at low levels of absorption is approximately the same for all particle sizes. The remission coefficient at low absorption levels is a measure of the scattering power of the material. Figure 22 displays the remission coefficients from Figures 18 and 21. The line labeled "sheet" has been scaled based on what the coefficient would have

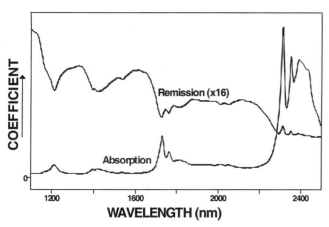

Fig. 17. Absorption and remission coefficients for polyethylene plastic sheets.

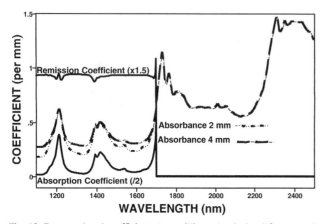

Fig. 18. Data used and coefficients (per millimeter) calculated for a sample of polyethylene particles having diameters on the order of 0.5 mm.

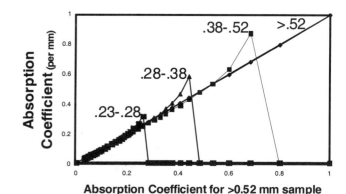

Fig. 19. Absorption coefficients (per millimeter) calculated for various size particles.

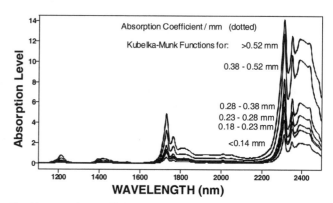

Fig. 20. Absorption coefficient (per millimeter) and Kubelka-Munk ($F[R_\infty]$) Equation 25 functions for polyethylene samples of various particle size.

been if the sheet had the same void fraction as the particulate samples. Notice the inverse correlation between particle size and scatter, as well as the approximate agreement between the coefficients obtained by the two methods for the same samples.

B. K-M Scattering

There have been essentially no results reported on scattering from small grains using the techniques described above to separate absorption and remission coefficients. The equations developed by Kubelka and Munk for analysis of diffuse reflectance also provide for computation of an absorption coefficient (K) and a scatter coefficient (S), and some experimental K-M coefficients have been obtained (Birth, 1986).

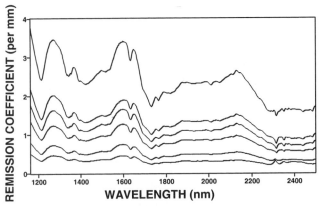

Fig. 21. Remission coefficients calculated from the Kubelka-Munk Equation 25 (smallest particle size at top, largest at bottom).

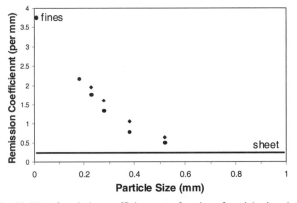

Fig. 22. Plot of remission coefficients as a function of particle size, determined from the Kubelka-Munk Equation 25 (dots) and from Equation 29 (diamonds).

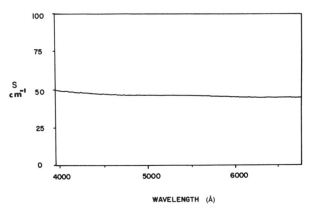

Fig. 23. Light scattering characteristics of pulverized glass having a particle size less than 0.125 mm. Data were computed with the Kubelka-Munk Equation 25 using reflectance and transmittance measurements. Standard deviation for these data is 2.43.

Data for a sample of pulverized green bottle glass was recorded to evaluate the procedure. The sample consisted of all particles that passed through a sieve having 0.0125-cm-diameter holes. Reflectance and transmittance measurements were recorded for five preparations of the sample. Each set of data was used to compute spectral curves of absorption and scattering coefficients. The five data sets were used to compute the average curve and the standard deviation. The results are shown in Figure 23. An important aspect of these results is the comparison between the absorption coefficient and the transmittance of the intact colored

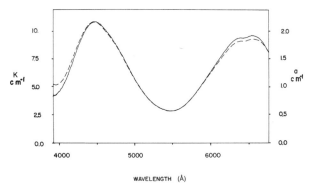

Fig. 24. Spectral absorption (K) characteristics of a pulverized green bottle glass (dashed line) and the absorptivity data (a) for an intact clear glass (solid line). Coefficient of variation for the pulverized glass was 4.3%.

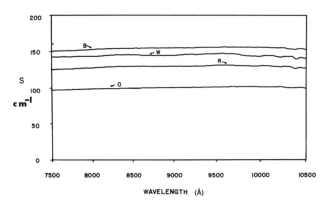

Fig. 25. Kubelka-Munk scatter coefficient (S) versus wavelength for four ground grains. Since there is relatively small wavelength dependence, a mean and standard deviation were computed for each grain. Results are given in Table II. Letters identify the specific grains.

Table II
Particle Size Distribution of Ground Grain

Grain	Sieve Size (mm)				
	1.0	**0.5**	**0.25**	**0.125**	**>0.125**
Oats	0.01[a]	0.11	0.38	0.31	0.19
Barley	0.02	0.14	0.31	0.19	0.33
Rye	0.01	0.06	0.22	0.24	0.47
Wheat	0.01	0.05	0.16	0.20	0.58

[a] Weight fraction of the sample remaining on the respective sieve.

Table III
Average Kubelka-Munk Scattering Coefficients (S) of Four Grains

Grain	Ground Grain			Intact Grain		
	Mean S (per cm)	**SD[a] (per cm)**	**d[b] (cm)**	**Mean S[a] (per cm)**	**SD[b] (per cm)**	**d[c] (cm)**
Oats	100	6.5	0.15	8.0	0.44	1.0
Barley	154	16.9	0.10	8.9	0.33	0.5
Rye	129	12.7	0.20	7.7	0.22	1.0
Wheat	145	15.9	0.20	6.9	0.13	1.0

[a] SD = Standard deviation.
[b] d = Sample thickness used for recording data.

glass expressed on an absorptivity scale. This comparison is illustrated in Figure 24. The agreement is good; however, the ratio between the two scales is about 5.0, whereas a value of 2.0 is predicted theoretically. Kortüm (1969) has shown results in which this ratio was 2.8. The difficulty of attaining the theoretical value was discussed above.

The scatter coefficient is not expected to have much wavelength dependence for a sample having particles several times larger than the wavelength of the radiation. However, the index of refraction does decrease with increasing wavelength, and this could explain a slight decrease in the scatter coefficient with increasing wavelength. The spectral scatter coefficients for four small grains (after grinding) are shown in Figure 25. Since the coefficients exhibit very little wavelength dependence, the mean

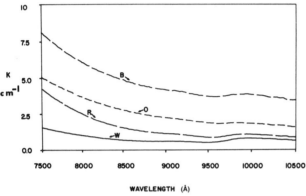

Fig. 26. Spectral absorption coefficient curves for the ground grain. Coefficients of variation for these data were barley, 5.1%; oats, 5.8%; rye, 8.5%; and wheat, 11.2%.

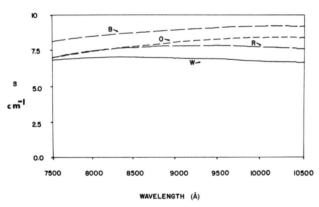

Fig. 27. Spectral scatter coefficient data of four grains (barley, oats, rye, and wheat) using intact grain. Averages and standard deviations are available in Table II.

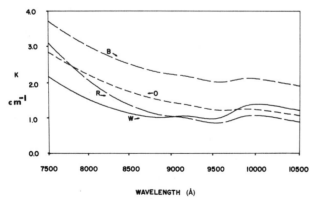

Fig. 28. Spectral absorption data of four grains using intact grain. The coefficients of variation for these samples were barley, 3.6%; oats, 5.7%; rye, 7.3%; and wheat, 2.8%.

values and standard deviations were computed. Those quantities are given in Table II, and the particle analysis is available in Table III.

The absorption coefficients for the ground grain are shown in Figure 26. The weak absorption near 1,000 nm is probably attributable to the presence of carbohydrates. The values for barley and oats are probably too high, but the value for wheat is reasonable. Since the absorption coefficients vary with wavelength, the standard deviation is expressed as a percentage of the mean, i.e., the coefficient of variation.

The K-M coefficients for intact grain are shown in Figures 27 and 28. These data indicate that the light scattering is less than $^1/_{10}$ of that for the ground samples. This is understood by recognizing that ground samples are composed of smaller particles than intact grain. Smaller particles have a higher surface area-to-volume ratio than larger ones, so that samples composed of small particles contain more scattering surfaces per unit mass than do large particles.

There is a large difference in the scatter coefficients of the ground grain and the intact grain, while the absorption for wheat is nearly the same for both conditions, being slightly higher for intact grain. The absorption curves should be nearly the same, since the same material was used for both samples. However, the quantity of material used was lower for the ground grain, so that sufficient radiation would be transmitted to be measurable. On the other hand, the value of K for ground barley was nearly twice the value obtained using intact grain. In addition to the differences in the size of the voids in the two sample preparations, the distribution of constituents is also much different; so although the two sets of curves are expected to be similar, one should not expect them to be the same.

VII. SUMMARY

Only one theoretical analysis of diffuse reflection has been described in this chapter: that of plane-parallel layers. (The K-M treatment is plane parallel, but uses continuous mathematics.) An alternate treatment is provided by Burger et al (1997). During the intervening years between the first and second editions of this book, considerable progress has been made in understanding the phenomena of diffuse reflectance. Even so, the approach most often used (and that gives usable results) is empirical. When the problem under consideration involves light scattering as well as absorption, the techniques described here to determine absorption and remission coefficients for a sample will be very beneficial. It is likely that, in the next several years, considerations such as those presented here will become more prevalent in compositional analysis.

ACKNOWLEDGMENTS

The data presented in Figures 17–22 were collected by Karl Norris, who also was primarily responsible for selecting the samples used to test our theory. We appreciate the kindness of Marie-Francoise DeVaux and Dominique Bertrand in making available their wheat–rape seed data, which we found very useful for understanding the effects of particle size distribution on absorbance. We appreciate the support of Foss/NIRSystems, in whose laboratories our experimental work was conducted.

LITERATURE CITED

Benford, F. 1946. Radiation in a diffusing medium. J. Opt. Soc. Am. 36:524.

Birth, G. S. 1978. The light scattering properties of foods. J. Food Sci. 43:916-925.

Birth, G. S. 1982. Diffuse thickness as a measure of light scattering. Appl. Spectrosc. 36:675-681.

Birth, G. S. 1986. The light scattering characteristics of ground grains. Int. Agrophys. 2:59-67.

Birth, G. S., Fyhn, P.-G., and Frank, J. 1988. Instrumentation to measure the diffuse thickness of scattering dispersions. Opt. Eng. 27:403-408.

Bohren, C. F., and Huffman, D. R. 1983. Absorption and Scattering of Light by Small Particles. John Wiley & Sons, New York.

Bouguer, P. 1760. Essai d'optique sur la gradation de la lumière. In: Traite d'optique sur la gradation de la lumiere. P. Bouguer and N. L. de la Caille, eds. De l'imperimerie de H. L. Guerin & L. F. Delatour, Paris. English translation by W. E. Knowles Middleton. 1961. In: Optical Treatise on the Gradation of Light. University of Toronto Press, Toronto, Canada.

Burger, T., Ploss, H. J., Kuhn, J., Ebel, S., and Fricke, J. 1997. Diffuse reflectance and transmittance spectroscopy for the quantitative determination of scattering and absorption coefficients in quantitative powder analysis. Appl. Spectrosc. 51:1323-1329.

Dahm, D. J., and Dahm, K. D. 1999a. Bridging the continuum-discontinuum gap in the theory of diffuse reflectance. J. Near Infrared Spectrosc. 7:47-53.

Dahm, D. J., and Dahm, K. D. 1999b. Representative layer theory for diffuse reflectance. Appl. Spectrosc. 53:647-654.

Dahm, D. J., and Dahm, K. D. Discontinuum theory of diffuse reflection. In: Handbook of Vibrational Spectroscopy. Vol. 2. J. M. Chalmers and P. R. Griffiths, eds. John Wiley & Sons, Ltd., London. In press.

Dahm, D. J., Dahm, K. D., and Norris, K. H. 2000. Test of the representative layer of diffuse reflectance using plane parallel layers. J. Near Infrared Spectrosc. 8:171-181.

DeVaux, M.-F., Nathier-Dufour, N., Robert, P., and Bertrand, D. 1995. Effects of particle size on the near-infrared reflectance spectra of wheat and rape seed meal mixtures. Appl. Spectrosc. 49:84-91.

Gates, D. M., Keegan, H. J., Schleter, J. C., and Weidner, V. R. 1965. Spectral properties of plants. Appl. Opt. 4(1):11-20.

Gausman, H. W., Allen, W. A., and Cardenas, R. 1969. Reflectance of cotton leaves and their structure. Remote Sens. Environ. 1(1):19-22.

Hecht, H. G. 1976. The interpretation of diffuse reflectance spectra. J. Res. Natl. Bureau Standards 80A:567-583.

Kortüm, G. 1969. Reflectance Spectroscopy: Principles, Methods, Applications. Springer-Verlag, New York.

Kortüm, G., and Vogel, J. 1958. Über reguläre und diffuse reflexion an pulvern und ihre abhängigkeit von der korngröbe. Z. Phys. Chem. N.F. (Frankfurt) 18:230-241.

Kubelka, P., and Munk, F. 1931. Ein beitrag zur optik der farbanstriche. Z. Tech. Phys. 12:593-601.

Kumar, R., and Silva, L. 1973. Light ray tracing through a leaf cross section. Appl. Opt. 12(12):2950-2954.

Lommel, E. 1887. Die photometrie der diffusen Zuruckwerfung. Sitzungsber. Munch. Akad. II. Kl. 17:95.

Loyalka, S. K., and Riggs, C. A. 1995. Inverse problem in diffuse reflectance spectroscopy: Accuracy of the Kubelka-Munk equations. Appl. Spectrosc. 49:1107-1110.

Melamed, N. T. 1963. Optical properties of powders. J. Appl. Phys. 34:560-570.

Pokrowski, G. I. 1924. Zur theorie der diffusen lichtreflexion. I. Z. Phys. 30:66-72.

Pokrowski, G. I. 1925. Zur theorie der diffusen lichtreflexion. II. Z. Phys. 35:34-37.

Pokrowski, G. I. 1926. Zur theorie der diffusen lichtreflexion. IV. Z. Phys. 36:472-476.

Seeliger, R. V. 1888. Zur Photometrie zerstreut reflectirender Substanzen. Sitzungsber. Munch. Akad. II. Kl. 18:201-248.

Simmons, E. L. 1972. Relation of the diffuse reflectance remission function to the fundamental optical parameters. Opt. Acta 19:845-851.

Simmons, E. L. 1975a. Diffuse reflectance spectroscopy: A comparison of the theories. Appl. Opt. 14:1380-1386.

Simmons, E. L. 1975b. Modification of the particle-model theory of diffuse reflectance properties of powered samples. J. Appl. Phys. 46:344-349.

Stokes, G. G. 1860-1862. Proc. Royal Soc. (Lond.) 51:545.

Weast, R.C., ed. 1984. Handbook of Chemistry and Physics. CRC Press, Cleveland, OH.

Wendlandt, W. W., and Hecht, H. G. 1966. Reflectance Spectroscopy. Interscience, New York.

Willstätter, R., and Stoll, A. 1918. Untersuchungen uber die Assimilation der Kohlensause. Springer-Verlag, Berlin.

Wolfe, W. L., and Zissis, G. J., eds. 1978. The Infrared Handbook. The Infrared Information and Analysis (IRIA) Center, Environmental Research Institute of Michigan, Ann Arbor, MI.

CHAPTER 2

Chemical Principles of Near-Infrared Technology

CHARLES E. MILLER
DuPont Engineering Technology
Houston, TX
U.S.A.

I. INTRODUCTION

A. Name Dropping

Many current users of the spectral region of 780–2,500 nm might take for granted that the name of the measurement technology they use is "near-infrared (NIR) spectroscopy." This name would seem logical, because the technology that exploits the neighboring spectral region of 2,500–25,000 nm (or 4,000–400 cm^{-1}) has long been called "infrared (IR) spectroscopy" or "mid-infrared spectroscopy." Those who might question, or even resent, this linkage to a more established analytical measurement technology might suggest other names for spectroscopy in the 780–2,500-nm range, such as "far-visible spectroscopy" or "overtone vibrational spectroscopy."

There are some who might call the name "near-infrared" a euphemism because it leads one to believe that it refers to a slightly modified version of "infrared spectroscopy," which has been shown to be a very reliable primary technique for both qualitative and quantitative analysis. In addition, perhaps some associate the name "near-infrared" with some of the practical advantages of the technology discussed elsewhere in this book, such as the large penetration depth that enables bulk material sampling and the wide availability of reasonably priced optical materials, including fiber optics. In contrast, other possible names that are more associated with the chemistry and spectroscopy of the technique, such as "overtone vibrational spectroscopy" and "anharmonic vibrational spectroscopy," might lead one to realize that the chemical roots of the technology are rather complicated, in that they arise from nonidealities in the phenomenon of molecular vibration, which is discussed below. Although it is natural to want to avoid these complicated roots, they are very effective at explaining the nature of what many NIR method developers consider to be unexplainable: the NIR spectrum.

If one is armed with a basic understanding of the source of NIR spectra, one is more capable of optimizing empirically derived NIR methods, improving the confidence of existing NIR methods, and more effectively interpreting results of NIR feasibility assessments. This chapter discusses the essential aspect of NIR spectroscopy that the salesperson did not want you to know: the chemical basis of the technology.

B. The Size and Speed of NIR

In Chapter 1, a discussion of the physics of NIR spectroscopy dealt with the interaction of light with matter at the micron scale. In contrast, this chapter discusses physical light–matter interaction phenomena that occur at the atomic and molecular levels. Not only do these interaction phenomena occur over very small distances, they also occur much too quickly to be observed by most laboratory equipment. As a result, one must rely heavily on theory

to understand these phenomena, which constitute the chemical basis for NIR spectroscopy. Although this might seem uncomfortable to the typical NIR user, who is accustomed to developing methods empirically rather than theoretically, it is the key to understanding NIR spectra.

In general, molecules can be thought of as having two types of properties: static properties and dynamic properties. The static properties include atomic composition, conformation, isomeric structure, stereochemistry, and morphology. The dynamic properties, such as molecular translation, rotation, and vibration, can also be considered as forms of molecular energy.

Of course, the users of NIR technology are most often interested in the static properties of molecules, rather than their dynamic properties. However, it is the dynamic properties that are directly responsible for the existence of NIR spectra. Fortunately, a molecule's dynamic properties are greatly influenced by its static properties, and this provides the basis for NIR analysis, as well as many other analytical techniques. In light of this, the attempt to understand the origin of NIR spectra requires two different tasks: (i) to understand the mechanisms by which NIR light can be used to detect dynamic properties (most often, molecular vibration), and (ii) to understand the relationship between a material's static properties and its molecular vibration properties.

II. THE SPECTROSCOPY OF NIR

The word "spectroscopy" is derived from the Latin root *spectrum* (appearance, image) and the Greek root *skopia* (to view). This definition is rather descriptive of the spectroscopic measurement itself: to view a light image coming from a specimen. An explanation of the *nature* of images observed from a specific sample, however, requires much more description. From the law of conservation of energy, we can be certain that the specimen does not create the light form itself. Instead, it must produce the light form by selective *absorption* of light entering the sample or by *emission* of certain light forms as a result of the sample losing some of its internal energy. In more general terms, spectroscopy involves energy transfer between light and matter.

Because most practical NIR methods to date involve absorption (i.e., the observation of the change in light properties as a result of interaction with a specimen), the discussion below is focused mainly on absorption methods.

A. Light Energy

Light can be considered to have both particle properties and wave properties. The concept of light energy results from the combination of the particle and wave properties of light: the energy of a single light particle, or photon (E_t), is defined as

$$E_t = h\nu = hc\overline{\nu}^* = hc/\lambda \tag{1}$$

in which h is Planck's constant, ν is the frequency of the light wave, $\bar{\nu}^*$ is the wavenumber of the light, λ is the wavelength of the light wave, and c is the speed of light (2.998×10^8 m/sec in a vacuum). Note that the energy of the light particle depends on its wave properties (wavelength, wavenumber, or frequency). Of course, these three wave properties are directly related to one another, as shown by

$$\nu\lambda = c \tag{2}$$

$$\bar{\nu}^* = 1/\lambda \tag{3}$$

It is worth noting that the speed of light (c) is not a constant, but rather a parameter, which depends on the refractive index of the medium within which the light is traveling. As a result, the relationship involving the frequency and wavelength (Equation 2) actually varies as the refractive index of the medium changes.

The type of radiation referred to as near-infrared is commonly defined to be all wavelengths between 780 and 2,500 nm. This range corresponds to frequencies (in a vacuum) between 3.84×10^{14} and 1.20×10^{14} Hz (or sec^{-1}) and to wavenumbers between 12,820 and 4,000 cm^{-1}. To convert from nanometers to wavenumbers, simply multiply the reciprocal of the wavelength in nanometers by 10^7.

$$\text{wavenumbers} = 10^7/\text{nanometers} \tag{4}$$

B. Vibrational Molecular Energy

As mentioned above, molecular energy can take many different forms, such as translational, rotational, electronic, and vibrational. Only vibrational and electronic energy forms greatly influence spectroscopy in the NIR spectral range, although rotational energy can influence the NIR spectra of gas-phase samples. Of these, vibrational energy is the predominant form and is the primary focus of this chapter. However, a short discussion on electronic energy and its relationship to NIR spectroscopy will be made, and a prime example of rotational energy effects in gas-phase spectra will be illustrated.

Vibrational energy refers to the oscillations of atoms through their bonds in a molecule, much like two balls could oscillate if they were connected by a spring. The energy of a system consisting of two atomic "balls" attached by a bond "spring" is

$$E = \frac{h}{2\pi}\sqrt{\frac{k}{\mu}} \tag{5}$$

in which k is the force constant of the bond between the two atoms (which is used to express the bond strength) and μ is the reduced mass, as shown by

$$\mu = \frac{m_1 m_2}{m_1 + m_2} \tag{6}$$

in which m_1 and m_2 are the masses of the two atoms (Fig. 1). Molecular vibrations can be effectively described using the harmonic oscillator model (Herzberg, 1950; Wilson et al, 1955; Chang, 1971; Sverdlov et al, 1974; Murray and Williams, 1987; Colthup et al, 1990; Duncan, 1991; Bonanno et al, 1992), which assumes that the potential energy of a vibrating system (V) at any given time is a quadratic function of the displacement of the atoms involved in the vibration, as shown by

$$V = \frac{1}{2}kx^2 \tag{7}$$

in which x represents the displacement of the atoms from their equilibrium position and k is the restoring force constant. Figure 1 shows the potential energy as a function of displacement for the ball and spring model discussed above.

Although the classical ball and spring model can be used to describe the concept of vibrational molecular energy, a quantum theory model is needed to determine the specific energy levels that are possible for a particular vibration. If this is done (Herzberg, 1950; Wilson et al, 1955; Chang, 1971; Sverdlov et al, 1974), one finds that the vibrating system does not have a continuum of vibrational energy levels but rather a set of discrete, quantized energy levels, defined by the equation

$$E_v = (v + 1/2)h\nu \tag{8}$$

in which v is the vibrational quantum number for the vibration, E_v is the energy of the vth quantum level of that particular vibration, and ν is the *fundamental frequency* of the vibration (equal to $1/2\pi \sqrt{k/\mu}$).

Although the diatomic molecule is particularly useful for demonstrating the concept of vibrational energy, most real molecules contain more than two atoms and are capable of more than one type of vibration per bond. For example, in the benzene molecule, the CH groups can stretch, bend in the plane of the ring, or bend out of the plane of the ring (Fig. 2). In addition, vibrations of different bonds in a polyatomic molecule cannot always be considered independent of one another, especially if adjacent bonds have comparable strengths and reduced masses. If one were to actually construct a prototype polyatomic molecule with balls and springs, one would most likely find that each bond in the molecule does not vibrate independently, but can be influenced by other vibrations in the molecule. As a result, many real molecular vibrations are described in terms of coupled vibrations of two or more bonds in the molecule. Although coupled vibrations are not discussed here in detail, two general rules regarding them are useful to remember:

1. bonds with hydrogen atoms tend to vibrate independently of the rest of the molecule, because the hydrogen atom is usually so much lighter than any of the other atoms in the molecule; and
2. the vibrations of adjacent bonds of identical or similar bond strength and reduced mass tend to be coupled with one another.

Some common examples of coupled vibrations are the vibrations labeled "amide I" and "amide II" for amino acids, polypeptides, and proteins, which involve simultaneous vibrations of the C=O, C-N, and N-H bonds (Fig. 3).

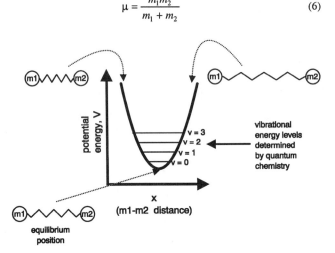

Fig. 1. Schematic of the harmonic oscillator model; potential energy versus atomic displacement for a diatomic molecule m_1–m_2.

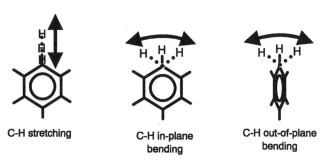

C-H stretching   C-H in-plane bending   C-H out-of-plane bending

Fig. 2. Three different vibrations of the CH group in benzene.

C. Vibrational Spectroscopy—Made Simple

ENERGY COUPLING

Vibrational spectroscopy, in general, exploits the coupling of light energy with vibrational molecular energy. According to quantum theory (Herzberg, 1950; Chang, 1971), a molecule is rather selective about the light that it will absorb: only light with energy that corresponds to the difference between two of its energy levels can be absorbed. Put mathematically, the energy of the light photon (Equation 1) must equal the energy of a vibrational transition, which equals the energy difference between two vibrational states. If harmonic vibrations are assumed (i.e., if Equations 7 and 8 hold), this energy difference (ΔE) is

$$\Delta E = E_{v2} - E_{v1} = \Delta v h \nu \qquad (9)$$

in which Δv is the change in the vibrational quantum number for the vibrational transition. Note that, in this case of the harmonic oscillator, the energy difference between any two adjacent vibrational levels is constant.

Because light energy is directly related to the frequency or wavelength of the light (Equation 1), it can be concluded that only certain light frequencies or wavelengths can be absorbed by a particular molecule. It is this very selective nature of light–molecule interaction, imposed by quantum theory, that makes analytical spectroscopy possible. A classical analogy to this selective interaction is the concept of physical resonance. For example, consider a long beam spanning two supports (Fig. 4). Although any periodic force could be applied to the beam by pushing down and then releasing at the beam center, only those periodic forces that are applied at a resonance frequency of the beam will be absorbed by the beam and cause it to oscillate. The possible resonance frequencies of the beam are dictated by the properties of the beam itself (its length, width, thickness, elasticity, etc.) and *not* by the frequency or intensity of the applied forces.

The matching of light energy with molecular energy is necessary, but not sufficient, for absorption; there must also be a "handle" by which the energy can be transferred. In the beam analogy above, this "handle" can be the direct physical contact of someone's hand on the center of the beam. If the beam were constructed of a magnetic material, a time-varying applied magnetic field could also cause it to oscillate. In the case of molecular vibrations, this "handle" is molecular polarity. Because the electrical charge density of a molecule is seldom distributed equally over the molecule, there can be several dipolar sites on the molecule, which can be "grabbed onto" by an oscillating electric field. As a result, only those transitions between energy states of molecular vibrations that correspond to a change in the molecule's polarity, or *dipole moment*, cause an absorption of radiation. For example, carbon dioxide, a linear triatomic molecule, has several possible vibrational modes (Fig. 5). However, the symmetric stretching vibrational mode (ν_1) cannot absorb a photon to rise to a higher energy level because this vibration does not involve a change in the dipole moment of the molecule. In contrast, the bending and asymmetric stretching vibrations (ν_2 and ν_3) involve changes in the molecular dipole moment and thus can absorb light to go to a higher energy level. For larger and more complex molecules that do not possess a high degree of symmetry, every vibration has at least a small dipole moment change associated with it, and thus every vibration has some capability of experiencing energy transitions by absorbing light photons. However, the *degree* of dipole moment change with the vibrational transition has a large impact on the *intensity* of light absorption, as discussed in a later section.

QUANTUM THEORY AND GROUP THEORY

For molecules that are not too complicated, one can apply quantum theory to determine whether a specific vibrational transition is "allowed" or "forbidden." According to quantum theory, a vibrational transition is allowed only if the following expression is nonzero:

$$\mu_{nm} = \int \Psi_n \mu \Psi_m d\tau \qquad (10)$$

in which μ_{nm} is called the transition electric dipole moment integral, Ψ_n is the wave function of the original vibrational state, μ is the electric dipole moment, and Ψ_m is the wave function of the final vibrational state (Herzberg, 1950; Chang, 1971). For rotational motion of a molecule, μ is constant. However, for vibrational motion of a bond in the molecule, μ is a function of the bond length, r, as shown by

$$\mu(r) = \mu_0 + \left(\frac{d\mu}{dr}\right)_0 q + \frac{1}{2}\left(\frac{d^2\mu}{dr^2}\right)_0 q^2 + \dots \qquad (11)$$

in which $q = r - r_e$, and r_e is the equilibrium distance of the bond. Using only the first two terms of Equation 11 as an approximation, we get

$$\mu_{nm} = \int \Psi_n \left[\mu_0 + \left(\frac{d\mu}{dr}\right)_0 q\right] \Psi_m d\tau \qquad (12)$$

Since Ψ_n and Ψ_m must be mathematically orthogonal to one another and μ_0 is a constant, it follows that $(d\mu/dr)_0$, or the dipole moment change with the vibration, must be nonzero in order for the whole expression to be nonzero and allow the transition.

This "selection rule" is strictly defined in terms of quantum theory, which might be rather intangible to many practical NIR users. However, it is not necessary to know the exact wave func-

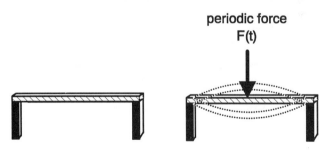

Fig. 4. The oscillation of a beam, as a result of a periodic applied force.

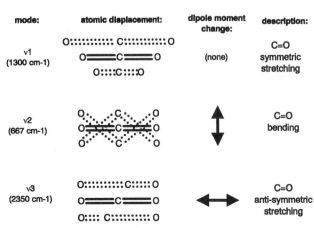

mode:	atomic displacement:	dipole moment change:	description:
ν_1 (1300 cm-1)		(none)	C=O symmetric stretching
ν_2 (667 cm-1)			C=O bending
ν_3 (2350 cm-1)			C=O anti-symmetric stretching

Fig. 5. The three fundamental vibrational modes of the CO_2 group: their geometric representation, their dipole moment change, and their common description.

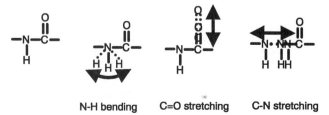

N-H bending   C=O stretching   C-N stretching

Fig. 3. Coupled vibrations in the amide group (the "amide I" and "amide II" vibrations), in terms of individual functional group vibrations. "Amide I" vibration (~1,650 cm^{-1}): ~80% C=O stretching; "amide II" vibration (~1,550 cm^{-1}): ~60% N-H bending and ~40% C-N stretching.

tions of the vibrational states to determine whether a transition is allowed or forbidden, but only the *symmetry properties* of the wave functions. Because of this, the problem of determining whether a transition is allowed or forbidden can be attacked by using a mathematical tool that deals only with symmetry properties of functions, called *group theory*.

When quantum and group theories are applied to cases in which all molecular vibrations are assumed to be harmonic (i.e., can be described by Equations 7 and 8), two main conclusions can be made about vibrational spectroscopy (Herzberg, 1950; Wilson et al, 1955; Chang, 1971):

1. vibrational transitions that involve a change in vibrational quantum number greater than one are forbidden; and
2. vibrational transitions that involve a change in vibrational quantum number equal to one, but do not satisfy symmetry criteria outlined by group theory, are also forbidden.

If it is also assumed that all molecular vibrational transitions start at the lowest possible vibrational energy states (i.e., $v = 0$), which is most often the case, then conclusion 1 can be modified to read "*only transitions from the $v = 0$ to 1 state, or the 'fundamental' vibrational transitions, are allowed.*" Because these fundamental transitions for most functional groups are in the energy range corresponding to light in the mid-IR range (4,000–400 cm^{-1} or 2,500–25,000 nm), this would essentially eliminate the possibility for vibrational absorptions in the NIR range. We all know, however, that this is not the case.

ABSORPTION INTENSITY

Most of the discussion above focused on the energies or positions of vibrational absorption bands, rather than on the intensities of the bands. Once again, quantum theory can be used to derive an expression for the intensity of a band corresponding to a vibrational transition:

$$\text{Band intensity} \propto \left(d\mu/dQ\right)_0^2 \qquad (13)$$

in which μ is the dipole moment and Q represents what is called the *normal coordinate* of the vibration (Herzberg, 1950; Chang, 1971). The concept of normal coordinate is particularly useful in the quantum theory explanation of molecular vibrations. Although the vibration of a molecule can be rather complex, any vibration can be expressed as a combination of "basis set" vibrations. There is one particular basis set of vibrations that enables molecular oscillation where each nucleus in the molecule moves in phase, and this is the set of *normal modes*. A normal coordinate can be thought of as simply a linear combination of different geometric properties of the molecule (e.g., bond angles and bond lengths). In the case

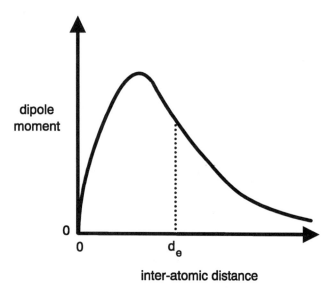

Fig. 6. Dipole moment as a function of interatomic distance for a typical heteronuclear atomic pair.

of the linear CO_2 molecule discussed above, the three normal modes of vibration are schematically expressed in Figure 5. In the case of diatomic molecules, the normal coordinate of the stretching vibration collapses down to a well-understood spatial coordinate, which is the interatomic distance. A more detailed discussion of normal coordinates can be found in several references (Herzberg, 1950; Wilson et al, 1955; Chang, 1971). For this chapter, it is simply sufficient to know that the magnitude of a normal coordinate represents the "progress" of a molecular vibration. With this in mind, one can see that Equation 13 essentially states that the intensity of a particular vibration is proportional to the change in the dipole moment as the vibration progresses, evaluated at the equilibrium position of the atoms.

D. Vibrational Spectroscopy—Made Complicated

ANHARMONIC VIBRATIONS

Vibrational spectroscopy made complicated is NIR made possible. The whole existence of vibrational overtone and combination bands, and thus NIR vibrational spectroscopy itself, relies on nonidealities or deviations from the assumptions regarding molecular vibrations discussed above. More specifically, the presence of vibrational overtone and combination bands depends on two main deviations of the behavior of *real* molecular vibrations from the assumptions discussed above:

1. *mechanical anharmonicity*: most real molecules undergo anharmonic, rather than harmonic, vibrations; and
2. *electrical anharmonicity*: for all heteronuclear diatomic pairs in a molecule, the dipole moment is not exactly a linear function of interatomic distance.

Mechanical anharmonicity is a necessary consequence of the fact that atomic nuclei, when pressed sufficiently close together, experience a strong repelling force and, when separated far enough apart, eventually dissociate. In this case, the potential energy of the molecule at any given time is not simply a quadratic function of the displacement, as in the harmonic oscillator model (Equation 7), but can be better approximated by using higher-order terms of displacement:

$$V = k_1 x^2 + k_2 x^3 + k_3 x^4 + \dots \qquad (14)$$

in which x once again represents the displacement of the atoms from their equilibrium positions. It is important to note that, according to Equation 14, higher-order terms become especially important for large displacements of the atoms from their equilibrium positions.

Electrical anharmonicity is a reflection of the nonlinear relationship between dipole moment and atomic displacement. For a heteronuclear diatomic molecule, the dipole moment at the point of coincidence of the atoms should be zero and should also be zero when the atoms are infinitely separated (Fig. 6). Between these two extremes, the dipole moment goes through a maximum value. In all known cases, the maximum in the dipole moment occurs at an atomic separation slightly smaller than the equilibrium atomic separation.

CONSEQUENCES OF ANHARMONICITY

There are several important consequences of mechanical and electrical anharmonicity that make NIR possible:

1. overtone transitions, which involve an increase in a vibrational quantum number of greater than one, are allowed to occur;
2. combination modes, which involve a simultaneous increase in the vibrational quantum numbers of two or more different vibrations from absorption of a single photon, are allowed to occur; and
3. the separation of the vibrational energy states of a given vibration are *no longer equal*, as in the harmonic oscillator case (Equation 8).

Detailed discussions regarding these consequences of anharmonicity, in terms of both classical and quantum mechanics, can be found in several references (Herzberg, 1950; Wilson et al,

1955; Chang, 1971; Murray and Williams, 1987). Only a brief and general discussion is given here.

The higher-order terms in the potential energy function (Equation 14), caused by mechanical anharmonicity, cause the wave functions of the vibrational states to be altered such that the transition probability (Equation 10) becomes nonzero for two vibrational states that are separated by more than one quantum level. In the simplest of cases, that of the diatomic molecule, the potential energy function can be fitted by a Morse function of the form

$$V = D_e \left[1 - e^{-a(r-r_e)} \right]^2 \qquad (15)$$

instead of Equation 7. In Equation 15, a is a constant for a given molecule in a specific electronic state, D_e is the spectroscopic dissociation energy, r_e is the equilibrium separation of the two atoms in the molecule, and r is the separation of the two atoms in the molecule. When quantum theory is applied to this potential function, the resulting vibrational energy levels are

$$E = h\nu(v + 1/2) - x_m h\nu(v + 1/2)^2 \qquad (16)$$

in which x_m is called the anharmonicity constant of the vibration, which is most often in the range of 0.005–0.05 (Murray and Williams, 1987). Like ν in Equations 8 and 16, the anharmonicity constant x_m is characteristic of the specific vibration and is observed to vary substantially between different vibrations and different molecules. Figure 7 shows the potential energy function and quantum theory-derived vibrational levels for an anharmonic oscillator. Note that the separation of adjacent vibrational levels is no longer constant.

Furthermore, for polyatomic molecules, the higher-order anharmonic terms in Equation 14 cause the individual vibrations to no longer be independent, in that they can now interact with one another. This is reflected in the total vibrational energy function, which now contains cross-terms from different vibrations in the molecule (Duncan, 1991):

$$E_v = \sum h\nu_r(v_r + 1/2) + \sum \sum hx_{rs}(v_r + 1/2)(v_s + 1/2) + \dots \qquad (17)$$

$$(r \le s)$$

in which ν_r and v_r are the fundamental vibrational frequency and vibrational quantum number of vibrational mode r, respectively, and x_{rs} is the anharmonicity constant for the interaction of vibrational modes r and s. The main consequence of this is that the normal vibrations can oscillate at frequencies that are integer multiples and summations of the fundamental vibrational frequencies (i.e., overtone and combination bands are allowed).

It is important to note that, even if a vibration is mechanically harmonic (i.e., vibrations could not interact and each obeyed Equation 7), electrical anharmonicity would still cause the presence of

overtone and combination bands (Herzberg, 1950), through similar higher-order and interaction terms in the molecular dipole moment function.

The frequencies of overtone bands are approximately equal to integer multiples of the frequency of the fundamental vibrational transition. For functional groups with higher-energy fundamental bands (for example, CH, NH, and OH groups), such overtone bands appear in the NIR spectral region. The frequency of a combination band is approximately the summation of the frequencies of the different vibrations that make up the combination. Combinations of fundamentals with overtones are possible, as well as combinations involving more than two different vibrations. However, not just any two vibrations in a molecule can produce a combination band; the vibrations must involve the same functional group and have the same symmetry. As a result, combination bands involving vibrations of two different functional groups (e.g., CH and OH) are not possible, and even two vibrations of the same functional group might not combine (Bonanno et al, 1992).

A good example of the symmetry rule for combination bands is given by Bonanno et al (1992). The methylene (-CH_2-) group has several different vibrational modes (Fig. 8). However, only two vibrations that have the same symmetry can make a combination band. The -CH_2- symmetric stretch (2,870 cm^{-1}) and -CH_2- bend (1,460 cm^{-1}) are symmetric vibrations in which all three atoms remain in the same plane. As a result, these two vibrations can combine to create a combination band at approximately 4,310 cm^{-1} or 2,320 nm, which is a characteristic NIR band of many hydrocarbons, polymers, and natural products. It should be noted, however, that the symmetry of an overtone is not necessarily the same as the symmetry of the corresponding fundamental. Therefore, such a "visual" symmetry approach to combination band activity assessment cannot be done in the case of combinations involving overtones.

Mechanical anharmonicity also causes the spacing of adjacent vibrational energy levels to no longer be constant. From Equation 16, the difference in energy between vibrational energy level 1 and vibrational energy level j is

$$E_{1 \text{ to } j} = h\nu(j-1) - h\nu x_m(j-1)^2 \qquad (18)$$

As a result, the frequencies of overtone bands are not exact integer multiples of the fundamental vibrational frequency, but must be adjusted due to anharmonicity. Although exact values for x_m have been calculated for only a few vibrations, two qualitative rules can be helpful:

1. vibrations involving hydrogen atoms (C-H, N-H, and O-H) tend to be very anharmonic; and
2. asymmetric stretching vibrations tend to be more anharmonic than symmetric stretching vibrations of the same group (Hollas, 1987; Bonanno et al, 1992).

In general, the intensities of overtone and combination bands are at least one order of magnitude weaker than fundamental vi-

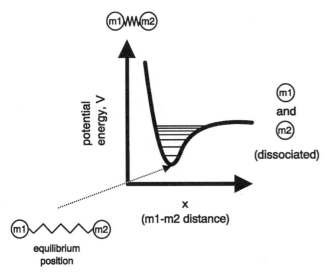

Fig. 7. Schematic of the anharmonic oscillator model; potential energy versus atomic displacement for a diatomic molecule m_1–m_2.

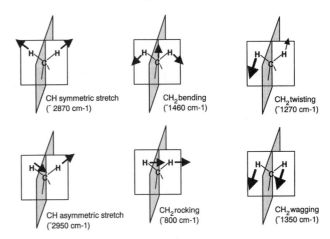

Fig. 8. The different fundamental vibrational modes of the -CH_2- group: their geometric representation, their approximate frequencies, and their common description.

brational bands (observed in the mid-IR spectrum). Furthermore, these bands tend to decrease drastically in intensity as the order of the overtone or combination (i.e., the overall or cumulative change in the vibrational quantum number or numbers) increases. This is the case because higher-order approximations in the potential energy function (Equation 14) are required to get higher-order overtone and combination bands. This effect is well illustrated in the NIR and mid-IR spectra of *n*-heptane (Fig. 9). As one goes to higher-order overtones and combinations, the sample pathlength must be increased to get absorbance bands of sufficient magnitude.

Unlike the intensities of fundamental vibrational bands, the intensities of overtone and combination bands depend on the *degree of anharmonicity* as well as the dipole moment change associated with the vibration. As a result, some of the most intense overtone and combination bands involve the stretches of CH, NH, and OH groups, which are very anharmonic. However, for symmetric molecules, there are even cases in which only the even- or odd-numbered overtones are allowed and others are not allowed, thus resulting in "alternating" overtone bands (Herzberg, 1950). Likewise, many combination bands are not allowed because the contributing vibrational modes do not have the same symmetry, even if they belong to the same functional group.

COMPLICATING AND DECOMPLICATING EFFECTS: RESONANCE AND THE LOCAL MODE PHENOMENON

Anharmonicity enables not only the presence of overtone and combination bands but also the occurrence of resonance between different vibrations within the same functional group. Such resonances are made possible by cubic, quartic, and higher-order terms in the potential energy functions (Equation 14) and can occur if the vibrational modes have the same symmetry. Fermi resonance, a direct consequence of cubic terms in the potential energy functions, can occur whenever the energy of a fundamental transition is close to the energy of an overtone or combination transition in the same molecule. Under "normal" (nonresonance) circumstances, the fundamental band is much more intense than the overtone or combination band. However, if Fermi resonance occurs, the intensities of the two bands become very similar and their frequencies are more separated than expected. In essence, the overtone or combination mode "borrows" intensity from the fundamental mode, thus resulting in two bands of more comparable intensity. A more detailed discussion of Fermi resonance can be found in several references (Herzberg, 1950; Chang, 1971; Bonanno et al, 1992).

One of the most definitive cases of Fermi resonance is observed in the Raman spectrum of carbon dioxide. The fundamental vibrations of CO_2, and their corresponding frequencies, are shown in Figure 5. Note that the first overtone of the v_2 mode would be expected at approximately $2 \times (667 \text{ cm}^{-1}) = 1,334 \text{ cm}^{-1}$, which is close to the frequency of the fundamental v_1 mode $(1,300 \text{ cm}^{-1})$. Furthermore, when group theory is applied, it is found that the first over-

tone of v_2 (or $2v_2$) and v_1 have the same symmetry. (It should be noted that, because these modes do not involve a change in dipole moment with the vibration, they are not allowed in the IR absorption spectrum but are observed in the Raman spectrum.) As a result, instead of a strong fundamental band at $1,300 \text{ cm}^{-1}$ and a weak overtone band at $1,334 \text{ cm}^{-1}$, two bands of comparable intensity in the Raman spectrum are observed at $1,285$ and $1,388 \text{ cm}^{-1}$, due to Fermi resonance between the overtone $2v_2$ and the fundamental v_1. These bands represent the resonance-weakened fundamental C=O symmetric stretching and resonance-enhanced C=O bending overtone band, respectively. A similar discussion of Fermi resonance effects of the -CH_2- group vibrations in the IR spectra of hydrocarbons has also been given (Bonanno et al, 1992).

Darling-Dennison resonances (Duncan, 1991; Bonanno et al, 1992) are similar to Fermi resonances, except that they involve interactions between different overtone transitions in a molecule, and consequently rely on quartic and higher-order terms in the potential energy function (i.e., higher orders of anharmonicity). The observation of such resonances is particularly evident for X-H vibrations, because the interacting energy levels lie close to one another and the vibrational anharmonicity is large. For Darling-Dennison resonances, the interacting levels tend to remain at similar separations at each level of excitation, but the interaction parameters tend to increase with increasing excitation (due to increased anharmonicity). As a result, as the level of excitation increases, the vibrational levels become more displaced from their expected positions and the description of the levels in terms of normal modes becomes less accurate. The most common examples of Darling-Dennison resonances are found in the spectrum of water (Duncan, 1991; Bonanno et al, 1992).

For the NIR user, the main practical consequence of Fermi and Darling-Dennison resonances, with regard to the interpretation of NIR spectra, is that one must be wary of the possibility of such effects causing the presence of two bands where only one is expected. This, of course, can further complicate NIR spectra, especially for lower overtone and combination bands in the region of 1,600–2,500 nm. However, there is actually a "decomplicating" effect in NIR spectroscopy that becomes apparent at higher overtones and combinations (typically, 700–1,600 nm). The origin of this effect is currently the subject of debate, but it appears to be related to what is called the "local mode" phenomenon (Henry et al, 1980; Duncan, 1991; Bonanno et al, 1992).

The spectrum of polyethylene from 850–2,500 nm (11,765–4,000 cm⁻¹) can be used to illustrate the complicating effects of resonance and the decomplicating effects of the local mode phenomenon (Fig. 10). For a polyethylene with sufficiently high molecular weight, one would expect to observe only overtones and combination bands involving the -CH_2- stretching and bending vibrations. In the mid-IR region, one clearly observes separate fundamental vibrational peaks for the symmetric and asymmetric -CH_2- stretching modes and for the bending and wagging -CH_2- modes (Fig. 8). In other words, if only fundamental vibrational bands in the mid-IR region are considered, one can associate an absorbance band with a *specific vibration* of a functional group. In the "first combination band region" (Fig. 10A) in the NIR spectrum (2,200–2,500 nm or 4,545–4,000 cm⁻¹), one observes two prominent bands, as well as a rather broad absorbance (most likely the result of multiple overlapped bands) at higher wavelengths. The prominent band at approximately 2,310 nm can be assigned to the specific combination of the -CH_2- symmetric stretch and the -CH_2- bend, as discussed earlier. However, the overlapped higher-wavelength bands cannot be definitively assigned to such a combination. In principle, these bands could be "pure" overtones of lower-frequency -CH_2- bending and wagging modes. However, because the fundamental bending and wagging vibrational transitions correspond to rather low frequencies (Fig. 8), any bands in this NIR region would have to be higher-order overtones, which would not be expected to show significant intensity relative to the combination band at 2,310 nm. It is more likely, however, that these bands are the result of Darling-Dennison resonance effects between these high-order overtone modes and the more intense combination mode. This pattern is also observed in the "first-overtone C-H region" (Fig. 10B), where two

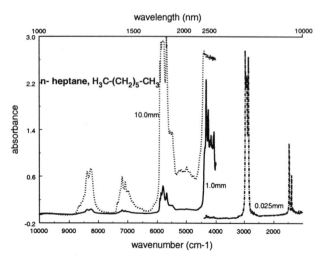

Fig. 9. Near-infrared spectra of liquid *n*-heptane at three different pathlengths: 0.1, 1.0, and 10.0 mm.

prominent (and slightly resolved) first-overtone bands corresponding to the -CH₂- symmetric and asymmetric stretching vibrations are seen, with a series of weaker overlapped bands at lower wavenumbers due to resonance between different overtone and combination modes. Note that the lower-wavelength overtone band for the asymmetric -CH₂- stretch is significantly stronger than the neighboring overtone band for the symmetric stretch. This is believed to be the case because this vibration is more anharmonic than the symmetric stretch. Note also that the NIR spectrum in this area is slightly less complicated than the "first-combination region" (Fig. 10A).

As one progresses to higher-order overtone and combination bands, located at the lower-wavelength NIR regions (850–1,300 nm, 11,765–7,100 cm⁻¹) (Fig. 10C and D), the spectra appear to become less and less complicated. At the "third-overtone region" (850–1,000 nm), there appears to be only a single major band. It is believed that this simplifying effect is the result of the decoupling of the two hydrogen atoms in the -CH₂- group as one moves to higher and higher overtones. One way of looking at this is to consider the vibrations of these high overtones in terms of *local*, rather than *normal*, modes, in which the two C-H bonds in the group do not move "in concert" but move more independently of one another. In this case, the C-H band becomes broader (in terms of energy or wavenumber) as ones goes from first, second, to third overtone (Fig. 10B, C, and D) because of a broader range of possible phases between the almost-uncoupled C-H motions (C. Marcott, *personal communication*). However, this simplifying effect for polyethylene has also been explained in terms of the improved "favorability" of

the asymmetric -CH₂- stretch relative to the symmetric -CH₂- stretch as one moves to higher energies or higher overtones (P. R. Griffiths, *personal communication*). In this case, the symmetries of the higher-overtone asymmetric stretching modes do not exactly match those of the bending mode, and, as a result, the possibility of spectral complication though Darling-Dennison resonances is reduced.

III. CHEMICAL FACTORS AFFECTING VIBRATIONAL SPECTRA

In an earlier discussion, it was stated that the position of an absorption band depends on the force constant(s) of the bond(s) involved and the masses of the atoms involved (Equations 5 and 6) and that the intensity of a band depends on the change in dipole moment corresponding to the vibration (Equation 13). In addition, it was stated that the symmetries of vibrations greatly influence the spectral activity of vibrational transitions, as well as the probability of observing combinations and resonance effects between vibrational modes. Therefore, any factor that affects atomic mass, bond strength, dipole moment, or symmetry has the potential to affect the NIR spectrum of a material.

A. The Primary Effect: Functional Group

The functional group effect is by far the most dominant of all effects in the NIR spectrum. According to Equation 5, the funda-

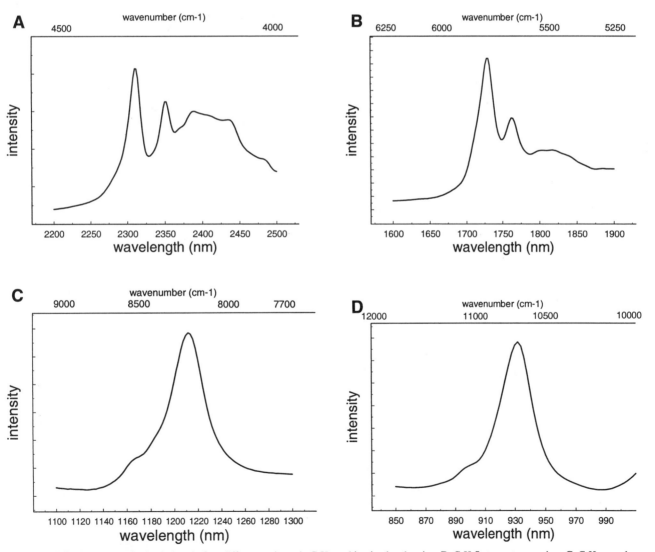

Fig. 10. Near-infrared spectra of polyethylene in four different regions. **A**, C-H combination band region; **B**, C-H first-overtone region; **C**, C-H second-overtone region; and **D**, C-H third-overtone region.

mental vibrational frequency increases as the reduced mass of the atoms involved in the vibration decreases and as the bond strength increases. When this is combined with the dipole moment change and anharmonicity requirements for overtone and combination bands, it becomes apparent that the most prominent bands in the NIR spectrum should involve C-H, N-H, and O-H stretching vibrations. Nonetheless, a significant number of other vibrations (C-H, N-H, and O-H bending and C-O, C-N, and N-O stretching) are also represented in the NIR spectrum to some extent (NIR correlation charts in the chapter appendix).

The three main attributes of functional groups that affect the positions and intensities of NIR bands are atomic mass, bond strength, and anharmonicity. For example, the frequencies of C-O stretching overtone bands for the carbonyl group (C=O) are higher than the corresponding bands for the carboxylate group (COO-), because the carbon–oxygen bond is stronger in the carbonyl group than in the carboxylate group. In addition, the first-overtone bands for C-H stretching are usually at higher frequencies than the first-overtone bands for N-H stretching, because the carbon atom is lighter than the nitrogen atom. These trends can be clearly observed in the NIR correlation charts.

Regarding the intensities of overtone and combination bands, the most dominant functional group factors are dipole moment, anharmonicity, and symmetry. The intensity of an overtone or combination band, like that of a fundamental vibrational band, is related to the change in dipole moment with the vibration. However, for symmetrical molecules, dipole moment change might not be sufficient to enable the presence of an overtone or combination band. Indeed, it is possible that an overtone of an IR-active fundamental is IR inactive, and, conversely, that an IR-inactive fundamental mode can contribute to an active overtone or combination band, if the molecule possesses some degree of symmetry (Herzberg, 1950). For example, the CO_2 molecule, because it has a center of symmetry, follows the "rule of mutual exclusion" regarding IR-active and inactive vibrations. As a result, only the odd overtones of the IR-active fundamental (ν_3, at 2,350 cm^{-1}) can be seen in the IR spectrum, and the even overtones cannot be observed in the IR spectrum.

An additional intensity factor for overtone and combination bands, which is not a factor for fundamental bands, is anharmonicity. For example, asymmetric stretching vibrations are generally more anharmonic than symmetric stretching vibrations of the same group (Hollas, 1987; Bonanno et al, 1992). It is thought that this effect causes -CH$_2$- asymmetric stretching overtone and combination bands to be more intense than corresponding -CH$_2$- symmetric stretching overtone and combination bands. This appears to be the case for the -CH$_2$- first-overtone stretching bands of polyethylene, discussed earlier (Fig. 10B). Another good example of anharmonicity-based intensity effects is the effect of hydro-

gen bonding on O-H stretching overtone bands, which is discussed later. Basically, the addition of a hydrogen bond acceptor atom to an OH group makes the O-H stretching vibration more harmonic, thus greatly reducing the intensity of stretching overtone bands in the NIR spectrum.

B. Secondary Effects

Even identical functional groups can produce very different spectra if they experience different environments or interactions. Such environmental factors can affect the spectra through changes in bond strength, symmetry, and dipole moment. Several secondary effects are common in NIR spectroscopy, and these are arranged according to their scale of observation.

THE ATOMIC LEVEL

The Neighboring Group Effect. Although the correlation charts in the chapter appendix can be used to obtain approximate positions of vibrational NIR bands for different functional groups, the exact positions of bands in specific materials can be influenced by the identity of a neighboring functional group. This is especially the case if the neighboring group is strongly electron withdrawing or electron donating, because it can then alter the bond strength and dipole moment of the functional group of interest. This effect can be seen through comparison of the NIR spectra of benzene and o-dichlorobenzene (ODCB) (Fig. 11). It is very unlikely that any C-Cl bands from ODCB can be observed in the NIR spectrum, because the reduced mass of this group is very large (Equations 5 and 6). As a result, the differences in the spectra in Figure 11 are mainly due to differences in aromatic C-H and C-C bands. The Cl groups in ODCB are highly electron-withdrawing and, therefore, cause a significant shift in the electron density distribution in the aromatic ring. This change in electron density distribution affects the strengths of the C-C and C-H bonds in the molecule (k in Equation 5), thus altering the positions of the NIR combination and overtone bands for these functional groups.

Hydrogen Bonding. Hydrogen bonding is a specific type of interaction that can have a profound influence on the vibrational spectra of materials. In fact, it is probably the most significant effect in NIR spectroscopy that does not involve a composition change! Hydrogen bonds form when a hydrogen atom that is electron poor (because it is attached to an electron-withdrawing oxygen or nitrogen "donor" atom) is attracted to a lone pair of electrons on an acceptor atom (Fig. 12). As a result of this interaction, an actual hydrogen bond is formed between the hydrogen atom and the acceptor atom, and the covalent bond between the donor atom and the hydrogen atom is weakened. In most cases, the hydrogen bond is much weaker than the donor-hydrogen covalent bond, but in some extreme cases, these two bonds can be of comparable strength (Pimental and McClellan, 1960; Schuster et al, 1976; Siesler and Holland-Moritz, 1980; Bonanno et al, 1992). Although the most common hydrogen bond donor groups are -OH and -NH, there is a significantly larger number of possible acceptor

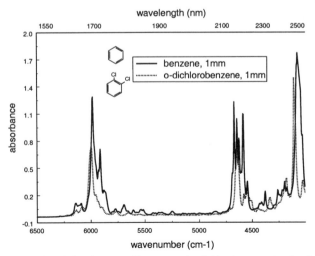

Fig. 11. Near-infrared spectra of benzene and o-dichlorobenzene, showing the neighboring group effect on C-C and C-H overtone and combination bands.

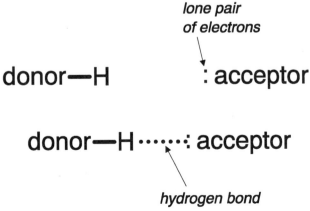

Fig. 12. Schematic of a typical hydrogen bond.

groups, which require only a lone pair of electrons (for example, OH, NH, C-O-C, C=O, and -N< groups).

At the atomic level, several aspects of hydrogen bond formation are very influential on the vibrational spectrum of the functional groups involved:

1. the strength of the donor-hydrogen bond is decreased, and the strength of bonds involving the acceptor atom (for example, C=O or C-N bonds) can even be decreased;
2. a new hydrogen bond is formed between the hydrogen and the acceptor atom; and
3. the donor-hydrogen stretching vibration becomes more harmonic (i.e., less anharmonic).

The increased harmonic nature of the donor-hydrogen stretching mode upon hydrogen bonding is intuitively obvious; in a linear hydrogen-bonded system, the donor-hydrogen stretch actually involves the oscillation of the hydrogen atom between the donor and acceptor atom, thus causing the potential energy of the system to go to infinity for both very small and very large displacements of the hydrogen atom from the donor atom (Fig. 13). As a result, the overall shape of the potential energy curve becomes more parabolic, which is indicative of a harmonic vibrational system.

Several consequences result from the aforementioned hydrogen-bonding effects that influence the vibrational spectroscopy of materials. The effect of band *frequency* and *shape* upon hydrogen bonding is more or less constant between fundamental (mid-IR) and overtone/combination (NIR) spectroscopy, with these results:

1. donor-hydrogen stretching bands decrease in frequency;
2. stretching bands of groups involving the acceptor atom decrease in frequency;
3. donor-hydrogen bending bands increase in frequency; and
4. donor-hydrogen stretching bands become broader.

Results 1, 2, and 3 are caused by changes in bond strengths that accompany hydrogen bond formation. The broadening of donor-hydrogen bands upon hydrogen bonding could be due to the multitude of slightly different hydrogen-bonded states present in the material, each of which would produce a slightly different band frequency.

Because the anharmonicity of donor-hydrogen stretching is greatly altered by hydrogen bond formation, the effect of hydrogen bonding on band *intensity* is quite different for the mid-IR and NIR regions:

1. hydrogen bonding causes an *increase* in the intensity of the *fundamental* donor-hydrogen stretching band, which does not depend on anharmonicity to exist; but
2. hydrogen bonding causes a *decrease* in the intensity of the *overtone* and *combination* bands involving the donor-hydrogen stretching vibration, which depend greatly on anharmonicity to exist.

As a result, the overtone and combination bands for OH and NH groups in the NIR spectrum are more intense for the "free" (non-hydrogen-bonded) forms than for the hydrogen-bonded forms. A good example of this is observed in the NIR spectra of alumina (Miller and Yin, 1989) and sucrose (Davies and Miller, 1988) (Fig. 14A and B, respectively). Crystalline sucrose contains both free and hydrogen-bonded OH groups (Davies and Miller, 1988), and alumina contains both free surface OH groups and hydrogen-bonded surface OH groups. In each case, the higher-frequency overtone stretching band for the free OH groups are significantly more intense than the lower-frequency bands for the hydrogen-bonded OH groups, even though the amount of free OH groups is relatively low compared with the amount of hydrogen-bonded OH groups. More detailed discussions about hydrogen bonds and their effects on vibrational spectroscopy can be obtained from several references (Pimental and McClellan, 1960; Schuster et al, 1976; Siesler and Holland-Moritz, 1980; Bonanno et al, 1992).

THE MICROSCOPIC LEVEL

The effects of functional group, neighboring group, and hydrogen bonding mentioned above can be directly observed only at the atomic level. However, some interactions of molecules observable in the microscopic domain can influence the vibrational spectroscopy of a material as well.

Arguably, the most common microscopic property that is of practical relevance in NIR analysis is crystallinity. Crystallinity refers to long-range molecular order in a material. Materials can be fully crystalline, fully amorphous (no molecular order), semicrystalline (some molecular order in an amorphous matrix—most often limited to polymeric materials), or a mixture of crystalline and amorphous (Mandelkern, 1964; Samuels, 1974; Siesler and Holland-Moritz, 1980; Hiemenz, 1984). Crystalline structures are characterized by specific intermolecular interactions and molecular arrangements, whereas an amorphous structure is characterized by more random intermolecular interactions and more random molecular arrangements. From an individual functional group's point of view, these states represent different potentials for interaction and different molecular environments and thus should influence its vibrational spectroscopy.

Because the crystalline state involves a regular pattern of discrete intermolecular interactions and a specific molecular order, crystalline materials tend to show sharper and more discrete vibrational absorption bands than their amorphous counterparts. These sharper bands represent the few discrete environments or interaction states experienced by the functional groups in the material. For example, the NIR spectrum of crystalline sucrose (Fig. 15) shows several overlapped, yet discrete, bands from the different states of OH and CH groups in the crystalline structure of

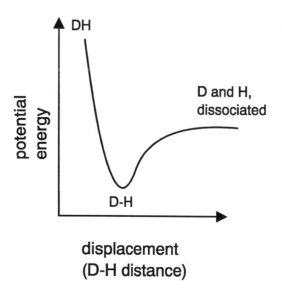

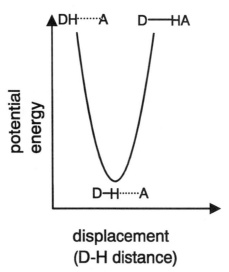

Fig. 13. Effect of hydrogen bonding on the anharmonicity of an X-H vibration. Top, *no* hydrogen-bonding case; and bottom, hydrogen-bonding case.

sucrose. In contrast, the spectrum of amorphous sucrose shows several relatively broad bands, each of which represents a random assortment of OH or CH groups in the amorphous structure. Note that the most dramatic crystallinity effects in the NIR spectrum of sucrose involve the first-overtone O-H bands (1,400–1,600 nm). This is the case because hydrogen-bonding effects are particularly strong in the NIR spectrum, as discussed earlier. In particular, the sharp 1,440-nm O-H band that is characteristic for crystalline sucrose represents a non-hydrogen-bonded form of OH group that is "frozen" into the crystal structure (Davies and Miller, 1988).

In the case of semicrystalline materials, such as poly(ethylene terephthalate) (PET), the degree of crystallinity can be manipulated by varying the material's thermal history. Upon an increase in the degree of crystallinity, very small changes in the spectrum of PET can be observed (Fig. 16A and B) (Miller and

Eichinger, 1990). On the molecular level, an increase in crystallinity generally involves an increase in the degree of molecular order, from a more random orientation of polymer chains to a more or-

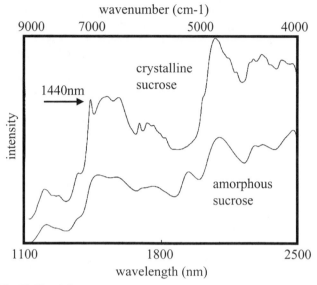

Fig. 15. Near-infrared spectra of crystalline and amorphous sucrose. (Reprinted with permission from Davies and Miller, 1988.)

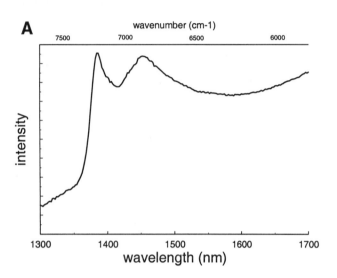

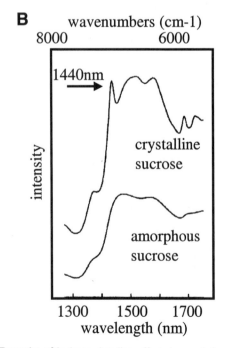

Fig. 14. Examples of hydrogen-bonding effects in near-infrared spectra. **A,** The first-overtone O-H stretching region of surface hydroxyl groups and surface water in alumina powder; **B,** the first-overtone O-H stretching region of the crystalline sucrose spectrum compared with the spectrum of amorphous sucrose. In both cases, the sharp band at lower wavelengths corresponds to the free OH groups, and the broader band at higher wavelengths corresponds to hydrogen-bonded OH groups. (**B,** Reprinted with permission from Davies and Miller, 1988.)

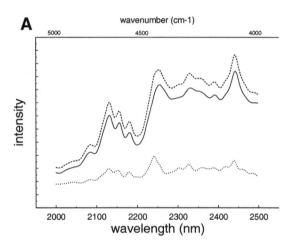

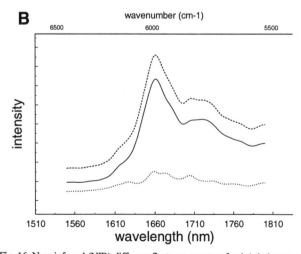

Fig. 16. Near-infrared (NIR) diffuse reflectance spectra of poly(ethylene terephthalate), showing the effect of crystallinity on the NIR spectrum of a semicrystalline material. **A,** 2,000–2,500 and **B,** 1,550–1,800 nm. Solid line = scan before annealing (lower crystallinity), dashed line = scan after annealing (higher crystallinity), and dotted line = subtraction of the two scans (scaled for clarity).

dered arrangement of these chains. Since PET contains no hydrogen bond donor groups, such increased molecular order is expressed in the NIR spectrum through either changes in intermolecular symmetry or changes in relatively weak intermolecular interactions. In Figure 16A and B, subtle differences are observed in the spectra across the entire NIR spectrum, indicating that the vibrations of both the ethylene -CH₂- and aromatic terephthalic acid molecular segments in the polymer are affected by the increased molecular order. The largest single effect seems to occur around 2,240 nm (Fig. 16B), which most likely corresponds to a -CH₂- stretching combination band for the ethylene group in the polymer. Because the ethylene group is the main element that enables "turns" in the PET polymer chain, and because molecular conformation (discussed below) can have a large impact on vibrational spectra, the NIR bands of this group are expected to be rather sensitive to changes in molecular order.

Many materials can possess several different crystalline forms, each of which involves a unique molecular arrangement. As a result, each state is expected to produce a unique and distinct pattern of intermolecular interactions and molecular environments that should be detectable by vibrational spectroscopy. However, it should be noted that any NIR spectral effects corresponding to changes between crystalline forms would be most noticeable in C-H bands or in the bands of weakly hydrogen-bonded OH and NH groups, because the bands of strongly hydrogen-bonded OH and NH groups are relatively broad and weak. The NIR spectra of two crystalline forms of glycine are shown in Figure 17 (Miller and Honigs, 1989). The weak peaks on top of the broad baseline inflections are mostly -CH₂- bands, and these appear to shift slightly and change in shape upon transition between the two crystalline forms. In addition, changes in N-H absorptions from changes in hydrogen bond structure are observed. However, the only N-H effect that is readily visible is the 2,147-nm band for α-glycine, which represents a weakly hydrogen-bonded form of the NH group observed only in the α-crystalline form. All of the other N-H effects involve changes in the shapes of the inflection points at approximately 1,270, 1,600, and 2,050 nm, which are *not* really baseline shifts but rather broad and overlapped bands of strongly hydrogen-bonded NH groups.

An additional microscopic phenomenon, which is very common for heterogeneous systems and block copolymers, is called phase separation. This phenomenon involves the segregation of molecular or functional group species into microscopic domains, according to their polarity or interaction potential. These domains are characterized by special interactions (often hydrogen bonding) between functional groups, which do not occur in the bulk material. As a result, the spectra of a single functional group in the bulk material and in the phase-separated domain can be quite different. In one specific study (Miller et al, 1990), the phase separation of a polyurethane block copolymer containing urethane/urea and polyether molecular segments was studied using NIR C=O overtone bands in

the region of 1,900–2,100 nm. As the polymer is annealed by heating at 100°C, these molecular segments phase separate into poly(urethane/urea)-rich and polyether-rich domains. As this occurs, the intensity of the NIR absorbance at 2,036 nm continually decreases (Fig. 18A and B). This absorbance was assigned to the "free" (non-hydrogen-bonded) form of the urea C=O groups in the polymer, based on model compound studies. This result suggests that phase separation causes the urea C=O groups in the polymer to leave the polyether-rich environment (which contains no hydrogen bond donors) and enter a urethane/urea-rich domain (where they can hydrogen bond to NH groups). It should be noted that other changes in the NIR spectra around 1,870–2,000 nm are most likely caused by changes in the hydrogen-bonding scheme of the urethane C=O groups in the polymers, with some possible interference from absorbed moisture.

For large molecules and polymers, molecular conformation can also have a significant effect on vibrational spectroscopy. The classical example of molecular conformation is the "chair"/"boat" transformation of cyclohexane. Because the intramolecular environments of the -CH₂- bonds in cyclohexane are different in the two forms, the vibrational spectroscopy of these bonds can also be different for the two forms. In the case of polymers, significant changes in the vibrational spectroscopy of functional groups in the polymer are often observed when going from a *trans* to a *gauche* conformation in the polymer chain (Fig. 19). A *gauche* linkage tends to cause

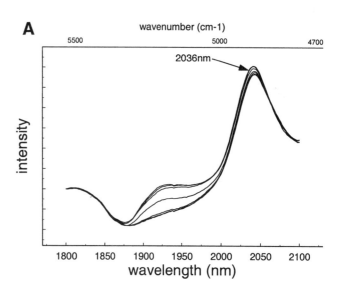

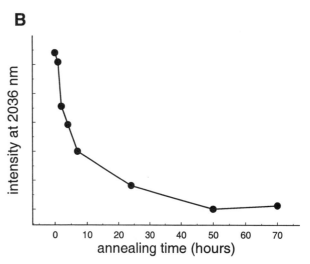

Fig. 18. Effect of phase separation on the near-infrared (NIR) spectrum of a polyurethane. **A,** NIR spectra in the region of the free urea C=O absorbance at 2,036 nm; and **B,** intensity at this wavelength during the phase separation process.

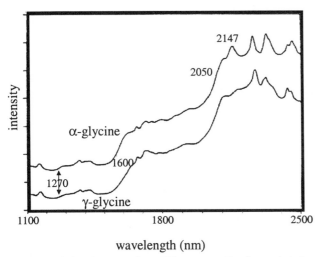

Fig. 17. Near-infrared spectra of two different crystalline forms of glycine. (Reprinted with permission from Miller and Honigs, 1989.)

a bend in the polymer chain, and a *trans* linkage tends to keep the chain pointing in the same direction. Work by Snyder (1967) discusses in detail the effects of conformation on the fundamental vibrational spectra of -CH₂- groups in long-chain hydrocarbons. In general, however, conformational changes tend to cause slight shifts in the frequencies of -CH₂- vibrational bands.

THE MACROSCOPIC LEVEL

Macroscopic changes can generally be classified into two groups: thermal and mechanical. For agricultural materials, the effect of temperature is by far the more important, although mechanical effects cannot always be ignored.

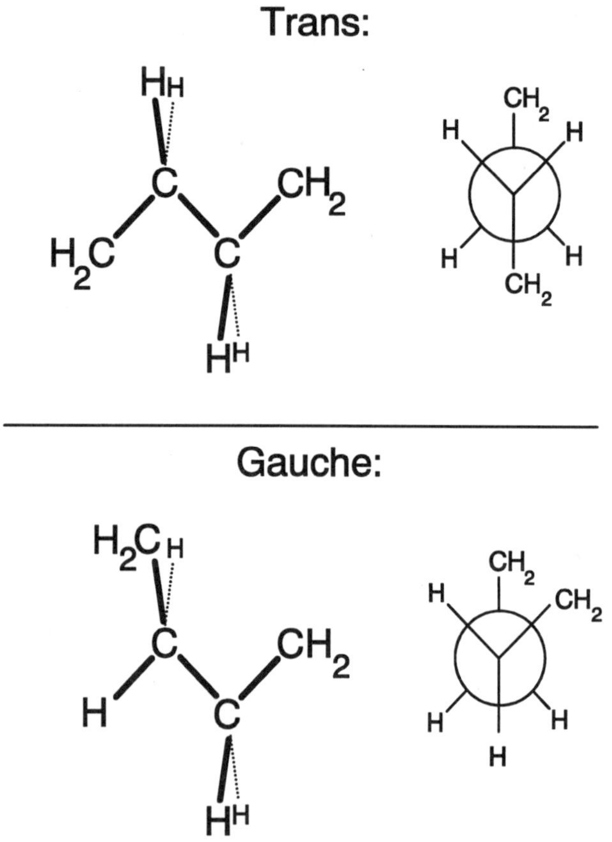

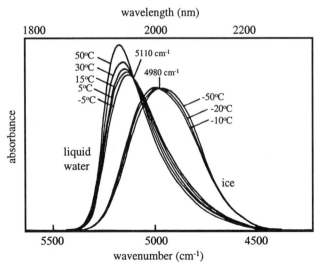

Fig. 19. Representations of the *trans* and *gauche* conformations of a hydrocarbon chain.

Thermal and Phase Effects. From a statistical thermodynamics point of view, a group of molecules can assume several different configurations, each of which corresponds to different energies. As the temperature is increased, more higher-energy configurations can be assumed at the expense of lower-energy configurations. Such configurations might correspond to different hydrogen-bonded states or conformational states of a molecule. Because changes in such states can significantly affect the vibrational spectrum of a material, one should expect the NIR spectrum of a material to be affected by temperature. However, the degree and nature of this effect depends on the specific molecular vibration and the nature of the different molecular states that are possible.

For materials that can hydrogen bond, the bands corresponding to hydrogen-bonding groups (i.e., N-H and O-H bands) are expected to be highly influenced by temperature. Because hydrogen bond formation typically involves an increase in enthalpy and decrease in entropy, increasing temperature favors higher-energy (i.e., weaker hydrogen bond) configurations over lower-energy configurations. When the effects of hydrogen bond strength on NIR spectra (discussed earlier) are also considered, it is expected that increased temperature causes O-H and N-H overtone and combination bands to shift to higher frequency, decrease in width, and increase in intensity. This expectation is confirmed by the observed spectra of liquid water and ice at different temperatures (Fig. 20) (Fornes and Chaussidon, 1978).

In the case of molecules that cannot hydrogen bond but can have several different conformational states (e.g., long-chain hydrocarbons and polymers), increasing temperature causes the population of higher-energy conformational states to increase at the expense of lower-energy conformational states. However, there is no universal rule relating energies of conformational states to specific conformations (i.e., the *gauche* conformation is not always at a higher energy than the *trans* conformation). In the case of the ethylene chain, found in straight-chain hydrocarbons and polyethylenes (Snyder, 1967), the *gauche* conformation is generally a higher-energy state than the *trans* conformation. As a result, it is expected that temperature affects the spectrum of non-hydrogen-bonding materials through band shifts that are caused by conformational changes. In most observed cases, however, these shifts are very small.

Temperature also affects the spectrum of a material through its phase. The phase of a material affects its spectroscopy through differences in available molecular configurations and interactions. For example, water molecules exhibit very different behavior in the vapor, liquid, and solid states. In the vapor state, the molecules have enough energy to avoid intermolecular interaction (i.e., hydrogen bonds) and are therefore allowed to rotate. As a result, the O-H combination (stretching + bending) band of water is at its highest frequency in

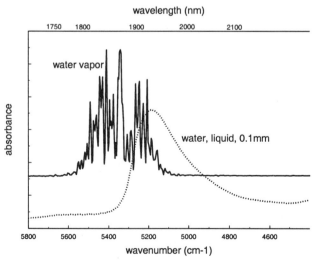

Fig. 20. The near-infrared spectra of water and ice at different temperatures. (Reprinted with permission from Fornes and Chaussidon, 1978. Copyright 1978 American Institute of Physics.)

Fig. 21. Comparison of the near-infrared spectra of water in the gas and liquid phases. The fine structure in the vapor-phase spectrum is caused by transitions between different rotational energy states in the gas-phase water molecules.

the vapor state, and a great deal of fine structure due to transitions between molecular rotation states can be seen (Fig. 21). In the liquid state, the molecules are still in movement, but they do not have enough energy to avoid intermolecular hydrogen bonding, and many different hydrogen-bonded states are possible. As a result, the O-H combination band is relatively broad and at lower frequency relative to the gas-phase case (Fig. 21) (Fornes and Chaussidon, 1978).

Interestingly, if one inhibits hydrogen bonding by dissolution into a non-hydrogen-bonding solvent, such as carbon tetrachloride, the spectrum reverts to a single, sharper band at the high-frequency end of the bulk liquid water band (Fig. 22), indicating that hydrogen bonding is either very weak or nonexistent. Note that, unlike the vapor-state spectrum (Fig. 21), the spectrum of water in CCl_4 shows only a single band, indicating the lack of any rotational transitions in the water molecules.

In the solid state (ice), the molecules assume a specific low-energy configuration that involves strong intermolecular hydrogen bonds of specific geometry and strength. In the case of ice, all of these different hydrogen bonds in the crystal structure are, on average, stronger than the hydrogen bonds in liquid water. As a result, one observes the summation of several weaker, broader, and lower-frequency bands, which, in this case, appears as a single broader, weaker, lower-frequency band in the spectrum (Fig. 20) (Fornes and Chaussidon, 1978). In other cases, in which relatively weak hydrogen bonds are "frozen" into the crystal structure, it might be possible to visually detect distinct peaks from individual hydrogen-bonded groups in the crystal structure, because these weakly hydrogen-bonded groups should produce stronger and sharper bands (e.g., Figs. 14 and 17).

For materials that cannot hydrogen bond, phase transitions are accompanied by changes in weaker intermolecular interactions and possibly by changes in intramolecular conformation. As discussed above, either of these changes can affect the NIR spectrum, although the observed effect is probably less than that observed for molecules that can hydrogen bond.

Mechanical Effects. For some materials, changes in the distribution of intramolecular states, and even changes in crystallinity, can be induced by applying mechanical stress. For example, the application of tensile stress to a solid polymer film can cause a change in the distribution of conformational states in the polymer molecules, as well as a change in the overall orientation of the molecules. Although it is much more common to observe this effect through dichroic spectroscopy (i.e., spectroscopy with polarized light, which actually looks at the alignment of oscillating dipoles in the sample) (Siesler and Holland-Moritz, 1980), small changes in peak *position* are also detectable, as a result of conformational changes. Figure 23 shows the NIR spectrum of a polyethylene film before and after uniaxial stretching. Even though randomly polarized light was used for these measurements, significant spectral changes are observed in the difference spectrum as a result of changes in molecular conformation (and

possibly crystallinity) (Miller, 1993). Such effects can cause considerable interference whenever highly oriented film or fiber samples are analyzed.

IV. ELECTRONIC NIR SPECTROSCOPY

Although most practical NIR applications exploit vibrational absorption bands, the presence of electronic NIR bands cannot be overlooked. When atoms are combined to make molecules, the electrons of the atoms are distributed between newly defined molecular orbitals, each of which corresponds to a different electronic energy level. Although these orbitals are strictly defined in terms of quantum theory, they can be qualitatively referred to as bonding, nonbonding, and antibonding. Electronic spectroscopy simply involves the change in the electronic state of a molecule (i.e., the movement of an electron between different energy levels) as a result of interaction with light energy.

For any given molecule, electronic transitions are generally of higher energy than vibrational transitions and are thus observed predominantly in the higher-energy visible and ultraviolet regions of the spectrum. The most dominant electronic transition is from the highest-energy molecular orbital that contains an electron to the lowest-energy molecular orbital that does not contain an electron. The energy of this transition corresponds to the difference in energy of these two electronic levels. For most molecules, the energy of this transition is in the visible or ultraviolet region of the spectrum. However, there are some molecules for which this transition is in the NIR spectral region (i.e., these electronic energy states are unusually close together). Some notable examples of such materials are highly conjugated organic molecules such as hemoglobin (Oda et al, 1991; Hall and Pollard, 1995; Norris and Kuenstner, 1995) and conducting polymers (Kaeriyama et al, 1989), as well as some inorganic materials such as rare earth oxides (Weidner et al, 1986).

Like vibrational NIR bands, electronic NIR bands are affected by intermolecular interactions and sample state. For example, changes in the NIR spectrum of whole blood upon a change in deoxy-hemoglobin level (Fig. 24A) involve changes in relatively broad bands in the region of 700–1,100 nm. In fact, these electronic bands are so broad that their change could be easily mistaken as a localized baseline shift in the 700–1,100-nm range. However, the spectrum of a mixture of solid rare earth oxides (Fig. 24B) exhibits rather sharp peaks across the whole spectral range, which are indicative of discrete molecular interactions associated with the crystalline state of these materials.

V. THE NIR COMPLICATION FACTOR

Even if an organic compound has only one functional group that can produce vibrational bands in the NIR spectral region, there can

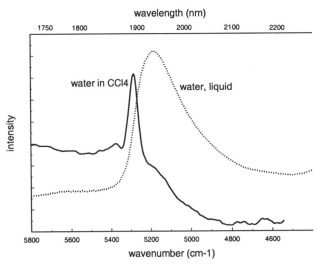

Fig. 22. Comparison of the near-infrared spectra of liquid water and a dilute solution of water in carbon tetrachloride. The spectra are scaled for clarity.

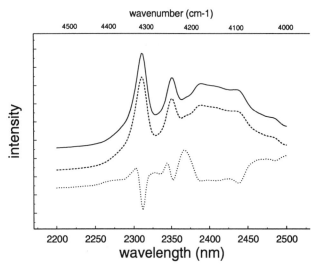

Fig. 23. Effect of mechanical strain on the near-infrared spectrum of a polyethylene film. Solid line = spectrum of polyethylene film before stretching, dashed line = spectrum of film after stretching, and dotted line = subtraction of the two spectra.

still be several different vibrational modes of that group, each of which corresponds to a single fundamental vibrational transition frequency. Then, as the many different possible combinations, overtones, and even combinations of overtones are considered, it becomes apparent that the NIR spectrum can become very complicated. This complicating effect is particularly apparent in the NIR spectrum of chloroform ($CHCl_3$). Chloroform is not only a rather simple molecule, but it also contains only one bond (C-H) that would have fundamental vibrational frequencies that are high enough to enable the presence of overtone and combination bands in the NIR spectral region. In fact, only six different fundamental vibrational modes of chloroform exist; they are listed in Table I (Murray and Williams, 1987). However, these six fundamental modes combine to produce an abundance of active overtone and combination bands in the NIR region, as shown in Table II. In the NIR spectrum of *n*-heptane (Fig. 9), a few dominant overtone and combination bands are observed in several spectral regions. However, if one looks more closely, the spectrum also contains a large number of weaker bands in other regions and, undoubtedly, several weaker bands that are obscured by the dominant bands. These examples illustrate the complicating effect of multiple overtones and combinations on the NIR spectrum, even for relatively simple molecules.

When one considers that most molecules of practical interest are considerably more complex than chloroform and *n*-heptane, it quickly becomes apparent that definitive separation and identification of individual NIR bands in more common materials is usually impossible, for several reasons:

1. the multitude of overtone and combination bands produced from only a few vibrations, discussed above;
2. the relatively large number of NIR-active groups (e.g., CH, NH, OH, and C=O), each of which contributes its own set of overtone and combination bands;
3. the possibility of resonances between vibrational modes, which results in bands that cannot be assigned to "pure" vibrations in the molecule; and
4. the possibility of several molecular configurations, each of which could produce a slightly different spectrum.

Although the local mode phenomenon discussed earlier tends to simplify the NIR spectrum at higher wavenumbers (lower wavelengths), these complicating factors are simply too overwhelming. However, such complication does not mean that practical implementation of NIR spectroscopy is impossible, as the large amount of NIR application literature demonstrates. In fact, the most notable NIR applications most often seem to be those that are the most demanding in terms of chemical complication. For example, the first

A

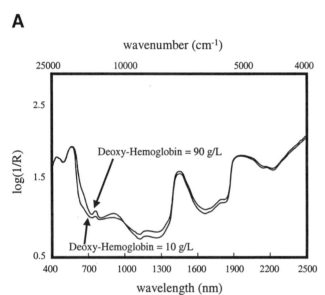

B

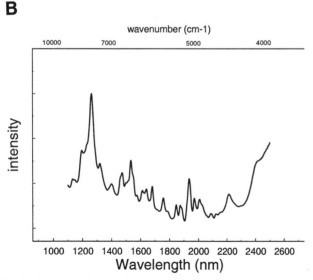

Fig. 24. Examples of electronic near-infrared (NIR) bands: **A,** the effect of hemoglobin oxygenation on the NIR spectrum; and **B,** the NIR spectrum of a mixture of rare earth oxides. (**A,** Reprinted from Hall and Pollard, 1995.)

TABLE I
Fundamental Vibrational Frequencies for the Chloroform Molecule

Vibrational Mode	Frequency (wavenumbers)
ν_1	3,040
ν_2	667
ν_3	370
ν_4	1,214
ν_5	755
ν_6	261

TABLE II
Vibrational Combination and Overtone Bands of Chloroform

Combination or Overtone Mode	Calculated Position (wavenumbers)	Calculated Position (nm)	Observed Position (nm)
$\nu_1 + \nu_4$	4,254	2,351	2,370
$\nu_1 + \nu_4 + \nu_6$	4,480	2,232	2,234
$\nu_1 + \nu_4 + \nu_3$	4,589	2,179	2,181
$4\nu_4$	4,856	2,059	2,107
$\nu_1 + \nu_4 + \nu_2$	4,886	2,047	2,049
$\nu_1 + \nu_4 + \nu_5$	4,974	2,010	2,008
$\nu_1 + 2\nu_4$	5,433	1,841	1,843
$\nu_1 + 2\nu_4 + \nu_6$	5,687	1,758	1,757
$\nu_1 + 2\nu_4 + \nu_3$	5,796	1,725	1,735
$2\nu_1$	6,080	1,645	1,692
$\nu_1 + 2\nu_4 + \nu_2$	6,093	1,641	1,655
$2\nu_1 + \nu_6$	6,171	1,620	1,619
$\nu_1 + 2\nu_4 + \nu_5$	6,181	1,618	1,618
$2\nu_1 + \nu_3$	6,280	1,592	1,590
$2\nu_1 + \nu_2$	6,577	1,520	1,525
$\nu_1 + 3\nu_4$	6,640	1,506	1,517
$2\nu_1 + \nu_5$	6,665	1,500	1,494
$2\nu_1 + \nu_4$	7,124	1,404	1,405
$2\nu_1 + \nu_4 + \nu_6$	7,378	1,355	1,357
$2\nu_1 + \nu_4 + \nu_3$	7,487	1,336	1,341
$2\nu_1 + \nu_4 + \nu_2$	7,784	1,285	1,290
$2\nu_1 + \nu_4 + \nu_5$	7,872	1,270	1,270
$2\nu_1 + 2\nu_4$	8,331	1,200	1,210
$3\nu_1 + \nu_6$	8,941	1,118	?[a]
$3\nu_1 + \nu_3$	9,050	1,105	1,109
$3\nu_1$	9,120	1,097	1,152
$3\nu_1 + \nu_2$	9,347	1,070	1,068
$3\nu_1 + \nu_5$	9,435	1,060	1,059
$3\nu_1 + \nu_4$	9,894	1,011	1,018
$3\nu_1 + 2\nu_4$	11,080	903	908
$4\nu_1 + \nu_2$	11,990	834	830
$4\nu_1$	12,160	822	883
$4\nu_1 + \nu_4$	12,534	798	800
$5\nu_1$	15,200	658	721

[a] ? = Band was not observed.

practical applications of NIR spectroscopy involved the determination of constituents in chemically complex agricultural samples, such as wheat (Massie and Norris, 1965). Many more-recent industrial and on-line applications of NIR involve rather complex samples as well (Weyer, 1985; Stark et al, 1986; Miller, 1991).

The extensive success of NIR applications demonstrates that spectral complication refers only to the overall spectrum, while the validity of an analytical method depends on its ability to detect *relevant variability* in the spectrum. As a result, even if a spectrum is very complicated, it is still quite possible to construct a useful NIR method. However, such methods most often need to be constructed empirically because it is often not possible to confidently isolate relevant effects in the spectrum based on knowledge of the sample's chemistry and spectroscopy only. Nonetheless, such knowledge can certainly be used to improve the confidence of empirically constructed NIR methods, as well as to assist in the development of NIR methods. Once again, the multiplicity of NIR application literature illustrates this.

VI. NIR CORRELATION CHARTS

This chapter demonstrated that there are several different chemical phenomena that affect the NIR spectrum of materials. However, a simple summary of the most prominent effect, that of the functional group, can be a very useful reference for both experienced and inexperienced users of NIR technology. The correlation chart, which has been a very useful reference to mid-IR users for years, can also be utilized in this case. However, because of the complicated nature of NIR spectra discussed earlier, most NIR band assignments were not made from fundamental studies of simple molecules, but rather from empirical NIR method development. In other cases, it was possible to estimate band positions of a functional group in the NIR region from known band positions of the same functional group in the IR spectrum.

The NIR correlation charts in the chapter appendix, labeled Charts I–VI, were constructed through the consideration of several sources of information:

1. a Sadtler atlas of NIR spectra, containing the spectra of about 1,000 different chemicals (Hirschfeld and Zeev-Hed, 1981);
2. a detailed correlation chart of the fundamental vibrational (mid-IR) region (Shriner et al, 1980), from which approximate locations of overtone and combination bands for different functional groups were calculated; and
3. assorted NIR feasibility studies, in which detailed band assignments were made during the preparation of empirically derived NIR methods.

It is important to note that the band positions represented in these charts are only approximate and were compiled from a limited amount of experimental data. In some cases, actual NIR spectra were not available for reference, and approximate band positions were calculated from the frequencies of fundamental vibrational bands listed in a mid-IR correlation chart (Shriner et al, 1980). In addition, some of the bands listed in the charts refer to secondary effects of the functional group listed. Despite these limitations, the charts should serve as useful quick references for NIR users.

VII. CONCLUSION

Although NIR spectra might appear to be rather complicated, there is certainly a fundamental basis for their existence. This chapter demonstrated that NIR spectra are not a random mix of unintelligible data but rather the result of fundamental molecular vibration mechanisms that are both well understood and very reproducible. In fact, this fundamental basis actually *predicts* the complicated nature of NIR spectra. In addition, with current multivariate calibration capabilities, it is not necessary to understand the entire NIR spectrum to construct a method, because such calibration procedures are often capable of extracting only the relevant spectral effects. As a result, an understanding of the sources of NIR spectra is not necessary to construct a useful method, but it can be used to improve the efficiency of method development and validate existing methods.

APPENDIX: CORRELATION CHARTS

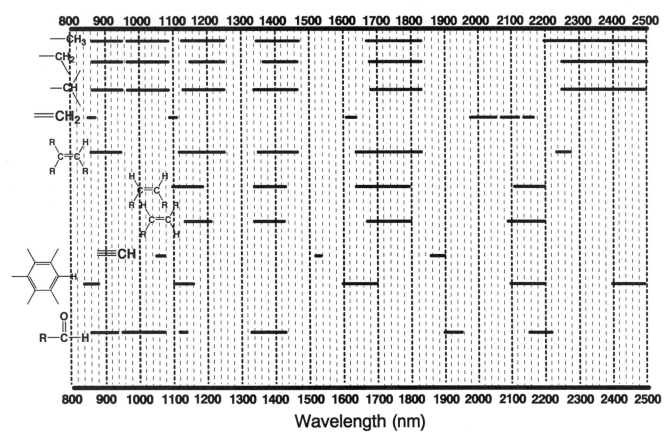

Wavelength (nm)

Chart I.

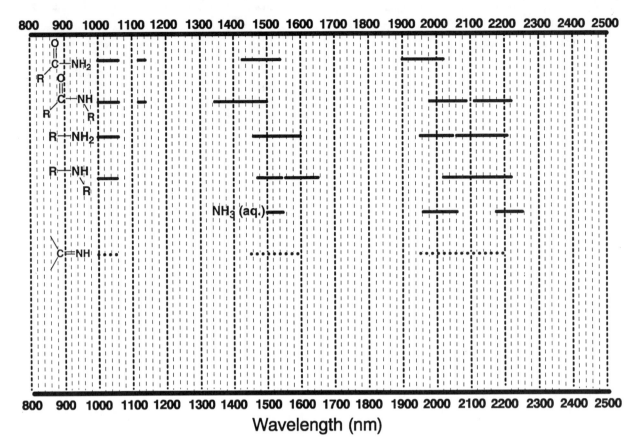

Chart II.

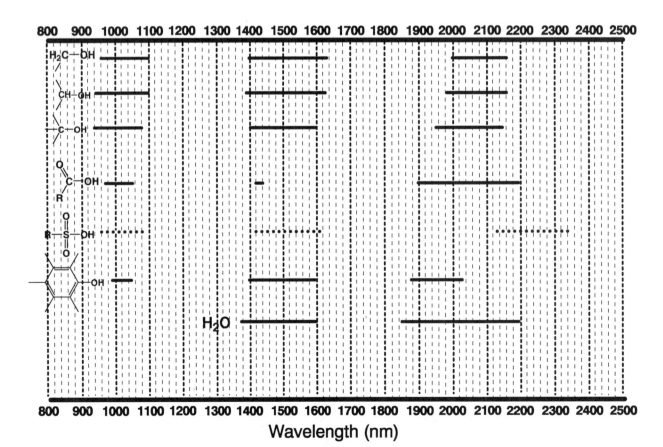

Chart III.

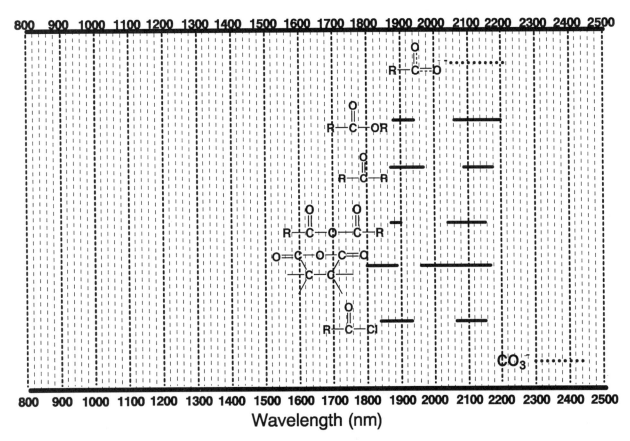

Chart IV.

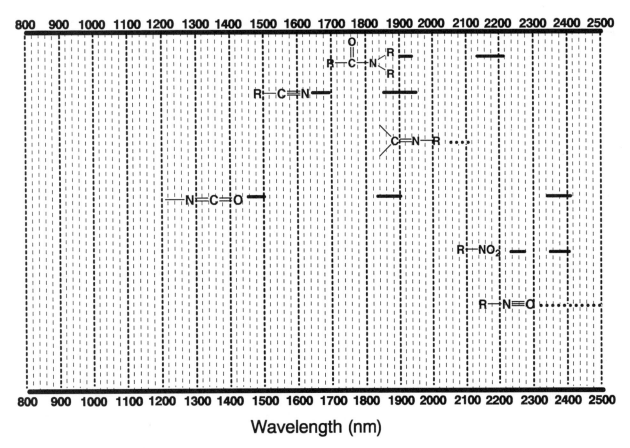

Chart V.

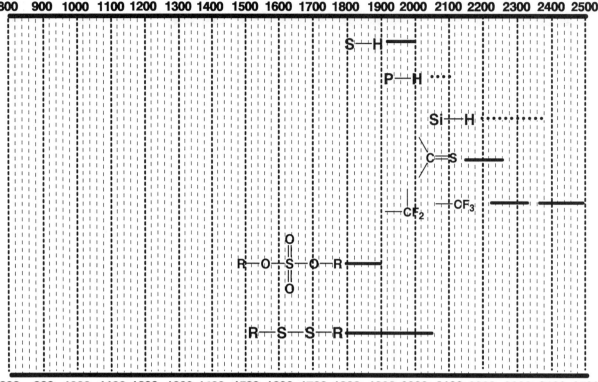

Chart VI.

LITERATURE CITED

Bonanno, A. S., Ollinger, J. M., and Griffiths, P. R. 1992. The origin of band positions and widths in near-infrared spectroscopy. Pages 19-28 in: Near-Infrared Spectroscopy, Bridging the Gap Between Data Analysis and NIR Applications. K. I. Hildrum, T. Isaksson, T. Naes, and A. Tandberg, eds. Ellis-Horwood, New York.

Chang, R. 1971. Basic Principles of Spectroscopy. McGraw-Hill, New York.

Colthup, N. G., Daly, L. H., and Wiberly, S. E. 1990. Introduction to Infrared and Raman Spectroscopy. 3rd ed. Academic Press, New York.

Davies, A. M. C., and Miller, C. E. 1988. Tentative assignment of the 1440 nm band in the near-infrared spectrum of crystalline sucrose. Appl. Spectrosc. 42:703.

Duncan, J. L. 1991. The determination of vibrational anharmonicity in molecules from spectroscopic observations. Spectrochim. Acta 47A(1): 1-27.

Fornes, V., and Chaussidon, J. 1978. An interpretation of the evolution with temperature of the n2+n3 combination band in water J. Chem. Phys. 68:4667-4671.

Hall, J. W., and Pollard, A. 1995. Near-infrared spectroscopy brought to life. Pages 421-430 in: Leaping Ahead with Near-Infrared Spectroscopy. G. D. Batten, P. C. Flinn, L. A. Welsh, and A. B. Blakeney, eds. Royal Australian Chemical Inst., Melbourne.

Henry, B. R., Hung, I. F., MacPhail, R. A., and Strauss, H. L. 1980. A local mode description of the CH-stretching overtone spectra of the cycloalkanes and cycloalkenes. J. Am. Chem. Soc. 102:515.

Herzberg, G. 1950. Molecular Spectra and Structure. II. Infrared and Raman Spectra of Polyatomic Molecules. D. Van Norstrand Co., Inc., Princeton, NJ.

Hiemenz, P. C. 1984. Polymer Chemistry. Marcel Dekker, New York.

Hirschfeld, T., and Zeev-Hed, A. 1981. The Atlas of Near-Infrared Spectra. Sadtler Research Laboratories, Philadelphia.

Hollas, J. M. 1987. Modern Spectroscopy. John Wiley & Sons, Chichester, United Kingdom.

Kaeriyama, K., Tanaka, S., Sato, M. A., and Hamada, K. 1989. Structure and properties of polythiopene derivatives. Synth. Metals 28:C611.

Mandelkern, L. 1964. Crystallization of Polymers. McGraw-Hill, New York.

Massie, D. R., and Norris, K. H. 1965. The spectral reflectance and transmittance properties of grain in the visible and near-infrared. Trans. Am. Soc. Agric. Eng. 8:598.

Miller, C. E. 1991. Near-infrared spectroscopy of synthetic polymers. Appl. Spectrosc. Rev. 26:275-337.

Miller, C. E. 1993. Use of near-infrared spectroscopy to determine the composition of high-density/low-density polyethylene blend films. Appl. Spectrosc. 47:222.

Miller, C. E., and Eichinger, B. E. 1990. Determination of crystallinity and morphology of fibrous and bulk poly(ethylene terephthalate) by near-infrared diffuse reflectance spectroscopy. Appl. Spectrosc. 44:496.

Miller, C. E., and Honigs, D. E. 1989. Discrimination of different crystalline phases using near-infrared diffuse reflectance spectroscopy. Spectroscopy 4:44.

Miller, C. E., and Yin, T.-K. 1989. Near-infrared reflectance analysis of poly(octadecyl methacrylate) adsorbed on alumina. J. Mater. Sci. Lett. 8:467.

Miller, C. E., Edelman, P. G., Ratner, B. D., and Eichinger, B. E. 1990. Near-infrared spectroscopic analysis of polyether(urethaneurea) block copolymers. II. Phase separation. Appl. Spectrosc. 44:581.

Murray, I., and Williams, P. C. 1987. Chemical principles of near-infrared technology. Pages 17-34 in: Near-Infrared Technology in the Agricultural and Food Industries. 1st ed. P. Williams and K. Norris, eds. Am. Assoc. Cereal Chem., St. Paul, MN.

Norris, K., and Kuenstner, J. T. 1995. Rapid measurement of analytes in whole blood with NIR transmittance. Pages 431-436 in: Leaping Ahead with Near-Infrared Spectroscopy. G. D. Batten, P. C. Flinn, L. A. Welsh, and A. B. Blakeney, eds. Royal Australian Chemical Inst., Melbourne.

Oda, I., Ito, Y., Eda, H., Tamura, T., Takada, M., Abumi, R., Nagai, K., Nakagawa, H., and Tamura, M. 1991. Non-invasive hemoglobin oxygenation monitor and computed tomography by NIR spectrophotometry. Pages 284-293 in: Proc. SPIE Vol. 1431, Time-Resolved Spectrosc. Imaging Tissues. B. Chance and A. Katzir, eds. SPIE (International Society for Optical Engineering) Press, Bellingham, WA.

Pimental, G., and McClellan, A. C. 1960. The Hydrogen Bond. W.H. Freeman, San Francisco.

Samuels, R. J. 1974. Structured Polymer Properties: The Identification, Interpretation and Application of Crystalline Polymer Structure. Wiley-Interscience, New York.

Schuster, P., Zundel, G., and Sandorfy, C. 1976. The Hydrogen Bond. II. Structure and Spectroscopy. North-Holland, Amsterdam.

Shriner, R. L., Fuson, R. C., Curtin, D. Y., and Morrill, T. C. 1980. The Systematic Identification of Organic Compounds. 6th ed. John Wiley & Sons, New York.

Siesler, H. W., and Holland-Moritz, K. 1980. Infrared and Raman Spectroscopy of Polymers. Marcel Dekker, New York.

Snyder, R. G. 1967. Vibrational study of the chain conformation of the liquid n-paraffins and molten polyethylene. J. Chem. Phys. 47:1316-1360.

Stark, E., Luchter, K., and Margoshes, M. 1986. Near-infrared analysis (NIRA): A technology for quantitative and qualitative analysis. Appl. Spectrosc. Rev. 22:335-399.

Sverdlov, L. M., Kovner, M. A., and Krainov, E. P. 1974. Vibrational Spectra of Polyatomic Molecules. John Wiley & Sons, New York.

Weidner, V. R., Barnes, P. Y., and Eckerle, K. L. 1986. A wavelength standard for the near-infrared based on the reflectance of rare earth oxides. J. Res. Natl. Bur. Stand. (U.S.) 91(5):243-253.

Weyer, L. G. 1985. Near-infrared spectroscopy of organic substances. Appl. Spectrosc. Rev. 21:1.

Wilson E. B., Decius, J. C., and Cross, P. C. 1955. Molecular Vibrations. McGraw-Hill, New York.

CHAPTER 3

Data Analysis: Wavelength Selection Methods

WILLIAM R. HRUSCHKA
Instrumentation Research Laboratory
Agricultural Research Service
U.S. Department of Agriculture
Beltsville, MD
U.S.A.

I. INTRODUCTION

The success and acceptance of near-infrared (NIR) analysis as a qualitative and quantitative tool are, in large part, the result of increases in the quality of the spectra and the power of the mathematics used in their analysis. This chapter discusses the basic factors involved in calibration and wavelength selection and describes mathematical solutions that, in commercial bench instruments, reduce to combinations of spectral measurements at a few ($\leq$20) wavelengths. Chapter 4 focuses on mathematical techniques that use measurements at all of the available wavelengths.

Typical NIR diffuse reflectance spectra are shown in Figures 1–3. Although the micrometer (μm) is the preferred wavelength unit in infrared work, most NIR work has been reported in nanometers, as we do here. The NIR region is composed of radiation with wavelengths of 700–3,000 nm. Wave numbers, also called reciprocal centimeters (cm^{-1}), are more often used in the mid- and far-infrared. They are units of frequency rather than wavelength. The conversion is cm^{-1} = 10,000/μm.

The vertical axes units, log(1/reflectance)—abbreviated as log(1/R) (log = log$_{10}$)—are related to absorption; a higher log(1/R) value means that more radiation has been absorbed (less reflected) by the sample at that wavelength. These units are used instead of percent reflectance because there is an almost linear relationship between the concentration of an absorbing component and its contribution to the log(1/R) value at the wavelength absorbed. Other units, such as Kubelka-Munk, for the vertical axis have been proposed but are not in general use by commercial NIR instruments. The log(1/R) function is used throughout this chapter with the understanding that it may be replaced by some other function of reflectance or transmittance without affecting the text, so long as attention is paid to the linearity question.

Figure 1 shows the spectra of water and ground wheat with two levels of moisture. The main features of the water spectrum are the two absorption bands that peak at 1,450 and 1,940 nm. Water is a strong absorber, and its absorption band at 1,940 nm is isolated from the bands of other absorbers, so the effect of increased moisture on the ground wheat spectrum is quite clear.

The other two major components of wheat are protein and starch (Fig. 2). These spectra have more bands, and the bands are not isolated. Not only do the protein bands overlap each other, but they overlap the starch bands. Protein absorbs NIR radiation much more weakly than does water, and the difference between the high-protein and low-protein wheat spectra is far less noticeable than that between wheats differing in moisture content. In fact, most of the differences in Figure 2 are traceable to small differences in the particle size distributions. The effect of large differences in particle size distributions is shown in Figure 3, in which the same

sample was ground to three different mean particle sizes. A change in particle size causes a change in the amount of radiation scattered by the sample, which causes a baseline shift (additive effect). The larger particles do not change the direction of the radiation as often, so more is absorbed before leaving the sample, resulting in a higher log(1/R) value. There is also a multiplicative effect, with the strong absorbers showing more change with particle size than the weak.

When differences in particle size are ignored, the differences caused by protein in Figure 2 are only small changes in the shape of some of the peaks (e.g., at 2,180 nm). Although these differences are barely visible, the various mathematical methods described in this chapter enable the use of these small changes to determine protein concentration with a standard error of less than 0.2% protein.

A lot of the research work in the NIR has involved analysis of spectra like those in Figures 1–3. The usual features are overlapping absorption bands with large spectral changes due to uncontrolled radiation scattering conditions masking the small spectral changes that are used to make measurements.

One set of data will be used throughout this chapter to illustrate the principles and techniques. The data are characterized in Table I and described in detail by Hruschka and Norris (1982) and Williams et al (1978, 1983). A set of 100 whole grain, hard red spring wheat lots was compiled that had small within-lot variability and large between-lot variability with respect to protein and moisture content. One sample was drawn from each lot and 20 additional samples were drawn from one lot for a total of 120 samples. The 20 replicates are called the reproducibility set in Table I. The other 100 samples were divided into two sets, each of 50 samples, which are called the calibration and validation sets. This division provided a roughly equal distribution of protein and moisture content in each set.

From each of the 120 samples, 20 subsamples were drawn and ground in duplicate to a mean particle size of 200 $\pm$ 30 μm. Sixteen of the subsamples were analyzed for protein by the Kjeldahl method over a 3-month period, and the average of these results is called the "definitive Kjeldahl" (KjD) for that sample. NIR reflectance spectra of the other four subsamples were obtained by an in-house-designed, computerized spectrophotometer (Williams et al, 1983), recorded as log(1/R), and averaged to form a single spectrum. The spectra are 1,001 points spaced 1.6 nm apart in the region of 1,000–2,600 nm with a resolution of 7 nm. The KjDs are reported on a moisture-free basis, with moisture data being obtained from the NIR spectra (Norris and Hart, 1965) and then combined with the KjDs to form the "as-is" or moisture-free protein data. The 120 spectra, with the corresponding chemical data, were stored on linc-tape as 16-bit integers for computerized data analysis. Figure 4 shows four of these samples with extreme protein and moisture values.

The wide variety of methods presented here, and the use of full spectra in the examples, obscure the fact that many calibrations have been developed using data from instruments that, in practice, have only a few preselected wavelengths available and use linear regression on one to six spectral values. Sections II–V cover the mathematics used in at least 90% of these applications. To help bridge the gap between this chapter and the mathematically more demanding Chapter 4, the conclusion contains an overview with simplified descriptions of all the mathematics used in both chapters.

II. CALIBRATION, MEASUREMENT, AND VALIDATION

Measurement is a general term for quantification. The technical literature, however, uses the term only for physical quantities such as reflectance or temperature and uses the term *determination* for chemical quantities such as percentage of protein in a wheat sample. Samples are *analyzed* for a component. When the analysis is quantitative, the component is said to be *determined*, and when the analysis is qualitative, the component is said to be *detected*. In the NIR literature, the term *prediction* is often used in place of measurement or determination. Although this term is routinely used by statisticians in the same way, it has connotations of looseness of method for nonstatisticians and will probably not be an acceptable term when NIR terminology becomes standardized. In this chapter, *measurement* is used for physical quantities, *determination* for chemical quantities, and *measurement* for the general case.

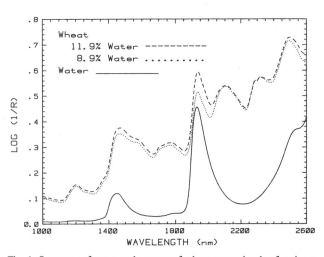

Fig. 1. Spectrum of water and spectra of wheat at two levels of moisture with constant protein content (12.4%).

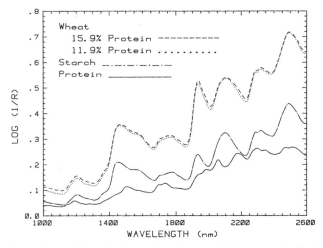

Fig. 2. Spectra of starch and protein and spectra of wheat at two levels of protein with constant moisture content (8.0%).

A. Calibration

NIR instruments determine protein and other components by measuring $\log(1/R)$ values that must then be related to the amount of the component as determined by some other method called a *reference* or *standard* method. Establishing this relationship by using a set of samples of known composition is called *calibration* of an NIR method, whereas using the relationship to determine the amount of a component in a new sample is called an *NIR determination*. The relationship between the $\log(1/R)$ values and the reference method values is expressed as an approximation and always involves some form of regression equation. (Sections IV and V contain a review of linear regression.) The regression equation has regression constants (the Y-intercept and regression coefficients), independent variables, and one dependent variable (the reference method value). The independent variables are mathematical combinations of $\log(1/R)$ values at various wavelengths. These combinations can be so complicated that they are better thought of as a series of steps, so we use the term *data treatment* to mean any mathematical process that combines $\log(1/R)$ values into independent variables for use in a regression equation.

For this discussion, a distinction is made between calibration and developing a calibration model. Each data treatment has data treatment constants (such as the derivative parameters or the amount of smoothing). Developing a calibration model involves testing different data treatments, data treatment constants, or sets of wavelengths. *Calibration* means finding the regression constants that go into the approximation once the form of the approximation, the data treatment constants, and the wavelengths have been decided upon. The following four examples illustrate the distinction.

1. One of the simplest examples of an approximation is the familiar single-term (one independent variable) regression equation

$$Y = a + bX + e$$

or

$$\hat{Y} = a + bX$$

in which Y and X are the reference and $\log(1/R)$ values, respectively, a and b are regression constants, $\hat{Y}$ is an approximation to Y, and $e = Y - \hat{Y}$ is the error in the approximation for a single sample. The data treatment here is simply to take the $\log(1/R)$ value as it stands, so that there are no data treatment constants. If this form of approximation were to be used, then developing the calibration model would mean finding the wavelength that provided the best approximation for all samples. If the wavelength had already been chosen, then calibration would mean finding the regression constants that gave

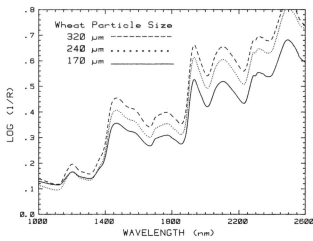

Fig. 3. Spectra of wheat (9.5% water, 11% protein) at three different particle sizes.

the best approximation for all samples. An example is the use of the log($1/R$) value at 1,680 nm to measure particle size index.

2. The most common example of an approximation is a multiterm linear regression with the equation

$$Y = a + b_1 x_1 + b_2 x_2 + \ldots + b_k x_k$$

in which the x's are log($1/R$) values at k different wavelengths. The data treatment is the same as in the first example, and there are no data treatment constants. Here, developing the calibration model involves deciding how many wavelengths to use and selecting an optimal set of wavelengths. Calibration means finding the regression constants to correspond with the set of wavelengths chosen. Numerous examples can be found using three to six wavelengths.

3. A more complicated case, but involving only single-term regression, occurs when x in the above equation is the quotient of two first derivatives (Section VI). The data treatment constants are the derivative parameters and the degree of smoothing. If this form of approximation were to be used, developing the calibration model would involve finding the best wavelengths, derivative parameters, and degree of smoothing. Calibration would involve finding a and b in the regression. An example is given by Williams et al (1983).

4. The most complex situations involve multiterm regressions in which each independent variable is a combination of the log($1/R$) values at all the available wavelengths. These combinations may be expressed as Fourier coefficients (Giesbrecht et al, 1981), curve fit coefficients (Hruschka and Norris, 1982), factor scores (Martens and Naes, 1983), etc., and developing a calibration model involves deciding which and how many combinations to use. Calibration in this case again means finding the regression constants.

There are also mixtures of the cases. For instance, Shenk et al (1981) and Marten et al (1985) use multiterm regression, in which each independent variable is a quotient of first or second derivatives. And it is possible for some independent variables in a regression to be NIR data and others to be physical or chemical data, e.g., sample temperature, grinder, variety, location, etc.

The left half of Figure 5 outlines the calibration process, using protein as an example. Several samples (the *calibration set*) should be analyzed by both the reference method and the NIR method, although, in practice, the two procedures usually analyze different subsamples of each sample. The reference method values are then regressed against the data treatment values to produce calibration constants. The number of samples required and the number of NIR wavelengths measured depend on whether one is developing a calibration model or actually calibrating. Development of a calibration model requires more samples and more available wavelengths than calibration itself. A simple rule of thumb is to use a minimum of 5–15 samples for each regression and data treatment constant and for any parameter of the data treatment (such as wavelength) that is allowed to vary. Examples are given in Section V.

The efficiency of a regression approximation for a set of calibration samples can be reported as the *standard error of calibration* (SEC), the *correlation coefficient* (r), or the *coefficient of determination* (r^2). The SEC is the standard deviation (s_e) of the individual approximation errors (e) and is also called the *standard error of estimate* (SEE) for the calibration regression. The relationship between r, r^2, and the SEC is illustrated in Table II, in which the standard deviation of the reference data (s_Y) was assumed to be 1.0 and the SEC was calculated assuming a single-term linear regression with 50 samples. The basic relationships to notice are that the SEC decreases as r increases, r is always larger in absolute value than r^2, $0 \le r^2 \le 1$, and $0 \le$ SEC. The last column of the table gives another way of understanding the correlation. In calculating its values, two things are assumed: first, the range of the reference data is about six times its standard deviation (a reasonable ratio for sample sizes larger than 50 when the reference data are normally distributed); and second, two samples can be correctly called different 95% of the time if their values are more than twice the SEC when apart. The correlation then permits division of the range into $6s_Y/(2\text{SEC})$ separate segments, and each sample is assigned to one of the segments by the calibration equation.

For example, suppose the Kjeldahl protein data for 50 samples had a standard deviation of 1.5% and a range of 9% (e.g., 10–19%). For a regression equation with a correlation of $r = 0.95$ ($r^2 = 0.90$), the corresponding entry in the fourth column of Table II

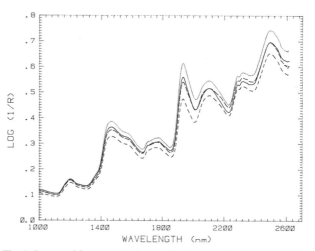

Fig. 4. Spectra of four samples from calibration set of Table I.

TABLE I
Reference Data for Illustrative Hard Red Spring Wheat Samples

Set	Protein (%) As Is	Protein (%) Dry Basis	Moisture (%)
Calibration (50 samples)			
Mean	14.15	15.98	11.47
Standard deviation	2.35	2.62	1.24
High	18.75	21.32	13.29
Low	10.13	11.35	8.97
Validation (50 samples)			
Mean	13.74	15.55	11.68
Standard deviation	2.11	2.39	1.43
High	17.71	20.05	14.32
Low	9.66	11.08	8.54
Reproducibility (20 samples)			
Mean	14.00	15.92	11.99
Standard deviation	0.061	0.065	0.273
High	14.09	16.06	12.78
Low	13.88	15.83	11.66

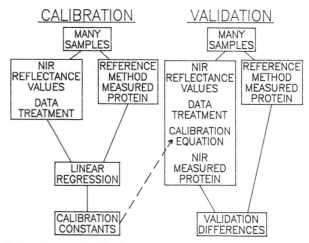

Fig. 5. Flow diagram of calibration and validation process.

(9.5) then shows that the range of 10–19% could safely be divided into nine segments and one could be 95% sure of not being more than one segment off when assigning a sample to a segment. Another way of looking at sample differences is in terms of the SEC. The SEC corresponding to $r = 0.95$ (0.315 from column 3 of Table II) is multiplied by the reference data standard deviation of $s_Y = 1.5\%$, because Table II is normalized to $s_Y = 1.00$. This gives an SEC of 0.47%. Then 95% of the NIR determinations will be within twice the SEC (0.94) of the reference values, and 99% should be within 3SEC = 1.41% of the reference values.

A second example uses the data in Table I. The as-is protein range of the calibration set is $18.75 - 10.13 = 8.62\%$ protein, which is 3.67 times the protein s_Y of 2.35% protein, indicating that the protein data are more uniformly than normally distributed. Because each KjD value is the average of 16 individual Kjeldahl values, the sampling error is greatly reduced, permitting very high correlations with data treatments of $\log(1/R)$ values. The SEC values of 0.15% protein reported by Williams et al (1983) correspond to $r = 0.998$ and $r^2 = 0.996$, because $0.15 = 0.064$ times the reference s_Y of 2.35%. Because the range is not $6s_Y$, in Table II, the fourth column entry of 47 segments must be adjusted to $(3.67/6) \times 47 = 29$ segments.

At the other extreme, correlation coefficients below $r = 0.6$ are not useful in this work because only three to four separate segments are statistically different. "Analysis" could be performed just as well by not using the regression equation at all but by declaring the determination of each unknown to be the mean value of the calibration samples. (You would still be sure of not being more than one segment off.) The situation is worse when the reference data are not normally distributed. For example, if they are uniformly distributed, then the range is approximately $3s_Y$ instead of $6s_Y$, and the values in column 4 of Table II must be divided by 2 to get the number of segments per range.

B. Measurement and Validation

Once an NIR instrument has been calibrated against a reference method, it can be used to determine the percentage of a constituent in different samples (called *unknowns*) or to measure some physical quantity of these samples. This *NIR measurement* should have a measurement error roughly equal to the SEC. However, the NIR measurement error may, in practice, be significantly larger than the SEC. Comparison of NIR measurements and reference method measurements on a new set of samples provides a basis for calculation of the true measurement error. This comparison is called *validation* (or verification) of the calibration.

Referring again to Figure 5 (right half), the mean ($\overline{D}$) and standard deviation (s_D) of the differences (D) between the NIR-measured and reference values for several samples (called the validation set, verification set, or prediction set) estimate the systematic and random errors, respectively, of the NIR method. The estimate $\overline{D}$ is called the BIAS and s_D is called the *standard deviation of validation differences* (SDVD = $\sqrt{[\Sigma(D - \overline{D})^2/(n - 1)]}$, in which n is the number of validation samples).

The SDVD, which is a term invented for this chapter, has been called the SEP (standard error of prediction or of performance), the SEP (C) = SEPBC (bias-corrected standard error of prediction), the SDP (standard deviation of prediction), and the SED (standard error of differences). I have not found any standard term in the statistical literature. Because the terms SEP and SED have also been used for estimators different from the SDVD, readers of the literature are cautioned to look for definitions in each paper, and authors should quote definitions until this area becomes standardized.

There is some confusion about the correct way to combine the BIAS and the SDVD and how to interpret a combination. The confusion stems from two possible interpretations of the BIAS. If the BIAS is interpreted to mean a constant systematic error that is not expected to change, then a correct statement is "If the samples used to determine the BIAS and SDVD are representative of the population to be measured, then 95% of future measurements will be within 2SDVD of (reference value plus BIAS) and 99% will be within 3SDVD of (reference value plus BIAS)." For example, if an instrument is calibrated with samples at one temperature and then is used to measure samples at a second temperature, the measurements may be biased. This bias is estimated by obtaining reference values for several samples, subtracting them from the corresponding NIR measurements, and finding the mean ($\overline{D}$) of the differences (D). Future measurements of samples at the second temperature should then be corrected for this systematic difference. If the BIAS is interpreted to mean a constant systematic error, it should not be combined with the SDVD, except in statements such as the above.

On the other hand, the BIAS $\overline{D}$ may be interpreted as being caused by special characteristics of a set of samples or measurement conditions, which may change from day to day. Considered this way, $\overline{D}$ is expected to change with each new validation set and is not, by itself, an estimate of a constant systematic error as in the previous paragraph. Under certain assumptions, the *root mean square deviation* (RMSD = $\sqrt{[\Sigma D/n]}$) is then a more appropriate measure of the random error than the SDVD. To discuss this, stricter terminology than was used in the last paragraph is needed. The set of validation errors (D) is assumed to be representative of the "population" of all such errors. The mean (μ_D) of the error population is the true systematic error and the standard deviation (σ_D) is the true random error. The parameters μ_D and σ_D were estimated in the previous paragraph by the error sample mean ($\overline{D}$) and standard deviation (s_D), respectively. Some researchers assume that the measurement has no systematic error, which is the same as saying that $\mu_D = 0$. This assumption can be postulated, justified by previous experience with the measurement process or similar sample sets, or justified by showing that the sample mean is not significantly different from zero. With the as-

TABLE II
Relationship of Measures of Correlation
for Single-Term Linear Regression Using 50 Samples[a]

r	r^2	SEC	Segments per $6s_Y$
0.0000	0.0000	1.010	3.0
0.2000	0.0400	0.990	3.0
0.4000	0.1600	0.926	3.2
0.6000	0.3600	0.808	3.7
0.8000	0.6400	0.606	4.9
0.9000	0.8100	0.440	6.8
0.9500	0.9025	0.315	9.5
0.9800	0.9604	0.201	14.9
0.9900	0.9801	0.143	21.0
0.9950	0.9900	0.101	29.7
0.9980	0.9960	0.064	47.0
0.9990	0.9980	0.045	66.4
0.9995	0.9990	0.032	93.9
0.9998	0.9996	0.020	148.4
0.9999	0.9998	0.014	209.9

[a] r = correlation coefficient, r^2 = coefficient of determination, SEC = standard error of calibration, and $s_Y = 1.00$ = standard deviation of reference data.

TABLE III
Relationship of BIAS, SDVD, and RMSD for SDVD = 1.00[a]

BIAS	RMSD	BIAS^2/SDVD^2
0.00	1.00	0.00
0.20	1.01	0.04
0.40	1.08	0.16
0.60	1.17	0.36
0.80	1.28	0.64
1.00	1.41	1.00
2.00	2.24	4.00
4.00	4.12	16.00
8.00	8.06	64.00

[a] BIAS = mean of validation differences, SDVD = standard deviation of validation differences, and RMDS = root mean square of validation differences.

sumption that $\mu_D = 0$, the RMSD is the appropriate estimator for the random error because it is an unbiased estimator for $\sqrt{(\sigma^2_D + \mu^2_D)}$.

So the decision whether to use the RMSD or the SDVD and BIAS is governed by assumptions about the systematic error of the measurement. If it is assumed to be zero, then the RMSD estimates the random error. If it is assumed to be different from zero, then the BIAS estimates the systematic error and the SDVD estimates the random error. For large n, RMSD $\cong \sqrt{(SDVD^2 + BIAS^2)}$.

RMSD is not a standardized term. It has also been used to mean $\sqrt{[\Sigma D^2/(n-1)]}$, which is the above RMSD with a divisor of $n - 1$ instead of n. Because this expression is not an unbiased estimator for the random error (σ_D), it is not a useful estimator, except in the sense that, for large n, it is very close to $\sqrt{(\Sigma D^2/n)}$. The RMSD is seldom found in statistics books, except as a term for the ordinary sample standard deviation.

The terms *precision* and *accuracy* have been avoided in this discussion because they have not been standardized. In this monograph, precision will mean the reproducibility of a measurement under specified conditions. Accuracy will mean the closeness of the measurement to the "true" value as determined by an accepted reference method.

C. Developing a Calibration Model

Development of a calibration model is dependent on the variety and flexibility of the data treatments available and the experience of the developer. The simplest situations involve regression on $\log(1/R)$ values, in which the only variables are which and how many wavelengths to use from a set of six. The most complex occur in research laboratories that have most of the methods in this chapter and Chapter 4 available.

The first step is recording the spectra and reference data on sufficient numbers of calibration and validation samples. For calibration, the number of samples considered sufficient is discussed in Section V. For validation samples, no consensus exists on a minimum number, and validation sets of from 6 to over 100 samples have been quoted. It is important that the validation samples cover the range of variability anticipated in future samples. This, however, is usually impossible to guarantee, and additional validations are often done routinely to detect changing situations and revise calibrations.

The second step should be the determination of the sampling and reference method errors discussed in Section III. The *reference-to-reference correlation, reproducibility,* or *laboratory error* (LE) may include sampling error and can be measured by the standard deviation of reference measurements on replicate samples. The tests should duplicate both sample preparation and analysis. An NIR method cannot correlate to a reference method better than the reference method correlates to itself, so the LE is the practical lower limit for the SEC. The LE should also be compared with the standard deviation of the reference data (s_Y). Table II can be used to convert the minimum expected SEC to the maximum expected absolute r by finding the corresponding normalized SEC = LE/s_Y in the third column. For example, if the LE is about one third of the reference standard deviation, then the fraction 0.33 gives a maximum expected absolute r of somewhat less than 0.95. Alternatively, because $r^2 \cong 1 - SEC^2/s_Y^2$ and LE $\leq$ SEC, then $r^2 \leq 1 - LE^2/s_Y^2$ gives a formula for a realistic maximum possible r^2.

The third step is to choose data treatments that provide acceptable SECs. "Acceptable" usually means from one to two times the LE. The NIR reproducibility should also be checked for any data treatment chosen.

The fourth step is to validate the acceptable calibrations to test for *overfitting*, which happens when calibration regression approximate nonrepresentative features of the particular samples used for calibration. Several different indications of overfitting are common.

1. The SEC is much lower than the LE. This is usually the result of using too many terms in the calibration regression or too few samples.
2. The BIAS is significant compared with the SDVD. The comparison should be the fraction $BIAS^2/SDVD^2$, be-

cause errors compare only when squared. For example, if the SDVD = 1.00, then a bias of 0.40 contributes only 16% of the variance, but a bias of 0.80 is significant. (Table III gives other examples.) If the BIAS looks significant, then further validation should be done to decide whether it is appropriate to compensate for the BIAS or to treat it as long-term random error.
3. The BIAS is negligible, but the SDVD is more than 2SEC. In this situation, it may be possible to get a good NIR measurement from a bad calibration by averaging the results of several NIR measurements on subsamples (essentially by averaging out sampling error).

The fifth step in developing a calibration model is to try to understand why it works: why the wavelengths are chosen and how the data treatment helps combine the $\log(1/R)$ values in a meaningful way. Sometimes this may be as simple as noticing that a protein calibration uses a wavelength at which radiation is absorbed by protein. A calibration that has not been justified spectroscopically is less likely to stand up in the field even though it has been validated.

Different data treatments must be validated on separate randomly selected sample sets for the validation results to be statistically comparable, although this is rarely done in practice because of the expense of reference measurements.

A table listing the individual calibration or validation errors is useful for learning about a data set, especially for detecting outliers and transcription errors. Table IV illustrates this. The number of asterisks beside each error represents the absolute size of the error in terms of the SEC. A reasonable calibration error distribution would have one third of the errors with one asterisk, 4% with two, and 1% with three. A similar distribution for validation errors would indicate no overfitting and no obvious outliers. A glance at the asterisks in Table IV suggests that the data for samples 4 and 5 have been interchanged. A reasonable error distribution results if the reference values of samples 4 and 5 are changed to 13.5 and 12.5, respectively.

III. SOURCES OF ERROR

The error in an NIR measurement is dependent on the interaction of many sources, which we will examine under three broad categories: sampling error, reference method error, and NIR method error. With high quality instruments, sampling error is usually the largest component of differences between reference and NIR measurements.

Errors are not necessarily a matter of imperfect machines or people, but they can be a result of using different measurement methods. In some ways, *difference* is a better word than *error*. For example, in the measurement of protein, the NIR method measures molecular bonding, whereas the Kjeldahl method determines total nitrogen content, which is not perfectly correlated to molecular bonding. The buyer wants to know the protein level because it relates to bread-making quality or resale price, neither of which is perfectly correlated to either protein or nitrogen level. These considerations suggest that it may be neither possible nor useful to improve the NIR-Kjeldahl correlation beyond a certain point.

TABLE IV
Portion of Table of Near-Infrared (NIR) Method and Reference Method Values

Sample No.	NIR Value	Reference Value	Error
1	9.6	9.5	0.1
2	10.7	10.5	0.2
3	11.4	11.5	−0.1
4	13.8	12.5	1.3******[a]
5	12.3	13.5	−1.2******
6	14.5	14.5	0.1
7	15.1	15.5	−0.4**
8	16.8	16.5	0.3*
9	17.4	17.5	−0.1
10	18.6	18.5	0.1

[a] Asterisks show errors in terms of standard error of calibration (SEC = 0.2).

A. Sampling Error

Sampling error is caused by lack of homogeneity in the material being sampled. For example, if two 1-g samples are drawn separately from a well-mixed, 1-kg batch of ground wheat, their protein content as measured by the Kjeldahl method may vary by 0.2% protein. Most of this variation results not from lack of precision of the Kjeldahl method, but from actual protein differences between the samples. Such variation is often the effective lower limit on the SEC in an NIR calibration. Table V shows common sources of sampling error for wheat and forages.

Sampling errors occur at several places in the calibration-validation process outlined in Figure 5.

1. The reference and NIR methods usually analyze different subsamples.
2. The methods may look at different amounts of the material. If a method used four times as much material, it will usually have about half the sampling error. In Table VI, the reproducibility for various protein determination methods are compared with the sample size used by the methods. The inverse relationship between sample size and error accounts for some of the precision differences between the methods.
3. The NIR measurement can introduce a sampling error because the radiation penetrates less than 2 mm into a sample that may be as deep as 10 mm, so only a portion of the sample is being measured. Reloading the same sample or a different subsample back into the cup several times gives a basis for calculating the loading (or reload or repack) error, which is primarily a sampling error; it may, however, include such sources of error as differences in sample compression, which affects reflectance, and differences in static electricity or particle size, which may cause certain types of materials to rise to the surface and thus introduce a bias in the measurement. Rotating the sample cup is like reloading, because a different part of the sample is observed. Using diffuse transmittance instead of reflectance can reduce the errors that are caused by only part of the sample being analyzed, but other sampling errors can then occur. For example, transmittance through whole kernel grain may overemphasize the outer layers of each kernel, which may not be representative of the whole kernel, and instruments using several small, filtered sources of radiation may have different wavelengths interacting with different parts of the sample.

Although it is rarely practicable to measure accurately all the kinds of sampling errors in a particular application, it is useful to know where they occur and the approximate contribution of each source to the total sampling error. This can result in savings by indicating where to repeat the sampling. For example, suppose sampling error at the boxcar level is 0.4% protein, and reload error on subsamples, after a single sample has been ground, is 0.2%. The boxcar + reload error is then $\sqrt{(0.4^2 + 0.2^2)} = 0.45$. Averaging the spectra of four reloads from the single ground sample halves the reload error so that the combined error from these two sources is $\sqrt{(0.4^2 + 0.1^2)} = 0.41$. But averaging the spectra of four resamplings from the boxcar would theoretically halve both errors, giving a combined error of $\sqrt{(0.2^2 + 0.1^2)} = 0.22$. In this kind of calculation, the cost of the various levels of resampling must be balanced against the benefit from the reduced error.

If the various stages of a sampling process are replicated, then the relative importance of the errors introduced at each stage can be estimated by the analysis of variance, which can also be used to measure other sources of error. I illustrate this here by using a personal communication from P. C. Williams on ground samples. Some data on NIR reflectance-determined protein variability are presented in Table VII. From each of four cargoes, six lots were drawn and thoroughly mixed. From each lot, two samples were drawn and two subsamples were ground from each sample. From each grind, two repacks of the sample cup were used for duplicate NIR protein determinations. The sum of squares column shows the relative amount of variance arising from each stage of the process. The root mean square column shows the error in terms of percentage of protein. The between-cargo variability dominates the total variation, with significant variability occurring between subsamples within a cargo lot and between duplicate tests (reloads). The variance due to individual cargo lots was insignificant, because sampling of cargoes is fully automatic and continuous, with low variability between lots relative to variance between cargoes. Subsampling of the cargo lot samples is manual. Variance due to grinding was also insignificant. Grinding of samples for NIR analysis at Canada's Grain Research Laboratory involved a cyclone grinder with a sample flow regulator. The ground sample is mixed very thoroughly in the sample grinder receiving jar and also in the sample tin, to minimize the grinding error. Loading the NIR instrument cell inevitably incurs some stratification, which is reflected in the variance of duplicate testing. The relative root mean square values show that most of the variance came from the cargoes themselves. Cargo lots showed roughly twice the variance of the subsamples, which in turn reflected about twice the variance due to grinding and reloading.

B. Reference Method Error

Every reference method has a measurement error (LE, reproducibility, or standard error of a single test [SET]) that is defined as the standard deviation of measurements on replicate samples. Several of these standard deviations can be combined using root mean square averaging. A different definition (which is $\sqrt{2}$ times the above definition) is the standard deviation of differences between measurements on duplicate samples. The literature uses both definitions, often without stating which. In this chapter, we use the first definition because it is more directly comparable to the SEC and SDVD.

For the as-is protein data in Table I, the LE using duplicate samples drawn after grinding and mixing 25 g of wheat is 0.12% pro-

TABLE V
Sources of Sampling Error in Two Commodities

Commodity	Sources of Sampling Error
Wheat	Type of sample used
	Foreign material
	Blending of withdrawn sample
	Sample storage
	Sample identification
	Subsampling for preparation for analysis
	Blending of prepared (usually ground) sample
	Subsampling for actual laboratory analysis
Straw and forage	Physical composition
	Dimensions (length, etc.)
	Leaf/stem ratio
	Stage of maturity
	Moisture content
	Subsampling for sample preparation
	Blending of withdrawn sample
	Sample identification
	Blending of prepared (usually ground) sample
	Subsampling for actual laboratory analysis

TABLE VI
Comparison of Methods for Determining Protein in Wheat

Method	Reproducibility (% Protein)	Sample Size (g)
Proton activation	0.189	1
Thermal decomposition	0.185	1
Kjeldahl nitrogen	0.154	16[a]
Near-infrared reflectance	0.141	16[b]
Neutron activation	0.074	100

[a] Average of 16 1-g replicates.
[b] Average of four 4-g replicates.

tein. This error contains very little sampling error because a small amount of wheat after grinding can be made very homogeneous. The KjDs, on the other hand, should have very little reference method error, because the average of 16 Kjeldahl determinations should have an LE of $0.12/\sqrt{16} = 0.03\%$ protein. The standard deviation of 0.06% protein for the reproducibility set is thus a measure of the sampling error between 1,200-g portions drawn from the original lots of grain.

There are also differences between what is measured by NIR and what is measured by the reference method. In the case of the Kjeldahl method, total nitrogen is determined and multiplied by a factor to give protein. This factor, called the protein-to-nitrogen ratio, varies between commodities and can vary within a commodity (Jones, 1931; Tkachuk, 1977; Coblentz, 1982). NIR does not determine nitrogen, but it measures vibrations of molecular bonds, which are more directly correlated to total protein. Thus, a further limit on the NIR-Kjeldahl standard error of calibration, if a constant factor is used for all samples, is the variability of the protein-to-nitrogen ratio.

In other measurements, what the reference method is measuring may not be a direct result of chemical composition. It could be a subjective value, such as a taste panel score or a physical property such as loaf volume. In cases like these, the NIR-reference method error will have a lower limit, although the molecular basis of the limit may be unknown.

As discussed in Section II, NIR results can often be used to detect errors in the reference method, especially errors involving transcription or sample mix-up. This is done by treating the calibration samples as a validation set and looking for abnormally large validation errors.

C. NIR Method Error and Smoothing

NIR method errors are caused by spectral measurement errors, lack of intrinsic correlation between spectral and reference methods data, and poor choice of data treatment. A detailed discussion of the detection, measurement, and correction of errors in the measurement of spectra is given in Chapter 9, and the lack of intrinsic correlation has been discussed earlier in this chapter. The relationship of specific data treatments to various types of error will be mentioned in later sections.

Instruments used in NIR reflectance spectroscopy have been designed to have very low errors in both wavelength and log(1/R) values because, in the applications developed for NIR reflectance instruments, the spectral variation related to variation in composition of samples is very low. For instance, Figure 2 shows that a 4% change in protein is equivalent to a 0.01 change in log(1/R) value at 2,180 nm. A 0.2% change in protein then corresponds to a 0.0005 change in log(1/R) value. Using the slope at 2,180 nm, we can calculate roughly that a wavelength error of 0.3 nm also corresponds to a 0.0005 error in log(1/R) value, which would again correspond to a protein determination error of 0.2% protein. Thus, the usual specification of log(1/R) and wavelength reproducibility as 0.0001 log(1/R) units and 0.05 nm, respectively, en-

sures that the instrument-caused error is well below the usual sampling error of 0.15–0.2% protein.

Figure 6 shows a wheat spectrum with added noise of 0.0001, 0.001, and 0.01 log(1/R). The 0.001-log(1/R) noise is at least 10 times the maximum normal noise. Even though it is barely visible to the eye, it causes a doubling of the calibration and validation errors when added to the data sets of Table I.

Although the usual noise specifications are adequate for most NIR reflectance applications, software packages developed for research scanning instruments usually have routines for mathematically smoothing spectral data. These generally use either a running mean or a Savitzky-Golay smooth, although several other methods could be used: binomial (Marchand and Marmet, 1983), Fourier, and Gaussian (Section VII).

A *running mean* (or boxcar smooth) simply replaces the value at each point by the mean of the values in a wavelength interval surrounding it. The interval is preferably specified in nanometers but is sometimes specified in number of points. The interval is centered at the given point, resulting in an odd number of data points per mean. In reading the literature, one should make sure of an author's definition of the running mean. If the number of points is $n = 2m + 1$, in which m is the number of points on either side of the central point, then sometimes the "number of points" quoted is m. The advantage of this method is its ease of calculation.

The principle of the Savitzky-Golay smooth is to fit the spectrum in a wavelength interval with a polynomial by least-squares methods, and the parameters are the degree of the polynomial and the number of points to fit (Savitzky and Golay, 1964; Morrey, 1968). The smoothed spectrum and its derivatives are then given by the polynomial and its derivatives. (Section VI has a review of derivatives.) The method is used primarily where the bandwidth or band motion to be detected is on the order of the distance be-

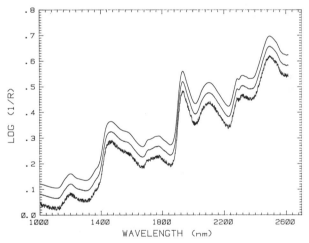

Fig. 6. Wheat spectra with added, uniformly distributed noise of 0.0001 (top), 0.001 (middle), and 0.01 (bottom) log(1/R) units. Spectra are displaced for clarity.

TABLE VII
Analysis of Variance of Wheat Cargo Analysis for Protein by Near-Infrared (NIR) Reflectance

Source	Sum of Squares	Degrees of Freedom	Mean Square	Variance Ratio	Probability	Root Mean Square
Cargoes	38.350	3	12.783	175.1	0.001	3.58
Lots	0.829	5	0.166	2.3	NS[a]	0.41
Error A	1.091	15	0.073			
Samples	0.057	1	0.057	3.0	NS	0.24
Error B	0.440	23	0.019			
Grinds	0.012	1	0.012	0.5	NS	0.11
Error C	1.125	47	0.024			
Duplicate NIR reflectance	0.015	1	0.015	3.9	0.5	0.12
Error D	0.359	95	0.0038			
Total	42.278	191				

[a] NS = not significant.

tween adjacent wavelengths or smaller, where the data are very noisy, or where it is important to preserve spectral band shape.

Figure 7 illustrates the effects of the two smoothing methods on computer-generated Gaussian curves with added noise. Notice that the running mean smooth lowers the peak height of the narrow band but does not distort the wide band. The remainder of this section contains the details of the Savitzky-Golay calculation.

The Savitzky-Golay calculation involves the solution of simultaneous equations. If λ_j is the wavelength at which the smoothed or derivative value is desired, n is the degree of the fitting polynomial, and $2m + 1$ is the number of points to fit (giving an interval symmetrical about λ_j), then the $n + 1$ simultaneous equations are

$$\sum_{i=-m}^{m} i^r y_i = \sum_{k=0}^{n} \left(b_{nk} \sum_{i=-m}^{m} i^{r+k} \right)$$

one for each r, in the $n + 1$ unknowns b_{nk}, in which $r = 0, 1, ..., n$, and y_i is the log($1/R$) value at wavelength λ_{y+i}. The value b_{nk} is the coefficient of the kth term of the nth degree polynomial, b_{n0} is the smoothed value at λ_j, and the value of the sth derivative at λ_j is $s!b_{ns}$.

The calculation can be thought of as a weighted average of the log($1/R$) values in the interval $(\lambda_{j-m}, \lambda_{j+m})$. The weights can be extracted during the calculation if a smoothed or derivative value for many points is desired. In that case, the weights for calculating the sth derivative at λ_j are $s!$ times the elements of the $(s + 1)$th row in the matrix $A^{-1}B$, in which

$$A_{r+1,k+1} = \sum_{i=-m}^{m} i^{r+k} \, (r, k = 0, 1, ..., n)$$

and

$$B_{r+1,i+m+1} = i^r \, (r = 0, 1, ..., n; \; i = -m, ..., 0, ..., m)$$

The zero-order derivative is the Savitzky-Golay smooth. Savitzky and Golay (1964) give hints for simplifying the calculations, as well as tables for computing $s!b_{ns}$ for various given s and n. Users of these tables should consult the corrections by Steinier et al (1972).

IV. SINGLE-TERM LINEAR REGRESSION AND THE CORRELATION PLOT

This section gives a brief introduction to the concepts involved in single-term linear regression, the statistical procedure that answers the following question: "Given a set of data with one independent variable X and one dependent variable Y (Table VIII) and the corresponding scatterplot of Y against X (Fig. 8), what is the straight line that best fits the data?" The answer is the straight line with the equation

$$\hat{Y} = a + bX$$

in which $\hat{Y}$ is an approximation to Y, and a and b are given by the formulas in the appendix to this chapter. Figure 9 shows the geometric meaning of Y, a, and b. For a given data point (X,Y), $\hat{Y}$ is the point on the line immediately above or below the data point. The Y-intercept (a) is the Y value at the point where the line crosses the y axis. The slope (b) is the steepness of the line, measured by the increase in Y for each unit increase in X.

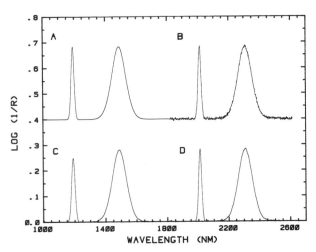

Fig. 7. Effect of smoothing: **A**, raw data, computer-generated Gaussian curves; **B**, normally distributed noise (standard deviation 0.001 log[$1/R$]) added to **A**; **C**, running-mean smooth of **B**; and **D**, Savitzky-Golay smooth of **B**.

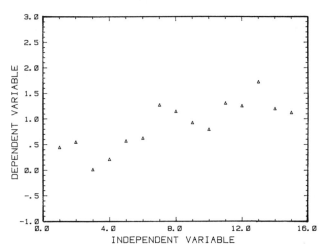

Fig. 8. Scatterplot of data in Table VIII.

TABLE VIII
Data Generated Randomly About Line $Y = 0.0 + 0.1X$[a]

Independent Variable	Dependent Variable	Residual
1.00	0.445	0.345
2.00	0.547	0.347
3.00	0.013	−0.287
4.00	0.213	−0.187
5.00	0.569	0.069
6.00	0.626	0.026
7.00	1.273	0.573
8.00	1.151	0.351
9.00	0.926	0.026
10.00	0.797	−0.203
11.00	1.312	0.212
12.00	1.258	0.058
13.00	1.723	0.423
14.00	1.200	−0.200
15.00	1.125	−0.375

[a] X = independent variable, Y = dependent variable, standard deviation of residuals = 0.289, and mean of residuals = 0.079.

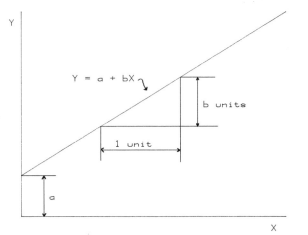

Fig. 9. Geometric meaning of constants in straight line $Y = a + bX$.

The best-fitting line is called the regression of Y on X or Y regressed against X. The regression constant a is called the constant term, and the regression constant b is called the regression coefficient. The vertical distance from a data point to this line is the residual or regression error for that point (Fig. 10), and the standard deviation of all the residuals is the SEC (or SEE). The correlation coefficient (r), which is related to the SEC, ranges from −1 to 1. If $r = 1$, the data are exactly on the regression line and are perfectly correlated. If $r = 0$, then there is no correlation, and if the slope of the regression line is negative, then r is negative, giving a perfect negative correlation if $r = -1$. Figures 11–14 give examples of scatterplots of data with four levels of correlation.

In developing a calibration model using single-term linear regression, when one does not yet know the best wavelength to use, one normally finds the r value at every available wavelength. The wavelength giving the highest r value is then used for the actual calibration and subsequent validation. However, in practice, this simple approach seldom gives an adequate SEC, and a more complex calibration is usually needed. The result of successively more complex mathematical treatments in single-term regression can be illustrated with the correlation plot, which is a graph of r against wavelength.

Figure 15 shows the correlation plot for the 50 calibration samples of Table I, in which the r values are calculated with as-is protein content regressed against the $\log(1/R)$ value at each of 1,000 wavelengths. The fact that the maximum r in Figure 15 is only 0.26 indicates that a more complicated mathematical treatment is necessary to obtain useful results.

In this example, the X in $Y = a + bX$ is the $\log(1/R)$ value at one wavelength. One way to improve the correlation is to let X be

the difference between $\log(1/R)$ values at two different wavelengths. The two wavelengths can be found by an iterative process. First, the single wavelength giving the best correlation is found; then, a second wavelength is found so that the difference between $\log(1/R)$ values at the first and second wavelengths gives

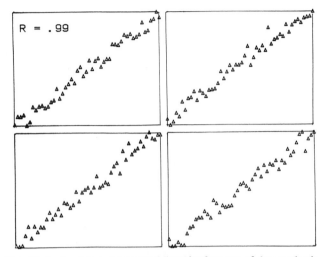

Fig. 12. Scatterplots and regression lines for four sets of data randomly generated about line $Y = 0.0 + 1.0X$ ($r = 0.99$).

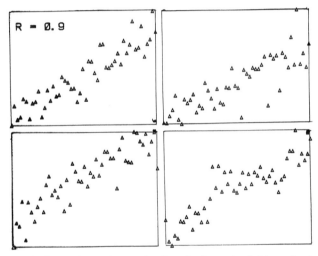

Fig. 13. Scatterplots and regression lines for four sets of data randomly generated about line $Y = 0.0 + 1.0X$ ($r = 0.9$).

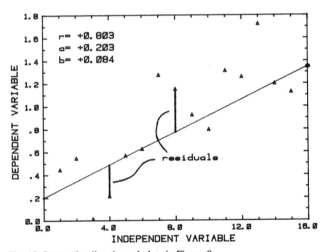

Fig. 10. Regression line through data in Figure 8.

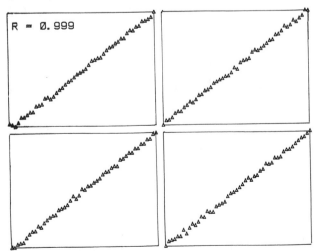

Fig. 11. Scatterplots and regression lines for four sets of data randomly generated about line $Y = 0.0 + 1.0X$ ($r = 0.999$).

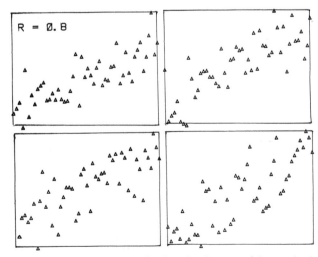

Fig. 14. Scatterplots and regression lines for four sets of data randomly generated about line $Y = 0.0 + 1.0X$ ($r = 0.8$).

the best correlation. The first wavelength is then replaced with a third wavelength whose difference with the second gives the best correlation, and so on until the process converges, which means that each of two wavelengths pairs best with the other. When this technique was applied to the calibration samples of Table I, the maximum *r* value rose to 0.92, with the general level of the correlation plot (Fig. 16) becoming much higher. An iterative procedure does not necessarily produce the pair of wavelengths whose difference provides the highest correlation. It only provides wavelengths that would choose each other. The same process can be used with quotients instead of differences (Fig. 17) and with the quotients of differences (Fig. 18). In this last case, there are various ways of iterating the process when selecting the four wavelengths whose $\log(1/R)$ values are put into the expression $(A - B)/(C - D)$. The various methods do not all yield the same choice of wavelengths.

These selection methods do not require a correlation plot, especially when only a few wavelengths are available; but the plot or a listing of the correlations often gives information as to why certain wavelengths are selected. Many researchers are reluctant to recommend wavelengths that are selected unless there is some spectroscopic justification. Of the various justifications offered by computer-selected wavelengths, the most common are that a chosen wavelength is (i) at the peak absorbance of the component to be determined; (ii) at a peak absorbance of a component whose concentration is highly correlated with that of the component to be determined; and (iii) part of a difference or quotient expression and serves to normalize the spectra to one level of scatter, particle size, temperature, etc. In this last case,

the wavelength may be called a *reference wavelength* (Sections V and VIII).

The correlation plot also shows which wavelengths may be unsuitable when transferring calibrations from a laboratory situation to the field. If the plot shows a sharp spike at a wavelength (Fig. 18, 1,850 nm), then a measurement based on that wavelength requires an instrument with very high wavelength accuracy and precision. The correlation plot can be used with multiterm regression, as discussed in Section V.

V. MULTITERM LINEAR REGRESSION

A. Basic Properties

Multiterm linear regression uses the information at a number of wavelengths to isolate the effect of a single absorber and to normalize the baseline. This technique is more difficult to describe geometrically than the single-term case, and the formulas for the regression coefficients are more complex. Given a set of data with one dependent variable (Y) and k independent variables $X_1, X_2, ..., X_k$, the problem is to find the set of constants $a, b_1, b_2, ..., b_k$ so that the function

$$\hat{Y} = a + b_1 X_1 + b_2 X_2 + ... + b_k X_k$$

is the best fit to the data. If $k = 1$, this function is the straight line over the X_1 axis as discussed in the previous section. If $k = 2$, the function is a plane over the $X_1 X_2$ plane. If $k = 3$, then there is no easily visualized function and no simple formula.

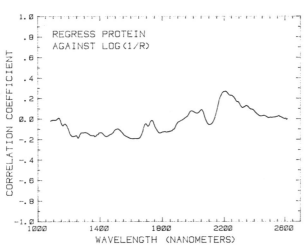

Fig. 15. Correlation to $\log(1/R)$ value at each wavelength using 50 wheat samples.

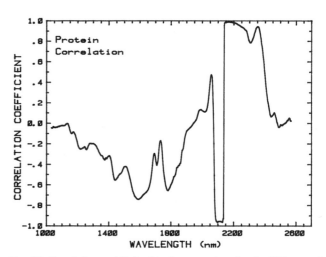

Fig. 17. Correlation to A/B for 50 wheat samples. $A = \log(1/R)$ at each wavelength, and $B = \log(1/R)$ at 2,138 nm.

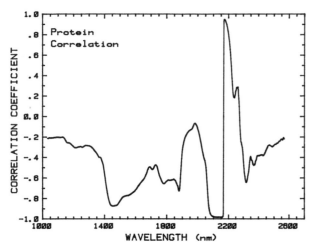

Fig. 16. Correlation to $A - B$ for 50 wheat samples. $A = \log(1/R)$ at each wavelength, and $B = \log(1/R)$ at 2,166 nm.

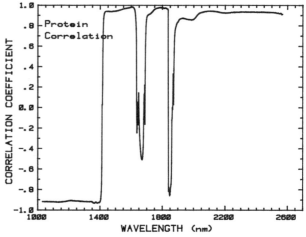

Fig. 18. Correlation to $(A - B)/(C - D)$ for 50 wheat samples. A, B, and $D = \log(1/R)$ at 1,982, 1,907, and 1,838 nm, respectively; and $C = \log(1/R)$ at each wavelength.

Usually the X's are log$(1/R)$ values at different wavelengths. But they can be anything that will help the measurement. Examples are derivatives, quotients of derivatives, Fourier coefficients, curve fit coefficients, principal component scores, and other types of scores, all discussed in later sections and Chapter 4. Auxiliary variables can also be added. For instance, the time a spectrum was measured, the instrument operator, the instrument, the wheat variety, etc., could be used in the regression. In most of this section, the term *wavelength* or *log(1/R) value* can be replaced by any other variable without changing the intended meaning.

There are various ways of choosing the wavelengths to use in a multiterm linear regression. The step-up or forward-stepwise procedure picks the wavelength giving the best single-term calibration as the first independent variable, and then finds the best wavelength to add as a second variable in a two-term regression, and so on until some stopping criterion is met. The step-down procedure starts with a multiterm linear regression using all available wavelengths and eliminates variables by some criterion. The all-possible-combinations procedure tests all possible linear regressions on all subsets of the available wavelengths and reports the subsets giving the lowest SEC. This procedure is usually limited to all subsets containing only two or three wavelengths. There are also combinations of these methods. For example, the all-possible-combinations method can select two or three wavelengths, and then the step-up method can be used to add wavelengths. Alternatively, each step in the step-up method can be followed by one step of the step-down method, to check for wavelengths that can be safely eliminated when a new wavelength is added. It is important to use some criteria for selection of subsets of available variables to avoid overfitting (Draper and Smith, 1966; Hocking, 1970). On the other hand, the question of the validity of the various stopping techniques in the face of the multicollinearity of the log$(1/R)$ values at several wavelengths requires further research.

Nonlinear regression is not in common use by NIR workers and will not be discussed here.

A k-term regression can be reduced to a $(k-1)$-term regression by "constraining" or "forcing" the regression coefficients to sum to zero. This constraint, which is one approach to the problem of shifting baseline, is equivalent to using one of the wavelengths as a reference wavelength, as in the previous section, or to normalizing the spectra (Section VIII), as can be shown as follows.

The constraint $\sum_{i=1}^{k} b_i = 0$ can be written as $b_1 = -\sum_{i=2}^{k} b_i$. Then

$$\sum_{i=1}^{k} b_i X_i = \left(-\sum_{i=2}^{k} b_i\right) X_1 + \sum_{i=2}^{k} b_i X_i = \sum_{i=2}^{k} [b_i(X_i - X_1)]$$

and X_1 is the reference wavelength log$(1/R)$ value.

A finite-difference derivative can be considered as a multiterm linear regression reduced to a single-term regression by a constraint on the regression coefficients. For example, the finite-difference second derivative can be considered to be a three-term regression on the log$(1/R)$ values at evenly spaced wavelengths, in which the regression coefficients must be in the ratio 1:−2:1. This constraint not only takes care of shifting baselines, but also separates overlapping absorption bands. Another advantage of such constraints is that they reduce the number of independent variables and thereby reduce the number of samples required for calibration.

It was mentioned above that an adequate calibration usually requires a minimum of 10 samples for each estimated regression constant and for each varied parameter (such as wavelength). The factor 10 is based on experience with validation rather than any statistical proof. Some researchers are content with as few as 5; some demand as many as 15. Following are a few examples using the factor 10.

1. A bias adjustment on an instrument is a change in the constant term used in the regression. Only one parameter is changed, so 10 samples are enough to make the adjustment.
2. A slope adjustment of a multiterm regression consists of multiplying all regression coefficients by the same number. Since essentially only one new number is determined, 10 samples are again sufficient.

3. A three-term calibration using three preselected wavelengths would require 40 samples (including 10 for the constant term).
4. Developing a 10-term calibration requires 210 samples (1 constant term + 10 coefficients + 10 wavelength selections = 21 parameters × 10 samples per parameter).
5. Developing a single-term calibration consisting of a quotient of second derivatives in which the wavelengths and gaps for the numerator and denominator derivatives (Section VI) are optimized would require 60 samples (10 × [2 regression constants + 2 wavelengths + 2 derivative parameters]).

The SEC, r, and r^2 have the same meaning in multiterm linear regression as in the single-term case, except that r is not negative. The correlation plot should be studied after the addition of each variable. For example, Figure 19 shows the correlation plots when the first, second, and third wavelengths are found. Notice that the maximum r in the first plot is the minimum r of the second.

B. Calculation

If there are two independent variables in a regression, the regression constants are given by

$$D = \sum x_1^2 \sum x_2^2 - \left(\sum x_1 x_2\right)^2$$

$$b_1 = \left(\sum x_2^2 \sum x_1 y - \sum x_1 x_2 \sum x_2 y\right)/D$$

$$b_2 = \left(\sum x_1^2 \sum x_2 y - \sum x_1 x_2 \sum x_1 y\right)/D$$

$$a = \overline{Y} - \left(b_1 \overline{X}_1 + b_2 \overline{X}_2\right)$$

in which the lowercase letters are the "centered" variables $y = Y - \overline{Y}$, $x_1 = X_1 - \overline{X}_1$, etc.

If $k \geq 3$, the a and b_j are best given as the solution to a set of simultaneous equations as follows. If Y_i are the Y values for n samples, and X_{ij} are the corresponding X_1 values for $j = 1, 2, ..., k$, then the approximation desired is given by the least-squares solutions $a, b_1, b_2, ..., b_k$ of the set of simultaneous equations

$$Y_1 = a + b_1 X_{11} + b_2 X_{12} + ... + b_k X_{1k}$$

$$Y_2 = a + b_1 X_{21} + b_2 X_{22} + ... + b_k X_{2k}$$

$$. \quad . \quad . \quad . \quad . \quad .$$

$$. \quad . \quad . \quad . \quad . \quad .$$

$$. \quad . \quad . \quad . \quad . \quad .$$

$$Y_n = a + b_1 X_{n1} + b_2 X_{n2} + ... + b_k X_{nk}$$

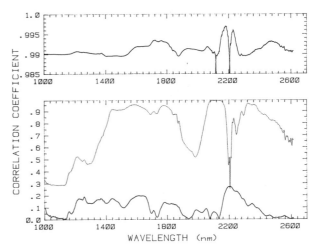

Fig. 19. Correlation plots for 50 wheat samples using one-, two-, and three-term correlations, bottom to top, respectively.

A simpler set of equations results from using the centered variables ($x = X - \overline{X}$).

$$y_1 = b_1 x_{11} + b_2 x_{12} + \ldots + b_k x_{1k}$$

$$y_2 = b_1 x_{21} + b_2 x_{22} + \ldots + b_k x_{2k}$$

$$\cdot \quad \cdot \quad \cdot \quad \cdot \quad \cdot$$

$$\cdot \quad \cdot \quad \cdot \quad \cdot \quad \cdot$$

$$\cdot \quad \cdot \quad \cdot \quad \cdot \quad \cdot$$

$$y_n = b_1 x_{n1} + b_2 x_{n2} + \ldots + b_k x_{nk}$$

Once these are solved for the b_i, the values of a, r, r^2, and the SEC are given by the formulas at the end of this chapter.

The multiterm case is more compactly stated in matrix notation. If $\mathbf{y}$ is a column vector of n centered data points for the dependent variable Y, and $\mathbf{x}$ is the n-by-k matrix whose rows are the independent variable values corresponding to these data points, with each column containing the data for one independent variable, and $\mathbf{b}$ is the column vector of regression coefficients, the set of simultaneous equations with centered variables shown above can be written as $\mathbf{y} = \mathbf{xb}$, and the least-squares solution is $\mathbf{b} = (\mathbf{x}^T\mathbf{x})^{-1}\mathbf{x}^T\mathbf{y}$ (T = transpose, -1 = inverse). The values of a, r^2, r, and the SEC are then found using the formulas at the end of this chapter.

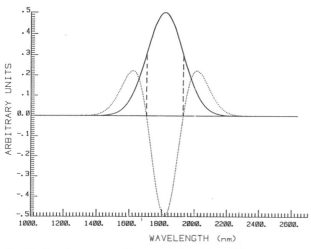

Fig. 20. Gaussian curve and its second derivative with respect to wavelength.

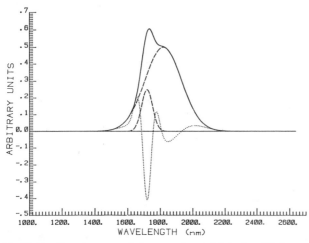

Fig. 21. Two Gaussian curves, their sum and second derivative of their sum.

VI. THE DERIVATIVE

A. Basic Properties

Derivatives are an approach to addressing two of the basic problems with NIR spectra: overlapping peaks and large baseline variations. The effect of derivatives is most clearly seen with the second derivative of a spectrum, which is able to separate overlapping absorption bands because it is related to curvature and has the same sign as the curvature of a spectrum. Figure 20 shows a single absorption band and its second derivative. The left and right portions of the band are curved upward, so the second derivative is positive there. The middle portion is curved downward, giving a negative second derivative with a minimum at the peak value of the spectrum where the downward curvature is greatest. The useful effect here is that the bandwidth (at half maximum) in the second-derivative spectrum is less than that of the original absorption band. The derivative will thus help resolve overlapping bands (Fig. 21).

The second effect of the second derivative is to remove baseline shifts. Because the curvature of a straight line is zero, the derivative of a spectrum-plus-a-straight-line-baseline will be the same as the derivative of the spectrum (Figs. 22 and 23). The quotient of two derivatives can be used to remove multiplicative as well as additive scatter effects.

The value of the second derivative in NIR work has stemmed from these two effects: resolution of overlapping peaks and re-

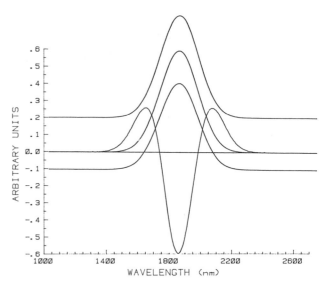

Fig. 22. Three Gaussian curves, identical except for displacement, and their identical second derivatives.

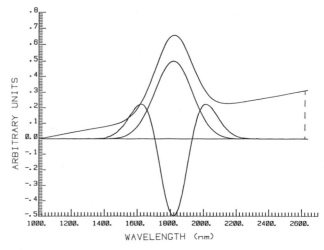

Fig. 23. Two Gaussian curves, identical except for addition of linear baseline, and their identical second derivatives.

moval of linear baselines. The first derivative also has these two effects, but to a lesser extent, and has a geometric interpretation as the slope of the spectrum at each wavelength (Fig. 24). Higher-order derivatives have the same two basic effects and will resolve overlapping absorption bands better than the lower-order derivatives, and they will remove gently curved background as well as linear background shifts. However, they are more sensitive to noise and generate more artifacts than the lower-order derivatives. They do not have any easily visualized geometric interpretation and have not been widely used in NIR work. Derivatives of orders higher than two have not been shown to have an advantage in calibration or validation.

The drawbacks of derivatives are the side lobes (Fig. 20), which can give false information about an absorption band, and the increased sensitivity to spectral noise and wavelength instability.

In developing a calibration model using a derivative, the same search procedures are used as with linear regression methods, except that the wavelength search is done using the derivatives of the calibration spectra rather than the original spectra. Figure 25 shows the second derivatives of the spectra in Figure 4. Figure 26 shows a correlation plot for protein regressed against the second-derivative spectra of the calibration samples of Table I together with the second derivative of a wheat protein spectrum. Notice the coincidence of correlation plot peaks with protein second-derivative peaks.

Software using more complex search techniques on multiterm quotients of derivatives in which each derivative can be of a different order and have a different gap and amount of presmoothing has been developed by Shenk and Westerhaus (Marten et al, 1985).

B. Calculation

Three common methods of calculating derivatives are the finite-difference, Fourier transform (FT), and Savitzky-Golay methods. There is no simple relationship between the derivatives as calculated by the three methods, and each has its advantages and disadvantages. The Savitzky-Golay method (Section III) fits the spectrum with a polynomial and then takes the derivative of that polynomial. It is most useful with very sharp absorption bands, with high noise spectra, in cases in which the value of the derivative at wavelengths between adjacent measured wavelengths is desired, and in detecting very small wavelength shifts. The FT method (Section VII) is most useful when the FT of the spectrum is already available for other purposes. The finite-difference method is easiest to calculate but is more sensitive to noise than the Savitzky-Golay method, because the latter includes a polynomial smooth. Both the FT and finite-difference methods are usually combined with some form of smoothing for

best results. The formulas for the finite-difference calculation is as follows.

If S is a spectrum defined for evenly spaced wavelengths (λ_n) and $n = 0, 1, ..., N - 1$, then the first derivatives S_n' at point n is defined by

$$S_n' = S_{n+g} - S_{n-g}$$

in which g is an integer called the gap or derivative size and S_n is the $\log(1/R)$ value at point n. Similarly, the second and third derivatives (S_n'' and S_n''', respectively) are defined by

$$S_n'' = S_{n+2g} - 2S_n + S_{n-2g}$$

$$S_n''' = S_{n+3g} - 3S_{n+g} + 3S_{n-g} - S_{n-3g}$$

Strictly speaking, these three equations represent first-, second-, and third-order *differences* rather than derivatives, as developed in formal numerical analysis. More important, the definitions vary from laboratory to laboratory, even between computer programs in the same laboratory. The principal variations involve whether division by some power of $\lambda_{i+g} - \lambda_{i-g}$ is applied, and whether g is replaced by $g/2$. Transferring calibrations or replicating results between laboratories requires knowledge of the definitions in use by the systems compared. Often, the definitions are not quoted in published papers. Authors, while retaining the slightly incorrect term *derivative* for *difference*, should quote the definition being used when publishing and should quote g in nanometers rather than numbers of points. The examples in this section used the finite-difference method preceded by a 5-point (8 nm) smooth and with a gap of 10 points (16 nm).

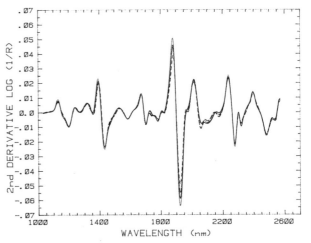

Fig. 25. Second derivatives of spectra in Figure 4.

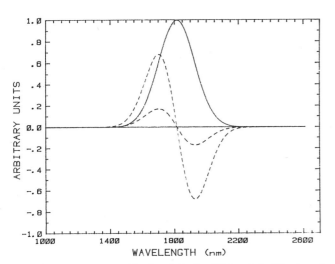

Fig. 24. Gaussian curve and its first derivative. Higher dashed line is multiple of lower.

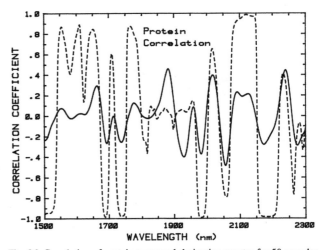

Fig. 26. Correlation of protein to second-derivative spectra for 50 samples (dotted) and second derivative of protein (solid).

VII. THE FOURIER TRANSFORM

Although the FT requires quite complicated calculations compared with those used in the methods discussed so far, its geometric interpretation permits an appreciation of how it has been used in NIR work. Bracewell (1965) wrote a basic reference work on the FT; Brigham (1974) wrote a basic introduction to the "fast FT," which is a method of efficiently organizing the calculation of the transform; Bell (1972) and Griffiths (1975) wrote basic texts on FT spectrophotometers, which are used primarily in mid-infrared work.

A. Basic Properties

The FT changes a spectrum into another kind of spectrum that we will call the *Fourier spectrum*. As an example, consider the two waves, a constant function, and their sum in Figure 27. This is meant to be a very simplified example of a spectrum with two peaks and some high-frequency noise. The FT of this sum consists of the three points in Figures 28 and 29. The point above zero in the Fourier spectrum is the FT of the constant function, and its height is proportional to the height of the constant function. The second point is the FT of the low-frequency wave, and its height is proportional to the height of that wave. The third point is the FT of the high-frequency wave, and its height is again proportional to the smaller amplitude of the high-frequency wave. Figures 28 and 29 are slightly oversimplified, but they illustrate the essential feature of the FT. The FT expresses each spectrum as a sum of waves and can be plotted as a graph whose horizontal axis is the wave frequency and whose vertical axis is the wave height.

The FT is reversible; the inverse FT of Figure 28 is Figure 27. Figure 30 is a flowchart showing how the FT is used. The Fourier spectrum is modified in some way and then the inverse FT is performed, giving a NIR spectrum that is different from the original one in some desirable way. For example, to remove the high-frequency noise from the spectrum in Figure 27, compute the Fourier spectrum, eliminate the point corresponding to the noise (giving a modified Fourier spectrum), and then compute the inverse FT of the modified Fourier spectrum. This will give the sum of the constant and the low-frequency wave without the high-frequency noise (Fig. 31).

So far the FT has been described in simplified form. The FT actually gives a pair of values for each wave frequency. The pair of values (F_k) is called a complex number; one is the real part, $\text{Re}(F_k)$, and the other is the imaginary part, $\text{Im}(F_k)$. The two values can be combined into the modulus $|F_k| = \sqrt{[\text{Re}^2(F_k) + \text{Im}^2(F_k)]}$, which is the value graphed in Figure 28. $\text{Re}(F_k)$ and $\text{Im}(F_k)$ are called Fourier coefficients, and the set of $|F_k|$ is called the power spectrum. The term *Fourier spectrum* is used here to mean the power spectrum except where an understanding of the application requires separation of the real and imaginary parts.

Figures 32 and 33 give a wheat spectrum of 1,024 points and the real part of its corresponding Fourier spectrum. The Fourier spectrum is symmetrical about the point 512, and point 0 is the zero-frequency point, which corresponds to the average value of the wheat spectrum. The high-frequency values are located near point 512 and the low-frequency values are located near points 0 and 1 (or near point 1,023, which is a reflection of point 1). (The symmetry occurs because the original spectrum is real, and the

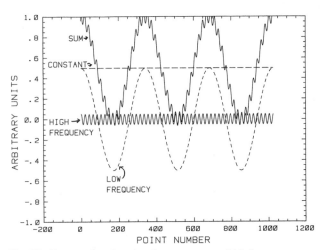

Fig. 27. Constant function, low-frequency wave, high-frequency wave, and sum of three functions.

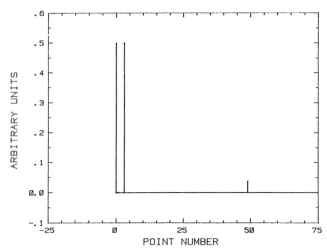

Fig. 29. Figure 28 expanded.

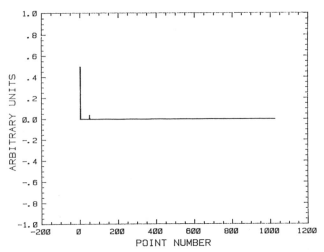

Fig. 28. Fourier transform of sum function in Figure 27.

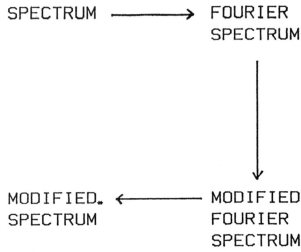

Fig. 30. Flowchart of general use of Fourier transform.

symmetry is actually more accurately expressed in the statement "F_{N-k} = the complex conjugate of F_k for $k \neq 0$.")

B. Applications

Of the many applications of the FT, seven that have been used with NIR spectra will be described: data compression, correlation to Fourier coefficients, smoothing, normalizing, calculation of derivatives, convolution, and deconvolution. Each involves a different kind of modification of the Fourier spectrum before taking the inverse transform to get the modified spectrum, and some dispense with the inverse transform altogether.

DATA COMPRESSION

The Fourier spectrum in Figure 32 can be modified by reducing the points near 512 to a value of zero (called "throwing out the high-frequency noise") without disturbing the essential information in the spectrum. Various measures of the lack of disturbance have been considered. One is the maximum of the difference between the spectrum and the modified spectrum. If this maximum is of the same order as the instrument noise, then probably only random instrument noise has been deleted. Another measure is the effect on calibration and validation from the resulting modified sample spectra. Zeroing all but the first 50 complex points will not disturb calibration or validation results for the samples of Table I.

Thus, 50 points of the Fourier spectra can be saved instead of 1,024 points of the original spectrum, conserving computer storage space. For example, if the original spectrum had 1,024 16-bit words (2,048 bytes) and the modified Fourier spectrum has 50 double-precision complex numbers (800 bytes), the space saving would be about 60%. If the original spectrum had 1,024 32-bit reals (4,096 bytes) and the modified Fourier spectrum has 50 single-precision complex numbers (400 bytes), the saving would be 90% (Giesbrecht et al, 1981).

CORRELATION TO FOURIER COEFFICIENTS

Sample composition can also be correlated directly to the Fourier coefficients rather than to values in the modified wavelength spectra. If only the first 10 (low-frequency) coefficients are retained and treated as 19 separate variables in a step-up linear regression (the zero-frequency coefficient plus nine real parts plus nine imaginary parts), sugar and nicotine in tobacco can be determined satisfactorily (McClure et al, 1984).

SMOOTHING

In the simplified example in Section VII A, it was demonstrated that smoothing could be accomplished with the FT. With an actual NIR spectrum and its FT, as discussed above, the high-frequency points were reduced to zero. Instead, the Fourier spectrum can be multiplied point by point with a weighting function of

the Fourier point numbers that increases the values at the low-frequency points and diminishes those at the high-frequency points. The modified Fourier spectrum is then inversely transformed into a smoothed wavelength spectrum. The degree of smoothing and amount of band distortion are controlled by the size and form of the weighting function.

NORMALIZING

Zeroing the constant coefficient of the Fourier spectrum results in a modified wavelength spectrum whose average value is zero. This normalization can reduce the particle size effect when applied to a set of samples (Section VIII).

DERIVATIVES

If the weighting function is designed to diminish the low frequencies and enhance the high frequencies, then the modified wavelength spectrum can show more detail. A specialized example is the Fourier method of calculating derivatives, which uses the relationships

$$\mathcal{L}\left(S^{(n)}\right) = k^n \times F_k \left(k \leq N/2\right)$$

and

$$\mathcal{L}\left(S^{(n)}\right) = \left(N - k + 1\right)^n \times F_k \left(k \geq N/2\right)$$

in which the superscript (n) denotes the nth derivative and the multiplication on the right is point by point.

CONVOLUTION

If each point in a spectrum is replaced by a weighted mean of it and its surrounding points, the spectrum is said to be convoluted with the weighting function. A simple example is the boxcar

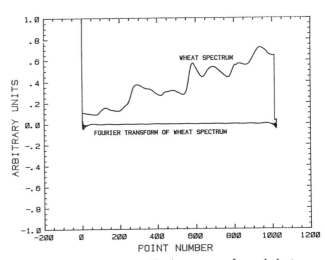

Fig. 32. Wavelength spectrum and Fourier spectrum of ground wheat.

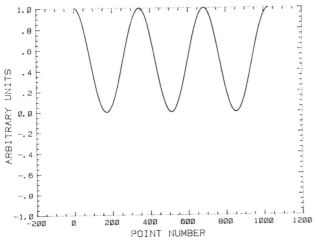

Fig. 31. Sum function of Figure 27 with high-frequency wave removed by Fourier method.

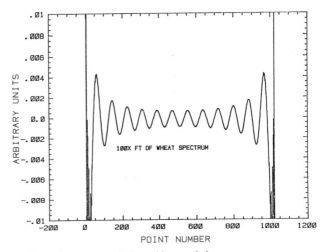

Fig. 33. Fourier spectrum of Figure 32 expanded.

smooth (Section III C), in which the weighting function is constant. A convolution is written as

$$S^M = S(*)W$$

in which S is the original spectrum, S^M is the modified spectrum, W is the weighting function, and (*) denotes convolution. The FT can be used to accomplish convolution because of the relationship

$$\pounds[S(*)W] = \pounds(S) \times \pounds(W)$$

in which the multiplication on the right is point-by-point of the FT of the spectrum and the weighting function. For example, since the FT of a Gaussian is a Gaussian, a Gaussian smooth can be accomplished by multiplying the Fourier spectrum point for point by a Gaussian, and then taking the inverse FT.

Most of the examples in this section could be discussed as some kind of convolution performed by FT methods.

DECONVOLUTION

Deconvolution is the restoration of a convoluted spectrum to its preconvoluted set of values. The convolution equation given in the preceding subsection can be reversed to read

$$\pounds(S) = \pounds[S(*)W] / \pounds(W)$$

in which the slash denotes point-by-point division. An example of the use of convolution occurs when the slit function and the experimental spectrum are known and it is necessary to find the theoretical spectrum. Deconvolution then removes the smoothing effect of the slit function.

VIII. OTHER METHODS

A. Component Spectrum Reconstruction

Honigs et al (1984a) describe the reconstruction of the spectrum of a component of a mixture from the spectra of several mixtures with different levels of the component. Success in this would be particularly valuable in mixtures in which the components are known to interact, cannot be isolated, or can change their spectra when isolated. The formula for the reconstructed spectrum is

$$S_n^R = \left(\sum xy\right)\left(1 - \bar{Y}\right)/\sum y^2 + \bar{X}_n$$

in which S_n^R is the reconstructed spectrum at point n, $x = X_n - \bar{X}_n$ is the centered log($1/R$) value at point n, and $y = Y - \bar{Y}$ is the centered fractional concentration of the component whose spectrum is to be reconstructed. A closer examination of the formula shows that it essentially adds a multiple of the slope of the regression line (of the log[$1/R$] value against the component level) at each wavelength to the average of all the spectra at each wavelength.

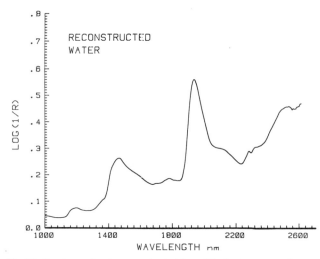

Fig. 34. Spectrum of water reconstructed from 50 wheat spectra and water data.

The water spectrum (Fig. 34) reconstructed from the spectra and moisture values for 50 calibration samples of Table I shows the strengths and limitations of the method. The reconstructed spectrum is remarkably like that of pure water. The exceptions are places (e.g., 2,280 nm) where water is probably correlated to starch and some other places (e.g., 2,100 nm) where the reconstructed spectrum may be that of bound water.

B. Fast Correlation Transform

Honigs et al (1984b) describe a method sometimes called the "fast correlation transform," which can be used to select a minimal calibration set of wavelengths and/or samples from a set of data such as that described in Table I. The benefit is to reduce the number of samples needed for calculation and to select wavelengths with the most information. The method for wavelength selection is as follows.

1. Decide on the number (k) of wavelengths to select.
2. Start with the first k available wavelengths.
3. Form a matrix with $k + 1$ columns, the first k columns being the spectral data at the k wavelengths, the last column being the chemical data.
4. Row reduce the matrix on the wavelength information as described by Honigs et al (1984b), so that the largest element in each column is placed on the main diagonal before eliminating other elements of that column.
5. Using the data in the first k rows of the reduced matrix, regress the chemical data against the spectral data.
6. On repetition of steps 3–7, save these wavelengths and r values (and discard previously saved wavelengths) if r is larger than previously saved r's.
7. Discard the wavelength with the largest regression coefficient (absolute value) and replace it with the next available wavelength.
8. Repeat steps 3–7 until all available wavelengths have been tried.
9. Using the final matrix as the starting point, with all the other wavelengths being "available," repeat steps 3–8 (second pass).
10. The wavelengths saved by step 6 are then the optimal set.

C. Normalizing Spectra

Normalization of spectra has been mentioned briefly in Sections V–VII. In general, it means changing a group of spectra so that the resulting ones have more features in common; or, equivalently, so that unwanted sources of variability (e.g., radiation scattering) are suppressed. This helps the visual understanding of the spectra and can reduce the complexity of a data treatment necessary to develop a calibration from the spectra. The simplest example (Section III) was subtracting the log($1/R$) value at a single wavelength (called the reference wavelength) in a spectrum from all the spectral values. The result is a set of spectra with value zero at the reference wavelength. Forcing the sum of the regression coefficients to zero essentially accomplishes the same thing as was seen in Section VI. The first derivative also implies a similar normalization because it reduces a constant baseline to zero.

A more complicated normalization is accomplished by adding a different straight-line baseline to each spectrum,

$$S_n^M = S_n + a + bn$$

so that the resulting spectra agree at two points instead of one (n is the point number of the wavelength). This normalization can be extended to the addition of in which degree polynomial in the variable n, so that the modified spectra agree at more than two points. ("Mathematical ball-milling" [Murray et al, 1981] has been used to describe the linear case where a and b are adjusted for each spectrum so that every resultant spectrum has the same slope and Y-intercept as the average spectrum, calculated by regressing log[$1/R$] values against wavelength.)

A similar normalization replaces S_n by some polynomial function of S_n, in which the polynomial constants are the same for all $\log(1/R)$ values in a single spectrum but change from spectrum to spectrum. The equation here is

$$S_n^M = a + b_1 S_n + b_2 S_n^2 + \dots$$

in which the polynomial variable is S_n rather than n. This is more theoretically sound than the previous method, since the scatter effect on a $\log(1/R)$ spectrum is dependent on absorbance rather than wavelength. Normalizing with a first-degree polynomial removes both the additive and multiplicative scatter effect. Figure 35 shows the four spectra of Figure 4 normalized by a second-degree polynomial to agree at the indicated points. Another normalization is discussed in Chapter 4 (Section V B: Multiplicative Scatter Correction).

D. Discriminant Analysis

The technique of discriminant analysis is like linear regression in that it uses a linear combination of variables, but the combination is used to classify a material rather than to analyze it. Mark and Tunnell (1985) give examples.

E. Neural Networks

Since the late 1980s, several investigators have used neural networks for NIR data analysis. For this method, "NEURAL NETWORK TRAINING" would replace "LINEAR REGRESSION" in the flow diagram of Figure 5, with all other parts of the calibration and validation processes remaining the same. The work of Hertz et al (1991) can serve as an introduction to the mathematics of the various types of neural networks and as a guide to the history and literature of the subject. Because most applications in NIR analysis have used only one type of network, the following (patterned after the discussion by Naes et al [1993]) outlines the mathematics involved in feed-forward back-propagation (FFBP) networks.

Recall the linear regression equation

$$\hat{Y} = a + \sum_i b_i X_i$$

in which the X_i are the result of some data treatment of $\log(1/R)$ values, the a's and b's are regression constants, and $\hat{Y}$ is the predicted value of the independent variable. An FFBP network can be considered a generalization of this equation, and in its simplest form becomes

$$\hat{Y} = f\left(a + \sum_i b_i X_i\right)$$

The X_i are called "inputs" to the network, $\hat{Y}$ is called the "output" of the network, the b's are called "weights," a is called the "bias," and f denotes a "transfer function" of the sum in parentheses. The collection of X_i is also referred to as the "input layer" and f(sum) is called the "output layer."

If f is the identity function, then the equation reduces to the linear regression equation. However, the transfer function is usually the sigmoid

$$f(\text{sum}) = 1/[1 + \exp(-\text{sum})]$$

and serves to introduce a nonlinearity into the equation and to force the outputs ($\hat{Y}$) to be between zero and one. So in this section, we have to consider the original Y values to have been scaled so that $0 \leq Y \leq 1$.

One or more "hidden layers" can be added to introduce a different kind of nonlinearity. For example, a network with one hidden layer has the equation

$$\hat{Y} = f\left[a^2 + \sum_j b_j^2 f\left(a_j^1 + \sum_i b_{ij}^1 X_i\right)\right]$$

with X's being the input layer, the interior f(sum)s of the input layer being the nodes of the hidden layer, and the exterior f(sum)

transferring the outputs of the hidden layer to the final output $\hat{Y}$. Here, the superscripts keep track of the weights and biases in the different layers and are not meant to be exponents.

This type of network is called feed forward because the output is calculated from the input in a forward direction without any recursion. The name back propagation refers to the method of calculating the weights and biases. This is done iteratively, starting from a random set of weights and biases. As each sample is presented to the network, the error, or difference between the desired output (or reference value) and the actual output (feed-forward value) is used to adjust the weights. Thus, in each iteration, the error is back-propagated through the network, changing all the weights, hopefully for the better in the sense of reducing the error. The formulas for weight adjustment may be found in Hertz et al (1991).

Various inputs have been or could be used, including raw reflectance values, few or many $\log(1/R)$ values, derivatives, principal component scores, etc. As with any method, choices have to be made and validation has to be used to check performance. The choices with neural networks include the learning rate (the portion of the error to use in back propagation), the transfer function(s), the number of iterations to use in calibration, and the number and size of hidden layers.

IX. CONCLUSION

The statistical processes of calibration, measurement, and validation have been described as used in NIR spectroscopy. Error sources have been discussed, and the mathematics required for several data treatments were presented. It would seem natural at this point to compare the performances of the various data treatments in more detail. However, although researchers continue to develop and evaluate data treatments, there are too few published comparisons to permit much more than accounts of preferences and experiences. Few laboratories have the resources to devote to extensive comparisons of mathematical methods, and applications researchers have achieved success by selecting one method and using it with as much flexibility and experience as possible. Because the main sources of error when using commercially available instruments are usually neither the data treatment nor the instrument error, one should raise the priority level for dealing with sampling error and reference method error and for developing a clear definition of the component or the physical property to be determined.

Chapter 4 is more complex mathematically than the material discussed up to now because it requires a familiarity with matrix algebra, which is reviewed in its appendixes. An overview of the mathematics discussed in both this chapter and Chapter 4 is now presented, so that the reader who is unprepared for the higher level of mathematics used in Chapter 4 can at least get an intuitive feel for the material.

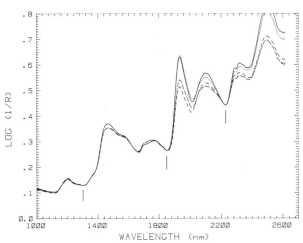

Fig. 35. Normalized spectra of those shown in Figure 4.

TABLE IX
Data Compression Methods

Method	Latent Variables (Loadings)	Scores
Fourier transform	Sine and cosine waves	Fourier coefficients
Curve fitting	Spectra of chemical components or representative samples	Curve fit coefficients
Principal components analysis	Artificial spectra that statistically account for spectral variability	Scores
Factor analysis	Linear combinations of principal components that approximate spectra of pure components	Scores
Partial least squares	Artificial spectra that account for spectral and chemical variability	Scores
Extended Beer's Law	Artificial spectra that account for spectral and chemical variability and approximate spectra of pure components	Scores

In Section II, several examples of linear regression approximations were presented. All the mathematics of both this chapter and Chapter 4 can be expressed in terms of such approximations. The basic form of the approximation is

$$\hat{Y} = a + b_1 X_1 + b_2 X_2 + \ldots + b_k X_k$$

in which Y is the variable to be measured, such as protein concentration, loaf volume, etc.; the X_1, X_2, ..., X_k are k independent variables, each a combination of one or more NIR spectral values; the b_1, b_2, ..., b_k are k regression coefficients; and a is the regression constant term (intercept). In calibration, a set of X's and known Y's are used to estimate the b's, whereas in measurement and validation, a set of X's and the estimated b's are used to estimate an unknown Y. The main differences between the various methods discussed in this chapter are in how the NIR spectral values are combined into the X_1, X_2, ..., X_k. These methods are now considered in this context.

As emphasized in Section I, the most common method is for the X's to be log($1/R$) values at a few preselected wavelengths. Sections II–V are concerned with this method. Some other function of R values at a few preselected wavelengths may also be used, such as the Kubelka-Munk $(1 - R)^2/2R$ or the empirical scatter-correcting functions discussed in Chapter 4. In what follows, the terms *spectrum* and *spectral values* refer to any of these functions of R values or to auxiliary variables. With any of the methods, spectral noise may be removed before the approximation by some form of smoothing (Section III), and X's can be indicator variables (Section V).

The next simplest method is for the X's to be derivatives, quotients, differences, or other mixtures of the spectral values at a few wavelengths (Sections IV–VI).

The following methods (data compression) are much more complex, but only in the sense that the X's require lengthier calculations. The calibration, measurement, and validation procedures are the same no matter how the X's are calculated. Each method begins by expressing a spectrum as the sum of elementary spectra (called loadings or latent variables), the definition of *elementary* being different for each method. Each elementary spectrum is multiplied by a "score" before being included in the sum, in order to make the sum of the elementary spectra as close as possible to the original spectrum. The scores (some or all) are then used as the X's in the approximation. What follows is summarized in Table IX.

In the FT procedure (Section VII and Chapter 4), the elementary spectra are sine and cosine waves and the scores are called Fourier coefficients. In curve fitting (Chapter 4), the elementary spectra are actual measured spectra (of pure materials or of representative samples), and the scores are called curve fit coefficients. In principal components analysis (Chapter 4), the elementary spectra (called principal components) are statistically estimated to account systematically for the spectral variation. Factor analysis is similar to principal components analysis, except that the elementary spectra are adjustments of the principal components to make them approximate actual spectra. The partial least-squares method (Chapter 4) is similar to the principal components analysis, except that the reference data as well as the spectral data are used in the estimation of the elementary spectra. Finally, the extended Beer's Law method (Chapter 4) is similar to the partial least-squares method in that the chemical data are used in the estimation of the elementary spectra. However, the components are designed to approximate actual spectra of pure chemical components, and the model is able to make use of assumptions about the statistical distributions of the spectral and chemical data.

Each method from either this chapter or Chapter 4 can be thought of as having been developed from a regression in which the independent variables (X's) can be very simple or very complex combinations of NIR spectral values or auxiliary variables. Computer packages usually calculate these X's in a way largely transparent to the investigator, who is then free to consider only the problems of overfitting, sampling, and the quality of spectral and chemical data.

ACKNOWLEDGMENTS

I thank Lois Harris, Dave Hopkins, Elaine Lanza, Howard Mark, Harald Martens, Dave Massie, Tom Nolan, Karl Norris, Dave Osten, John Shenk, Ed Stark, Mark Westerhaus, and Phil Williams for invaluable help in preparing this chapter.

APPENDIX

The statistical notation follows Steel and Torrie (1980), except for the use of lowercase x and y for deviations from sample means and the use of $\mathbf{M}^T$ for matrix transpose. The FT notation was adapted from several sources. Alternative notation and terminology in current use is enclosed in parentheses.

Mathematical Symbols

$\neq$	Not equal
$\leq$	Less than or equal
$\geq$	Greater than or equal
$\cong$	Approximately equal
$\times$	Multiplication
$\sqrt{}$	Square root
$n!$	n factorial
$\lvert X \rvert$	Absolute value of X
log	Base 10 logarithm
Σ	Sum of
ΣX_i	$X_1 + X_2 + \ldots + X_n$

Abbreviated versions of summation are used when i and n are clear from the context.

$$\sum X = \sum X_i = \sum_i X_i = \sum_{i=1}^{n} X_i \qquad \sum X^2 = \sum \left(X_i^2 \right)$$

$$\sum XY = \sum \left(X_i Y_i \right) \qquad \sum (X - Y)^2 = \sum \left[(X_i - Y_i)^2 \right]$$

Spectroscopic Symbols

R	Reflectance
nm	Nanometer
S	Spectrum of log($1/R$) values
S_n	log($1/R$) value at spectral point number n (not wavelength n)
S'	First derivative of S with respect to wavelength
S'_n	First derivative at point number n
S''	Second derivative of S with respect to wavelength
S^M	Modified spectrum
λ_n	Wavelength at spectral point number n

Statistical Symbols

X	Independent variable
Y	Dependent variable
$\hat{Y}$	Estimate of dependent variable
X_i, Y_i	Individual observation
e, e_i	Individual calibration or validation error (regression error)
D, D_i	Difference between paired observations
$\overline{X}, \overline{Y}$	Sample mean (average)
μ, μ_X	Population mean
x, y	Centered variable (deviation from sample mean)
x_i, y_i	Individual centered observation (deviation)
s^2, s_X^2, s_Y^2	Sample variance
s, s_X, s_Y	Sample standard deviation (SEST, standard error of a single test)
σ, σ_X	Population standard deviation
n	Sample size
k	Number of variables in regression
a	Y-intercept, constant term (b_0)
b	Slope in a single-term linear regression (regression coefficient) (b_1)
b_i	Regression coefficient, not including the constant term
r	Correlation coefficient
r^2	Coefficient of determination
$\mathbf{M}$	Matrix
M_{ij}	Matrix element in row i, column j
$\mathbf{M}^{-1}$	Inverse of $\mathbf{M}$
$\mathbf{M}^T$	Transpose of $\mathbf{M}$
$\mathbf{b}$	Column vector of regression coefficients $b_1, ..., b_k$
$\mathbf{x}$	Matrix of centered independent observations (rows = observations, columns = variables)
x_{ij}	ith observation of jth variable centered on mean of jth variable (element in ith row, jth column of $\mathbf{x}$)
$\mathbf{X}_j$	Column of data for independent variable j
$\mathbf{y}$	Column vector of centered data for dependent variable
$\overline{X}_j$	Mean of one variable

Statistical Abbreviations

BIAS	Bias (mean difference)
CV	Coefficient of variation or variability
RMSD	Root mean square difference (RMS, root mean square)
SDVD	Standard deviation of validation differences (Section II has discussion of several other terms, including verification, prediction measurement, analysis, SEP, SED, and SDP)
SE	Standard error (used verbally to mean SEE, SEC, or SDVD when meaning is clear from context)
SEC	Standard error of calibration (SEE in a calibration regression)
SEE	Standard error of estimate (standard deviation of regression errors)

Statistical Formulas

$\overline{X}$	$(\Sigma X)/n$
x	$X - \overline{X}$
s_X^2	$\Sigma x^2/(n-1) = \left(\Sigma X^2 - (\Sigma X)^2/n\right)/(n-1)$. The divisor $n-1$ is used when the population mean is estimated by the sample mean. The divisor is actually n minus the number of parameters estimated by sample statistics. For example, if the population mean is assumed to be zero, no parameters are estimated and the divisor is n. In a multiterm linear regression with k independent variables, $k+1$ parameters are estimated and the divisor is $n-1-k$.
CV	$100 s_X/\overline{X}$

In any regression:

e_i	$Y_i - \hat{Y}_i$
SEE	s_e (formula depends on type of regression)
r^2	$(\Sigma y\hat{y})^2 / \left(\Sigma y^2 \Sigma \hat{y}^2\right) = 1 - s_e^2/s_Y^2$

In a single-term linear regression calibration:

b	$\Sigma xy/\Sigma x^2$
a	$\overline{Y} - b\overline{X}$
r	$\Sigma(xy)/\sqrt{\left(\Sigma x^2 \Sigma y^2\right)}$
SEC	$\sqrt{\left\{\left[\Sigma y^2 - (\Sigma xy)^2/\Sigma x^2\right]/[n-2]\right\}}$
	$= s_y\sqrt{\left[(n-1)(1-r^2)/(n-2)\right]} = \sqrt{\left[\Sigma e^2/(n-2)\right]}$

In a multiterm linear regression calibration:

$\mathbf{b}$	$\left(\mathbf{x}^T\mathbf{x}\right)^{-1}\mathbf{x}^T\mathbf{y}$
a	$\overline{Y} - \Sigma b_i \overline{X}_i$
r^2	$\Sigma_i\left(b_i\Sigma_j x_{ji}y_j\right)/\Sigma y^2 = \left(\mathbf{b}^T\mathbf{x}^T\mathbf{y}\right)/\left(\mathbf{y}^T\mathbf{y}\right)$
SEC	$s_Y\sqrt{\left[(n-1)(1-r^2)/(n-1-k)\right]} = \sqrt{\left[\Sigma e^2/(n-1-k)\right]}$

In a validation, X = NIR method value, Y = reference method value (Section II C has alternative terminology):

D_i	$X_i - Y_i$
SDVD	$\sqrt{\left\langle\left\{\Sigma[X-Y]^2 - [\Sigma(X-Y)]^2/n\right\}/(n-1)\right\rangle}$
	$= s_{(X-Y)} = s_D = \sqrt{\left[\Sigma(D-\overline{D})^2/(n-1)\right]}$
BIAS	$\overline{X} - \overline{Y} = \Sigma D/n = \overline{D} = \Sigma(X-Y)/n$
RMSD	$\sqrt{\left[\Sigma(x-Y)^2/n\right]} = \sqrt{\left[\Sigma D^2/n\right]} \cong \sqrt{\left(\text{SDVD}^2 + \text{BIAS}^2\right)}$

Fourier Transform Notation

j	$\sqrt{(-1)}$
exp	Exponential function
π	Ratio of circumference to diameter of circle
N	Number of points in spectrum
$\pounds$	Fourier transform
$\pounds^{-1}$	Inverse Fourier transform
S	Spectrum of $\log(1/R)$ values
S_n	$\log(1/R)$ value at point number n (not wavelength n) (points numbered $0, 1, 2, ..., N-1$)
F	$\pounds(S)$ = Fourier spectrum of S
F_k	Fourier spectrum at point k (points numbered $0, 1, 2, ..., N-1$)
S	$\pounds^{-1}(F)$ = inverse Fourier transform of F
(*)	Convolution
Re	Real part
Im	Imaginary part
$\|F_k\|$	Modulus of F_k

Fourier Transform Formulas

$$F_k = \sum_{n=0}^{N-1}\left[S_n \exp(-j2\pi nk/N)\right]$$

$$S_n = \left\{\sum_{k=0}^{N-1}\left[F_k \exp(j2\pi nk/N)\right]\right\}/N$$

LITERATURE CITED

In this list and in the additional reference list that follows, the statistical references all presume a "first course" in that subject, for which there are numerous equivalent tests. Neter et al (1978) being only one example.

Bell, R. J., ed. 1972. Introductory Fourier Transform Spectroscopy. Academic Press, New York.

Bracewell, R., ed. 1965. The Fourier Transform and Its Applications. Electrical and Electronic Engineering Ser. McGraw-Hill, New York.

Brigham, E. O., ed. 1974. The Fast Fourier Transform. Prentice-Hall, Englewood Cliffs, NJ.

Coblentz, W. K. 1982. A study in the near infrared reflectance spectra of proteins and related compounds. M.S. thesis. Pennsylvania State University, University Park.

Draper, N., and Smith, H. 1966. Applied Regression Analysis. John Wiley, New York.

Giesbrecht, F. G., McClure, W. F., and Hamid, A. 1981. The use of trigonometric polynomials to approximate visible and near infrared spectra of agricultural products. Appl. Spectrosc. 35:210-214.

Griffiths, P. R. 1975. Chemical Infrared Fourier Transform Spectroscopy. John Wiley, New York.

Hertz, J., Krogh, A., and Paimer, R. G. 1991. Introduction to the Theory of Neural Computation. Addison-Wesley, New York.

Hocking, R. R. 1970. The analysis and selection of variables in linear regression. Biometrics 32:1-49.

Honigs, D. E., Hieftje, G. M., and Hirschfeld, T. 1984a. A new method for obtaining individual component spectra from those of complex mixtures. Appl. Spectrosc. 38:317-322.

Honigs, D. E., Hieftje, G. M., and Hirschfeld, T. 1984b. Number of samples and wavelengths required for the training set in near-infrared reflectance spectroscopy. Appl. Spectrosc. 38:844-847.

Hruschka, W. R., and Norris, K. H. 1982. Least-squares curve fitting of near infrared spectra predicts protein and moisture content of ground wheat. Appl. Spectrosc. 36:261-265.

Jones, D. B. 1931. Factors for converting percentages of nitrogen in foods and feeds into percentages of proteins. U.S. Dep. Agric. Circ. 183.

Marchand, P., and Marmet, L. 1983. Binomial smoothing filter: A way to avoid some pitfalls of least-squares polynomial smoothing. Rev. Sci. Instrum. 54:1034-1041.

Mark, H. L., and Tunnell, D. 1985. Qualitative near-infrared reflectance analysis using Mahalanobis distances. Anal. Chem. 57:1449-1456.

Marten, G. C., Shenk, J. S., and Barton, F. E., II 1985. Near infrared reflectance spectroscopy (NIRS): Analysis of forage quality. U.S. Dep. Agric., Agric. Handb. 643.

Martens, H., and Naes, T., eds. 1983. Food Research and Data Analysis. Applied Science Publishers, London.

McClure, W. F., Hamid, A., Giesbrecht, F. G., and Weeks, W. W. 1984. Fourier analysis enhances NIR diffuse reflectance spectroscopy. Appl. Spectrosc. 38:322-329.

Morrey, J. R. 1968. On determining spectral peak position from composite spectra with a digital computer. Anal. Chem. 40:905-914.

Murray, I., Jessiman, C., and Kenley, H. 1981. Use of a near infrared reflectance (NIR) spectrocomputer for forage evaluation. Pages 65-72 in: Annu. Meet. Eur. Soc. Nucl. Methods Agric., 12th. European Society of Nuclear Methods in Agriculture (ESNA), Aberdeen, United Kingdom.

Naes, T., Kvaal, K., Isaksson, T., and Miller, C. 1993. Artificial neural networks in multivariate calibration. J. Near Infrared Spectrosc. 1:1-11.

Norris, K. H., and Hart, J. R. 1965. Direct spectrophotometric determination of moisture content of grain and seeds. Pages 19-25 in: Principles and Methods of Measuring Moisture in Liquids and Solids, Vol. 4. A. Wexler, ed. Reinhold, New York.

Savitzky, A., and Golay, M. J. E. 1964. Smoothing and differentiation of data by simplified least squares procedure. Anal. Chem. 36:1627-1638.

Shenk, J. S., Westerhaus, M. O., and Hoover, M. R. 1978. Infrared reflectance analysis of forages. Pages 242-244 in: Proc. Int. Grain and Forage Harvesting Conf. Am. Soc. Agric. Eng., St. Joseph, MI.

Shenk, J. S., Landa, I., Hoover, M. R., and Westerhaus, M. O. 1981. Description and evaluation of a near infrared reflectance spectrocomputer

for forage and grain analysis. Crop Sci. 21:355-358.

Steel, R. G. D., and Torrie, J. H. 1980. Principles and Procedures of Statistics: A Biometrical Approach. 2nd ed. McGraw-Hill, New York.

Steinier, J., Termonia, Y., and Deltour, J. 1972. Comments on smoothing and differentiation of data by simplified least square procedure. Anal. Chem. 44:1906-1909.

Tkachuk, R. 1977. Calculation of nitrogen to protein conversion factors for food legume breeders. Pages 78-81 in: Nutritional Standard and Methods of Evaluation. J. H.. Hulse, K. O. Rachie, and L. W. Billingsley, eds. IDRC-TS7E. Int. Dev. Res. Center, Ottawa, Canada.

Williams, P. C., Norris, K. H., Johnsen, R. L., Standing, K., Fricioni, R., MacAffrey, D., and Mercier, R. 1978. Comparison of physicochemical methods for measuring total nitrogen in wheat. Cereal Foods World 23:544-547.

Williams, P. C., Norris, K. H., Gehrke, C. W., and Bernstein, K. 1983. Comparison of near-infrared methods for measuring protein and moisture in wheat. Cereal Foods World 28:149-152.

ADDITIONAL REFERENCES

American Society for Testing and Materials. 1982. Compilation of ASTM Standard Definitions. Philadelphia.

Davies, A. M. C., Gee, M. G., and Foster, P. W. 1984. A colour graphics display system to aid the selection of "best-pair" wavelengths for regression analysis of near infrared data. Lab. Pract. 38:78-80.

Frank, I. E., and Kowalski, B. R. 1982. Chemometrics. Anal. Chem. 54: 232R-243R.

Judd, D. B., and Wyzsecki, G. 1963. Color in Business, Science, and Industry. 2nd ed. John Wiley, New York.

Landa, I., and Moen, R. 1979. Reduction of predictable errors in NIR measurements. (Abstr.) Cereal Foods World 24:460.

Martens, H. A. 1985. Multivariate calibration. Dr. Techn. thesis. Technical University of Norway, Trondheim.

Mosteller, F., and Tukey, J. W. 1977. Data analysis and regression. Addison-Wesley, Reading, MA.

Neter, J., Wasserman, W., and Whitmore, G. A. 1978. Applied Statistics. Allyn and Bacon, Boston.

Norris, K. H. 1983. Extracting information from spectrophotometric curves: Predicting chemical composition from visible and near-infrared spectra. Pages 95-114 in: Food Research and Data Analysis. H. Martens and H. Russwurm, Jr., eds. Applied Science Publishers, London.

Norris, K. H., and Williams, P. C. 1984. Optimization of mathematical treatments of raw near-infrared signal in the measurement of protein in hard red spring wheat. I. Influence of particle size. Cereal Chem. 61: 158-165.

Snedecor, G. W., and Cochran, W. G. 1967. Statistical Methods. 6th ed. Iowa State University Press, Ames.

Snee, R. D. 1977. Validation of regression models: Methods and examples. Technometrics 19:415-428.

Wendlandt, W. W., and Hecht, H. G. 1966. Reflectance Spectroscopy. John Wiley, New York.

Williams, P. C. 1975. Application of near infrared reflectance spectroscopy to analysis of cereal grains and oilseeds. Cereal Chem. 52:561-576.

CHAPTER 4

Multivariate Calibration by Data Compression

H. MARTENS
Norwegian Computing Center
Oslo
Norway

T. NAES
Norwegian Food Research Institute
Aas
Norway

I. INTRODUCTION

The current chapter is, apart from an extra section (Section IX), almost identical to the contribution with the same title published in the first edition of this book. Instead of changing or modifying the chapter, we have decided to keep the presentation as it was written. The reasons for this is that the aspects covered are still important today and it has methodological details not published anywhere else. The chapter gives an impression of the status of the field in the mid-1980s, and, for being the first overview paper on multivariate calibration by full-spectrum methods, it has a certain innocent charm, in the authors' own view.

Since the late 1980s, several developments have taken place. Among the most prominent with specific interest for calibration are

- nonlinear methods such as neural networks and locally weighted regression for handling multivariate nonlinear relationships;
- methodology for suggesting how to select calibration samples the best possible way;
- data compression methodology based on wavelets for efficient data reduction for extremely large data tables;
- empirical validation methods for partial least-squares (PLS) regression and other types of regression based on the principle of resampling (e.g., cross-validation and jack-knifing);
- methods for removing irrelevant or detrimental wavelength channels from the calibration model, based on the same resampling methods;
- calibration transfer methods for transferring data between instruments; and
- three-way methodology for handling spectral data with matrix structure in them.

These are all topics of interest to the practitioner. Description of relevant methodology and major references can be found in the

works by Martens and Naes (1989), Martens and Martens (2001), and Naes et al (2001).

The new addendum in Section IX addresses two more developments, the selection of calibration samples and the nonlinearity problem.

A. Multivariate Calibration and Validation

The purpose of a near-infrared (NIR) instrument is to determine the concentration of chemical variables, such as protein content, quickly and precisely from spectrophotometric measurements. It can thereby replace slower, more expensive, or more imprecise methods for assessing the desired chemical constituents. To do this, the NIR instrument must first be calibrated, i.e., a mathematical relationship for converting the NIR optical measurements to the desired quantity (the prediction equation) has to be established.

Diffuse reflectance spectrometry in the NIR wavelength region yields nonspecific data due to interference from strongly overlapping constituents' spectra and from light-scatter variations. To minimize these interferences, the analytical equation of NIR instruments has to be multivariate, i.e., measurements on many different wavelengths must be combined. Since some of these interferences cannot be physically isolated and measured directly, the calibration of NIR instruments has to be done indirectly. A set of "representative" calibration samples with known chemical composition is used to program the computer of the NIR instrument to recognize future "unknown" samples. This approach is different from that met in conventional transmission spectroscopy of clear solutions in which the spectra of pure constituents can be measured and used for direct calibration.

A number of indirect multivariate calibration methods are available. In Chapter 3, a review was given of the established methods for calibration and analysis currently used in commercial NIR instruments. This chapter summarizes some important approaches to calibration. In contrast to the established methods, they utilize as many relevant wavelengths as possible in the analysis and are, therefore, here called *full-spectrum calibration methods*. This chapter aims at presenting the different techniques in a unified model framework. Some of the methods will be illustrated on artificial and some on real data sets. In general, the important approaches are attempts at using the available data more efficiently by combining important statistical regression methods with an understanding of the physical phenomena involved.

Sections IX A and B are reproduced from "Multivariate Calibration 2nd edition" by Martens and Naes (1989) with permission from John Wiley & Sons, Ltd.

Section IX C is reproduced in portions from "Extended Multiplicative Signal Correction and Spectral Interference Subtraction: New Preprocessing Methods for Near Infrared Spectroscopy" by Martens and Stark (1991) in the Journal of Pharmaceutical and Biomedical Analysis by permission from Elsevier Science.

Figure numbers within these three sections of text have been adjusted to reflect the figure numbers within this chapter.

B. Calibration

Calibrating an NIR instrument presents a number of problems. These are summarized in Table I. Figure 1 illustrates the basic steps required for solving them. Four different stages are required. The first two concern the preparation of the data and the last two, the statistical estimation. A calibration procedure must be reasonably simple and fast, so that it can run on small computers. Therefore, it is preferable to use some kind of linear mathematical model to relate NIR information to chemical information. Such models are easy to compute and fast to fit to the data and they can be used with a higher degree of complexity than nonlinear models. Secondly, linear statistical models are well understood theoretically.

The recorded electrical signals from the spectrophotometer are not linearly related to the chemical composition. Over a narrow range of concentrations, most curved relationships can be approximated by a straight line, but over a wider range, the error may become intolerable. In order to be able to use linear models over a wide variation range, some kind of initial data transformation may be required.

Experience has showed that it may be convenient to distinguish between two different data transformation ("data pretreatment") aspects and let the transformation consist of four stages.

1. The NIR data should first go through a response linearization step, ensuring that the NIR instrument responds linearly to changes in the chemical composition of the sample.
2. The second part of the transformation, the optical correction stage, concerns the elimination of optical NIR interferences such as light-scatter and specular reflectance. This step is needed to provide a model as simple as possible. Automatic warnings should be issued if abnormal levels of these optical phenomena are discovered for a sample. Both stages may yield estimates of certain constants.

After these transformations, the NIR data are suitable for linear modeling. Due to the high intercorrelation between the different NIR wavelengths (multicollinearity), the statistical estimation is difficult. One way to proceed that has been found to be useful consists of two stages: data compression and calibration regression.

3. The data compression stage concentrates the information from all of the available, relevant NIR wavelengths in the spectrum into a small number of basic variables that are termed *regression factors*. Data compression is important for statistical reasons (parsimony), but compressing the NIR data reduces computation time and computer storage requirements. Experience has shown that stage 2 can reduce the number of factors in stage 3. Data compression also gives certain estimated parameters, *NIR loading spectra*, that define how the values of the regression factors are to be calculated. It also gives various diagnostic checks that reveal abnormalities in the NIR data—NIR measurement errors and extreme samples or wavelengths. It also provides estimates of

TABLE I
Factors Involved in Calibrating Near-Infrared (NIR) Instruments

Chemical and optical interferences in the NIR data demand indirect multivariate calibration, i.e., it requires the determination of calibration coefficients for several different wavelengths.

The response of the NIR instrument to chemical variations is nonlinear, which makes linear modeling difficult.

Light scatter and other optical interference in NIR data is strong and nonlinear.

Measurement noise in chemical and NIR data creates danger of statistical overfitting.

Multicollinearity in NIR data creates statistical estimation problems.

A high number of calibration samples is desired, but is expensive.

Accurate chemical or physicochemical analysis is essential, but not always available.

Comprehensive sample selection is essential, but the samples are not always available.

the "normal" range of variation and the "normal" level of measurement noise in the NIR data, which help in revealing abnormalities.

4. The final stage, calibration regression, establishes linear regression relationships between the obtained regression factors and the chemical variables to be determined. It produces estimated parameters, *chemical loadings*, that define how the chemical variables are determined from the regression factors, as well as various regression diagnostics that reveal abnormalities in the chemical data in the calibration set. It may also provide estimates of the "normal" range of compositional variation and the "normal" error level when chemical data are being determined by calibrations based on the NIR data.

The NIR and chemical loadings combine mathematically to yield the calibration coefficients for the analysis of unknown samples. The calibration coefficients define the weights given to the different wavelengths in a linear analytical equation. This equation is a simple formula, and it will, therefore, be used as a common way of expressing the analytical results from all of the different linear calibration methods treated.

Some of the methods to be presented in this chapter perform each of these four stages separately, while others perform two or more stages at the same time. Methods also exist that encompass all of the stages simultaneously. Transformations are considered in Section V and the stages 3 and 4 are considered in Section III.

C. Validation and Analysis

Analysis consists of determining the chemical composition of a sample from NIR measurements using the calibration equation. During analysis, the NIR data must pass through the same four main stages as during calibration.

1. The measurements from the NIR instrument are first submitted to response linearization.
2. The data are then passed through the optical correction stage yielding quantitative information about the light scatter in each sample.

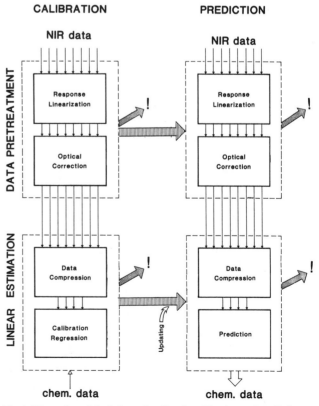

Fig. 1. Illustration of the information flow in calibration and prediction.

3. In the data compression stage, the values of the regression factors, which are linear combinations of the spectrum, are computed. This stage may also give various warning diagnostics for samples that appear abnormal.

4. In the final stage, the desired chemical variables are determined linearly from the computed regression factors using the chemical loadings found in the calibration. Diagnostics for controlling the chemical results may also be obtained.

In this part of the chapter, standard statistical language and terminology are used with the modification that Greek letters are avoided whenever possible. Capital bold letters represent matrices, lowercase bold letters represent column vectors (or row vectors, if explicitly noted), and lowercase italic letters represent scalars.

The following definitions are used, given a matrix X.

X' = the transpose of X
X^{-1} = inverse of X, i.e., such that $XX^{-1} = I$, the identity matrix

The following running index system is used.

$i = 1, 2, ..., I$: sample number
$j = 1, 2, ..., J$: chemical constituent number (protein, starch, etc.)
$k = 1, 2, ..., K$: NIR wavelength number
$a = 1, 2, ..., A$: regression factor number

If both a data set for calibration and a data set for validation analysis are available, I_c may be used to denote the number of samples in the calibration set and I_p the number of samples in the validation set.

A problem exists for describing multivariate calibration in standard statistical notation: regression is performed both over calibration samples and over wavelengths. The following chemometric convention is used in this part of the chapter.

$X = \{x_{ik}\}$: matrix of NIR data for sample i at wavelength k
$C = \{c_{ij}\}$: matrix of chemical data for sample i and chemical constituent j.

The superscript ("hat") denotes an estimated value, e.g., $\hat{C}$. Symbols and abbreviations used are summarized in Appendix A. Readers who are not familiar with matrices are referred to Appendix B.

II. LINEAR PREDICTION AND ALTERNATIVE WAYS TO FIND THE CALIBRATION COEFFICIENTS

A. Linear Analytical (Prediction) Equation

The main emphasis in this part of the chapter is on calibration resulting in predictors that are linear functions of the spectrum x. For constituent j in sample i, the linear analytical equation may be written as

$$\hat{c}_{ij} = \hat{b}_{0j} + \sum_{k=1}^{K} x_{ik}\hat{b}_{kj} \qquad (1)$$

in which $\hat{b}_{0j}$ and $\hat{b}_{kj}$ are the calibration coefficients, estimated statistically by some calibration method. The constants $\hat{b}_{kj}$ represent proportionality constants for the chemical constituents versus wavelength k, and $\hat{b}_{0j}$ is the intercept term.

On vector form, the model is

$$\hat{c}_{ij} = \hat{b}_{0j} + x_i\hat{b}_j \qquad (2)$$

in which the NIR spectrum x_i is a K-dimensional row vector and the calibration constant vector $\hat{b}_j$ is a K-dimensional column vector as illustrated in Figure 2.

For two or more chemical constituents, $(J > 1)$ becomes

$$\hat{c}_i = b_0 + x_i\hat{B} \qquad (3)$$

in which the chemical concentration vector $\hat{c}_i$ and the intercept vector $\hat{b}_0$ are J-dimensional row vectors and the matrix $\hat{B}$ is a $K \times J$ matrix. For analyses on several samples, the equation can be written as

$$\hat{C} = 1b_0 + X\hat{B} \qquad (4)$$

in which the matrix of predicted concentrations $\hat{C}$ is an $I \times J$ matrix and the NIR matrix X is an $I \times K$ matrix. Here, 1 is an I-dimensional column vector with elements equal to 1.

There exist several ways to estimate the calibration coefficients in the linear analytical equation. In, for example, transmission spectroscopy for clear solutions, the calibration coefficients can be determined from measurements of the absorbance spectra of the constituents present. This is called *direct calibration*. In NIR reflectance and diffuse transmission, this kind of calibration is complicated by strong interference from unknown noise factors due to light scatter and unknown constituents. Furthermore, some constituents do not have the same absorbance spectra in biological tissue as in their pure state. Therefore, the calibration coefficients must, in this case, be estimated statistically from a set of representative calibration samples with measured chemical and NIR data. This is called *indirect calibration*.

B. Multiple Linear Regression as a Calibration Method to Determine the Calibration Coefficients

The calibration coefficients $\hat{b}_j$ for each constituent may be estimated by multiple linear regression (MLR) using the following model.

$$c_j = 1b_{0j} + Xb_j + f_j \qquad (5)$$

in which c_j and f_j are I-dimensional column vectors. Notice that while c_i denoted the observations of the J chemical constituents in sample i, c_j denotes the observations for constituent j in all I samples. The ordinary least-squares (LS) estimator for $(b_{0j}, b_j')'$ is

$$\left[(1, X)'(1, X)\right]^{-1}(1, X)'c_j \qquad (6)$$

However, the matrix to be inverted in Equation 6 is $(1 + K) \times (1 + K)$ dimensional. If K is large, inverting such large matrices requires large computer memory and long computation time. Numerical accuracy problems may also make such an operation difficult.

In this method, each calibration coefficient is treated as an "independent" statistical parameter to be estimated. Therefore, even if a solution to Equation 6 could be obtained numerically for a certain calibration set, analysis of unknown samples with the resulting calibration coefficients may give inaccurate results, unless the number of calibration samples is very high and all of the data

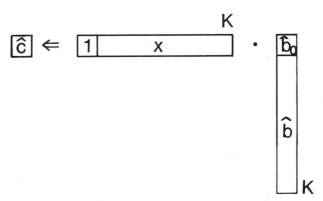

Fig. 2. The linear prediction equation, yielding concentration $\hat{c}$ from the measured spectrum x and the estimated calibration coefficients $\hat{b}_0$ and $\hat{b}$ from constituent j.

very precise. The reason is sometimes called *statistical overfitting*, when too many independent parameters are estimated with too few calibration samples (Fig. 3).

The data obtained at the different wavelengths are usually highly intercorrelated. The huge light-scatter variation introduces a strong correlation between all the wavelengths, unless some optical correction is applied on the NIR data. In addition, if, for example, 100 different wavelengths are measured for each sample, there are usually less than 100 different independent ways in which these wavelengths vary from sample to sample. This is certainly true if the number of calibration samples is less than 100! Hence, some of the wavelengths are bound to be strongly correlated to the other wavelengths.

This simultaneous intercorrelation between many different NIR wavelengths is termed *multicollinearity* and makes it difficult to obtain good calibration by the MLR method using all available wavelengths. Some type of data compression is required to correct for multicollinearity.

C. Different Classes of Calibration Methods

Several calibration methods can be used for NIR instruments to correct for multicollinearity. Many of them give similar results, but some methods behave much better than others under certain circumstances. These methods can be classified in different ways.

One way of classifying calibration methods concerns whether they rely on selection or weighting of NIR information to "solve" the multicollinearity problem. The calibration methods described in Chapter 3 select only a small subset of the available NIR wavelengths, ignoring the rest. In contrast, full-spectrum methods define the regression factors as weighted linear combinations of all the relevant wavelengths, and they use different ways to obtain the weights that define these linear combinations of wavelengths. The spectra of certain calibration samples may be selected as wavelength weights; in other methods, the weights are obtained from predefined mathematical formulas, and in still others, the weights are estimated statistically. The latter estimation is usually based on the NIR data alone, but sometimes even the chemical data are used to estimate the wavelength weights.

Another way of classifying the methods concerns the way they utilize information about the statistical distribution of the chemical composition in the population of samples to be analyzed. Some methods use this type of information, either implicitly or explicitly, while others do not (Brown, 1982).

In the conventional NIR calibration methods (Section III B), the general calibration model is

$$\text{Chemical data} = f(\text{spectral data}) + \text{errors} \qquad (7)$$

in which $f(\ )$ means "function of." Here, the chemical data are modeled as a regression on spectral data plus errors. The conditional distribution of the chemical data given the spectral data is implicitly present in the model. Calibration methods developed in this model will also depend on this kind of information.

The direct calibration methods sometimes employed for multicomponent analysis in transmission spectroscopy in clear solutions can be extended to indirect versions suitable for NIR calibration. Here, the model is

$$\text{Spectral data} = f(\text{chemical data}) + \text{errors} \qquad (8)$$

For methods developed in this model, one may choose explicitly to what extent distribution of the chemical compositions should be incorporated in the predictor.

Another class of methods is based on latent regression factors, e.g., Fourier regression, principal component regression, and PLS regression (PLSR); the model is

$$\text{Spectral data} = f(\text{latent regression factors}) + \text{errors}$$

or $\qquad (9)$

$$\text{Chemical data} = g(\text{latent regression factors}) + \text{errors}$$

In the first equation, spectral data are modeled as a function of latent factors. In the second equation, the chemical data are modeled as a regression on the same latent factors. The methods developed in this model contain information about the population of samples used in the calibration and are easiest to justify if the calibration samples are representative for the distribution of future samples.

The effect of the different usage of distributional information has not been fully studied statistically, but theoretical considerations indicate that it is of some relevance. Sometimes a reasonably small analytical error over a maximum range of future saple qualities is required. This is obtained by not incorporating distributional assumptions. In other cases, an absolutely minimal error over a smaller, but more important, range of sample variation is necessary. This is obtained by incorporating distributional information. Naes (1985a) discusses this distinction in more detail.

A third way of classifying methods is based on the orthogonality properties of the regression factors. If the regression factors are calculated mathematically so that they are uncorrelated (orthogonal), then certain mathematical operations become simpler, i.e., no matrix inversion is needed. This is discussed later.

III. STATISTICAL CALIBRATION METHODS FOR MULTICOLLINEAR NIR DATA

Several indirect calibration methods are presented for multicollinear data (after a possible transformation) in the unified model structure introduced in Section I B. The model applies for all the methods covered, except for the Beer's model methods to be discussed later.

For those methods in which the chemical data in some way are used during the data compression stage, it is probably advantageous to calibrate for each chemical constituent separately, but possibilities of calibrating for several constituents simultaneously also exist (Sundberg, 1982). For those methods that do not use chemical data in the data compression stage, it is immaterial whether the calibration is performed for each chemical constituent separately or for several simultaneously ($J > 1$). Here, main attention is given to the calibration of one single constituent.

A calibration method that may be important, but is not covered herein, is ridge regression (Fearn, 1983). The formula is similar to Equation 6, but a diagonal matrix with the so-called ridge parameter on the diagonal is added to the matrix in brackets before inversion (Gunst and Mason, 1979). This reduces the multicol-

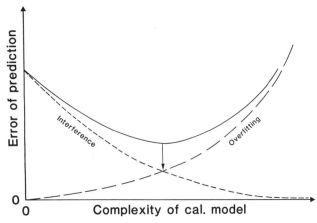

Fig. 3. Optimizing in indirect multivariate calibration. The good effect of modeling interferences and the bad effect of statistical overfitting affect the error of prediction in opposite directions as the complexity of the model increases. The complexity of the model is the amount of information built into the calibration model relative to the total amount of information available in the calibration data. The amount of information built into the calibration model roughly corresponds to the number of parameters estimated or selected, which increases with the number of regression factors, A.

linearity problem, but does not give much insight into the NIR data.

A. The General Model Framework

Instead of modeling only the chemical data as a function of spectral data, as is done in Equations 5 and 7, i.e.,

$$c = 1b_0 + Xb + f \qquad (10)$$

both the NIR data and the chemical data (one constituent) can be modeled in terms of a small set of common latent regression factors $t_1, ..., t_A$ (I-dimensional column vectors), as illustrated by Equation 9 and in Figure 4. The model is

$$X = 1\overline{x} + t_1 p_1 + ... + t_a p_a + ... + t_A p_A + E$$

and

$$c = 1\overline{c} + t_1 q_1 + ... + t_a q_a + ... + t_A q_A + f$$

in which p_a and q_a are the NIR and chemical loadings of regression factor a, E, and f are error terms after A factors, $\overline{c}$ is the mean of the constituent in question and $\overline{x}$ is a K-dimensional row vector of means of X over the I calibration samples.

Using full matrix notation and subtracting the means from both sides, this general calibration model can be written more simply as

$$(X - 1\overline{x}) = TP + E \text{ (data compression)} \qquad (11)$$

and

$$(c - 1\overline{c}) = Tq + f \text{ (calibration regression)} \qquad (12)$$

This notation, shown in Figure 5, will be used for illustrating the different calibration methods.

During calibration, the means $\overline{x}$ and $\overline{c}$ are calculated and subtracted from the X and c in order to eliminate general offset effects. The NIR loadings P, the latent regression factors T, and the chemical loadings q are then estimated, together with the errors E and f. The way P and T are estimated is different for the methods treated, but, in most cases, q is found by least squares using $\hat{T}$ and the model summarized in Equation 12. This implies that

$$\hat{q} = (\hat{T}'\hat{T})^{-1}\hat{T}'(c - 1\overline{c}) \qquad (13)$$

The NIR residuals $\hat{E}$ and chemical residuals $\hat{f}$ are found by subtracting $\hat{T}\hat{P}$ and $\hat{T}\hat{q}$ from the corresponding NIR and chemical observations.

The calibration model can be expressed with increasing degrees of complexity by increasing the number of regression factors, A. Figure 3 shows that testing of unknowns, cross-validation, or some other criterion with statistical significance is required to determine the optimal number of factors. If too few factors are used, the model is incomplete and some interferences remain uncorrected. If too many factors are used, the estimated constants $\hat{P}$

and $\hat{q}$ (and hence coefficients $\hat{B}$, which are constructed from $\hat{P}$ and $\hat{q}$) become noise contaminated. Both result in high analytical error.

The determination of the constituent concentrations in an unknown sample implies first a computation of the regression factors in that sample, $\hat{t}_i$ (A-dimensional row vector), from the NIR spectrum x_i (K-dimensional row vector), and the estimated NIR calibration mean spectrum $\overline{x}$ and loadings $\hat{P}$, using LS and the equation

$$x_i - \overline{x} = t_i\hat{P} + e_i \qquad (14)$$

The estimate of t_i, $\hat{t}_i$, is then

$$\hat{t}_i = (x_i - \overline{x})\hat{P}'(\hat{P}\hat{P}')^{-1} \qquad (15)$$

(An exception is the PLS method (Section III F), in which the $\hat{t}$'s are found by a slightly different procedure.)

This procedure also yields an estimate of the NIR residuals of the unknown sample:

$$\hat{e}_i = x_i - \hat{x}_i = x_i - (\overline{x} + \hat{t}_i\hat{P}) \qquad (16)$$

The obtained $\hat{t}_i$ and $\hat{e}_i$ may be used for detection of outliers of various types. The regression factors $\hat{t}_i$ (sometimes called *estimated scores* or only *scores*) are used to predict the chemical variable by

$$\hat{c}_i = \overline{c} + \hat{t}_i\hat{q} \qquad (17)$$

In all methods considered here, $\hat{t}_i$ is a linear combination of x_i and consequently $\hat{q}$ can be obtained from an ordinary linear prediction equation of the form

$$\hat{c}_i = b_0 + x_i\hat{b} \text{ (Equation 2)}$$

The vector $\hat{b}$ can be written as

$$\hat{P}'(\hat{P}\hat{P}')^{-1}\hat{q} \quad \text{and} \quad \hat{b}_0 = \overline{c} - \overline{x}\hat{b}$$

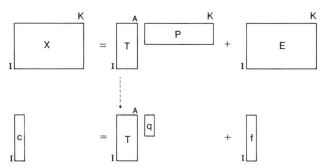

Fig. 5. General data model for calibration expressed in condensed matrix form. X and c are assumed centered.

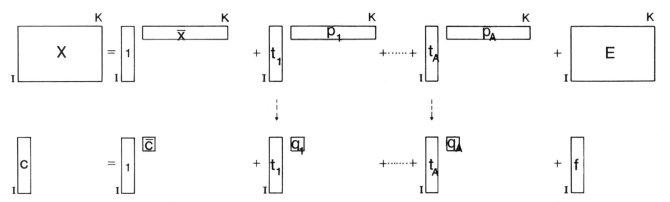

Fig. 4. General data model for the calibration of one chemical variable on K near-infrared (NIR) variables, using I calibration samples. The NIR variables are concentrated to the A regression factors $T = \{t_1, t_2, ..., t_A\}$, on which the chemical variable is regressed.

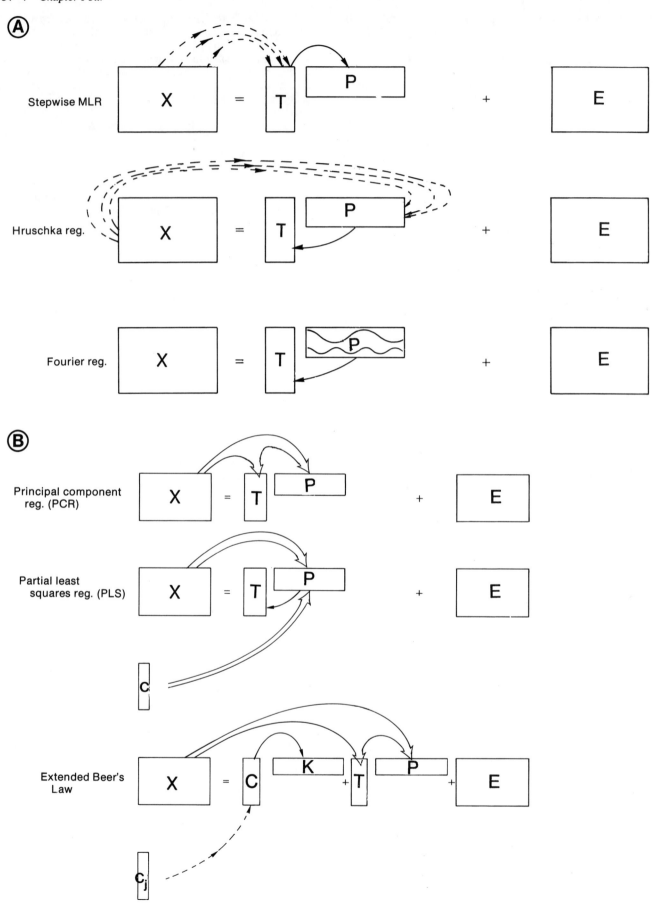

Fig. 6. **A,** The data compression stage for the calibration methods stepwise multiple linear regression (SMLR), Hruschka regression, and Fourier regression. Dotted arrows indicate selection. Solid arrows indicate least-squares fit. **X** is assumed centered. **B,** The data compression stage for the calibration methods principal component regression (PCR), partial least squares (PLS), and the extended Beer's model. Dotted arrows indicate selection. Solid arrows indicate least-squares fit. **X** and **c** are assumed centered.

Figure 6 illustrates the different data compression techniques. Figure 7 illustrates the subsequent use of the regression factors for the different methods to be treated.

B. Conventional NIR Calibration Methods: Selecting the "Best" Wavelengths

Figure 6 illustrate
linear regression (SM
be termed *stepwise*
general model frame
but they are still pos:
of Section III A.

The SDLR metho
$\log(1/R)$, in which R
for calculating first d
tions) and second de
with respect to wavel
from other chemical
terms serves as an op

The data compres
lecting a combinatior
of derivative terms (
can be done in a vari
different combinatior
"best" or most reason

The wavelengths
gression factor matrix
the chemical loading
residuals $\hat{f}$ are found
tives are used, some
quired (Norris, 1983)

For determination
sample, the same wav
stage to find $\hat{t}_i$. This $\hat{t}$
by the equation

For the wavelength
the calibration coeffici

An estimate of E
However, an extra re
tional methods to pro
used NIR wavelength
lengths put into $\hat{T}$).
of the NIR loadings a

$$\hat{P} = $$

in which X represent
are estimated by

$$\hat{E} = X$$

The regression factor:
to detect outliers in th

Outlier detection i
of the unused NIR wa

$$\hat{e}_i = x$$

C.
Selecting tl

If there are many wavelengths, K, but only a limited number, A, of relevant, independent ways in which the NIR spectra may vary, it is possible to use A selected calibration sample's NIR spectra as the NIR loading matrix. These A samples must be selected in such a way that they include all the variation types present in the other $I - A$ calibration samples, as well as all future prediction samples (Hruschka and Norris, 1982; Martens et al, 1983a).

In analogy to the previous method in which a subset of wavelengths were selected and put into the regression factor matrix $\hat{T}$, Figure 6 shows how the spectra of a subset of samples from the calibration set is selected and put into the loading matrix $\hat{P}$. In the data compression step, the loading matrix, $\hat{P}$, is selected and used to estimate the corresponding scores T. Ordinary multiple linear regression (MLR) over the K wavelengths was used by and Norris (1982), i.e.,

$$\hat{T} = \left(X - 1\bar{x}\right)\hat{P}'\left(\hat{P}\hat{P}'\right)^{-1} \qquad (22)$$

ely, a weighted least-squares (WLS) estimator might be der to improve the precision of the $\hat{T}$ estimates, i.e.,

$$\hat{T} = \left(X - 1\bar{x}\right)V^{-1}\hat{P}'\left(\hat{P}V^{-1}\hat{P}'\right)^{-1} \qquad (23)$$

$V = \text{diag}(v_k)$ is a diagonal matrix and v_k is the expected l in wavelength k.

calibration regression, $\hat{T}$ is used as regressor matrix in ry linear regression equation estimating the chemical

analysis of an unknown sample, the same NIR loading is used to compute $\hat{t}_i$, i.e.,

$$\hat{t}_i = \left(x_i - \bar{x}\right)\hat{P}'\left(\hat{P}\hat{P}'\right)^{-1} \qquad (24)$$

als $\hat{e}_i$ are obtained as in Equation 16 and can be used for tection. The desired chemical constituents are deter-n Equation 17.

ection of samples whose NIR spectra are used as NIR ling matrix $\hat{P}$ can be done according to several differ-a. If they are chosen as described by Hruschka and 82), according to their ability to reduce the NIR cali-iduals, $\hat{E}$, this is analogous to Ottestad's (1975) com-alysis. They could also be chosen according to their model the chemical data from NIR data, i.e., by mini-In order to avoid overfitting, this could preferably be ediction testing or cross-validation.

ations in which it is important to limit the number of analyze chemically, the Hruschka method for compo-sis is interesting. Given cheap NIR data but no chemical arge set of samples, a small subset of samples spanning n NIR variation types can be obtained. This small sam-then be analyzed chemically by a slower, more expen-nce method to provide a good set of data for calibration hka, *personal communication*; Martens et al, 1983b). uschka regression method requires precise NIR spectra. measurements contain significant noise, the following e expected to yield somewhat improved analyses.

urier Transform Regression: Concentrating e NIR Data to the Main Spectral Features

ethod is analogous to the former, but uses mathemati-ed rather than experimentally measured loading spectra, vs the concentration of the information and the isolation

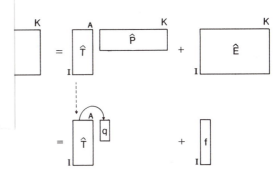

Fig. 7. The calibration regression stage in all the methods covered, except those based on Beer's model. Estimated regression factors $\hat{T}$ are used for the least-squares estimation of q. X and c are assumed centered.

of certain types of noise. Each NIR spectrum is transformed trigonometrically from the usual wavelength "domain" to a new type of spectrum of feature "rhythms" assumed to contain all of the valuable information in the NIR spectra. This procedure can eliminate redundancy and erratic features that are suspected to represent random noise (the narrow peaks or high-frequency "ripple").

Figure 6 illustrated the use of Fourier transforms for multivariate calibration. During calibration, the Fourier series corresponding to 1, 2, 3, 4, ... full sine and cosine functions over the length of the relevant wavelength range can be computed according to

$$\hat{p}_{ak} = \sqrt{\frac{2}{K}} \sin\left(\frac{2\pi ak}{K}\right), \quad a = 1, 2, ..., \frac{A}{2}; \quad k = 1, 2, ..., K$$

$$(25)$$

$$\hat{p}_{\frac{A}{2}+a,k} = \sqrt{\frac{2}{K}} \cos\left(\frac{2\pi ak}{K}\right), \quad a = 1, 2, ..., \frac{A}{2}; \quad k = 1, 2, ..., K$$

The rows of $\hat{P}$ are then orthogonal, i.e., $\hat{P}\ \hat{P}' = I$.

Since factors for high a only represent noise, they are eliminated (i.e., $^A/_2$ is selected such that $^A/_2 \ll {}^K/_2$). The number of Fourier terms to be included, A, may be determined by, for example, significance testing, analytical testing, or cross-validation. Other types of "wavelengths" than the "harmonic" ones in the standard Fourier approach can be envisioned (Damsleth and Spjøtvoll, 1982).

Each observed NIR spectrum, x_i, may first be adjusted so that it meets the infinite Fourier transform requirement that the first and the last wavelength have the same value. In the data compression stage, the loading matrix $\hat{P}$ and the actual NIR data are used for each sample separately to estimate $\hat{T}$, according to Equation 22. (For faster operation, the fast Fourier transform algorithm may be used instead.) "Zero filling" can be used for higher resolution (Rabiner and Gold, 1975). In the calibration regression stage, $\hat{T}$ is used as regressor matrix in a MLR equation for estimating the chemical loadings, $\hat{q}$ (Equation 13).

During analysis of an unknown sample, the first step "slope-adjusts" the NIR data x_i (if slope-adjustment is used) and sets up the feature "rhythm" matrix $\hat{P}$ with the desired number of factors, A. The regression factors and the NIR residuals for outlier detection are then found by, for example, using a similar formula such as Equation 16. The chemical variables c_i are determined as by Equation 17. This corresponds to a calibration coefficient matrix

$$\hat{b} = \hat{P}'\left(\hat{P}\hat{P}'\right)^{-1}\hat{q} = \hat{P}'\hat{q} \tag{26}$$

since $\hat{P}\ \hat{P}' = I$.

The Fourier technique is only applicable for "continuous" or piecewise continuous NIR spectra, but for such data a great simplification can be obtained in terms of computer storage, computation time, and danger of overfitting. By so-called deconvolution prior to the back-transformation, some effects of wide bandpass slit effects, etc., can be removed. A combination of the Hruschka or Fourier regression with the subsequent estimated compression methods may prove optimal in terms of immunity to NIR noise and computer requirements.

To summarize the previous methods, in the conventional calibration, the regression factors $\hat{T}$ were selected more or less directly from the NIR data. In the two subsequent full-spectrum methods, the regression factors $\hat{T}$ were estimated by first selecting the NIR loading matrix $\hat{P}$; in the Hruschka regression, the NIR spectra of some calibration samples were chosen as loadings, while in the Fourier transform regression, the loading spectra consist of trigonometric functions. In the following three methods, the spectra P are estimated statistically.

The three methods differ in the way they use the chemical information in the data compression stage. In principal component regression (PCR), the chemical data are not used at all during the data compression; the regression factors are estimated statistically solely on the basis of the NIR data. This contrasts to PLSR, in which the chemical data are used for guiding the estimation of the factors from the NIR data in such a way that the NIR regression factors become better regressors for the calibration regression. In the last group of methods, based on the causal Beer's model, the chemical data are used most strongly in the data compression: the chemical loading matrix is estimated so as to represent the pure chemical constituents' spectra directly.

E. PCR: Concentrating the NIR Data to Their Most Dominant Dimensions

The PCR (Gunst and Mason, 1979) uses for data compression the singular value decomposition of the NIR data X (also called principal component analysis [PCA], eigenvector decomposition, and Karhunen-Leuwe decomposition), to estimate both T and P. Given the centered NIR data matrix $X - 1\bar{x}$, the t_1 (the first column of T, I-dimensional) and p_1 (the first row of P, K-dimensional) are estimated so that the product $\hat{t}_1\hat{p}_1$ accounts for as much as possible of the total variation in the centered X data. The vectors $\hat{p}_2$ and $\hat{t}_2$ are then chosen such that

$$\hat{t}_2'\hat{t}_1 = 0 \tag{27}$$

and

$$\hat{p}_2\hat{p}_1' = 0 \tag{28}$$

and such that $\hat{t}_2\hat{p}_2$ accounts for as much as possible of the variation in the NIR data remaining after the first factor. In this data compression stage, as much as possible of the variation in the NIR data in the calibration samples is described by as few orthogonal factors as possible. The last factors will mainly account for non-systematic measurement noise in the NIR data and are eliminated from the final calibration regression stage. Thereby the number of regression variables is reduced from the original K wavelengths in X to the A first PCA regression factors in $\hat{T}$.

A number of different algorithms exist for computing the loadings and scores. The NIPALS or NILES method (H. Wold, 1966) has been found to be particularly suitable. The method is detailed as follows.

CALIBRATION

The data matrix X is first centered, i.e.,

$$U = X - 1\bar{x} \tag{29}$$

For each new factor $a = 1, 2, ..., A$, perform steps 1, 2, and 3.

1. Estimate the NIR loading vector p_a (K-dimensional row vector) by LS from a preliminary estimate of the regression factor vector $\hat{t}_a$ (I-dimensional column vector) (or from some start value, e.g., a column in U) using the model

$$U = \hat{t}_a p_a + E \tag{30}$$

The solution is scaled to length 1 and is then

$$\hat{p}_a = \kappa \hat{t}_a' U \tag{31}$$

in which the scaling factor κ is obtained as the inverse of the length of the $\hat{p}_a$ vector obtained without scaling, i.e.,

$$\kappa = \left(\hat{t}_a' U U' \hat{t}_a\right)^{-1/2} \tag{32}$$

2. Estimate the NIR score vector t_a by LS from the previous estimate of $\hat{p}_a$ using the model

$$U = t_a \hat{p}_a + E \tag{33}$$

The LS solution is

$$\hat{t}_a = U\hat{p}_a' \tag{34}$$

Repeat steps 1 and 2 until convergence (typically 4–10 iterations). Convergence criterion is, for example, that the length of $\hat{t}_a$, $\hat{t}_a'\hat{t}_a$ no longer "changes."

3. Prepare new NIR residuals

$$U^{(new)} = \hat{E} = U^{(old)} - \hat{t}_a\hat{p}_a \tag{35}$$

and increase a by 1 and go to step 1.

The regression factors $\hat{t}_1, ..., \hat{t}_A$ obtained in this way are used to estimate the vector of chemical loadings, $\mathbf{q}$, in the subsequent calibration regression stage. This can be done as in Equation 13 or, since the scores are orthogonal, by the simple formula

$$\hat{q}_a = \frac{\hat{t}_a(\mathbf{c} - 1\bar{c})}{\hat{t}'_a \hat{t}_a} \tag{36}$$

The number of regression factors A to be used in the final equation can be chosen in different ways. S. Wold (1978) gives a cross-validation method for finding the number of significant PCA factors based on $\mathbf{X}$ alone, and significance tests based on distributional assumptions also exist. It is more suitable to use the ability of determination of the chemical data $\mathbf{c}$ as a criterion, based on true analysis or on cross-validation in the calibration regression stage. Notice that PCR as described here deletes small eigenvalues. Other versions of PCR that delete eigenvectors due to predictive relevance are discussed by Joliffe (1982).

PREDICTION

During analysis of an unknown sample, $\hat{t}_i = (\mathbf{x}_i - \bar{\mathbf{x}})\hat{\mathbf{P}}'$, because $(\hat{\mathbf{P}}\hat{\mathbf{P}}') = \mathbf{I}$. The NIR residuals of this sample are

$$\hat{\mathbf{e}}_i = \mathbf{x}_i - \left(\bar{\mathbf{x}} + \hat{t}_i\hat{\mathbf{P}}\right) \tag{37}$$

and the concentration of the chemical constituent is determined from $\bar{c}$ and $\hat{\mathbf{q}}$ by

$$\hat{c}_i = \bar{c} + \hat{t}_i\hat{\mathbf{q}} \tag{38}$$

Alternatively, the same analysis step can be written as

$$\hat{c}_i = \hat{b}_0 + \mathbf{x}_i\hat{\mathbf{b}} \tag{39}$$

in which

$$\hat{b}_0 = \bar{c} - \bar{\mathbf{x}}\hat{\mathbf{b}} \quad \text{and} \quad \hat{\mathbf{b}} = \hat{\mathbf{P}}'\hat{\mathbf{q}} \tag{40}$$

F. PLSR: Concentrating the NIR Data to Their Most Relevant Dimensions

The PLSR is a relatively new approach to linear regression, and the algorithm for finding the estimates is a generalization of the NIPALS algorithm for PCA. The method and its extensions are especially suitable for calibration on a small number of samples with experimental noise in both chemical data and NIR data. It is also simple to compute, involving only a series of simple linear regressions. The method used for calibration was initially developed by S. Wold et al (1983) on the basis of the general PLS method (H. Wold, 1982). It has been used for NIR data (Martens and Jensen, 1983), fluorescence data (Jensen et al, 1982; Jensen and Martens, 1983), X-ray computer tomography data (Vangen and Martens, 1982; Martens et al, 1983b), and sensory data (Hildrum et al, 1983; Martens et al, 1983c). Computer programs for PLS analysis are commercially available. The works by Sjöström et al (1983) and Lindberg et al (1983) can be referred to for applications of the method.

In both PCR and PLS, the regression factors are linear combinations of the NIR wavelengths, estimated from the NIR spectra, but their estimation is different. In PCR, the data compression stage is performed independently of the calibration regression and in such a way that the principal components account for a maximum of the total variation in the NIR data alone. In PLS, the two stages can be considered tied together, and the PLS factors primarily describe the NIR variation types that are relevant for modeling the variations in the chemical data. The resulting PLS factors describe important variations in the NIR data themselves, but at the same time are relevant for the determination of the chemical data. If the different variation types in the NIR data have the same relative importance for modeling the NIR data as for modeling the chemical data, then the PCR and the PLS factors are nearly identical. An early comparison of PCR and PLS can be found in the work by Naes and Martens (1985).

If the NIR data instead contain several major, irrelevant variation types (e.g., due to light scatter) and only minor variations relevant for modeling the chemical data, then the PCR solution may mistakenly regard the minor variations as noise and ignore them, since they come in as late factors (large a). The PLSR has the possibility to detect these minor, relevant variations in early factors, which are clearly distinguished from irrelevant factors and the data noise in the later factors. This may be expected to lead to more efficient data compression and thus better calibration.

The basic PLSR method related one chemical variable at a time ($J = 1$) to the K NIR variables (PLS1). It can be extended in many different ways; for instance, it can be modified to calibrate for many different chemical constituents at a time (PLS2) (Martens and Jensen, 1983; S. Wold et al, 1983; Sjöström et al 1983). The algorithms for both PLS1 and PLS2 are described in detail. Some extensions will also be discussed. More detail may be found in the work by, e.g., Martens and Naes (1989) and Martens and Martens (2001).

Two different PLS1 algorithms exist, one that yields orthogonal scores and one that does not. Preliminary investigations indicate that the two algorithms yield the same final results. First, the basic version with nonorthogonal scores (Martens, 1985) will be shown in order to illustrate the PLS philosophy compared with that of PCR. The other algorithm, which is less illustrative from a statistical point of view, will be described afterwards and used in the computations in the subsequent sections of the paper because of its computational simplicity.

Like the NIPALS PCA algorithm, the PLS algorithms can accept missing data, but this is not shown here. In addition, the method can be used even if $I < K$.

BASIC PLS ALGORITHM (FIG. 8)

Calibration. The data matrix $\mathbf{X}$ and the vector $\mathbf{c}$ are first centered, i.e.,

$$\mathbf{U} = \mathbf{X} - 1\bar{\mathbf{x}} \tag{41}$$

$$\mathbf{v} = \mathbf{c} - 1\bar{c} \tag{42}$$

For each new factor $a = 1, 2, ..., A$, perform steps 1, 2, 3, and 4.
1. Use the chemical residuals $\mathbf{v}$ and estimate the NIR loading vector $\mathbf{p}_a$ by LS from the model

$$\mathbf{U} = \mathbf{v}\mathbf{p}_a + \mathbf{E} \tag{43}$$

The solution is scaled to length 1, which gives

$$\hat{\mathbf{p}}_a = \kappa\mathbf{v}'\mathbf{U} \tag{44}$$

in which κ is the scaling factor that makes the length of $\hat{\mathbf{p}}_a$ equal to 1. If desired, this estimated loading vector $\hat{\mathbf{p}}_a$ can be modified (e.g., smoothed, as described in Section III F, Modifications of the PLSR Method) prior to the scaling to length 1.

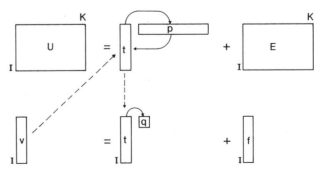

Fig. 8. The basic PLS1 algorithm. The noniterative one-regress and PLS1-regression illustrated for the process of estimating the parameters of partial least-squares (PLS) factor a ($\mathbf{p}_a$, $\mathbf{t}_a$, and q_a).

2. Estimate the regression factors t_a by LS from the estimated NIR loadings $\hat{p}_a$ using the model

$$\mathbf{U} = \mathbf{t}_a\hat{\mathbf{p}}_a + \mathbf{E} \qquad (45)$$

The LS solution is

$$\hat{\mathbf{t}}_a = \mathbf{U}\hat{\mathbf{p}}'_a$$

Steps 1 and 2 are the same as in NIPALS PCA, except that $\mathbf{v}$ is used as the start value in step 1.

3. Estimate the chemical loadings for the previous a factors, $\mathbf{q}$, by MLR (Equation 13) from the regression factors $\hat{\mathbf{T}} = \{\hat{\mathbf{t}}_1, ..., \hat{\mathbf{t}}_a\}$, using the model summarized in Equation 12. Equation 36 in the PCR-NIPALS algorithm cannot be used, since the scores are not orthogonal in this PLS version.

4. Prepare new residuals, i.e.,

$$\mathbf{v} = \mathbf{c} - \mathbf{1}\bar{c} - \hat{\mathbf{T}}\hat{\mathbf{q}} \qquad (46)$$

and

$$\mathbf{U} = \mathbf{X} - \mathbf{1}\bar{x} - \hat{\mathbf{T}}\hat{\mathbf{P}} \qquad (47)$$

and go to step 1 if $a < A$.

The number of factors to be accepted in the final calibration, A, may be determined, for example, by prediction testing or cross-validation, like in PCR. Notice that a new vector $\hat{\mathbf{q}}$ is computed by MLR for each a.

Analysis. In this basic PLS version, the analysis of an unknown sample from its NIR spectrum $\mathbf{x}_i$ proceeds just like in PCR. Equation 37 yields its NIR residuals $\hat{\mathbf{e}}_i$, and Equation 38 or 39 yields the concentration of the chemical constituent, $\hat{c}_i$.

Just like the NIPALS algorithm, this basic PLS algorithm yields orthogonal NIR loading vectors $\hat{\mathbf{p}}_1, ..., \hat{\mathbf{p}}_A$, obtained by projecting the NIR residuals $\mathbf{U}$ onto each loading vector $\hat{\mathbf{p}}_a$ and subtracting this projection $\hat{\mathbf{t}}_a\hat{\mathbf{p}}_a$ before proceeding to the next factor. Thereby the information concerning the loading vector $\hat{\mathbf{p}}_a$ is removed from the NIR data before $\hat{\mathbf{p}}_{a+1}$ is estimated, and hence orthogonality between rows in $\hat{\mathbf{P}}$ is ensured.

In the NIPALS algorithm, a projection of $\mathbf{U}$ on the vector $\hat{\mathbf{t}}_a$ was done prior to subtracting this effect, and hence orthogonal columns in $\hat{\mathbf{T}}$ are also obtained. But, in the basic PLS version given above, no such projection on $\hat{\mathbf{t}}_a$ occurs, and hence the columns $\hat{\mathbf{t}}$ are nonorthogonal.

PLS ALGORITHM
WITH ORTHOGONAL REGRESSION FACTORS (FIG. 9)

A second version of the PLSR (S. Wold et al, 1983) has been designed in order to gain the computational advantage that orthogonality in the regression factors gives. The reason is that the MLR is avoided in step 3 above. Only simple regressions are necessary. The algorithm is very similar to the PLS described above, with the exception of the introduction of a second type of NIR loadings (loading weights), $\mathbf{w}$.

Calibration. The data are first centered as in Section III F, Basic PLS Algorithm. For each new factor $a = 1, 2, ..., A$, perform the following steps.

1. Estimate the NIR weights $\mathbf{w}_a = (w_{a1}, ..., w_{ak})$ from the chemical residuals $\mathbf{v}$ using LS and the model

$$\mathbf{U} = \mathbf{v}\mathbf{w}_a + \mathbf{E} \qquad (48)$$

The solution is scaled to length 1. The estimate is

$$\hat{\mathbf{w}}_a = \kappa\mathbf{v}'\mathbf{U} \qquad (49)$$

in which κ scales $\hat{\mathbf{w}}_a$ to length 1. If desired, this estimated loading vector $\hat{\mathbf{w}}_a$ can be modified (e.g., smoothed), as described in Section III F, Modifications of the PLSR Method, prior to the scaling to length 1.

2. Estimate the factor score vectors $\mathbf{t}_a$ by LS from the estimated NIR loading weights $\hat{\mathbf{w}}_a$ using the model

$$\mathbf{U} = \mathbf{t}_a\hat{\mathbf{w}}_a + \mathbf{E} \qquad (50)$$

The solution is

$$\hat{\mathbf{t}}_a = \mathbf{U}\hat{\mathbf{w}}'_a \qquad (51)$$

2-II. Estimate the NIR loadings $\mathbf{p}_a$ from the estimated factor score vectors $\hat{\mathbf{t}}_a$ using LS and the model

$$\mathbf{U} = \hat{\mathbf{t}}_a\mathbf{p}_a + \mathbf{E} \qquad (52)$$

The solution is

$$\hat{\mathbf{p}}_a = \frac{\hat{\mathbf{t}}'_a\mathbf{U}}{\hat{\mathbf{t}}'_a\hat{\mathbf{t}}_a} \qquad (53)$$

3. Estimate the chemical loading element for factor a from the estimated factor scores $\hat{\mathbf{t}}_a$ by LS using the equation

$$\mathbf{v} = \hat{\mathbf{t}}_aq_a + \mathbf{f} \qquad (54)$$

which gives the solution

$$\hat{q}_a = \frac{\hat{\mathbf{t}}'_a\mathbf{v}}{\hat{\mathbf{t}}'_a\hat{\mathbf{t}}_a} \qquad (55)$$

4. New NIR residuals are obtained by Equation 35 ($\hat{\mathbf{p}}$ is used, not $\hat{\mathbf{w}}$!).

$$\mathbf{U}^{(new)} = \mathbf{U}^{(old)} - \hat{\mathbf{t}}_a\hat{\mathbf{p}}_a$$

The new chemical residuals are found from

$$\mathbf{v}^{(new)} = \mathbf{v}^{(old)} - \hat{\mathbf{t}}_a\hat{q}_a \qquad (56)$$

Notice that Equation 51 projects $\mathbf{U}$ on the vector $\hat{\mathbf{w}}_a$ and Equation 53 projects $\mathbf{U}$ on $\hat{\mathbf{t}}_a$. Hence, orthogonal rows in $\hat{\mathbf{W}}$ and orthogonal columns in $\hat{\mathbf{T}}$ are obtained! Notice also that $\hat{\mathbf{p}}$ in the basic PLSR algorithm corresponds to $\hat{\mathbf{w}}$ in this orthogonalized version.

Analysis. In this second PLSR version, the analysis of an unknown sample from its NIR spectrum $\mathbf{x}_i$ proceeds as follows.

The first score, t_1, is found from the model

$$\mathbf{x}_i - \bar{\mathbf{x}} = t_{i,1}\hat{\mathbf{w}}_1 + \mathbf{e} \qquad (57)$$

The solution is

$$\hat{t}_{i,1} = (\mathbf{x}_i - \bar{\mathbf{x}})\hat{\mathbf{w}}'_1 \qquad (58)$$

The next score, t_2, is found from the model

$$[(\mathbf{x}_i - \bar{\mathbf{x}}) - \hat{t}_{i,1}\hat{\mathbf{p}}_1] = t_{i,2}\hat{\mathbf{w}}_2 + \mathbf{e} \qquad (59)$$

The solution is

$$\hat{t}_{i,2} = [(\mathbf{x}_i - \bar{\mathbf{x}}) - \hat{t}_{i,1}\hat{\mathbf{p}}_1]\hat{\mathbf{w}}'_2 \qquad (60)$$

The procedure continues until the Ath factor. Notice that the procedure corresponds to the estimation procedure in the calibration.

The chemical concentration is then predicted by Equation 38 or 39. In this case, vector $\hat{\mathbf{b}}$ can be written as

$$\hat{\mathbf{b}} = \hat{\mathbf{W}}'(\hat{\mathbf{P}}\hat{\mathbf{W}}')^{-1}\hat{\mathbf{q}} \qquad (61)$$

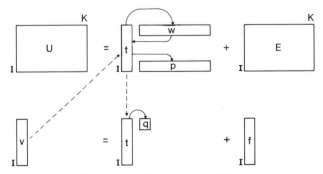

Fig. 9. The algorithm for partial least squares (PLS) with orthogonal regression factors.

instead of by Equation 40. (This formula was developed heuristically [Martens, 1985] and tested in practice, but has now been proven theoretically.)

MODIFICATIONS OF THE PLSR METHOD

In different calibration situations, different noise problems occur and different types of additional information can be used to counter this noise. The PLSR algorithm, especially in its orthogonalized version, is well suited for incorporating different kinds of additional information.

Using the Correlation Between Different Chemical Constituents. An iterative extension of the PLS algorithm given above when $J \geq 2$ (termed PLS2) (S. Wold et al, 1983; Martens and Jensen, 1983) can, in certain cases, be advantageous, e.g., when the signal-to-noise ratio in the chemical calibration data for an important constituent is so bad that the calibration is difficult. If precise chemical data are available for other chemical constituents, and these other constituents correlate more or less to the constituent in question, then these other variables can be used to improve the calibration of the important constituent.

In conventional regression, e.g., SMLR, these other chemical variables would have to be used directly as regression factors, just like the NIR wavelength, and hence would have to be measured chemically in every new verification sample as well. In contrast, the PLS2 method uses the additional chemical variables for stabilizing the statistical estimation process only during calibration.

However, if there are strong nonlinearities in the data, then different linear approximations may be required in order to give optimal calibration for different chemical constituents. In such cases, the one-constituent ordinary PLS1 method is probably preferable to the PLS2 method.

The PLS2 method requires a statistical weighting of the different chemical constituents' data in **C** to compensate for their differences in signal-to-noise ratio.

The PLS2 method requires data from two or more chemical constituents ($J \geq 2$; implying the matrix notation **C**, **F**, **V**, $\mathbf{q}_a$, and $\bar{\mathbf{c}}$ instead of **c**, **f**, **v**, q_a, and $\bar{c}$). The algorithm in Section III F, PLS Algorithm with Orthogonal Regression Factors, has two extra steps between steps 3 and 4.

3-II. Test if convergence has occurred, e.g., by checking that $\hat{\mathbf{t}}'_a\hat{\mathbf{t}}_a$ no longer increases.

3-III. If convergence is not reached, estimate temporary factor scores from the chemical loadings $\hat{\mathbf{q}}_a$ by LS using the model

$$\mathbf{V} = \mathbf{t}_a\hat{\mathbf{q}}_a + \mathbf{F} \qquad (62)$$

The solution is

$$\hat{\mathbf{t}}_a = \mathbf{V}\hat{\mathbf{q}}'_a(\hat{\mathbf{q}}_a\hat{\mathbf{q}}'_a)^{-1} \qquad (63)$$

Repeat steps 1, 2, 2-II, 3, 3-II, and 3-III until convergence (typically three to seven iterations).

Decreasing Sensitivity to Chemical Noise by Combined PLS/PCR. In contrast to the PCR method, the PLSR methods use the chemical residuals **v** when estimating the loading spectra $\mathbf{p}_a$. One would expect this to lead to more efficient data compression in PLS than in PCR, but at the expense of a slightly stronger tendency to overfitting due to errors in the chemical data. Since the PLS algorithm is a modification of the NIPALS algorithm, it appears possible to combine the two. For a given chemical constituent ($J = 1$), one could use PLS for the first, dominating factors to ensure their chemical relevance, and PCA for the last ones to eliminate the effect of chemical errors in the factor estimation. An intermediate algorithm can be used for the intermediate factors. In the PLS algorithm with orthogonal scores, a PCR corresponds to an iteration between Equations 51 and 49.

Thus, PLS factors are obtained as long as steps 1, 2, and 2-II are performed only once for each factor. PCR factors are instead obtained by repeating the above iteration until convergence, whereby all information about the chemical residual **v** is eliminated from the estimation of loading weights $\hat{\mathbf{w}}_a$. Regression

factors with properties between those in PLS and PCR are obtained by repeating the above iteration more than once, but less than until convergence; each iteration will make the influence of the chemical residuals **v** on the estimation of loading weights $\mathbf{w}_a$ progressively weaker. The method has been tested successfully on NIR data.

Application of Extra Information About NIR Loadings. The PLS loading vector $\hat{\mathbf{w}}_a$, obtained from Equation 48 by $\hat{\mathbf{w}}_a = \mathbf{v}'\mathbf{U}$, may be improved in different ways by modifying it according to additional information available. Care must be taken to retain the orthogonality of the loading vectors.

Smoothing—If the NIR data come from a continuously scanning instrument, the loading vector $\mathbf{w}_a$ should be a smooth function of wavelength. By applying digital filtering to $\hat{\mathbf{w}}_a$, much of the noise in the NIR data can be eliminated from the calibration process. A low-pass far-infrared filter (Rabiner and Gold, 1975) is probably one of the best methods for eliminating random noise. A median filter can be used to remove spikes in the NIR data first. This smoothing approach has given good results on a data set based on measurements from a data-tomograph (Martens et al, 1983b). Other smoothing algorithms can be seen in the work by Esbensen and Wold (1983).

Eliminating irrelevant NIR wavelengths—In conventional SMLR, only a few of the available NIR wavelengths are selected as regression factors. In contrast, the full-spectrum methods employ all the available relevant NIR wavelengths (although their relative importance can be selected by weighing them prior to the estimation).

A flexible intermediate method can be obtained in PLSR (and probably for other methods of this section) by replacing the estimated loadings $\hat{w}_{ak}$ for certain wavelengths with zero, thereby eliminating them from the estimation of that factor. If a sensible criterion is used for selecting which wavelengths to keep and which to eliminate, one would expect improved calibration results, due to the reduced number of statistical parameters to be estimated and the high information content in the remaining wavelengths.

Wavelength elimination in PLSR based on jack-knifing (Martens and Martens, 2000) is shown by Westad and Martens (1999) and Martens and Martens (2001). An analogue approach can be used for making the PLSR more or less like Hruschka regression by selecting samples instead of variables.

G. Calibration Based on Beer's Model for Mixtures

Because of the multicollinearity between the different wavelengths, the MLR is dangerous to use as a calibration method. The alternative methods described above use some type of data compression in order to reduce the number of independent variables, followed by a calibration regression stage consisting of an LS fit of the chemical constituent to the obtained regression factors.

Since c is used as a dependent variable in the calibration regression of these approaches (Equation 7), the methods developed will depend on the distribution of the c's in the calibration sample set. In many cases, it is natural to incorporate this information, but if for instance the calibration samples are selected in a way that gives no information about the population of samples, this type of information should not be utilized. These two cases are sometimes called *natural* and *controlled* calibration, respectively (e.g., Brown, 1982). It can be shown that if the calibration samples are selected according to their c values in a way other than of being representative for the total population of samples, these methods may be difficult to justify theoretically. Generally speaking, it may be said that it is safest to use samples selected to represent the variability in the population of samples. The set of calibration methods described next are based on the physical mixture model (Beer's model) and allows an explicit choice whether to utilize population information or not.

These calibration methods do not have a traditional calibration regression stage; therefore, the general model of Section III A will not be referred to in this section. Calibration of several constituents will be treated simultaneously ($J \geq 2$).

The Beer model for mixtures can be written as

$$X = 1v + CK + E$$

in which v is an intercept (K-dimensional row vector), C corresponds to the concentration of the constituents of interest, and E consists of measurement noise with expectation zero, unknown interference effects, and other model errors (Fig. 10, a simplified illustration). The error terms, e, for different samples are independent. A discussion of the model can be found in the work by Naes (1985b). Beer's model is used as an explaining concept for matrix algebra in Appendix B.

NIR spectra have been shown to contain phenomena causing systematic disturbances that may be approximated by a linear factor structure. It may, therefore, be appropriate to assume that

$$E = TP + E^*$$

in which the number of factors (rows of P) is small and the error terms in E^* for the different wavelengths are uncorrelated. The elements in E^* correspond to random measurement noise and TP corresponds to interferences such as unknown constituents and light-scatter effects. Thus, the total model (extended Beer's model), including the contribution from the chemical constituents CK and the systematic interferences TP, may be written as

$$X = 1v + CK + TP + E^* \tag{64}$$

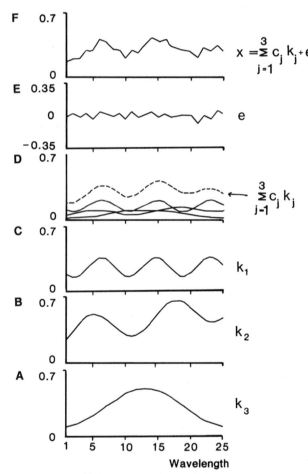

Fig. 10. Beer's model illustrated by artificial data. Three constituents have absorbance spectra A, k_1; B, k_2; and C, k_3. A mixture containing the three constituents in concentrations of 60, 20, and 20%, respectively, should yield D, the absorbance spectrum (dotted line), if no other interferences occurred. This spectrum is the sum of the three individual contributions drawn beneath it. But if E, the measurement process adds noise to the measurement, then F, the actually measured absorbance spectrum looks like this. If this spectrum were obtained from an unknown sample, the constituent concentrations and the noise spectrum could be estimated by some multivariate prediction method. The k spectra plotted are the ones used for the artificial data set in Section VI.

Four versions of the model are of interest, and these versions correspond to different ways of handling unknown c's and t's (Naes, 1985a). This is illustrated in Figure 11. The symbol t (row vector) is here the interference effect for one single sample, and c (row vector) is the corresponding chemical composition of the sample.

From above, it is known that, in many cases, it may be appropriate to assume that we have information in the calibration data set about the distribution of unknown c's and, in other cases, no such information exists, for instance because the calibration samples tell us nothing about future samples. In the two cases, it is natural to consider unknown c's as random variables and as fixed parameters, respectively. In the random case, the distribution of c corresponds to the distribution of c in the population of samples.

Concerning tP, it may sometimes be realistic to assume that both the matrix P and the distribution of future t's are well estimated by the calibration samples. In other cases, it is not so. It may for instance be realistic to assume that P, or the space spanned by the systematic interferences, is well estimated, but the distribution of future t's is not sufficiently well described to be used in the prediction. In such cases, we may want to consider future t's as unknown fixed parameters.

When c is considered random, $E(c)$ (the mean of c) is denoted by u and $cov(c)$ (the dispersion matrix of c) by H. When a future t is considered random, it is assumed that $E(t) = 0$ and a ratio of P and t is chosen such that $cov(t) = I$. When t is random, $cov(e) = cov(tP + e^*) = P'P + W$ is denoted by Σ, in which t and e^* are row vectors (corresponding to one single sample) and $W = cov(e^*)$.

In order to be useful for the analysis of unknown samples, calibration methods based on this extended Beer's model require information about its model parameters. In some applications of the Beer model (e.g., some types of transmission spectroscopy), some model parameters (e.g., K) are known or can be measured directly, while others (e.g., v and P) are assumed to be zero. Calibration in such cases is sometimes called *direct calibration*. This direct approach, however, cannot be used in diffuse NIR spectroscopy for the reason discussed in Section I. When the model parameters are unknown, indirect versions of the predictors derived from the model may be obtained by replacing the known parameters by estimates. The different methods based on Beer's model differ in the way these parameters are estimated and in the way they are used in the analysis.

The model parameters in Beer's model can be estimated in different ways dependent on, for example, robustness requirements and distributional assumptions. The maximum likelihood estimators (under normality assumption) for v, K, u, and H have been found useful (Naes, 1985b). Concerning the factor structure parameters P and W, these may be found by factor analyses of, for example, the maximum likelihood estimator for Σ (Naes, 1985b).

Calibration methods are given for the four different cases mentioned above and illustrated in Figure 11. Estimates of the model parameters are assumed to be available and denoted by the "hat."

Distributional assumption on c

		No	Yes
Distributional assumption on t	No	EGLS1	EBLUP
	Yes	EGLS2	EBLP

Fig. 11. Illustration of the different usage of distributional assumptions.

1. No distributional assumption on **c** and **t**: EGLS1. In this case, **u**, **H**, and Σ are of no interest. It is natural to use the generalized least-squares (GLS) estimator to estimate an unknown **c**, treating $(\hat{\mathbf{K}}', \hat{\mathbf{P}}')'$ as design matrix and $\hat{\mathbf{W}}$ as residual covariance matrix. In this case, the calibration coefficient matrix $\hat{\mathbf{B}}$ is the first J columns of

$$\hat{\mathbf{W}}^{-1}(\hat{\mathbf{K}}', \hat{\mathbf{P}}')\left[(\hat{\mathbf{K}}', \hat{\mathbf{P}}')' \hat{\mathbf{W}}^{-1}(\hat{\mathbf{K}}', \hat{\mathbf{P}}')\right]^{-1} \quad (65)$$

and

$$\hat{\mathbf{b}}_0 = -\hat{\mathbf{v}}\hat{\mathbf{B}}$$

The equation derived from Equation 65 is sometimes called EGLS (estimated GLS). Here, it is called EGLS1, to distinguish it from EGLS2 below.

2. Distributional assumption on **t** only: EGLS2. In this case, only **u** and **H** are uninteresting. The natural predictor to use is then the GLS estimator using $\hat{\mathbf{K}}$ as design matrix and $\hat{\Sigma}$ as covariance matrix, i.e.,

$$\hat{\mathbf{B}} = \hat{\Sigma}^{-1}\hat{\mathbf{K}}'(\hat{\mathbf{K}}\hat{\Sigma}^{-1}\hat{\mathbf{K}}')^{-1} \quad (66)$$

and

$$\hat{\mathbf{b}}_0 = -\hat{\mathbf{v}}\hat{\mathbf{B}}$$

Even this is an estimated version of GLS, but different from EGLS1. The difference is the design and covariance matrices used in the two cases.

3. Distributional assumption on both **c** and **t**: EBLP. If both **c** and **t** are assumed to be random variables, it is natural to base the determination of **c** on the estimated version of the best linear predictor (BLP with regard to [w.r.t.] mean squared error [MSE] $= E[\hat{c} - c]^2)$ for **c** with **K** as design matrix and $\hat{\Sigma}$ as error covariance matrix. In this case,

$$\hat{\mathbf{B}} = (\hat{\mathbf{K}}'\hat{\mathbf{H}}\hat{\mathbf{K}} + \hat{\Sigma})^{-1}\hat{\mathbf{K}}'\hat{\mathbf{H}} \quad (67)$$

and

$$\hat{\mathbf{b}}_0 = -(\hat{\mathbf{v}} + \hat{\mathbf{u}}\hat{\mathbf{K}})\hat{\mathbf{B}} + \hat{\mathbf{u}}$$

4. Distributional assumptions on **c** only: EBLUP. In this case, **t** is an unknown fixed parameter. It is then natural to determine unknown **c**'s by the estimated version of the best linear unbiased predictor for **c** (BLUP w.r.t. MSE; e.g., Henderson, 1975), treating **t** as a fixed effect. In this case,

$$\hat{\mathbf{B}} = (\hat{\mathbf{K}}'\hat{\mathbf{H}}\hat{\mathbf{K}} + \hat{\mathbf{W}})^{-1}\hat{\mathbf{K}}'\hat{\mathbf{H}} \quad (68)$$

and

$$\hat{\mathbf{b}}_0 = -(\hat{\mathbf{v}} + \hat{\mathbf{u}}\hat{\mathbf{K}} + \hat{\mathbf{t}}\hat{\mathbf{P}})\hat{\mathbf{B}} + \hat{\mathbf{u}}$$

in which $\hat{\mathbf{t}}$ is the GLS estimator of **t** in the model treating **c** as random (with estimated model parameters).

Practical applications of the methods in cases 2 and 3 can be found in the works by Naes (1985b), Naes (1986), Brown (1982), and Skrede et al (1983). The methods show a good performance in NIR and are multivariate generalizations of the classical and inverse methods of univariate calibration (e.g., Hoadley, 1970). In Section VII, an application of the method in case 1, which is originally treated by Martens (1981), was presented. The method in case 4 is not tested, but may be of practical interest. Although the methods treated here at first glance look fundamentally different from earlier methods, it can be shown that there is a great similarity between EBLP and, for example, PCR. They may both be considered as estimates of the BLP for case 3, and the difference is how this BLP is estimated.

Other kinds of calibration methods in Beer's Law may be found in the works of Haaland and Easterling (1982) and Saxberg and Kowalski (1979). These methods are, however, not so well suited for NIR data.

The practical significance of the different distributional assumptions on **c** and **t** have not been fully explored (Naes, 1985a; Lwin and Maritz, 1982). It is known, however, that if the variances in $\hat{\mathbf{w}}$ are very small, the methods will give similar results. In general, it can also be said that the approaches treating **c** as a constant unknown parameter are usually better suited than other methods for analysis outside the range of the calibration samples (extrapolation). Incorporating distributional information usually improves the precision of the analysis when the future samples are within the range of the calibration samples (interpolation). In addition, if calibration samples are representative for the population, incorporation of distributional assumptions gives the best average results (averaged over the samples analyzed).

Outlier detection is also possible in the methods presented here; the work by Naes and Martens (1987) can be referred to for some results.

IV. ANALYTICAL ABILITY AND OUTLIER DETECTION

A. Evaluating Analytical Ability

The equation described in Section II A can be applied to the samples in the calibration set as well as to new samples. By comparing the NIR concentrations with chemically determined values, the success of the calibration can be assessed (validation). Different types of analyses may be envisioned. Presented below are a few types with special importance in practice.

Let the chemical lack-of-fit residual, $\hat{f}_{ij}$, be defined by

$$\hat{f}_{ij} = c_{ij} - \hat{c}_{ij} \quad (69)$$

and for a set of I samples let

$$\hat{\mathbf{f}}_j = (c_{1j} - \hat{c}_{1j}, \ldots, c_{Ij} - \hat{c}_{Ij})' = \mathbf{c}_j - \hat{\mathbf{c}}_j \quad (70)$$

in which $\hat{\mathbf{f}}_j$, $\hat{\mathbf{c}}_j$, and $\mathbf{c}_j$ are I-dimensional column vectors.

The goal of a calibration is to obtain calibration coefficients that determine concentrations in future samples so that the residuals $\hat{f}_{ij}$ are as close to zero as possible. The success can be assessed statistically by absolute values (i.e., in the same unit as the concentration itself) or by relative values (i.e., relative to the total variation in the data).

ABSOLUTE VALUES

External Validation. The most straightforward assessment of a determination (validation) is obtained by testing on a set of new samples. The root mean square of performance (RMSP) for each chemical constituent is defined by

$$\text{RMSP}_j = \left(\frac{1}{I_p} \hat{\mathbf{f}}_j' \hat{\mathbf{f}}_j\right)^{1/2} \quad (71)$$

This measures the total analytical error, since it includes bias as well as variation around the mean. Hruschka and Norris (1982) used the corresponding standard error of performance (SEP), in which $\hat{\mathbf{f}}_j$ is corrected for the analytical bias. A discussion of this is in Chapter 3.

Internal Validation. When the predictor is tested on the calibration set itself, the root mean square of estimation (RMSE) is obtained in analogy to Equation 71, except with I_p replaced by I_c minus the degrees of freedom used in the calibration. This may sometimes be dangerous to use as a measure of the real analytical error, since it may become much too low if the calibration model is overfitted, i.e., measurement noise is drawn from the calibration data into the predictor. This may be the result of testing too many alternative wavelength combinations or using too many regression factors in the calibration regression.

A better internal validation criterion is the root mean square of cross-validation (RMSCV) (e.g., Stone, 1974). In cross-validation, one repeats the calibration N (e.g., four) times, each time treating one Nth part of the whole calibration sample set as "unknown" samples instead. In the end, all the calibration samples have been treated as unknown or analytical samples. RMSCV is then obtained in analogy to Equation 71. Alternatively, RMSCV can, in some cases, be computed by a leverage-corrected RMSE (Cook and Weisberg, 1982), which we term RMSEL.

RELATIVE MEASURES

A measure of the success used frequently is

$$R_j^2 = \frac{STOT_j^2 - RMS_j^2}{STOT_j^2} \tag{72}$$

in which RMS_j^2 represents the square of one of the measures RMSE, RMSP, RMSCV, or RMSEL. The term STOT is the total standard deviation of the constituent in the sample set, i.e.,

$$STOT_j = \left[\frac{1}{I} \sum_{i=1}^{I} \left(c_{ij} - \bar{c}_j \right) \right]^{1/2} \tag{73}$$

The quantity R_j^2 is closely related to the squared correlation between estimated and measured value (Searle, 1971). The relative ability of prediction (analysis) (RAP) (Hildrum et al, 1983) is probably a better measure of the quality of the analysis, because it takes into account the level of experimental error in the chemically measured data used as "truth," represented by the chemical noise standard deviation, SCHEM. RAP is defined as

$$RAP_j = \frac{STOT_j^2 - RMS_j^2}{STOT_j^2 - SCHEM_j^2} \tag{74}$$

This section has discussed how the quality of analysis can be based on $\hat{f}$. It should be emphasized here that other aspects are also important when selecting a calibration method. For instance, having a simple model is useful for interpretation.

B. The Importance of Outlier Detection

Measurements used for calibration and analysis usually contain random measurement noise. In general, such random noise can be compensated for by increasing the number of calibration samples or the number of wavelengths employed. However, abnormally large errors also sometimes occur that make the analytical error intolerable. Abnormalities in NIR data can be caused by a variety of reasons. These are discussed in a subsequent chapter.

NIR spectroscopy is often operated in routine environments. It is therefore important to have methods for automatic detection of such abnormalities or outliers, during both calibration and analysis. During calibration, outlier detection is important in order to obtain optimal calibration coefficients. When carrying out NIR analysis, outlier detection is primarily used as error warning.

Various types of data information are available for detecting outliers, in calibration as well as in analysis.

1. NIR residual, $\hat{e}_{ik}$: the lack-of-fit between NIR measurement and the calibration model. These residuals can be studied with respect to their general variation level as well as to their remaining spectral features.
2. Leverage: the position relative to the rest of the calibration sample set (Cook and Weisberg, 1982).
3. Chemical residual, $\hat{f}_{ij}$: the lack of fit between chemical measurements and NIR predicted values.
4. Criteria combining the variables above.

Each of these outlier information types will be discussed below, with main emphasis on their use in the general model framework of Section III A. Some of the statistical tests are based on rather heuristic arguments and the statistical properties are not fully explored. In addition, the tests are based on assumptions that are not always exactly fulfilled. Tests are therefore only approximate in many applications, but give useful warnings for potential problems. The final choice of what to do when a sample or a variable has been detected as an outlier must rest with the user; the distinction between an erroneous sample and a correct but extreme sample is a practical, not a statistical, problem.

Certain types of NIR outliers might also be detected in the response linearization or the optical correction stages, and grave errors in chemical calibration data may be found by inspecting the input data. This will not be treated here.

For a thorough presentation of outlier detection and influence in linear models based on MLR, the reader may refer to the works by Cook and Weisberg (1982) and Beckman and Cook (1983).

C. Analysis of NIR Residuals

The multivariate NIR instruments usually measure more wavelengths, K, than the number of regression factors, A, required for modeling the true variability of the calibration samples. Therefore, over-determined equations are obtained, yielding NIR residuals

$$\hat{e}_{ik} = x_{ik} - \bar{x}_k - \sum_{a=1}^{A} \hat{t}_{ia} \hat{p}_{ak} \tag{75}$$

These residuals, $\hat{e}_{ik}$, are available for every calibration sample and every unknown sample and can be used in different ways for a variety of purposes. During both calibration and analysis, they can be tested statistically to identify samples or wavelengths that do not fit the calibration model. The outliers' residuals can then be studied in more detail to identify the reason for the lack of fit, to tell if the problem stems from erratic measurement errors or from systematic errors due to some specific interference.

For scanning instruments, the presence of spectral peaks and valleys in the residuals of a sample indicates that systematic information remains in the NIR data. Residual features can be detected by plotting $\hat{e}_i$ versus wavelength, but, for routine operation, some automatic method is preferable. One method is to study the power spectrum of $\hat{e}_i$ generated by the Fourier transform (Rabiner and Gold, 1975). Random ("white") noise in $\hat{e}_i$ generates a flat power spectrum, while remaining systematic features yield abnormally high values for the "low-frequency" power terms. Fourier analysis can also be used to design optimal digital filters against erratic measurement errors in NIR data from scanning instruments. Cross-correlations of an outlier's NIR residual spectrum against the corresponding residuals from a data bank of NIR spectra may be used to identify the most probable cause of the lack of fit.

DETECTING ABNORMAL SAMPLES AND WAVELENGTHS

In the calibration set, containing I_c samples, the total residual standard deviation $s(\hat{e})$ after A factors is defined as

$$s(\hat{e}) = \left(\frac{1}{df} \sum_{i=1}^{I_c} \sum_{k=1}^{K} \hat{e}_{ik}^2 \right)^{1/2} \tag{76}$$

in which the degrees of freedom, df, depends on the calibration method and the number of factors used. In, for example, PCR and PLS, it may be approximated by

$$df = I_c K - K - \max(K, I_c) A \tag{77}$$

(This probably underestimates the df somewhat [Martens and Jensen, 1983], but the problem has not been solved theoretically [Mandel, 1971]).

The NIR standard deviation of an individual calibration sample is defined as

$$s(\hat{e}_i) = \left(\frac{I_c}{df} \sum_{k=1}^{K} \hat{e}_{ik}^2 \right)^{1/2} \tag{78}$$

in which df is the degrees of freedom used in Equation 77.

Correspondingly, for an unknown sample, it is defined as

$$s(\hat{e}_i) = \left(\frac{1}{K - A} \sum_{k=1}^{K} \hat{e}_{ik}^2 \right)^{1/2} \tag{79}$$

To detect abnormally high residuals for a sample, one may test $s(\hat{e}_i)$ against the total calibration NIR residual, i.e., by the quotient

$$s(\hat{e}_i)/s(\hat{e}) \tag{80}$$

A large value of this quantity shows that the sample is abnormal in some way. Martens and Jensen (1983) used this method successfully for detecting erroneous wheat and barley samples. It should be noted that, for calibration samples, the quantities $s(\hat{e}_i)$

and $s(\hat{\mathbf{e}})$ are not strictly independent. Another test for a sample is obtained by replacing $s(\hat{\mathbf{e}})$ with the total residual calculated without sample i, i.e.,

$$s(\hat{\mathbf{e}}_{-i}) = \left\{ \frac{1}{(I_c - 1)} \left[s(\hat{\mathbf{e}})^2 I_c - s(\hat{\mathbf{e}}_i)^2 \right] \right\}^{1/2} \qquad (81)$$

The distribution of this criterion is unknown. However, under certain assumptions, the distribution of the square of the quotient $s(\hat{\mathbf{e}}_i)/s(\hat{\mathbf{e}}_{-i})$ can be approximated by a Fisher distribution with the degrees of freedom corresponding to $s(\hat{\mathbf{e}}_i)$ and $s(\hat{\mathbf{e}}_{-i})$. This fact can, in practice, be used to give a good indication of when a sample is considered to be abnormal.

In the calibration samples, the residual standard deviations for the different wavelengths are defined by

$$s(\hat{e}_k) = \left(\frac{K}{\mathrm{df}} \sum_{i=1}^{I_c} \hat{e}_{ik}^2 \right)^{1/2} \qquad (82)$$

in which df is the degrees of freedom used in Equation 77. This can also be compared with $s(\hat{\mathbf{e}})$ or $s(\hat{\mathbf{e}}_{-k})$ (obtained in analogy to Equation 81) as above to reveal wavelengths with abnormally high noise.

Errors in individual data points, i.e., abnormal $\hat{e}_{ik}$, can be detected by testing against $s(\hat{\mathbf{e}})$, $s(\hat{\mathbf{e}}_{-i})$, or $s(\hat{\mathbf{e}}_{-k})$. Also, in this case, an F distribution approximation can be useful.

For calibration samples, Studentized NIR residuals may be used instead of the obtained $\hat{\mathbf{e}}$'s in the outlier tests. Studentized residuals are the $\hat{\mathbf{e}}$'s divided by estimates of their standard deviations (involving leverages) in order to get variables with equal variance. This is discussed by Cook and Weisberg (1982).

Generally, for procedures such as PCR and PLS, the data for each wavelength should have the same noise level in order to yield optimal prediction ability. Abnormally high $s(\hat{e}_k)$ indicates a need for a weighting of the NIR data prior to repeated calibration. If the different noise levels are unknown, the estimated $s(\hat{e}_k)$'s can be used. The weighting may be as follows.

$$x_{ik}^{(\mathrm{new})} = x_{ik}^{(\mathrm{old})} v_k \qquad (83)$$

in which v_k is the weight corresponding to the inverse of the estimated noise level of wavelength k, e.g., $v_k = 1/s(\hat{e}_k)$. The level of prior knowledge about the information content of the different wavelengths may also be drawn into the weights v_k.

D. Leverage: Position Relative to the Rest of the Calibration Sample Set

Leverage of an observation is a concept developed in ordinary linear regression theory and concerns the position of the observation's independent variables relative to the independent variables of other observations (e.g., Cook and Weisberg, 1982). Modifications of the original leverage definition are treated here, as applied to the general data model of Section III A.

The leverage of sample i, h_i, is here defined as

$$h_i = \hat{\mathbf{t}}_i \left(\hat{\mathbf{T}}' \hat{\mathbf{T}} \right)^{-1} \hat{\mathbf{t}}_i' \qquad (84)$$

in which $\hat{\mathbf{t}}_i = \{\hat{t}_{ia}, a = 1, ..., A\}$ is the row vector of regression factors for sample i, while $\hat{\mathbf{T}}$ is the regression factor matrix for the whole calibration sample set (dim $I \times A$). For calibration methods yielding orthogonal regression factors, this simplifies to

$$h_i = \sum_{a=1}^{A} \left(\frac{\hat{t}_{ia}^2}{\hat{\mathbf{t}}_a' \hat{\mathbf{t}}_a} \right) \qquad (85)$$

in which $\hat{t}_{ia}$ is the regression factor of the sample i for factor a, while $\hat{\mathbf{t}}_a$ is the ath score vector of all the calibration samples (I-dimensional column).

The leverage of a calibration sample describes its "uniqueness" or its potential contribution to the estimated calibration solution. It is an important tool in outlier detection that complements the residuals from the NIR data and the chemical data. It reveals cali-

bration samples that may have had very high importance for the estimation of the calibration coefficients $\mathbf{b}$. It is known that $0 \leq h_i \leq 1 - 1/I$, because $\mathbf{X}$ is centered (Cook and Weisberg, 1982).

A leverage of zero for a calibration sample indicates that this calibration sample has very little importance for the calibration solution. A leverage near 1 indicates a very high importance; this sample has "grabbed" a regression factor alone. This is dangerous if the sample contains much noise in its $\mathbf{X}$ or $\mathbf{c}$ data. The average value of h_i is approximated by A/I, and a warning may be given for, for example, $h_i > 2A/I$.

The leverage can also be computed for unknown samples by using the same $\hat{\mathbf{T}}$ in Equation 84, but using the $\hat{\mathbf{t}}_i$'s for these samples. In that case, it shows how the NIR data of the unknown samples are positioned relative to those of the calibration samples. This h_i is useful for detecting extreme samples, e.g., samples that are outside the normal concentration range. For unknown samples, a high leverage indicates a rather extreme, possibly erroneous set of concentration data.

The uniqueness of the different NIR wavelengths may be described by their leverages h_k defined by

$$h_k = \hat{\mathbf{p}}_k' \left(\hat{\mathbf{P}} \hat{\mathbf{P}}' \right)^{-1} \hat{\mathbf{p}}_k$$

in which $\hat{\mathbf{p}}_k$ is the column vector of NIR loadings corresponding to variable k. For methods with orthogonal $\mathbf{P}$ columns, h_k can be computed by

$$h_k = \sum_{a=1}^{A} \left(\frac{\hat{p}_{ak}^2}{\hat{\mathbf{p}}_a \hat{\mathbf{p}}_a'} \right) \qquad (86)$$

in which $\hat{\mathbf{p}}_a$ is the row vector of $\hat{\mathbf{P}}$ corresponding to regression factor a. For the PLS method with orthogonal scores, $\hat{\mathbf{w}}$ must be used instead of $\hat{\mathbf{p}}$.

A leverage of zero indicates that the wavelength also has had very little importance on the obtained calibration solution and may probably be eliminated as irrelevant. The average value of h_k is A/K; a warning for irrelevance may be given by, for example, $h_k < 0.2A/K$; for outliers by, for example, $h_k > 2A/K$.

E. Analysis of the Chemical Residuals

In analogy to the NIR residuals, $\hat{e}_{ik}$, the chemical residuals $\hat{f}_{ij}$ reveal certain types of outliers, e.g., samples with errors in the chemical data c_{ij} or samples that do not fit the chemical–NIR relationship of the other calibration samples for other reasons, and

$$\hat{f}_{ij} = c_{ij} - \hat{c}_{ij} = c_{ij} - \bar{c}_j - \sum_{a=1}^{A} \hat{t}_{ia} \hat{q}_{aj} \qquad (87)$$

A sample's $\hat{f}$ value for a certain constituent can be tested statistically to detect outliers, e.g., due to errors in its chemical data c_{ij}. To detect abnormal samples, the individual $\hat{f}_{ij}$ values can be divided by a lack-of-fit measure like RMSP or RMSCV and squared. This can be used as test criterion in an approximate F test (Section IV C).

The ordinary calibration error RMSE is less suitable for this purpose because it may be too small due to overfitting. However, if the $\hat{f}_{ij}$ values are "leverage-corrected" (similar to Studentized, Section IV C) prior to the calculation of RMSE, then this corrected RMSE can also be used for testing unknown samples. Cook and Weisberg (1982) give such a correction for residuals in conventional MLR. For the methods using an explicit calibration regression stage, an analogue correction appears to be directly applicable for the chemical calibration residuals:

$$\hat{f}_{ij}^{(\mathrm{corrected})} = \frac{\hat{f}_{ij}}{(1 - h_i)} \qquad (88)$$

The leverage-corrected RMSE (RMSEL) for the calibration sample set is defined by

$$\mathrm{RMSEL}_j^2 = \left\{ \frac{1}{I_c} \sum_{i=1}^{I_c} \left[\frac{\hat{f}_{ij}}{(I - h_i)} \right] \right\}^2 \qquad (89)$$

This expression is applicable for many methods, e.g., PCR. For PLSR, in which the c data are "used twice," Equation 89 underestimates RMSP. In unknown samples without chemical data c_{ij} available, the residual $\hat{f}_{ij}$ is undefined. But in certain cases when all the major chemical constituents in the sample have been calibrated for and determined, the fact that they should add up to 100% may be used for control. Let

$$\text{DIFF}_i = 100\% - \sum_{j=1}^{J} \hat{c}_{ij} \tag{90}$$

This can be tested against the expected mean and standard deviation of DIFF, estimated from the calibration sample set. The method is tested successfully by Hildrum et al (1984).

F. Combined Criteria

Expressions combining leverage and residuals may measure the influence (Cook and Weisberg, 1982) of an observation on the estimation and have the ability of detecting several different types of abnormalities at the same time. In ordinary regression, there are several such criteria, e.g., DFFITS and the Cook statistic, and analogues for the situation are immediately derived using Equation 85 or 86.

A new approach for combining criteria is proposed. The criterion is designed to differentiate between influential calibration samples that are good, neutral, or bad for their analytical ability.

In calibration, the leverage and the residuals may be used for detecting extreme samples. It is often difficult to determine if an extreme sample is good or bad for the final analytical ability. A new criterion is proposed that weighs the leverage contribution from each regression factor by the positive or negative change in the analytical ability caused by that factor.

For calibration methods with orthogonal regression factors, the following criterion is proposed. For sample i, the definition is

$$g_i = \sum_{a=1}^{A} \underbrace{\frac{\hat{t}_{ia}^2}{\hat{t}_a'\hat{t}_a}}_{\substack{\text{leverage}\\\text{contribution}}} \times \underbrace{\frac{\left[\text{RMSP}^2(a-1) - \text{RMSP}^2(a)\right] \times A}{\text{STOT}^2 - \text{RMSP}^2(A)}}_{\substack{\text{weighting by relative}\\\text{prediction}}} \tag{91}$$

(RMSCV or RMSEL may be used instead of RMSP.) The parenthesis behind RMSP indicates the number of regression factors incorporated. The criterion g_i is here termed *leverage relevance*. The difference $\text{RMSP}^2(a-1) - \text{RMSP}^2(a)$ gives positive or negative weighting for the leverage contribution of each factor dependent on whether the factor has good or bad influence on the analytical ability.

$$\frac{\left[\text{RMSP}^2(a-1) - \text{RMSP}^2(a) \times A\right]}{\text{STOT}^2 - \text{RMSP}^2(A)} \tag{92}$$

has an average value of 1. The "average" calibration sample thus has an analytical leverage, g_i, of A/I_c (same as h_i); good ones have $g_i > A/I_c$, while irrelevant or bad samples have $g_i < A/I_c$ (g_i may even be negative for exceptionally damaging samples!).

The leverage-relevance criterion has not been studied in detail. It has been found useful, however, in warning against possible detrimental calibration samples (Section VI). Also, $(g_i - h_i)$ compared with the ordinary leverage has been found useful, as will be shown in Section VI F.

Further extensions of the leverage-relevance criterion for calibration samples, in analogy to DFFITS, can be envisioned, e.g., combining the g_i criterion with both chemical and NIR residuals in some Studentized form.

V. DATA PRETREATMENT

The transformations of the diffuse spectroscopy measurements should ideally pass through two stages, response linearization and optical correction, before being used in linear calibration modeling as treated in Section III B. Other types of data pretreatment can be found in the work by Young (1981).

A. Response Linearization

The NIR measurements from diffuse reflectance or transmission spectroscopy must pass through a linearizing stage in order to be suitable for the linear analytical equation

$$\hat{c}_{ij} = \hat{b}_{0j} + \mathbf{x}_i \hat{\mathbf{b}}_j \tag{93}$$

over a wide enough concentration range. Different linearization functions are available. For reflectance data, R, we have, for example,

$$\text{Apparent absorbance}: A = \log(1/R) \tag{94}$$

and

$$\text{Kubelka - Munk transform}: K/S = \frac{(1-R)^2}{2R} \tag{95}$$

TABLE II
Percentage of Chemical Data in Artificial Samples[a]

Sample	Protein $j = 1$	Starch $j = 2$	Water $j = 3$	Sum	Comment
Calibration set					
$i = 1$	0	80	20	100	
2	30	50	20	100	Error-free samples with constant water content
3	50	30	20	100	
4	80	0	20	100	
5	17 (−3)	50 (−10)	20	87 (−13)	Random measurement errors in chemical data
6	50 (+10)	38 (−2)	20	108 (+8)	
7	60	20	20	100	Random noise added to near-infrared (NIR) data; error-free chemical data
8	50	50	0	100 ◆	
9	0	50	50	100	Error-free samples with abnormal water content
10	50	30	20	100	Abnormal light-scatter level added to NIR data; error-free chemical data
Prediction set					
$i = a$	70	10	20	100	Error-free sample at normal water content
b	45 (+5)	44 (+4)	20	109 (+9)	Random measurement errors in chemical data
c	30	30	40	100	Error-free sample with abnormal water content
d	30	50	20	100	Random noise added to NIR data; error-free chemical data
e	60	20	20	100	Abnormal light-scatter level added to NIR data; error-free chemical data

[a] Concentrations of three chemical constituents in calibration and prediction data. Simulated chemical errors added to obtain the data are given in parentheses. The simulated measurement errors added to the NIR spectra of samples 7 and d were random, normally distributed noise with zero expectation and a standard deviation of 5% of the theoretical NIR reading. The abnormal light scatter was introduced in samples 10 and e by the following, somewhat arbitrarily chosen, formula:

$$x_{ik}{}^{(\text{new})} = -\log\left(v_i^{-1} \times 10_{ik}^{-x}\right)$$

with $v_i^{-1} = 1.5$ and 0.8 for the two samples, respectively. (Section V has a further discussion on scatter.)

These are discussed in Chapter 3. They can be extended in different ways, e.g., by replacing the measured reflectance R by a reflectance corrected for internal and external specular reflection, $R*$, according to the Saunderson theory (Saunderson, 1942):

$$R* = \frac{R - k_1}{1 - k_1 - k_2 \times R} \qquad (96)$$

in which k_1 and k_2 are constants defining specular reflection and internal reflection loss, respectively.

B. Multiplicative Scatter Correction

With correct response linearization, all the chemical interferences can be expected to be well modeled by the methods in Section III. Optical interferences may require different linearizations than the chemical ones. In fact, for data linearized according to the Kubelka-Munk theory for chemical interferences, the huge effect of light scatter should affect the linearized data in a strictly multiplicative way that is impossible to compensate for in linear models as treated in this chapter.

Martens et al (1983a) developed a multiplicative scatter correction (MSC) to eliminate this optical interference:

$$x_{ik}^{(new)} = \frac{\left[x_{ik}^{(old)} - \hat{u}_i\right]}{\hat{v}_i} \qquad (97)$$

in which $x_{ik}^{(old)}$ is the NIR data after some response linearization, $\hat{u}_i$ is the estimated effect of specular reflection, and $(1/v_i)$ is the estimated scatter interference in sample i. The constants $\hat{u}_i$ and $\hat{v}_i$ were estimated from simple regressions of each spectrum $\mathbf{x}_i$ against the average NIR spectrum of the calibration sample set, $\bar{\mathbf{x}}$, over the K wavelengths or a selected subset of them, using the model

$$x_{ik} = u_i + v_i \bar{x}_k + e_{ik} \qquad (98)$$

This simple scatter correction was found very useful for samples with large scatter variations (wheat fractions, meat [Geladi et al, 1985]). Section VII shows applications of the method.

VI. ILLUSTRATION BY ARTIFICIAL DATA

A. Artificial Input Data

Data for 10 artificial calibration samples (termed 1–10) and five test samples (termed a–e) were constructed in order to illustrate some of the phenomena encountered in NIR reflectance analysis of wheat flour. For simplicity, the hypothetical wheat flour was assumed to consist of protein ($j = 1$), starch ($j = 2$), and water ($j = 3$) so that the sum of these three constituents always is 100%. Table II shows the artificial constituent concentration data chosen, together with a comment on what each particular sample is meant to represent.

The NIR data were assumed to be recorded in some absorbance unit (e.g., log[1/R]). To construct the spectra of the artificial samples, each sample was regarded as a linear combination of the pure constituents' spectra (Beer's model, Appendix B):

$$x_{ik} = \sum_{j=1}^{3} c_{ij} k_{jk} + e_{ik} \qquad (99)$$

with $i = 1, 2, ..., 15$ and $k = 1, 2, ..., 25$. The symbol e_{ik} is the noise added to sample i at wavelength k. Figure 10 illustrated Equation 99 for one of the samples ($i = 7$).

B. Graphical Study of the Input Data

Figure 12 shows the input NIR data for the calibration samples 1–10 and unknown samples a–e plotted for two NIR wavelengths expected to contain much information about protein and starch (sample 23 = high for protein, low for starch; sample 17 = low for protein, high for starch). The position of the three pure constituents are also given as the corners of the triangle inside which all

error-free samples should theoretically fall. The figure shows that samples 1–6, a, and b fall along a single straight line (samples b and 6 overlap). This is because water amounts to constantly 20% in these samples, so that protein and starch add up to 80%. These samples, therefore, only contain one type of chemical variability (protein versus starch) and thus form a straight line. Ideally, samples 7 and d should also fall on the line, but due to random measurement noise in the NIR data, they lie off the line. Samples 8, 9, and c have other water contents, and samples 10 and e have abnormal light scatter.

C. The Effect of Using Insufficient Range of Calibration Samples

Suppose for a moment that only the first seven samples were available for calibration; samples 8, 9, and 10 were unavailable.

If nothing was known about the chemical data of the different samples, some conclusions could still be drawn from the NIR data. The straight line for samples 1–7 in Figure 12 indicates that

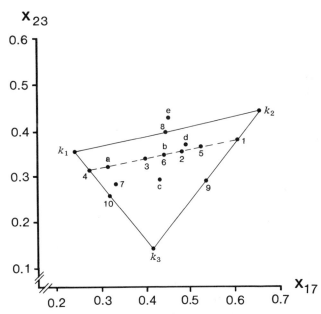

Fig. 12. Input near-infrared data $\mathbf{X}$ of artificial calibration and prediction samples plotted for wavelength 23 versus wavelength 17. The individual samples are marked according to Table II. Triangle corners represent spectra of the three pure hypothetical constituents protein ($\mathbf{k}_1$), starch ($\mathbf{k}_2$), and water ($\mathbf{k}_3$).

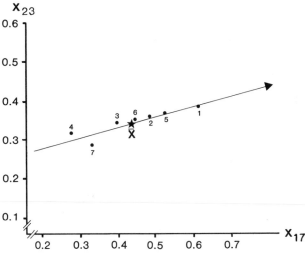

Fig. 13. Calibration samples 1–7 illustrated as in Figure 12 together with their mean, $\bar{\mathbf{x}}$, and their first principal component.

the data at the 25 wavelengths can be compressed into a few factors suitable for calibration regression. PCA will be used for this purpose.

A PCA of samples 1–7 yielded the following results. Figure 13 repeats the positions of the seven samples from Figure 12. In addition, $\bar{x}$ (the average of these seven samples) is given, together with the first principal component. Because of the errors in sample 7 the estimated component line does not exactly follow the theoretically correct path passing through samples 1, 5, 2, 6, 3, and 4, but it is very close. Figure 14 shows the full loading spectra of the mean spectrum $\bar{x}$ and the first two eigenvectors. The mean spectrum contains peaks from protein, starch, and water, while the

first factor's spectrum represents a difference spectrum between protein and starch. The second factor is strongly dominated by random noise. In this case, it is caused by noise in sample 7 (Fig. 10), since the other six calibration samples used were made to be error free.

The five unknown samples were then submitted to analysis based on the one-factor PCR solution obtained above. The NIR protein values $\hat{c}$ are, in Figure 15, plotted against the "measured" protein values, c. The figure shows that sample a was analyzed accurately ($\hat{c} = 70.2\%$, i.e., $\hat{f} = 0.2\%$). This was expected, since it belongs to the same type as calibration samples 1–7 (20% water). Sample d was also accurately analyzed ($\hat{f} = 0.4\%$); the random noise in its NIR data did not affect the protein determination much because it was averaged over all 25 wavelengths. Sample b shows a residual of $\hat{f} = 4.5\%$ that is similar to the 5% error actually present in its chemical protein data c. This exemplifies a sample whose NIR protein $\hat{c}$ was much better than its chemically measured reference data c. For these samples, the analytical data were acceptable. But the samples with abnormal water content and abnormal light scatter (samples c and e) yielded wildly erroneous protein results ($\hat{f} = 15$ and 25%) because their types of variability were not included in the calibration sample set.

D. Using a Complete Calibration Data Set

Samples c and e were not accurately analyzed when using only samples 1–7 for calibration. This can be improved by including more types of samples in the calibration set. A PCR calibration using all 10 calibration samples was performed. Four components were required to give full modeling of the NIR calibration data. Figure 16 shows the NIR loading spectra for the first four principal components. Vectors 1 and 2 show smooth features and correspond to linear combinations of the protein, starch, and water. Vectors 3 and 4 are both affected by the random noise in sample 7.

The RMSP for protein for samples a–e decreased to a minimum after four factors, so that four factors appeared to be optimal for PCR with these data. Figure 17 shows the individual NIR results for protein with this four-factor PCR solution. Samples a

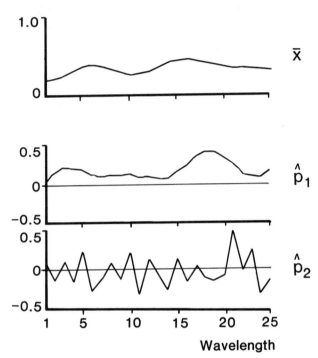

Fig. 14. Results of principal component analysis (PCA) for calibration samples 1–7. Average spectrum $\bar{x}$ and loading spectra of the first two principal components are plotted against wavelength.

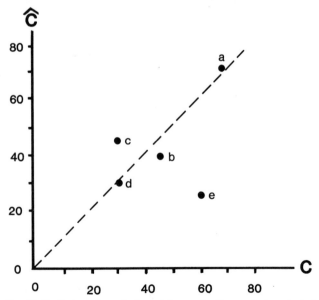

Fig. 15. Prediction of protein in samples a–e by the one-factor principal component regression (PCR) solution from calibration samples 1–7. The dotted diagonal represents the ideal prediction line.

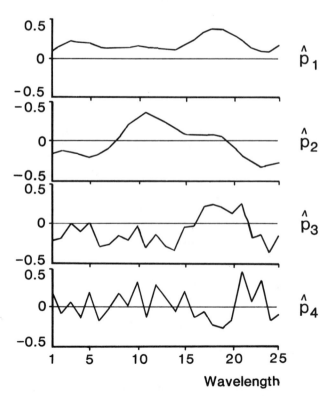

Fig. 16. Result of PCA for all 10 calibration samples. Loading spectra of the first four principal components plotted against wavelength.

and b were accurately analyzed, similar to the incomplete data set. Sample c with abnormal water content was also accurately analyzed, and so was the sample with abnormal scatter, sample e. Sample d, which had large random NIR noise, was now less accurately analyzed by the present four-factor solution than in the one-factor solution above. This was not unexpected, since there are more chances for four regression factors to find an incidental correlation to its NIR noise than for the one-factor solution used in Section VI C.

E. PLSR

The same 10 calibration samples were submitted to PLS calibration for protein (the PLS method with orthogonal regression factors was used). Like PCR, the PLS data compression required four regression factors to describe the NIR calibration data completely. The NIR loading weights, $\hat{w}$, are shown in Figure 18. The first two factors are somewhat similar to those for PCA (Fig. 16), except for having opposite signs. But in contrast to PCR, the PLS displaces the random noise contributions of sample 7 to the last of these four factors. This illustrates how PLSR concentrates the NIR variation that has relevance for the protein modeling into the first factors. It leaves the less relevant variation behind in the last factors. PCR, on the other hand, takes out NIR factors with respect to their NIR modeling ability in the data compression stage, irrespective of their protein relevance.

The significance of this difference between PCR and PLS is shown in Figure 19, which compares the RMSP for protein of the PCR and PLS methods on the same 10 calibration samples. The two methods show identical RMSP after zero factors (only the mean subtracted) and after four factors (full description of all NIR calibration data). But while PCR required four factors to give optimal RMSP, PLS had optimal analytical power after only three factors, and this RMSP is lower than that of PCR.

F. Outliers

Table III shows some outlier detection results obtained by the PLS method used in Section VI E, using samples 1–10 for calibration and samples a–e for determination of protein. Only the samples that yielded significant outlier detector values are given, and results for the obviously inadequate solutions (zero, one, or two factors) are ignored. For comparison, the average detector values of the calibration sample set are given.

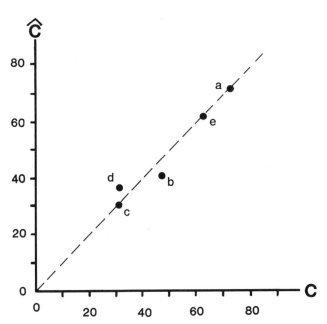

Fig. 17. Prediction of protein in samples a–e by the four-factor principal component regression (PCR) solution for calibration samples 1–10. The dotted diagonal represents the ideal prediction line.

CALIBRATION DATA SET (SAMPLES 1–10)

Calibration sample 7 (with strong random NIR noise) was detected as an outlier after three factors, both by its high NIR residual $s(\hat{e}_i)$ and by its large protein residual $\hat{f}_i$.

After four factors (the overfitted solution), all the NIR residuals were approximately zero, even for sample 7, in spite of its NIR noise. Sample 7 also showed zero protein residual. Instead, the sample was then detected as an outlier by its high leverage h_i and as a bad outlier by its low leverage relevance g_i, (shown as $g_i - h_i$). Its abnormally high leverage ($h_i = 0.900$, which is the maximum obtainable, $[1 - 1/I]$) proved that it had "grabbed" the fourth PLSR factor for modeling its own irrelevant noise, thereby contaminating the calibration coefficients. This overfitting effect was apparent in the erratic loading spectrum of factor 4 in Figure 18 and in the increased RMSP at four factors in Figure 19.

The calibration samples 5 and 6 (with errors of –3 and +10% added to their protein data) were detected as outliers with respect to $\hat{f}_i$ both after three and four factors; sample 6 was particularly conspicuous. Their estimated protein residuals $\hat{f}_i$ correspond closely to their true errors (–3.5 and +9.6% after three factors).

In addition to these apparently bad outliers, the calibration samples 8, 9, and 10 were detected as abnormal by their high leverages h_i. They were considered to be good outliers, however, because of their high predicting relevance g_i, which indicated that they were essential in defining regression factors important for analytical ability.

UNKNOWN OR ANALYTICAL DATA SET (SAMPLES a–e)

Samples b, d, and e were discovered as outliers of different types. Sample b (with 5% error in its protein value) was detected after three and four factors from the large residual $\hat{f}_i$ between its NIR and known, but erroneous, protein values. Sample d (with random NIR noise) was detected from its high NIR residual standard deviation $s(\hat{e}_i)$ after three and four factors. It also yielded abnormal NIR protein values (because of its random NIR noise). Sample e (with abnormal light scatter) was discovered as an outlier since its leverage h_i was exceptional. Some NIR information

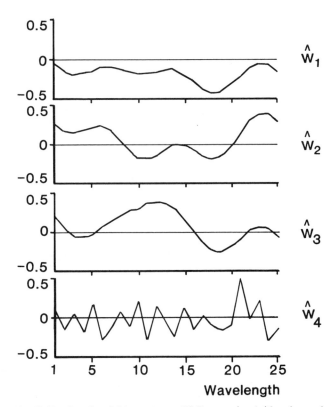

Fig. 18. Results of partial least-squares (PLS) regression (with orthogonal regression factors) for all 10 calibration samples. Near-infrared loading weights of the first four PLS factors are plotted against wavelength.

remained in this sample after three factors; this illustrates the effect of light scatter: The scatter information in calibration sample 10 was not sufficient to model that of sample e, because their light-scatter effects did not span exactly the same space. It was sufficient, however, to ensure acceptable protein prediction for sample e. This again illustrated that high leverage h_i in itself does not indicate an erroneous sample.

RECALIBRATION WITH OUTLIERS ELIMINATED FROM CALIBRATION DATA SET

Samples 6 and 7 were regarded as outliers and eliminated from the calibration data set before recalibration on the remaining eight samples; sample 6 was eliminated because of its extreme chemical

RMSP

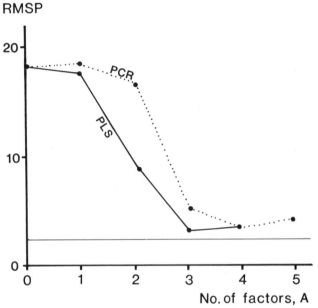

Fig. 19. Prediction error, root mean square of performance (RMSP), for protein predicted by principal component regression (PCR) and partial least-squares (PLS) regression, plotted against the complexity of the calibration model (A). Both curves were calculated over the prediction samples a–e, using the calibration coefficients estimated using samples 1–10. The horizontal line represents the theoretical lower prediction error limit caused by the protein error of 5% in sample b, resulting in a purely chemical RMSP of 2.2%. The minimal RMSP of PCR was 3.7%, while the corresponding value for PLS regression was 3.3%. Subtracting the purely chemical RMSP contribution, their "true" prediction errors came out as 1.5 and 1.1%, respectively.

residual $\hat{f}_i$ and sample 7 because of its NIR residuals $s(\hat{e}_i)$ and low leverage relevance g_i.

In order to compare with the previous NIR results, the detected outliers in the unknown set were not eliminated. The RMSP was now further reduced from 3.23 to 3.18% after three factors. In real situations, samples b, d, and e should have been measured again, together with samples 6 and 7, with respect to NIR and protein data. Their errors then could have been discovered and corrected.

G. Conclusions

In summary, this artificial data example illustrated how data compression can be accomplished by PCR and PLSR, concentrating the 25 intercorrelated NIR variables to three or four regression factors without losing information. It also shows similarities and differences between the relatively well-known PCR and the PLSR method for calibration, with the latter giving the simplest solution and best analytical results. The example shows how outliers can be detected from their NIR residuals $\hat{e}_i$, their leverage h_i, their leverage relevance g_i, and their chemical residuals $\hat{f}_i$. Finally, it illustrated that improved analytical results can be obtained by eliminating calibration outliers.

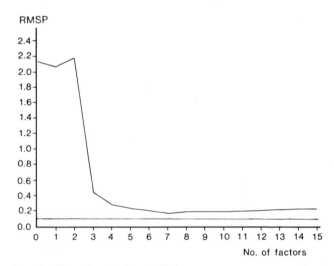

Fig. 20. Effect of overfitting. Prediction error root mean square of performance (RMSP) of protein for real wheat data plotted as a function of the number of partial least-squares (PLS) factors, A. The near-infrared data were in the $\log(1/R)$ unit and protein in the percentage of sample weight. The horizontal line represents the approximate level of errors in the chemical data themselves.

TABLE III
Detection of Bad Outliers Illustrated by Partial Least-Squares (PLS) Regression for Protein in the Artificial Data Set After Three and Four Factors

	Outlier detectors							
	Three PLS factors				Four PLS factors			
	$s(\hat{e}_i)^a$	h_i^b	$g_i - h_i^c$	$\hat{f}_i^d$	$s(\hat{e}_i)^a$	h_i^b	$g_i - h_i^c$	$\hat{f}_i^d$
Calibration set (samples $i = 1$–10)								
Average	1.1[e]	0.3	0.0	0.4[e]	0[e]	0.4	0.0	1.1[e]
Sample 5	0.02	0.1	−0.1	<u>−3.5</u>	0	0.1	−0.1	<u>−3.5</u>
Sample 6	0.5	0.0	0.0	<u>9.6</u>	0	0.0	−0.0	<u>9.0</u>
Sample 7	<u>2.8</u>	0.2	−0.1	<u>−3.8</u>	0	0.9	<u>−0.8</u>	0
Prediction set (samples $i = $ a–e)								
Sample b	0.4	0.0		<u>4.6</u>	0	0.0		<u>4.0</u>
Sample d	<u>2.4</u>	0.1		<u>−5.5</u>	<u>2.4</u>	0.1		<u>−6.2</u>
Sample e	<u>1.5</u>	<u>2.2</u>		−0.4	0	<u>2.5</u>		−3.0

[a] Near-infrared residual $s(\hat{e}_i)$ defined by Equations 78 and 79.
[b] Leverage h_i defined by Equation 85.
[c] Leverage relevance g_i defined by Equation 91.
[d] Protein residual $\hat{f}_i$ defined by Equation 87.
[e] Root mean square of the other seven calibration samples (1–4 and 8–10). The most extreme values are underlined.

VII. RESULTS FOR REAL DATA

A. The Real Data Sets

WHEAT SAMPLES WITH NORMAL VARIATIONS
(DATA SET 1)

One hundred representative samples of Canadian hard red spring wheat were analyzed at the U.S. Department of Agriculture (Beltsville, MD) in a Cary scanning spectrophotometer with about 1,000 readings between 1,000 and 2,600 nm. The Kjeldahl-N was measured in 16 replicates for each sample. The samples were split randomly in two representative sample sets of 50 samples each (calibration set "whelgb" and prediction set "whelga" [Hruschka and Norris, 1982]). A total of 155 wavelengths between 1,906 and 2,398 nm (every 3.2 nm) were chosen for a comparison of various calibration methods.

WHEAT SAMPLES WITH VERY LARGE VARIATIONS
(DATA SET 2)

A single batch of Canadian hard red spring wheat was submitted to abrasive milling, resulting in 10 successive fractions ranging from almost pure hull in large flakes via aleuron-rich fractions to almost pure endosperm, finely ground. Each of these fractions was sieved into four particle-size fractions (a few fractions gave too little material to be measured). The remaining pearled endosperm and the whole-kernel sample were also included in the set after conventional milling in a Udy cyclone mill. The reflectance of the 47 samples (23 calibration, 24 prediction) was measured in a Technicon InfraAlyzer 400 with 19 fixed bandpass filters between 1,445 and 2,348 nm. Six different chemical constituents (water, ash, protein, starch, fiber, and fat) were measured on the 50 fractions in at least two replicates. Some NIR calibration results on these data have been reported elsewhere (Martens and Jensen, 1983; Martens et al, 1983a).

B. Effect of Overfitting

A critical aspect of indirect multivariate calibration is the determination of the optimal number of regression factors, A. Using too few factors implies insufficient modeling of the chemical and optical interferences. Using too many factors implies estimating too many statistical parameters with the given calibration data. Both mistakes lead to unnecessarily large analytical errors. This is illustrated for wheat samples with normal quality range (data set 1) in log($1/R$) data in the 1,900–2,400-nm range.

Calibrating for protein in these 50 samples by unweighted PLS1 regression, using 155 different wavelength readings in this region (every 3.2 nm), yielded the results shown in Figure 20. The analytical error RMSP decreased sharply with the first four factors. It reached a minimum of 0.16% after seven factors, whereafter it increased again. This example illustrates the importance of selecting the optimal number of regression factors, A.

C. Comparison of Some Calibration Methods

Some different calibration methods were applied to data set 1. Figure 21 compares the analytical error RMSP for PLS (Section

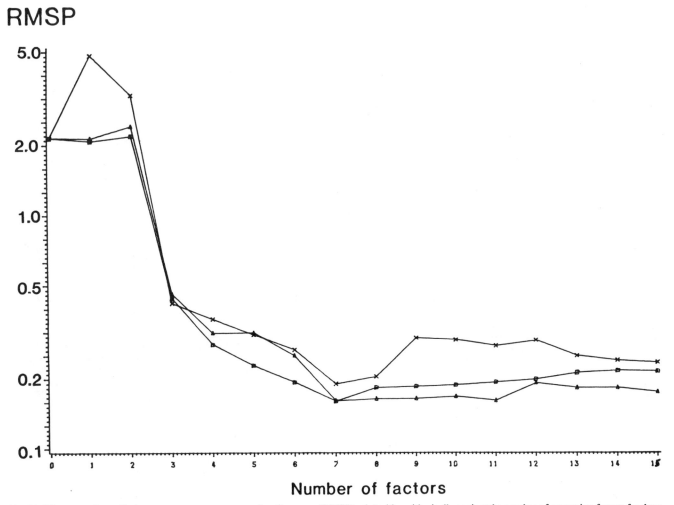

Fig. 21. Wheat protein prediction error, root mean square of performance (RMSP), plotted logarithmically against the number of regression factors for three different full-spectrum calibration methods: estimated generalized least-squares estimator (EGLS1) (×), principal component regression (PCR) (Δ), and partial least squares (PLS) (□).

III F) and with PCR (Section III E) and EGLS1 (Section III G) for data in log(1/R) units. To emphasize details, the RMSP is given on a logarithmic scale. The physically based EGLS1 predictor regresses **X** on **c** with no distributional assumptions, while the two PCR and PLS predictors instead regress **c** on **X** and incorporate information about the calibration sample distribution. The figure shows that all three methods gave the best results after seven factors for these data. PCR and PLS yielded slightly better data (RMSP = 0.16%) than did EGLS1 (RMSP = 0.19%). This indicates an advantage in using population information in obtaining optimal predictors (Sections II C and III G).

PLSR gave virtually the same optimal results as PCR (with minimal RMSP found after seven factors at 0.16%). The PLS analytical error approached minimum more rapidly than the PCR error did, just like for the artificial data (Fig. 19).

However, Figure 21 shows that a penalty of the more thorough use of the chemical data in PLS is a slightly increased sensitivity to overfitting, compared with PCR. The PCR-PLS intermediate method followed the PLS up to six factors, but its RMSP increased less than that of PLS and gave a minimum RMSP of 0.15% after 11 factors. It was thus less sensitive to overfitting than PLS, while retaining the advantage of PLSR at fewer factors.

D. Transformations of NIR Data

WHEAT DATA WITH NORMAL VARIATIONS
(DATA SET 1)

Three different response linearizations and three different optical corrections were tested for protein in data set 1. The PLSR of Section III F, Modifications of the PLSR Method, was used as a calibration method. The conventional log(1/R) linearization without optical correction yielded an optimal analytical error RMSP of 0.16% (Table IV). Paradoxically, the Kubelka-Munk linearization (K/S) gave much worse results (0.27%) in spite of being more founded on a physical modeling of the diffuse reflectance data. With the physical model further elaborated with Saunderson-corrected reflectance, using reasonably realistic parameters (specular reflectance $k_1 = 0.01$ and internal reflectance $k_2 = 0.5$), K/S gave still worse results (RMSP = 0.29%). A relatively complicated calibration solution was required in each case (five to seven PLS factors).

However, after optical correction, the picture changed drastically for all three linearizations. The MSC transformation without constant term improved the optimal analysis from 0.16–0.15% for log(1/R) data, from 0.27–0.16% for Kubelka-Munk data, and from 0.29–0.16% for Saunderson-corrected Kubelka-Munk data. Very good results (RMSP = 0.16%) were obtained after only four factors with MSC. Reasonable data (RMSP = 0.30%) were, in some cases, obtained after only two factors with MSC, contrasting with RMSP = 2.2% without.

The MSC without offset is analogous to the SDLR method of dividing one derivative term by another derivative term. The MSC with constant term added may allow more general applications. For the present data it allows a simpler modeling with fewer factors.

The strong effect of MSC on Kubelka-Munk data is not surprising, since the K/S linearization attempts to express scatter as a multiplicative effect, $K×(1/S)$.

WHEAT DATA WITH VERY LARGE VARIATIONS
(DATA SET 2)

NIR reflectance data measured in the fixed-filter instrument at 19 wavelengths were used to study the optimization of calibration on wheat with widely varying chemical composition and particle size. The results (Martens and Jensen, 1983) are summarized for protein in Table V, which shows that for these data

1. PLS gave better analyses than the conventional SMLR calibration method for log(1/R) data;
2. an optical correction consisting in dividing each log(1/R) spectrum by its sum gave improved prediction;
3. the MSC with offset term gave even better improvements; and
4. again, K/S and Saunderson-corrected K/S gave worse analytical accuracy than log(1/R) without MSC, but better than log(1/R) with MSC.

Optimizing the response linearization and the optical correction both improved the determination of protein in wheat. Together they brought the error for samples with abnormally large chemical and physical variabilities down toward that of normal wheat samples (data set 1). This indicates that the importance of standardizing the sample preparation may be reduced by data pretreatment.

INTERPRETATION
OF THE KUBELKA-MUNK MSC SOLUTION

Figure 22 shows the mean spectrum $\bar{x}$ and the standard deviation spectrum of the 50 calibration samples in data set 1 for the data in Kubelka-Munk units after MSC in the 1.9–2.4-µm range. The sharp water peak in the mean spectrum at 1,940 nm and the wide starch peak at 2,100 nm with its protein peak shoulders on both sides are, for instance, clearly visible. The water peak region at 1,940 nm showed a larger variability than that of the other wavelength regions. Figure 23 shows the six first NIR loading spectra (PLS).

The NIR loading weights of factor 1 show that the scores of factor 1 correlate negatively with the water peak in the 1.94 µm region and with the starch peak near 2.10 µm and positively at higher wavelengths. Factor 2 correlates positively with wave-

TABLE IV
Prediction Error Root Mean Square of Performance (RMSP)
for Protein in Percent of Sample Weight in Data Set 1[a]

Response Linearization	Optical Correction	RMSP (%)		
		$a = 2$[b]	$a = 4$[b]	Optimum (A)[b]
Log(1/R)	...	2.18	0.28	0.16 (7)
K/S	...	2.20	0.33	0.27 (7)
(K/S)[c]	...	2.20	0.34	0.29 (5)
Log(1/R)	MSC[d]	0.35	0.16	0.15 (7)
Log(1/R)	MSC[e]	0.30	0.28	0.15 (9)
K/S	MSC[d]	0.35	0.16	0.16 (4)
K/S	MSC[e]	0.43	0.32	0.16 (8)
(K/S)[d]	MSC[d]	0.35	0.17	0.16 (6)
(K/S)[d]	MSC[e]	0.71	0.25	0.16 (8)

[a] Total initial standard deviation was 2.39% and the standard deviation of the chemical Kjeldahl protein data was about 0.1% (Hruschka and Norris, 1982).
[b] Number of partial least-squares factors.
[c] Based on Saunderson-corrected reflectance ($k_1 = 0.01$, $k_2 = 0.5$).
[d] Multiplicative scatter correction (MSC) model: $x_{ik} = u_i + v_i \bar{x}_k + e_{ik}$.
[e] MSC model: $x_{ik} = v_i \bar{x}_k + e_{ik}$.

TABLE V
Prediction Error Root Mean Square of Performance (RMSP)
for Protein in Percent of Sample Weight in Wheat Data Set 2[a]

Response Linearization	Optical Correction	RMSP (%)	Calibration Method[b]	Reference
Log(1/R)	...	0.39	SMLR	Martens and Jensen, 1983
Log(1/R)	...	0.33	PLSR	Martens and Jensen, 1983
Log(1/R)	Divided by NIR[c] sum	0.25	PLSR	Martens and Jensen, 1983
Log(1/R)	MSC[d]	0.24	PLSR	Martens et al, 1983a
R	...	0.24	PLSR	Martens et al, 1983a
R	MSC[d]	0.23	PLSR	Martens et al, 1983a
K/S	...	0.47	PLSR	Martens et al, 1983a
K/S	MSC[d]	0.21	PLSR	Martens et al, 1983a
Saunderson & K/S	...	0.54	PLSR	Martens et al, 1983a
Saunderson & K/S	MSC[d]	0.20	PLSR	Martens et al, 1983a

[a] Total initial standard deviation was 2.35% and the standard deviation of the chemical Kjeldahl protein data was about 0.15% (Hruschka and Norris, 1982).
[b] SMLR = stepwise multiple linear regression, and PLSR = partial least squares regression.
[c] NIR = near-infrared
[d] Multiplicative scatter correction (MSC) model: $x_{ik} = u_i + v_i \bar{x}_k + e_{ik}$.

length just above the water peak at 1.94 µm and negatively with starch at 2.10 µm. Factor 3 correlates negatively with wavelengths just below the water peak and with 2.25 µm, which has tentatively been ascribed to cellulose (K. Norris, *personal communication*), and positively with protein peaks at 2.17 and 2.28 µm. Factor 4 correlates positively with wavelengths just below and negatively with wavelengths just above the water peak, and it may possibly be a factor that compensates for wavelength shifts in the water peak caused by, for example, temperature. Factor 5 correlates negatively with wavelengths below the water peak, coinciding with a region of strong interferences from atmospheric water vapor (K. Norris, *personal communication*). The inclusion of factor 5 in the calibration solution gave decreased analytical ability, which indicates the water vapor region to have no relevance. Factor 6 was likewise nonsignificant, and the reason is apparent in Figure 23; its loading

weights display a very erratic wavelength dependency that may reflect random NIR noise and possibly some nonlinearity compensation.

The effect of these factor loading weights on the resulting calibration coefficients $\hat{b}_j$ is shown in Figure 24. In Figure 24A, the effect of using an increasing number of factors ($A = 2$, 4, and 6) is illustrated. It shows that as the number of regression factors increases, the size of $\hat{b}_{kj}$ increases. Since the calibration constants amplify the random noise e_{ik} in the NIR measurements, it is probably preferable to have small coefficients. In addition, the detrimental effects of, for example, the noise factor ($a = 6$) is apparent. Figure 24B gives the optimal calibration coefficients for protein (four factors) in detail. It shows a positive relationship between protein and wavelengths at bands near 2.28, 2.18–2.21, 2.07, 1.98, and 1.94 µm, and a negative relationship at 2.25, 2.08–2.15, and 2.02 µm and at both ends of the spectral region.

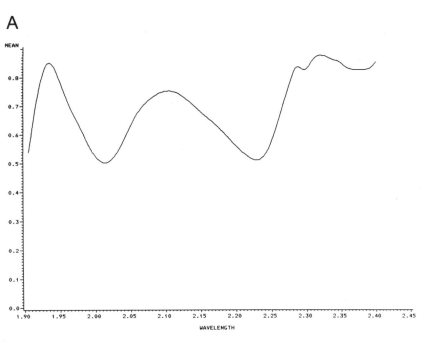

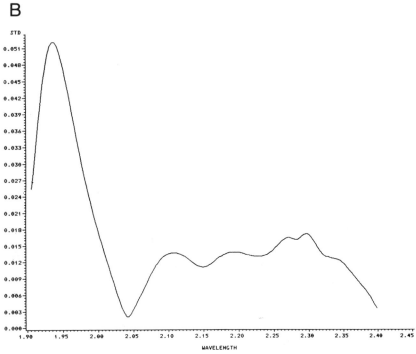

Fig. 22. The **A,** mean and **B,** standard deviation (Std) spectra for the 50 calibration samples in data set 1 after Kubelka-Munk and multiplicative scatter correction (MSC) transformation.

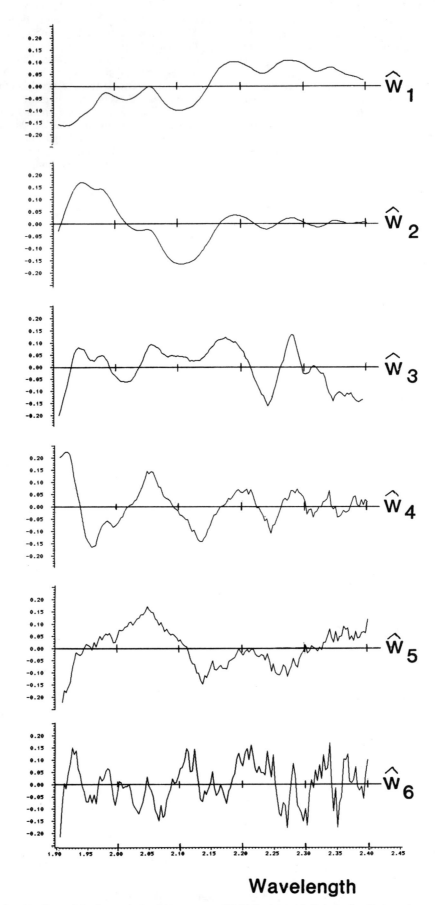

Wavelength

Fig. 23. The six first near-infrared loading weights for protein (partial least squares [PLS]) for data set 1. Only the four first ones improved the prediction ability.

E. Improvements of the PLS Calibration Method

WEIGHTING NIR DATA

In all the calibrations reported here, the NIR variables were used unweighted. Improved analysis could probably have been obtained by weighting down or eliminating the water vapor region at 1.90–1.93 μm and the apparently noisy region above 2.3 μm. An up-weighting of well-identified protein, starch, and cellulose peaks might possibly have given further improvements.

SMOOTHING

Theoretically, one would expect the loadings of the first factors to be more precisely estimated than those of later factors due to measurement errors. Figure 23 shows an increasing level of erratic ripple in the loading spectra of the later factors.

In an attempt at reducing this ripple in the estimates, a weighted moving average filter over ±3 neighbor variables (±9.6 nm) was applied to the loading spectra $\hat{w}_a$ inside the PLS algorithm, as described by Martens et al (1983a). The smoothing reduced the optimal protein RMSP from 0.164 to 0.160% after four factors.

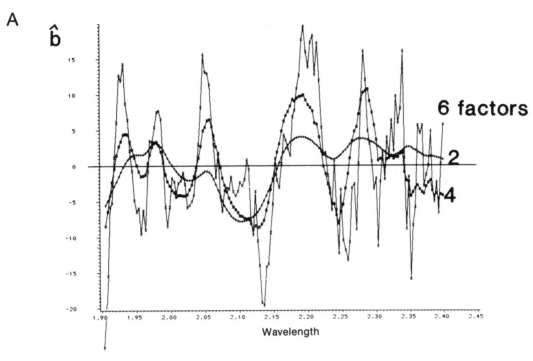

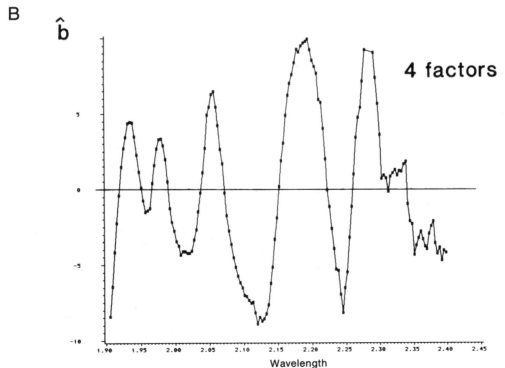

Fig. 24. **A,** Calibration coefficients $\hat{b}$ for protein from the partial least-squares (PLS) solution, for *K/S* data after multiplicative scatter correction (MSC), shown for two, four, and six factors, corresponding to prediction errors root mean square of performance (RMSP) of 0.35, 0.16, and 0.17%, respectively. **B,** The optimal solution, with four factors, is also plotted separately to show its details. Protein in a sample hence is predicted by multiplying the near-infrared measurement at each wavelength by the corresponding $\hat{b}$ value and summing these products plus and offset, i.e., $\hat{c} = b_0 + \mathbf{x}\hat{b}$.

Increasing the number of wavelength variables is another way of making the calibration more robust against random NIR noise. By increasing the number of wavelengths in the 1,904–2,400-nm range from 155 to 309, again using Kubelka-Munk linearization with MSC and smoothing over ±3 neighbor variables (±4.8 nm), the RMSP decreased further from the above mentioned 0.160 to 0.157% after four factors. Alternatively, the individual measured spectra x_i could have been smoothed.

TRUNCATED PLS VERSION

Section III F discussed how each NIR loading vector could be truncated by forcing $\hat{w}_{ak}$ of all wavelengths with small initial coefficients to zero. This was tested on the normal-range wheat data set 1 in log($1/R$) units, and was found to give a small improvement compared with the full-spectrum PLS solution. The RMSP of the initial $\hat{w}_a$ vector was for simplicity used as the cutoff limit. Likewise, each regression factor vector $\hat{t}_a$ could be truncated by forcing all samples with small initial scores to zero. This was tested on the same data and likewise found to give a minor improvement in RMSP compared with the full sample set method. For truncation, the score $\hat{t}_{ia}$ of all samples was forced to zero, except the one having the largest absolute initial score, and one term in the PLS algorithm was repeated.

VIII. DISCUSSION

A. The Statistical Calibration Methods

Two of the existing calibration methods (SMLR and SDLR) only use a few wavelengths in the final calibration. Thereby the information of all the other wavelengths available or potentially available is discarded in the data compression and left unused when analyzing new samples.

A number of new calibration methods that use the whole spectrum **x** to obtain the predictor are described (full-spectrum methods). All these also rely on a data compression stage to eliminate redundancy and reduce noise in the NIR data. The different methods presented can be regarded as approximations to the general MLR method for estimating the calibration coefficients $\hat{b}_0$ and $\hat{\mathbf{b}}$ in the prediction equation $\hat{c} = \hat{b}_0 + \mathbf{x}\hat{\mathbf{b}}$.

Figure 24 illustrated how the $\hat{\mathbf{b}}$ spectrum may evolve as the dimensionality of the calibration model increases, from too simple ($A = 2$) via an optimum complexity ($A = 4$) to overfitting ($A = 6$). The different calibration methods will be summarized and a subjective evaluation of them given.

The conventional data compression types (SMLR and SDLR) probably require very precise NIR data. These methods (as well as the wavelength-truncated version of PLSR) may be preferable if strong nonlinearities exist in the relationship between some elements of **x** and c. The least nonlinear wavelengths can then be chosen. By treating the unused NIR wavelengths as c variables, outliers can be detected even with these conventional methods.

The Hruschka regression method uses the spectra of selected calibration samples for data compression. The method may be simpler to understand than the subsequent ones and may possibly be advantageous when the number of calibration samples is very small, since few degrees of freedom are used in the determination of the factor loadings, $\hat{\mathbf{P}}$. But the subsequent methods that estimate the factor loadings statistically are probably better if the NIR data contain more noise.

The Fourier regression method has the advantage of allowing smoothing and data compression in the same step, but it is applicable only for scanning instrument data. The PCR and PLSR are similar to each other, but while the former is easier to understand theoretically, the latter is faster and appears more flexible and is the method mostly used in practice by the authors. The PLS method can easily be modified in a variety of ways to accommodate different types of restrictions or additional information, as illustrated by the multiconstituent version, the smoothed-loadings version, the truncated versions for ignoring irrelevant NIR information, and the PCR-PLS combination.

The extended Beer's model methods (EGLS1, EGLS2, EBLP, and EBLUP) have the advantage of being related to the physical model for the NIR signals. They allow the modeling of unknown interference factors as well, which the ordinary multiconstituent curve resolution method of some UV/visual-range transmission spectrophotometers does not. Although the basic models are statistically well founded, these estimated versions are not fully understood. But they allow a choice between the different usages of distributional assumptions that is theoretically appealing.

EGLS1 and EGLS2 may allow extrapolation outside the constituent concentration range of the calibration sample set. EGLS1 and EBLUP likewise seem to allow extrapolation with respect to the level of the unknown interference factors modeled during calibration. The EBLP method is expected to behave similarly to the PLS method when the error in the chemical calibration data is low.

In general, there is a great need for closer comparison of the different statistical calibration methods, both from a theoretical and a practical angle. However, for NIR data with very low levels of random noise, one should expect the methods to behave similarly. The apparent prediction error will often be dominated by the noise in the chemical data.

B. Factors Affecting Choice of Method

The actual choice of statistical calibration method and its dimensionality should, for a particular instrument and a particular product type, depend on many aspects. Here are some important ones.

1. Complexity of the calibration problem:
   - number of different types of compositional variation that interfere with the NIR measurements in the samples (e.g., for determining wheat protein, water, and starch variations constitute two interferences);
   - degree of overlap between spectra of the different constituents and interference (protein, starch, water, etc.);
   - degree of chemical interaction between constituents, and other phenomena modifying their spectra (e.g., water modified by protein salt and temperature); and
   - degree of optical scatter effect variations and other nonlinear interferences.
2. Noise level:
   - random noise levels and probability of outliers in the chemical calibration data;
   - random noise level and probability of outliers in the NIR calibration data; and
   - random noise level and probability of outliers in the NIR prediction data.
3. Sample distribution:
   - to what extent the calibration sample set is representative for future samples, in terms of average quality and quality variation concerning both the constituent calibrated for and the interferences.
4. Computer storage and processing capacity.
5. Need for automatic error warnings.

C. Data Pretreatment

When light scatter varies strongly, it appears that theoretically more exact response linearizations such as the Kubelka-Munk transformation can give better linear analyses than the empirical log($1/R$), but only after an optical correction. One such optical correction, the MSC, was tested and found beneficial and apparently yields useful estimates for the relative scatter coefficient.

None of the response linearizations tested give fully satisfactory linearity; new formulas should therefore be tested. The presented MSC method is likewise somewhat unsatisfactory, because it ignores chemical effects on the light scatter. Ideally, the data pretreatment should be optimized iteratively together with the linear estimation.

An alternative approach is to allow nonlinear estimation. For instance, several different nonlinear transformations of the same

NIR wavelengths can be used together as "x variables" in the data compression stage. In order to avoid estimating loadings for all these x variables, a wavelength selection method may then be desirable to eliminate irrelevant ones; wavelength-truncated PLS might be useful for this purpose.

Alternatively, nonlinear relationships between $\mathbf{c}$ and $\hat{\mathbf{T}}$ may be used in the calibration regression stage.

D. Error Detection

Outliers of different types can be detected in the calibration sample set as well as among unknown samples, using modern regression diagnostics to asses NIR residuals, chemical residuals, and leverages. Samples with abnormal light-scatter levels may be detected during the optical correction stage. The outlier detection methods need more testing in practice.

In the calibration stage, the elimination of outliers among the samples or among the variables should be compared with methods for down-weighting them (robust methods [e.g., Huber, 1981]).

E. Updating

Finally, the problem of updating calibration results should be mentioned. Advanced instruments such as NIR spectrophotometers require routine checking. These control data may be used for updating the calibration constants in order to improve accuracy (reduce estimation errors from the calibration, account for instrument drift, etc.). In some methods, the updating algorithm is relatively straightforward (e.g., for update of MLR [Cook and Weisberg, 1982]). In other methods, updating algorithms are not yet available. But at least the offset and general slope (i.e., the simple corrigible errors [Cardone, 1983]) can always be updated in order to minimize the bias. Since calibration methods may yield biased predictors, this appears to be important in practice. A weighted average of the old offset (Equation 10) and the new bias may be used:

$$\hat{b}_o^{(new)} = w\hat{b}_o^{(old)} - (1-w)(\bar{f}) \tag{100}$$

in which $\bar{f}$ is the bias estimated in a set of new control samples and w is a weight between 0 and 1 that balances the variance of the estimated bias against the variance of the old offset. A proportional update factor, common to the coefficient of all wavelengths, $\hat{b}$, can also be made. In this way, the information in the data of analytical control samples is put to efficient use.

However, such bias-and-slope updating is not the most informative way to update calibration models. In the following, an alternative approach to the updating of linear regression models is used. This is based on a bilinear representation of the information in the reduced-rank covariance matrices involved. The theory was described by Martens and Naes (1989, page 162). Here is a summary of the methodology. Assume that we had $N_{initial}$ (e.g., 200) initial calibration samples and that A factors (e.g., $A = 7$) were needed to generate a good initial PCR calibration model for $y = f(x)$.

The X-X and X-Y covariance "structure" estimated from the calibration data set $(\mathbf{X},\mathbf{y})_{initial}$ can be represented by $2A$ (e.g., 14) "samples" that span the modeled variabilities around the model centrum $(\bar{\mathbf{x}}, \bar{y})$. For each factor, this variability is described as plus and minus each factor's loadings $\mathbf{p}_a$ and q_a. Thus, each factor generates a pair of artificial "bilinear samples"

$$(\mathbf{x}, y) = (\bar{\mathbf{x}}, \bar{y}) + (\mathbf{p}_a, q_a)$$

$$(\mathbf{x}, y) = (\bar{\mathbf{x}}, \bar{y}) - (\mathbf{p}_a, q_a)$$

All valid factor $a = 1, 2, ..., A$ can thus be represented by the $2A$ "samples" (101)

$$(\mathbf{X}, \mathbf{y})_{blm+} = \mathbf{1}(\bar{\mathbf{x}}, \bar{y}) + (\mathbf{P}, \mathbf{q})$$

$$(\mathbf{X}, \mathbf{y})_{blm-} = \mathbf{1}(\bar{\mathbf{x}}, \bar{y}) - (\mathbf{P}, \mathbf{q})$$

in which $\mathbf{p}_a$ and q_a, the loadings for each factor in $\mathbf{P}$ and $\mathbf{q}$, have been *scaled to the relative importance of each other*, by multiplication of the original $\mathbf{p}_a$ and q_a by the factor's size, $\sqrt{(\mathbf{t}_a'\mathbf{t}_a)}$.

The final artificial sample set $(\mathbf{X},\mathbf{y})_{blm}$, centered around the original calibration set average and spanning the A factors according to their original importance, then consists of both $(\mathbf{X},\mathbf{y})_{blm+}$ and $(\mathbf{X},\mathbf{y})_{blm-}$.

This set of $2A$ bilinear "samples," $(\mathbf{X},\mathbf{y})_{blm}$ has the same mean $(\bar{\mathbf{x}}, \bar{y})$ and more or less the same relevant systematic covariance structures $\mathbf{s}_{xy}$ and $\mathbf{S}_{xx}$ as those obtained as with the original $N_{initial}$ samples, $(\mathbf{X},\mathbf{y})_{initial}$. Thus, if submitted to PCR or PLSR analysis alone, this set of $2A$ "samples" would give back almost the same bilinear model parameter estimates as those originally obtained for the initial calibration set.

More significantly, these $2A$ artificial samples' X and Y data can be merged with new X and Y data from N_{update} new updating samples, thereby allowing model updating without having to use all the original $N_{initial}$ samples' raw data. This can give storage and computation time advantages.

For an example in Figures 25 and 26 (Martens et al, 1991), the input data consisted of a total of 14 samples representing impure solvent mixtures and contained known levels of benzene in the concentration range of 0–16%, but unknown levels of other unknown solvents. The NIR data of these mixtures were measured in

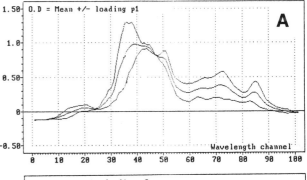

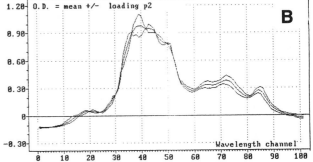

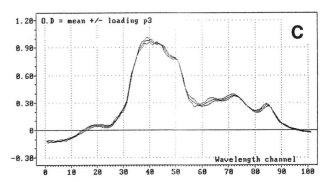

Fig. 25. Artificial bilinear sample spectra $\mathbf{X}_{blm}$ for factors **A**, 1; **B**, 2; and **C**, 3. The initial mean spectrum $\bar{\mathbf{x}}$ is included for comparison. (Reprinted from Martens et al [1990] with permission from Instrument Society of America. Copyright 1990. All rights reserved.)

a remote-sensing Guided Wave Model 200 process spectrophotometer (Guided Wave, Inc., El Dorado Hills, CA), using a single-strand optical fiber. These 14 samples served as an *initial calibration set*, $(\mathbf{X},\mathbf{y})_{initial}$.

An *updating set* of samples, containing the same solvents but at a different level, were similarly obtained. These contained known levels of benzene in the concentration range of 45–90%, in a total of 12 samples, $(\mathbf{X},\mathbf{y})_{update}$.

In addition, at a later stage, a *test set*, $(\mathbf{X},\mathbf{y})_{test}$, covering the full range from 0–90% benzene was generated, primarily in order to demonstrate the need and effect of updating.

Given the ever-present noise and nonlinearities in data, the calibration model based on the initial calibration set (<16% of analyte) did not give good predictions for the samples with much higher analyte levels (>45%) in the test set. The need for updating the model was therefore obvious.

One way to do the model updating would, of course, be just to put the two sets of raw data $(\mathbf{X},\mathbf{y})_{initial}$ and $(\mathbf{X},\mathbf{y})_{update}$ into one big $(\mathbf{X},\mathbf{y})$ set and recalibrate on this joint data set. In this current case, that would be simple, since the size of each of these sets is small. However, if the initial calibration set had been larger—many samples and many variables—this could be wasteful and possibly impractical with respect to computation time and memory size. Therefore, a way to simplify the updating process was sought.

For the $(\mathbf{X},\mathbf{y})_{initial}$ data, a three-factor PLSR model was found to give optimal prediction of percent benzene ($A = 3$).

The three factors $a = 1$, 2, and 3 in the initial PLSR calibration model represent all the valid information in the initial calibration data set.

On the basis of this information, the more compact data set of the "bilinear samples" $(\mathbf{X},\mathbf{y})_{blm}$ was created. Figure 25A, B, and C show the initial model center, $\bar{\mathbf{x}}$, alone and plus/minus the loading $\mathbf{p}_a$ for factors $a = 1$, 2, and 3, respectively. These six samples in $(\mathbf{X},\mathbf{y})_{blm}$ were merged with the new empirical data $(\mathbf{X},\mathbf{y})_{update}$, and the PLSR modeling was repeated.

When applied to the test set for which the benzene percentage was known, this updated calibration model shows good predictive ability (Fig. 26) throughout the benzene percentage range, as desired. Thus, the purpose of the updating was fulfilled, to establish a new multivariate calibration model with a wider application range and more precise performance.

The adaptive updating method can be modified in various ways. One may, for example, give different weights to the initial versus the updating sets in the recalibration. These updating weights can be different for the model center estimation ($\bar{\mathbf{x}}$, $\bar{\mathbf{y}}$) and for the model span (loadings $\mathbf{P}$ and $\mathbf{q}$). The method can also be made computationally more effective using A "sample" only. An advantage of the method version with 2A "samples" is that updating can then be performed with existing standard chemometric calibration software.

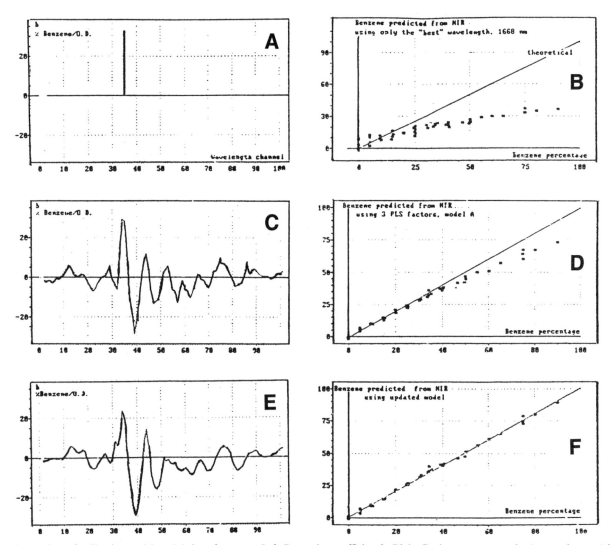

Fig. 26. Comparison of calibration models and their performances. Left: Regression coefficient **b**. Right: Pred. versus measured **y** (percent benzene) in test set. **A and B**, Univariate ($x_{35} = OD_{1,668\ nm}$), using initial calibration set only. **C and D**, Multivariate calibration, using initial calibration set only. **E and F**, Updated calibration, using bilinear summary plus the updating set. (Reprinted from Martens et al [1990] with permission from Instrument Society of America. Copyright 1990. All rights reserved.)

IX. MISCELLANEOUS TOPICS

A. Design Is Central in Calibration

Calibration for simple mixtures is usually easy and cheap, but calibration of complex measurements, such as diffuse reflectance of powders, slurries, or intact biological material, requires a lot of representative empirical data and careful multivariate calibration. This can be time-consuming and also quite expensive. So multivariate calibration is only worth the work for reasonably large series of analyses.

We first have to decide upon analytical procedures for measuring **Y** and **X**, and to consider their expected performances. Thus the question is: 'Given the noise in the reference method **Y**, and given the selectivity problems and noise in the **X**-data, is it at all possible to attain the required prediction ability?' If the answer is definitely 'no,' then there is no sense in wasting time trying to make a calibration.

If the answer to the preliminary assessment is 'yes' or 'maybe,' then the problem is: 'How should we calibrate?' This requires an assessment of what are the phenomena that have to be modelled—analyte and interferences, nonlinearities, interactions, etc. This can be based on prior knowledge about the object material and the instrumentation, or possibly on data analytic inspection of a large amount of available *X*-data.

On this basis follows the actual *experimental planning*—how to get good calibration data. We must decide which variables to measure, which experimental factors to control, how to combine them, and how many replicates to use. We must also decide whether to include a number of more or less *randomly* chosen observations in order to pick up uncontrollable factors and unexpected surprises. In some extreme cases in for instance NIR spectroscopy, so little information is available for design that a randomly selected sample may even be the best we can use. This is termed random or *natural* calibration as opposed to *controlled* calibration, which is usually to be preferred.

In multivariate calibration, experimental design is required in order to ensure a satisfactory description of the analyte–instrument relationship and of the instrument interferences in light of the unavoidable measurement noise and model errors. Good calibration plans can save us time and money. They increase the efficiency of our calibration work, by ensuring that the important information is obtained, while useless or unnecessary measurements are avoided.

Ideally, good planning does require good understanding of our instruments and good understanding of the type of objects to be calibrated for. But even when our understanding is very incomplete, elements of conscious planning can help us get useful calibrations. Thus we can learn while we work, through a cycle of planning—measurements—calibration—practical prediction—outlier detection and interpretation—more planning—more measurements—improved recalibration, etc.

IMPORTANT ASPECTS OF DESIGN

The design of a calibration experiment has two basic aspects.

1. To span the important types of systematic variability, including interferences, so that they can be assessed and hopefully modelled.

2. To minimize the effect of random noise of various sorts.

Both should be given attention.

Fortunately, unexpected interferences can often be detected later on through their abnormal leverage contribution and/or their abnormal spectral residual contribution during prediction. But to avoid recalibration or updating, it would have been better to include the various types of interference from the beginning.

It should be noted that the word *interferences* in this context means both chemical/physical phenomena affecting the *X*-data, and effects like nonlinear instrument responses and interconstituent interactions.

In general, the wider the range of objects calibrated for in one calibration model, the higher is the chance of running into diffi-

cult nonlinearity and heterogeneity problems. So from a mathematical model point of view, one should develop calibration models for *narrow* ranges of objects.

On the other hand, our local approximation models have parameters that need to be *estimated* statistically, and that calls for *wide* ranges of calibration objects to be modelled together. Sampling problems and measurement noise in **X** and **Y** can create damaging errors in these parameter estimates, depending on the signal/noise ratio in the calibration data. We can reduce the estimation error by bringing down the noise, improving our sampling and measuring techniques and/or by including more calibration objects. But this brings the cost of calibration up.

So in designing multivariate calibration experiments, we must strike a compromise between optimizing the model structure (narrow range) and optimizing the parameter estimation (wide range) and usually under constraints of cost, time, and availability of objects.

DEFINING THE TARGET POPULATION OF A CALIBRATION

Since a multivariate calibration only gives local approximations, it will give different predictive abilities for different types of objects, even within the normal ranges of analyte and interferents included in the calibration set.

For instance, in the case of NIR protein determinations in a certain type of wheat, we may want to calibrate for a typical protein range, say 10–14%. Even within this range our predictions would probably be most precise for the most typical protein levels calibrated for, say near 12%, and considerably more uncertain at 10 and 14%, and even more so for objects with 16% protein or with other qualities outside the normal range calibrated for.

The general goodness of a predictor $f(\mathbf{X})$ is a *statistically weighted average over the different types of future object qualities*. So it is dangerous to develop a calibration model and to specify its predictive ability if we do not know which distribution of future objects to target on.

Once the target population has been more or less identified, it is important to get calibration objects that together are representative for these future objects, both for model estimation and for validation assessments.

BASIC PRINCIPLES

First of all, it is important to note that since the *X*-variables in the linear regression equation and in bilinear regression are often empirical values impossible to specify in advance of an experiment (i.e., $\mathbf{X} \Leftarrow$ causal $g[\mathbf{Y}]$). So ordinary regression design theory as reported in, e.g., Box et al (1978) and in Deming and Morgan (1987), where exactly such control is assumed, may often be impossible to apply directly.

An important exception here is the linear mixture model where the *X*-variables are linear functions of the chemical constituents plus noise. In this case, each **x** can be written as

$$\mathbf{x}_i = \mathbf{k}_0 + \mathbf{K}\mathbf{y}_i + \mathbf{e}_i$$

and the combinations of *y*-values can be designed according to a traditional design scheme for precise estimation of $\mathbf{k}_0$ and **K** or precise predictions of future *y* (e.g., Ott and Myers, 1968).

A simple and efficient design which can be applied in the model above is a factorial design with two levels on each factor. This is a so-called orthogonal design making estimation of different parameters (for different constituents) statistically independent. If the number of factors to control is high, the number of calibration objects in the design can be reduced by sensible application of fractional factorial designs (see, e.g., Box et al, 1978). In case of closure among the constituents, the constituents cannot be varied independently, and modifications such as those presented in, e.g., Cornell (1981) must be applied.

In general, the causal mixture model represents oversimplifications and the calibration model has to be expanded, either by using more bilinear factors than the number of known constituents or by using additional factors in **e**.

If a bilinear model is used, the variation in **X** is approximated by $\hat{T}P'$, where $\hat{T}$ spans all possible types of systematic factors: chemical concentrations, interferences and nonlinear functions, and interactions of the chemical and physical factors. Some of these may be known and controllable, others not.

Experimental design for such models is complicated and is discussed in very few applications (e.g., Naes and Isaksson, 1989). One of the problems is that the $\hat{T}$-factors are abstract and normally it is impossible to identify one single factor as, e.g., protein content. Secondly, the 'dependent' *Y*-variable is generally part of the design and this creates problems with application of traditional regression design theory.

However, the essence of the general ideas used in the simpler cases of regression design carry over their importance and should be applied for safe calibration. The most basic point to know here is that only factors which are present and modelled (by $\hat{T}P'$) in the calibration can be compensated for in prediction. This means that an interferent not modelled in the calibration will create alias problems and therefore bad predictions. Thus, the following is of fundamental importance.

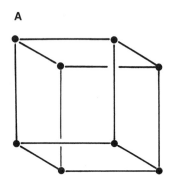

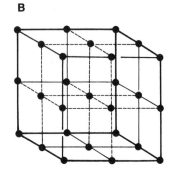

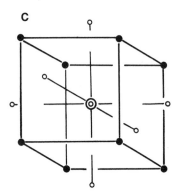

Fig. 27. Factorial designs for three factors (analytes or known interferents). **A,** Factorial design with two levels for each factor. **B,** Factorial design with three levels for each factor. **C,** Central composite design. (Reprinted from Martens and Naes [1989] with permission from John Wiley & Sons, Ltd. Copyright 1989.)

All phenomena that vary (in X) in the target population, must be spanned in the calibration set and described in the calibration model.

Each phenomenon should be spanned as well as possible, to be separated from noise and from each other. Remember, however, the dilemma telling us to consider model fit in the larger population as well.

What then about factors in $\hat{T}$ that represent nonlinear functions of chemical constituents or interactions between chemical and physical phenomena? To be sure that all combinations of them are well represented or that all factors corresponding to nonlinearities are properly spanned, each varying phenomenon should be represented at more than two levels. A general rule of thumb that can be used is that the number of levels should at least correspond to one plus the expected complexity of a polynomial approximation of the causal relationship (e.g., Naes and Isaksson, 1989). So, if it is important to keep the number of calibration objects low, and if for instance the functional relation between **x** on one hand and the varying interferences on the other can be approximated by a second degree polynomial with interactions, then it may suffice to use three levels of each factor, for instance a low, an intermediate, and a high value (Fig. 27).

This type of multilevel multifactor orthogonal design can give a 'combinatorial explosion,' resulting in unnecessarily expensive calibration experiments. One interesting way to simplify such multifactor factorial designs is to combine a simple two-factorial experiment with some extra points into a central composite design (Fig. 27C).

In case of closure, one can alternatively use a type of simplex lattice design as described in, e.g., Cornell (1981).

Thus, factorial designs in all varying phenomena with more than two levels for each factor are reasonable candidates for BLM designs, representing a type of even spread of points over the actual region.

This even spread of design points is also important from another point of view, namely for detecting nonlinear relations between **y** and **x** and to smooth out moderate nonlinearities. This is discussed in detail in Naes and Isaksson (1989).

In many cases, the target population has more objects near the centre than far from it. In such cases, it may be natural to reflect this in the design strategy by selecting more calibration objects from near the centre. The consequences of this are not properly analysed, but an illustration is given below.

Finally, remember from the above that in cases where information about interfering constituents and physical phenomena is incomplete, it may be of use to add some randomly selected objects as well.

ILLUSTRATION: THE IMPORTANCE OF GOOD DESIGN

The following illustration is a summary of Naes and Isaksson (1989) and concerns NIR reflectance determination of protein in mixtures of fish meal, starch, and water in different proportions. A high number of mixtures were generated and measured in a Technicon InfraAlyzer 400 with 19 wavelength channels. Various subsets of calibration objects were selected from these, each set having the same number of calibration objects ($I = 41$). All the calibrations were tested on the same 25 independent prediction objects. In addition, the calibrations were tested on two different subsets of the prediction set, i.e., the central points and the border points. The calibration method used was PCR with deletion of eigenvectors corresponding to the smallest eigenvalues. The data were scatter-corrected according to the multiplicative signal correction (MSC) to reduce the effect of physical phenomena on the spectral readings.

The results of the different calibrations are shown in Figure 28 and the results are given in RMSEP [root mean square error of prediction] units. In each of these five designs, the RMSEP corresponds to the result of PCR with optimal number of factors.

The most striking feature in Figure 28 is the clear demonstration of the need for spanning all dimensions properly. Calibration design iv only covers a small region of the target population and the predictions outside this region are very bad. Design v spans the variation of the constituent protein calibrated for, but as we see, this is not enough to ensure good prediction. To obtain this, one obviously needs calibration objects spanning the whole space of interest (here mainly water and carbohydrates in addition to protein).

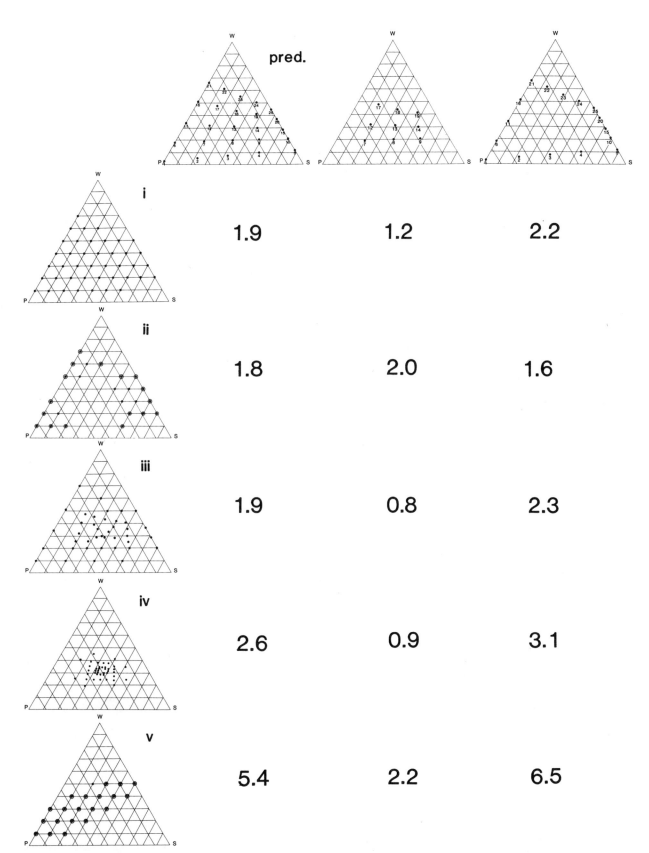

Fig. 28. Illustration of the effect of different design strategies for the mixture experiment. Vertically are given the designs of the five calibration sets i, ii, iii, iv, and v, and horizontally are given the designs of the three test sets, illustrated in the three-phase diagram of the three constituents. The resulting prediction error for the test sets is presented in root mean square error of performance (RMSP) units for each calibration model. Solid circles indicate two points. (Reprinted from Martens and Naes [1989] with permission from John Wiley & Sons, Ltd. Copyright 1989.)

As we also see, the performance difference between the other three calibration sets are not very clear. There is, however, a tendency of better results in the middle for design i and the 'natural population' case iii than for design ii. As discussed in the paper by Naes and Isaksson (1989), this can be due to the 'least squares effect' of best prediction ability near the centre. Alternatively, it can be due to a nonlinearity which is not properly handled in case ii. This becomes clearer when studying the individual prediction errors as done in the original paper.

Overall, it seems that the nonlinearity problem is very important in this type of application, especially when the region of interest is as large as here. To be able to handle this, it is necessary that all different parts of the actual target population are covered.

B. Linearity Problems

SPLITTING, NONLINEAR MODELLING OR PRETREATMENTS

If there are nonlinearities or heterogeneities in the *X-X* and/or *X-Y* relationships, the best way to obtain good linear fits to data is often to split the population of objects into more narrow subclasses, each with sufficient linearity to let us ignore the whole problem of nonlinear calibration. The splitting can be based on a priori knowledge about the objects, or on a preliminary data analysis step (e.g., splitting on the basis of visual inspection of the score plots from a preliminary bilinear modelling, or by initial cluster analysis [Mardia et al, 1979]).

It may, however, be difficult to know how to split the calibration set and there are other drawbacks as well: the splitting increases the need for calibration data, the data analysis takes more time, and having many different predictors may be confusing in practice. So, after having learned to master the linear calibration techniques, it may be useful to study the topic of nonlinearity.

In general, multivariate calibration concerns data approximation of continuous but possibly nonlinear unknown relationships. As in general Taylor series expansions, the linear simplification can give a good enough description of many different types of nonlinear relationships, provided it is developed over a narrow enough range. So for initial screening, for preliminary calibration results, and sometimes even for final results, linearization can be ignored. But for optimal results in terms of predictive ability and interpretation simplicity, linearity problems in data should be consciously addressed.

We refer to Martens and Naes (1989) for methods to detect nonlinearity. In some cases, nonlinear calibration methods may be necessary, but in most cases, a nonlinearity transformation followed by linear modelling is easier to apply and gives equally good results. This section is mainly devoted to such pretreatment of spectral and chemical data, intended to be used once for each data set. But it should be mentioned that a repeated iteration between nonlinearity transformation and linear modelling until convergence can be useful.

Care should be taken to avoid losing important information during the linearization process. And it is important to avoid overfitting by the linearization. Choosing the best out of many attempted alternative linearizations, or be estimating a lot of linearization parameters, may not give the best predictive ability.

DIFFERENT TYPES OF NONLINEARITY

The simplest and most desirable calibration situation with respect to linearity is illustrated in Figure 29A, i.e., the instrument signal *x* is linearly related to analyte concentration *y*. There are,

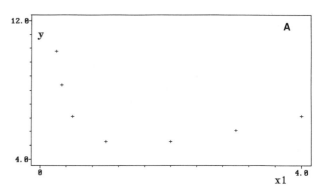

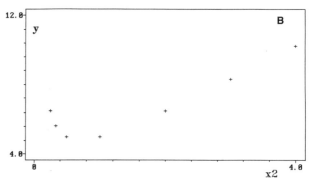

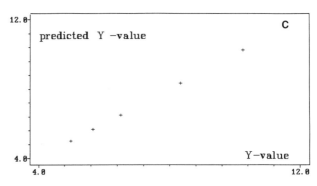

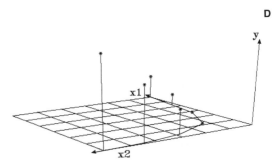

Fig. 30. Successful linear modeling of nonlinear relations. Simplified example with two *x* variables and one *y* variable. **A,** Raw data: *y* versus x_1. **B,** Raw data: *y* versus x_2 with two *x* variables and one *y* variable. **C,** Prediction $\hat{y} = 2.0 + x_1 \times 1.0 + x_2 \times 2.0$ plotted versus *y* (*b̂* was here estimated using two partial least-squares regression (PLSR) factors). **D,** *y* versus x_1 and x_2. The *Y-X* curvature forms a two-dimensional plane and can be modeled linearly by two factors when seen from a particular angle. Both these factors can be seen in the x_1 versus x_2 space (curved path). (Reprinted from Martens and Naes [1989] with permission from John Wiley & Sons, Ltd. Copyright 1989.)

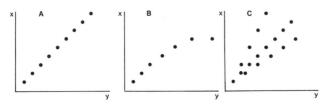

Fig. 29. Linear and nonlinear data. **A,** Linear instrument response. **B,** Curved instrument response. **C,** Multiplicative problem. (Reprinted from Martens and Naes [1989] with permission from John Wiley & Sons, Ltd. Copyright 1989.)

however, many analytical situations with different structure and two simple problems are well known.

Curvature Problems. If a detector saturates at high analyte concentrations, there will be a nonlinear response of the curvature type (Fig. 29B). Response curvature would also arise if x represents conventional spectrophotometric data in transmission T in-stead of absorbance units $\log(1/T)$. These cases would traditionally be regarded as 'bad analytical practice,' but it is not necessarily so. In general, some curvature must be expected in all instrument responses. Thus, for instance, Beer's 'Law' for spectroscopy and other conventional 'laws' are only valid under certain ideal conditions—but to limit ourselves to those ideal conditions is quite unnecessary!

Multiplicative Problems. In chromatography, the response varies in proportion to the total amount of sample applied to the column, and if this varies uncontrollably, then we have a nonlinear problem of the multiplicative type (Fig. 29C). The same type of problem arises in ordinary transmission spectroscopy if the curvette length changes (Lambert's 'Law'). In reflectance spectroscopy of, e.g., powders, changes in light scattering create a similar effect, as expressed by the Kubelka-Munk theory (e.g., Kortum, 1969).

LINEAR METHODS CAN HANDLE SOME NONLINEAR DATA STRUCTURES

Let us look at a simple illustration to show how this can work. Assume that we want to determine a single analyte y in objects $i = 1, 2, \ldots$, without any interferences, using two detectors x_1 and x_2. These detectors can give various types of nonlinear responses due to saturation, as illustrated in Figures 30A and 30B.

Still, y is linearly related to the linear combination of them, found by regressing y on both x_1 and x_2 (Fig. 30C) using the model

$$y = b_0 + x_1 b_1 + x_2 b_2 + f$$

This phenomenon is important and explains much of the reason why, e.g., linear calibration models have had such success in NIR reflectance spectroscopy. Using for instance PCR or PLSR in this application, it is found that quite many bilinear factors are needed (compared to the number of interferences) to obtain optimal prediction ability (typically between 5 and 10). The interpretation of this is that the high number of factors are needed to account for nonlinear effects between composition, interference, and individual spectral variables.

But it is easy to find examples where this is not the case, as illustrated in Figure 31. Again y is nonlinearly related to both x_1 and to x_2 (Fig. 31A and B), but in this case, even the best linear combination of these two input X-variables gave bad nonlinear relation to y (Fig. 31C).

In Figures 30D and 31D, we see that when plotted in the three-dimensional space $y - x_1 - x_2$ the points in both examples lie along a single curved path (a 'banana'). This could be due to a single

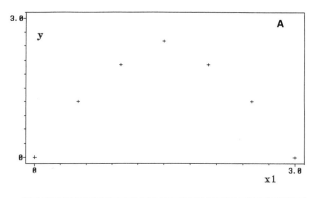

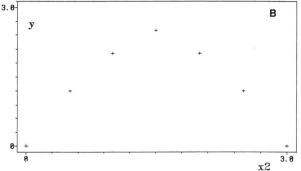

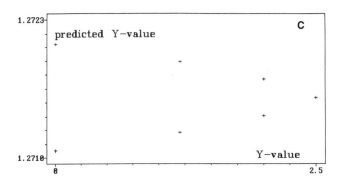

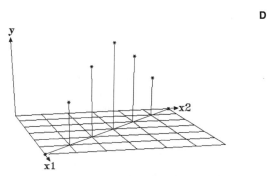

Fig. 31. Unsuccessful linear modeling of nonlinear relations. An example with two X variables and one Y variable. **A,** Raw data: y versus x_1. **B,** Raw data: y versus x_2. **C,** Predicted $\hat{y} = 1.27 + x_1 \times -0.002 + x_2 \times$ versus y ($\hat{b}$ was here estimated using one partial least-squares regression [PLSR] factor). **D,** y versus x_1 and x_2. The Y-X curvature forms a two-dimensional plane. However, only one of its dimensions can be seen in the x_1 versus x_2 space (straight path). (Reprinted from Martens and Naes [1989] with permission from John Wiley & Sons, Ltd. Copyright 1989.)

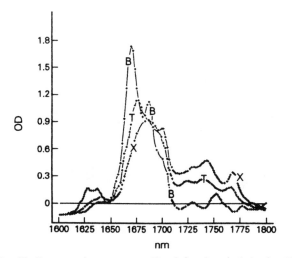

Fig. 32. Pure constituent spectra. Near-infrared optical density (OD) spectrum of the analyte toluene (T) and the two interferences benzene (B) and xylene (X). (Reprinted from Martens and Stark [1991] with permission from Elsevier Science.)

chemical phenomenon causing the variations in the *X*-data. The difference is in how the 'banana' is positioned relative to the *X*-space. In Figure 30D, we can recognize the curvature of the 'banana' in our two-dimensional *X*-space (the $x_1 - x_2$ plane) itself. Thus, we can use **X** to estimate a linear two-factor model—a plane in which the *Y-X* 'banana' shape is seen as a 'boomerang' shape. This adequately describes the curved *Y-X* relationship.

In Figure 31D, we cannot see the curvature of the 'banana' just from the *X*-data, because both *X*-variables show the same kind of

nonlinear response. Hence, we cannot estimate the second dimension in the 'boomerang' plane from the *X*-data alone, and the *Y-X* curvature remains unmodelled. In this case, something explicit has to be done to overcome the curvature problem. We refer to Naes et al (2001) for an overview of how this can be done.

C. Other Data Preprocessing Methods

For effective multivariate calibration modelling, it is important to combine *a priori* assumptions ("prior knowledge") and empirical data in a balanced way.

The present section assumes a flexible intermediate between hard modelling and soft modelling: *a priori* knowledge is applied in "hard" modelling during preprocessing, to simplify the structure in the spectral data. "Soft" modelling is then used for cleaning up empirically what the causal modelling could not explain. The goal is to obtain maximal understanding and maximal predictive reliability and relevance at minimum experimental and data analytic costs.

The section outlines two spectral preprocessing methods for improving the multivariate calibration of multichannel analytical instruments based on spectroscopic background knowledge: extended multiplicative signal correction (EMSC) is designed to improve the separation of light scattering and light absorbance, and spectral interference subtraction (SIS) for elimination of interferences with known spectral effects.

If the physical and chemical effects in the spectra are sufficiently different, they may be separated by multivariate statistical modelling. One method is the MSC technique (Section V B). This is now termed "multiplicative signal correction" [Martens and Naes, 1989, Chapter 7], as a generalization of the original term "multiplicative scatter correction" [Martens et al, 1983a], since it is also applicable to other types of data, e.g., correcting for varying amounts of sample applied to a chromatography column. MSC seeks to correct the baseline and amplification effects to the same "average" level in every spectrum.

Stark and Martens developed MSC into the extended multiplicative signal correction (EMSC) in order to attain a more effective separation of chemical and physical effects in light spectroscopy. The EMSC method employs knowledge about the spectra of the analytes and interference effects to improve the path length estimation.

When spectral information about the analyte and interferents is available, it is possible to reduce or eliminate the spectral effects of the interferents in the preprocessing stage. This further reduces the need for empirical calibration data. The technique presented here is called spectral interference subtraction (SIS). Its purpose is to filter out the effects of known constituents from the spectral data, with as little modification as possible of the effects of the unknown constituents and phenomena. The mathematical details are given in Martens and Stark (1991), where the current example is taken from.

EXAMPLE

Input Data: Fibre-Optic Process Near Infrared Spectra. Figure 32A shows the OD spectrum of the analyte, $k_{toluene}$, together with that of two interferents, benzene and xylene. The figure shows that the spectra are strongly overlapping. Wavelength channel 39 (1,676 nm) seems to be the single most typical wavelength for the analyte.

Figure 33 shows the NIR OD spectra of some mixture samples after different preprocessing, and Figure 34 shows the corresponding univariate and multivariate predictive performance when calibrating for the analyte toluene.

Raw Input Spectra. Figure 33A shows the raw input OD spectra prior to preprocessing. A large degree of variation is evident. Figure 34A shows virtually zero correlation between the "best" single wavelength (channel 39) and the toluene concentration for these data. Figure 34B shows that with multivariate PLSR calibration these 101 wavelengths together yielded a clearly improved predictive ability, but the relationship is not satisfactory.

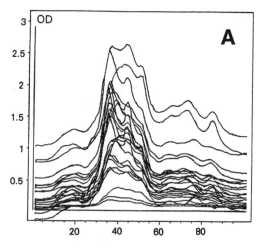

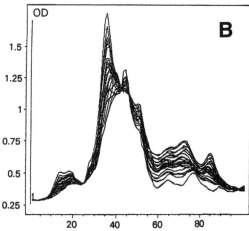

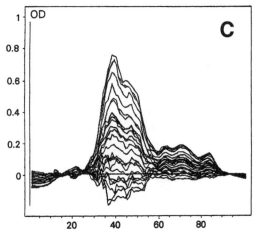

Fig. 33. Spectra for solvent mixtures. **A,** "Input spectra," illustrating uncontrolled random pathlength and baseline variations; **B,** same spectra after extended multiplicative signal correction (EMSC); **C,** and same spectra after EMSC and spectral interference subtraction (SIS). (Reprinted from Martens and Stark [1991] with permission from Elsevier Science.)

Extended Multiplicative Signal Correction. Figure 33B shows the same spectra after EMSC preprocessing. This normalizes all the spectra to an average estimated baseline level and an average estimated path length ("light scattering") level. The variability in the spectra is now much smaller. Figure 34C shows that satisfactory predictions can still not be attained using only one single wavelength, due to the spectral overlap between the analyte (toluene) and the interferences (benzene, xylene) and to "unknown interferences." However, when using all 101 wavelengths in multivariate PLSR calibration (Fig. 34D), an excellent predictive ability is attained.

Spectral Interference Subtraction. Figure 33C again shows the same EMSC-treated spectra, after an additional SIS preprocessing step to remove additive effects of the known interferents benzene and xylene. The variability in the spectra is now quite systematic and represents mainly increasing levels of the analyte toluene (cf. Fig. 32). However, some low-toluene samples now show negative OD (as opposed to the expected level close to zero). This is probably due to nonrepresentativity in the "known" component spectra, or to nonadditivity, for example, caused by constituent interactions of some kind. Such unexpected spectral phenomena have to be "cleaned up" by the subsequent multivariate calibration.

Figure 34E shows that good predictions can now be attained with a single wavelength. But the single-wavelength calibration is still not optimal, due to the spectral overlap between the analyte

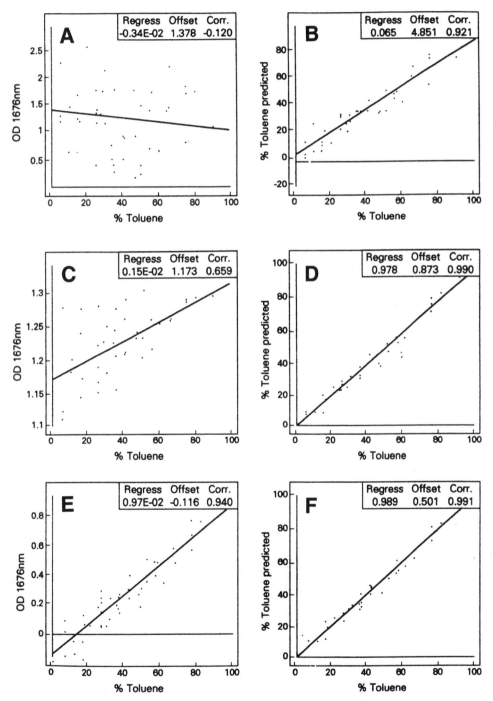

Fig. 34. Calibration performances for prediction of percent toluene (abscissa). Left side: Traditional univariate calibration, using the "best" single wavelength; ordinate = optical density at 1,676 nm, channel 39. Right side: Multivariate calibration, using a combination of four partial least-squares regression (PLSR) factors; ordinate = predicted toluene concentration as obtained in full cross-validation. **A and B,** "Input spectra." **C and D,** Same spectra after extended multiplicative signal correction (EMSC). **E and F,** Same spectra after EMSC and spectral interference subtraction (SIS). (Reprinted from Martens and Stark [1991] with permission from Elsevier Science.)

(toluene) and "unknown interferences" in the spectra. With all 101 wavelengths combined in multivariate PLSR (Fig. 34F), an excellent predictive ability is attained.

The unpreprocessed spectra contained interference problems that the PLSR calibration could not handle well enough. A four-factor model was required in order to obtain reasonable predictive ability, and this yielded rather high predictive error (root-mean-square error of prediction [RMSP] = ±8.9% toluene) (cf. Fig. 34B). The MSC-treated spectra had drastically improved predictive ability and needed only two factors to give good predictive ability, as expected for a three-constituent mixture system where the sum of the constituents is constant. The EMSC-treated spectra gave a further improvement. The PLSR loadings for MSC- and EMSC-treated spectra (not shown here) were somewhat difficult to interpret, since they represent two difference-spectra between the three solvents. A couple of minor factors gave slightly improved predictive ability (RMSP = ±3.2%

benzene after four factors, cf. Fig. 34D). Other studies have revealed that these effects represent interconstituent interactions of an optical or chemical kind, deviations from fully linear instrument response, etc.

The SIS-treated spectra gave good predictive ability already after one factor. The PLSR loadings of this factor (not shown here) was virtually identical to the spectrum of the analyte itself. Very good predictive ability was again attained after, for example, four factors (RMSP = ±3.1% toluene, cf. Fig. 34F).

This example has shown that EMSC preprocessing effectively removed most of the path length and baseline effects, allowing the subsequent additive PLSR to work better. EMSC gave a small, but clear improvement over the traditional MSC treatment. This difference is expected to be greater in systems where the absorbance spectra of the constituents differ more widely than in the present case (e.g., for mixtures containing water and displaying water temperature effects).

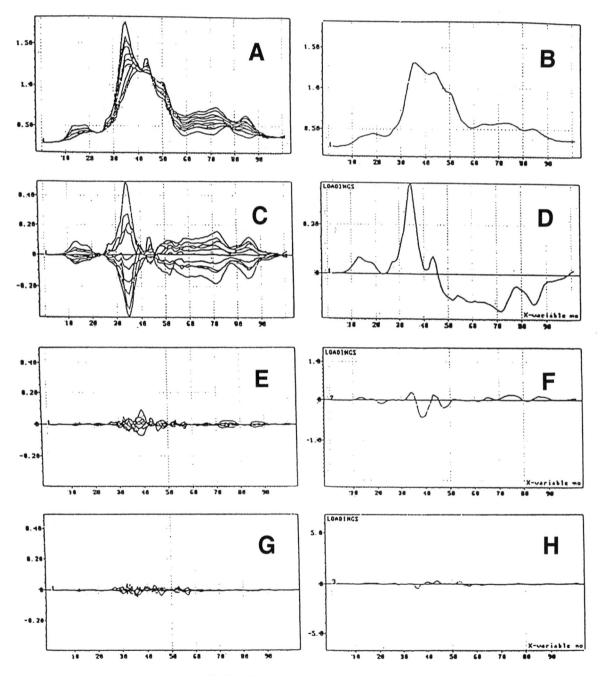

Fig. 35. Multivariate partial least-squares regression (PLSR) calibration modeling of near-infrared data of three-constituent mixtures. Left side: Spectra for the 10 first calibration samples. Right side: Model parameters estimated by PLSR. A, Input data $\mathbf{X}$ after extended multiplicative signal correction (EMSC) and spectral interference subtraction (SIS); B, $\bar{\mathbf{x}}'$; C, $\mathbf{E}_0 = \mathbf{X} - \mathbf{1}\,\bar{\mathbf{x}}$; D, loading $\mathbf{p}_1$, E, $\mathbf{E}_1 = \mathbf{E}_0 - \mathbf{t}_1\mathbf{p}_1$; F, loading $\mathbf{p}_2$; G, $\mathbf{E}_2 = \mathbf{E}_1 - \mathbf{t}_2\mathbf{p}_2$; and H, loading $\mathbf{p}_3$. (Reprinted from Martens et al [1991].)

The SIS preprocessing allowed the removal of most of the known additive interference, making the resulting PLSR model easier to interpret. This preprocessing can be advantageous in situations where it is difficult to generate real calibration samples to span the full variability of the expected future sample qualities, e.g., in on-line process control applications. Every mixture spectrum, both in the calibration set and in future unknown samples, can instead be made "immune" against certain expected future interference by SIS preprocessing.

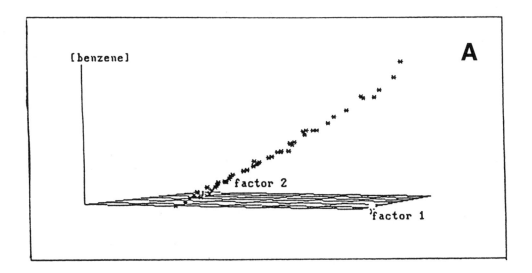

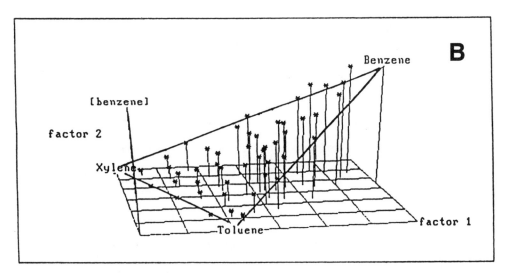

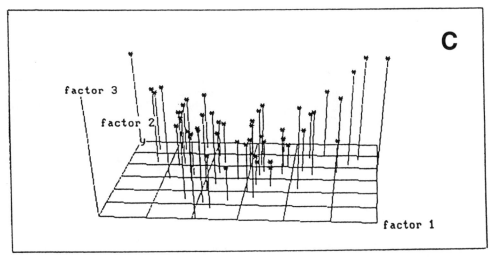

Fig. 36. **A,** Multivariate partial least-squares regression (PLSR) modeling of near-infrared (NIR) data of three-constituent mixtures: $\hat{y}$ versus NIR scores t_1 and t_2. **B,** Same as A, but seen from another angle. **C,** A look into **X** space scores: t_3 versus t_1 and t_2. (Reprinted from Martens et al [1991].)

D. Graphical Interpretation
of NIR Calibration Based on Soft Modeling

It has become evident that multichannel NIR instrumentation sometimes faces a credibility problem with traditional analytical chemists. It may, therefore, be important to show that multivariate calibration of NIR instruments is not a set of unscientific hocus pocus methods. Also, it is important to remind oneself that the flexible chemometric methods like PLSR should not be used as totally black boxes. One should always try to understand the causal basis for the results.

The current example shows how a PLSR model is developed and how it makes perfect sense from a traditional linear mixture perspective, while offering an additional flexibility for modeling the unexpected. The example is taken from the work by Martens et al. (1991).

Fifty mixtures of organic solvents (benzene, toluene, and xylene) in various known ratios were prepared. This sample set is intended to simulate samples from an industrial process stream of an impure solvent. Benzene is the solvent—the analyte to be calibrated for—while toluene and xylene are again to be treated as 'unidentified interferents.' The NIR data of these mixtures were measured in a Guided Wave Model 200-45 process spectrophotometer, using a 2-m single-strand optical fiber and a transmission probe configuration. The transmission (T) spectra were obtained at every 2 nm between 1,600–1,800 nm. The data were linearized by the conventional transform into optical density, $OD = -\log(T)$.

Analyte $\mathbf{y}$ = benzene percentage was calibrated for by PLSR using 27 calibration samples. The left side of Figure 35 illustrates how the bilinear modeling proceeded for 10 of the calibration samples, and the right side of the figure shows model parameters estimated. The PLSR loading parameters are scaled according to the relative importance of each factor to the modeling of the NIR spectra in the calibration set.

Figure 35A shows the NIR spectra of calibration samples, and Figure 35B shows their mean spectrum, $\bar{\mathbf{x}}$. After subtraction of $\bar{\mathbf{x}}$, the variation in the NIR data $\mathbf{X}$ looks like Figure 35C. The first bilinear factor's loading $\mathbf{p}_1$ (Fig. 35D) summarizes most of this variability. After subtraction of this first factor, Figure 35E shows the residual NIR variation left in $\mathbf{X}$. The second factors $\mathbf{p}_2$ (Fig. 35F) summarizes much of this remaining NIR variation. Figure 35G shows the residual NIR variation after subtraction of the second factor as well. Figure 35H gives the third factors' loading, $\mathbf{p}_3$. The fourth and the subsequent factors appeared to contain only random noise and are not shown. Thus, Figure 35 shows how various types of systematic variation in the $\mathbf{X}$ data are picked up and subtracted by the self-modeling PLSR method. No spectral interpretation was required at this stage, except a visual check for apparent random noise structure. Both the graphical inspection and the statistical cross-validation prediction testing indicated one major and two minor factors to be validly estimated from the current calibration data set. This calibration model was used in order to predict y from $\mathbf{X}$ in test set.

Figure 36 explains how the multivariate selectivity enhancement works geometrically. Each point now represent a test mixture.

Figure 36A shows corresponding benzene percentage $\mathbf{y}$ of both the calibration set (objects 1–27) and the test set (objects 28–50) plotted against their scores for the first two PLSR factors from the NIR data, $\mathbf{t}_1$ and $\mathbf{t}_2$. Seen from a certain angle (corresponding to the ratio between the Y loadings, q_1 and q_2) a highly selective relationship is seen between $\mathbf{y}$ and these two linear combination of NIR data $\mathbf{X}$.

Thus, the multivariate PLSR calibration has converted nonselective spectra $\mathbf{X}$ (Figs. 32, 35A) into very selective analyte predictions, without using any explicit information about the interferences!

For an analytical chemist, it may be strange to accept that good predictive ability can be attained in spite of serious nonselectivity in the raw data—particularly when the causes of the interferences are unknown. The reason why it works is that there is sufficient

systematic covariation information in the calibration data themselves to model the unidentified interference effects automatically. However, multivariate calibration—initial calibration as well as updating—should be regarded as a causal learning experience, not just a statistical black-box process. Bilinear calibration methods such as the PLSR give graphical access into the covariance matrices involved in the multivariate analysis.

Figure 36B shows the same as Figure 36A, $\mathbf{y}$ versus two first factors' scores, $\mathbf{t}_1$ and $\mathbf{t}_2$, but seen from another angle and with vertical lines indicating the three-dimensional coordinates of the various samples. The samples of pure benzene, toluene, and xylene were included in the test set, for illustration, and their positions are marked explicitly. Line segments are drawn between them. The figure shows that the samples fall inside the triangle that has the analyte benzene and the two "unknown" interferants at its corners. So, the self-modeling PLSR has been able to pick up and describe *both* the analyte variation *and* the variation in the unknown interferants. The first two factors thus automatically spanned the two-dimensional variation of a three-component mixture system (three components summing to 100%). Seeing this plane from the side gave the selectivity enhancement required (Fig. 36A), and seeing it from the top gave chemical interpretability (Fig. 36B).

What was the third, minor but valid, factor? In Figure 36C, this factor score vector $\mathbf{t}_3$ is plotted against the two first factors' scores, $\mathbf{t}_1$ and $\mathbf{t}_2$. The figure shows two things. i) The three-component mixtures generally lie down in a "pocket" or valley in the middle, with higher scores $t_{i,3}$ for the three pure solvents and the two-component mixtures. This nonlinear effect may be due to solvent interactions of some sort. By the third factor, the self-modeling bilinear PLSR has been able to account for some of the unknown nonlinear structure in the data. ii) The scores from the third factor seem generally more irregular and noisy than those for the first two factors. This is not unexpected, considering how small variations were available for its estimation (Fig. 35G and H). But this effect is small, so it was difficult to observe and model it precisely. On the other hand, since it is small, it would have had no grave consequence to leave it out from the calibration model.

X. CONCLUSIONS

This chapter has shown that indirect multivariate calibration is necessary in diffuse NIR spectroscopy, due to extensive chemical and optical interferences at all wavelengths. It has also been shown how the optimal dimensionality of such calibration methods has to be found in order to model these interferences, while avoiding the noise problems from overfitting (Fig. 3).

Several linear full-spectrum calibration methods that can be used in NIR spectroscopy have been described. One general group of methods relies on a data compression stage followed by a calibration regression stage, while another group relies on a joint data compression/calibration stage that allows alternative prediction methods (Beer's model methods). Methods for detection of errors in NIR data have been discussed and also of errors in chemical data and of abnormal sample qualities. The importance of proper linearization of the NIR data has been stressed, and a basic model of additive chemical interferences and multiplicative light-scatter interference has been shown to be useful.

The linear calibration methods and outlier detection methods discussed must be further tested in practice before a final evaluation can be made.

ACKNOWLEDGEMENTS

We would like to thank Ragnhild Norang and Unni Haugdahl for the patient and skillful typing of the manuscript. Karl Norris, William Hruschka, Lars Munck, and Sven Age Jensen are thanked for lending us data, and we thank them, as well as Svante Wold and Kim Esbensen, for valuable discussions.

APPENDIX A: ABBREVIATIONS AND SYMBOLS

Uppercase bold	matrix
Lowercase bold	vector (column vector unless otherwise stated)
Uppercase italic	upper limit for running index
Lowercase italic	scalar
^	"hat;" statistically estimated

$$\mathbf{1} = \begin{bmatrix} 1 \\ 1 \\ \vdots \\ 1 \end{bmatrix}$$ column vector on ones used for matrix formalism

a, A	index for regression factors, $a = 1, 2, ..., A$
$\mathbf{B}, \mathbf{b}, b$	calibration coefficients
b_0	calibration offset coefficient
BLP	best linear predictor
$\mathbf{C}, \mathbf{c}, c$	concentration of the chemical constituents
df	degrees of freedom
$\mathbf{E}, \mathbf{e}, e$	NIR residuals, residuals in Beer's model
e^*	uncorrelated residuals in Beer's model
EBLP	estimated best linear predictor
EBLUP	estimated best linear unbiased predictor
EGLS1	estimated generalized least-squares estimator (EGLS) with distributional assumptions on unknown interferences
EGLS2	EGLS without distributional assumptions on unknown interferences
$\mathbf{f}, f$	chemical residuals
g_i	leverage relevance
GLS	generalized least squares
h_i, h_k	leverage
$\mathbf{H}$	covariance matrix of $\mathbf{c}$
i, I	index for samples, $i = 1, 2, ..., I$
$\mathbf{I}$	identity matrix
j, J	index for chemical constituents, $j = 1, 2, ..., J$
k, K	number of NIR wavelengths, $k = 1, 2, ..., K$
K/S	Kubelka-Munk transform
LS	least squares
ML	maximum likelihood
MLR	multiple linear regression
MSC	multiplicative scatter (or signal) correction
MSE	mean squared error
$\mathbf{P}, \mathbf{p}, p$	loadings for interferences in Beer's model, NIR loadings
PCA	principal component analysis
PCR	principal component regression
PLS	partial least squares
$\mathbf{q}, q$	chemical loadings
R	reflectance
R^*	reflectance corrected according to Saunderson theory
RAP	relative ability of prediction
RMSCV	root mean square of error in cross-validation of calibration
RMSE	root mean square of error in calibrating estimation
RMSEL	leverage-corrected root mean square of error in calibrating estimation
RMSP	root mean square of performance (analysis)
$s(\hat{\mathbf{e}})$	standard deviation of $\hat{\mathbf{e}}$
SDLR	stepwise derivative linear regression
SMLR	stepwise multiple linear regression
Std	standard deviation?
STOT	total standard deviation of c prior to modeling
$\mathbf{T}, \mathbf{t}, t$	regression factor scores, interferences in Beer's model
$\mathbf{u}$	mean of $\mathbf{c}_i$
$\mathbf{U}$	symbol used in the PCA and PLS algorithms to replace $\mathbf{E}$
v	symbol used in the PCA and PLS algorithm to replace $\mathbf{f}$, intercept in Beer's model
$\mathbf{w}$	NIR loading weights in PLS with orthogonal scores
$\mathbf{W}$	covariance matrix of e^*
$\mathbf{X}, \mathbf{x}, x$	NIR observations
Σ	covariance matrix of $\mathbf{e}$ in Beer's model

APPENDIX B: MATRIX OPERATIONS ILLUSTRATED FOR MULTICOMPONENT ANALYSIS

Describing different calibration methods at sufficient precision and yet in a readable way requires matrix and vector notation. This appendix is written to help the reader who is not familiar with matrix algebra to be able to read at least the main sections of the chapter.

There are four basic matrix operations required, namely transpose, sum, product, and inversion, corresponding roughly to "turning a table 90°," "summing tables," "multiplying tables," and "dividing tables." Beer's model will be used here to illustrate these operations.

Vectors

The NIR spectrum of the sample i and K different wavelengths is a row vector $\mathbf{x}_i$ that can be written as

$$\mathbf{x}_i = \{x_{ik}\} = (x_{i1}, x_{i2}, ..., x_{ik}, ..., x_{iK}) \tag{B-1}$$

Assume that the NIR spectrum $\mathbf{x}_i$ is correctly modeled by the sum of the spectral contributions from water, starch, and protein in that sample:

$$\mathbf{x}_i = \text{water} + \text{starch} + \text{protein} \tag{B-2}$$

The spectral contribution of constituent j (e.g., water) is defined by its NIR spectrum multiplied by the concentration of the constituent in that sample. Its NIR spectrum $\mathbf{k}_j$ is, like the sample spectrum $\mathbf{x}_i$, a row vector:

$$\mathbf{k}_j = \{k_{jk}\} = (k_{j1}, k_{j2}, ..., k_{jk}, ..., k_{jK}) \tag{B-3}$$

Its concentration in sample i is simply one number (a scalar) termed c_{ij}:

$$c_{ij} \times \boxed{} \tag{B-4}$$

Hence, the spectral contribution of constituent j (e.g., water) in sample i is the product of its concentration c_{ij} and spectrum $\mathbf{k}_j$:

$$c_{ij} \times \mathbf{k}_j = \{c_{ij}k_{jk}\} = (c_{ij}k_{j1}, c_{ij}k_{j2}, ..., c_{ij}k_{jk}, ..., c_{ij}k_{jK}) \tag{B-5}$$

The sum of the contributions from the three constituents water, starch, and protein is then

$$\begin{aligned} \mathbf{x}_i = &\, c_{i1}\mathbf{k}_1 \\ &+ c_{i2}\mathbf{k}_2 \\ &+ c_{i3}\mathbf{k}_3 \end{aligned} \tag{B-6}$$

Matrices and Matrix Multiplication

In matrix algebra, Equation B-6 can be even more efficiently written. This is done by including all three concentration numbers c_{i1}, c_{i2}, and c_{i3} into one single row vector $\mathbf{c}_i$:

$$\mathbf{c}_i = \{c_{ij}\} = (c_{i1}, c_{i2}, c_{i3}) \qquad \boxed{} \qquad \text{(B-7)}$$

Likewise, all three constituent spectra $\mathbf{k}_1$, $\mathbf{k}_2$, and $\mathbf{k}_3$ are included into one single table or matrix $\mathbf{K}$:

$$\mathbf{K} = \{k_{ik}\} = \begin{bmatrix} \mathbf{k}_1 \\ \mathbf{k}_2 \\ \mathbf{k}_3 \end{bmatrix} \qquad \text{(B-8)}$$

The model can then be expressed more efficiently than in Equation B-6 as follows.

$$\mathbf{x}_i = \mathbf{c}_i \mathbf{K} \qquad \text{(B-9)}$$

This illustrates the use of vectors ($\mathbf{x}_i$ and $\mathbf{c}_i$) and matrices ($\mathbf{K}$). Matrix multiplication is done in this way. From given $\mathbf{c}_i$ and $\mathbf{K}$, the elements in the resulting vector $\mathbf{x}_i$ are obtained by

$$x_{ik} = \sum_{j=1}^{3} c_{ij} k_{jk} \qquad \text{(B-10)}$$

If there are a number of different samples $i = 1, 2, …, I$, then their spectra $\mathbf{x}_i$ can be collected into a single table or matrix $\mathbf{X}$:

$$\mathbf{X} = \{x_{ik}\} = \begin{bmatrix} \mathbf{x}_1 \\ \vdots \\ \mathbf{x}_i \\ \vdots \\ \mathbf{x}_I \end{bmatrix} \qquad \text{(B-11)}$$

Similarly, the concentration vectors $\mathbf{c}_i$ are collected in one single concentration matrix $\mathbf{C}$:

$$\mathbf{C} = \{c_{ij}\} = \begin{bmatrix} \mathbf{c}_1 \\ \mathbf{c}_2 \\ \vdots \\ \mathbf{c}_i \\ \vdots \\ \mathbf{c}_I \end{bmatrix} \qquad \text{(B-12)}$$

The model Equation B-10 for, for example, $I = 100$ different samples can be written simply as

$$\mathbf{X} = \mathbf{C}\mathbf{K} \qquad \text{(B-13)}$$

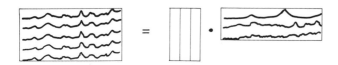

Matrix multiplication thus consists of summing up individual multiplication elements. The elements are chosen in this way.

$$\text{(B-14)}$$

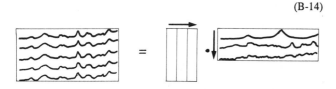

Hence, the number of columns in $\mathbf{C}$ must equal the number of rows in $\mathbf{K}$; in this case, this is three (water, starch, and protein).

Multicomponent Analysis by LS Fit

Suppose now that the pure constituents' spectra $\mathbf{K}$ were known and that the spectra of I samples, $\mathbf{X}$, had been determined. The concentrations, $\mathbf{C}$, can then be estimated. If the data behaved linearly, as for transmission spectroscopy absorbance in clear solutions, the model could then be

$$\mathbf{X} = \mathbf{C}\mathbf{K} + \mathbf{E} \qquad \text{(B-15)}$$

in which $\mathbf{E} = \{e_{ik}\}$ represents measurement errors. The LS fit, which averages the fitting over all the K wavelengths, would then be

$$\hat{\mathbf{C}} = \mathbf{X}\mathbf{K}'\mathbf{Z}^{-1} \qquad \text{(B-16)}$$

in which the matrix $\mathbf{Z}$ is equal to $\mathbf{K}\mathbf{K}'$ and $\hat{\mathbf{C}}$ means estimated $\mathbf{C}$.

Two new matrix operations are introduced in this formula: transpose $(\ )'$ and inverse $(\ )^{-1}$. These will now be treated.

Transpose of a Matrix, K′ (or K^T in Some Texts)

This means that $\mathbf{K}'$ represents another matrix with exactly the same elements as $\mathbf{K}$, but turned around so that rows become columns and columns become rows. If the matrix $\mathbf{K}$ is the three spectra with K wavelengths, i.e.,

$$\mathbf{K} = \begin{bmatrix} k_{11} & k_{12} & k_{13} \\ k_{21} & k_{22} & k_{23} \\ k_{31} & k_{32} & k_{33} \end{bmatrix} \qquad \text{(B-17)}$$

then $\mathbf{K}'$ is

$$\mathbf{K}' = \begin{bmatrix} k_{11} & k_{21} & k_{31} \\ k_{12} & k_{22} & k_{32} \\ k_{13} & k_{23} & k_{33} \end{bmatrix} \qquad \text{(B-18)}$$

Hence, the matrix product $\mathbf{Z} = \mathbf{K}\mathbf{K}'$ implies collapsing over the wavelengths; so that the result $\mathbf{Z}$ is the 3×3 cross-product matrix:

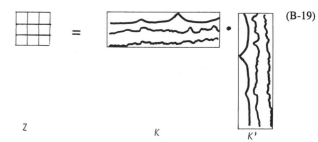

$$\text{(B-19)}$$

Z K K'

Inverse of a Matrix, Z^{-1}

In ordinary algebra, the inverse of a single scalar number z (say $z = 20$) can be written either $1/z$ or $z^{-1} = (1/20 = 0.05)$. One definition of this scalar inverse is that a number z multiplied by its inverse z^{-1} equals 1.

$$z \times z^{-1} = 1$$
$$20 \times 20^{-1} = 1 \qquad \text{(B-20)}$$

Matrix inversion is a straightforward extension of this; the inverse of a quadratic matrix (a matrix with the same number of rows as columns), say $\mathbf{Z}$, is defined such that

$$\mathbf{Z}\mathbf{Z}^{-1} = \mathbf{I} \qquad \text{(B-21)}$$

in which $\mathbf{I}$ is the "unit matrix" corresponding to the number 1 for scalars. For a 3×3 matrix $\mathbf{Z}$, the following could be obtained.

$$\mathbf{I} = \begin{bmatrix} 1 & 0 & 0 \\ 0 & 1 & 0 \\ 0 & 0 & 1 \end{bmatrix} \qquad \text{(B-22)}$$

Given the matrix $\mathbf{Z}$, a computer is required in order to obtain the inverse, $\mathbf{Z}^{-1}$, but this is rather easily done.

However, one problem should be remembered: just like the scalar inverse $1/z = z^{-1}$ is meaningless if $z = 0$, the inversion of a matrix $\mathbf{Z}$ can be meaningless or impossible if the matrix $\mathbf{Z}$ does not contain sufficient "information" for the computer inversion algorithm (its determinant is zero). This happens if the $J \times J$ matrix $\mathbf{Z}$ contains less than J different types of variation. In the example above, this could occur if starch and protein had the same NIR spectra ($\mathbf{k}_3 = \mathbf{k}_2$) or if the protein spectrum was equal to some combination of the water and starch spectra ($\mathbf{k}_3 = \alpha\mathbf{k}_1 + \beta\mathbf{k}_2$). The computer then cannot distinguish between protein information on one hand and water information on the other, and the inversion of $\mathbf{Z} = \mathbf{K}\mathbf{K}'$ becomes impossible. It can then be said that $\mathbf{Z}$ does not have full rank and that $\mathbf{K}$ does not have full row rank; the rows are so-called linearly dependent.

Provided that $\mathbf{K}$ has full row rank, the formula Equation B-16 for estimating the concentrations $\mathbf{C}$ can then be written as

$$\hat{\mathbf{C}} = \mathbf{X}\hat{\mathbf{B}} \qquad \text{(B-23)}$$

in which

$$\hat{\mathbf{B}} = \mathbf{K}'\mathbf{Z}^{-1} = \mathbf{K}'(\mathbf{K}\mathbf{K}')^{-1} \qquad \text{(B-24)}$$

Equation B-23 can also be written without matrix algebra:

$$\hat{c}_{ij} = \sum_{k=1}^{K} x_{ik}\hat{b}_{kj}, \quad i = 1, 2, \ldots, I; \quad j = 1, 2, \ldots, J \qquad \text{(B-25)}$$

How the calibration coefficients $\hat{\mathbf{B}}$ were estimated (Equation B-24), however, is difficult to express in traditional summation notation, while it is easy in matrix algebra, e.g., Equation B-24.

LITERATURE CITED

Beckman, R. J., and Cook, R. D. 1983. Outliers. Technometrics 25:119-163.

Box, G. E. P., Hunter, W. G., and Hunter, J. S. 1978. Statistics for Experimenters: An Introduction to Design, Data Analysis, and Model Building. John Wiley and Sons, New York.

Brown, P. J. 1982. Multivariate calibration. J. Roy. Stat. Soc. B 44:287-308.

Cardone, M. J. 1983. Detection and determination of error in analytical methodology. Part II. Correction for corrigible systematic error in the course of real sample analysis. J. Assoc. Off. Anal. Chem. 66:1283-1293.

Cook, R. D., and Weisberg, S. 1982. Residuals and Influence in Regression. Chapman and Hall, London.

Cornell, J. 1981. Experiments with Mixtures: Design, Models and the Analysis of Mixtures. J. Wiley and Sons, New York.

Damsleth, E., and Spjøtvoll, E. 1982. Estimation of trigonometric components in time series. J. Am. Stat. Assoc. 77:381-387.

Deming, S. N., and Morgan, S. L. 1987. Experimental Design: A Chemometric Approach. Elsevier Publishing Co., New York.

Draper, N., and Smith, H., Jr. 1981. Applied Regression Analysis. 2nd ed. John Wiley & Sons, New York.

Esbensen, K., and Wold, S. 1983. SIMCA, MACUP, SELPLS, GDAM, SPACE and UNFOLD: The way towards regionalized principal components analysis and subconstrained N-way decomposition—with geological illustrations. Pages 11-36 in: Proc. Nordic Symp.—Appl. Stat. Stokkand Forlag Publ., Stavanger, Norway.

Fearn, T. 1983. Misuse of ridge regression in the calibration of a near infrared reflectance instrument. Appl. Stat. 32:73-79.

Geladi, P., McDougel, D., and Martens, H. 1985. Linearization and scatter-correction for near-infrared reflectance spectra of meat. Appl. Spectrosc. 39:491-500.

Gunst, R. F., and Mason, R. L. 1979. Some considerations in the evaluation of alternate prediction equoting Technometrics 21:55-63.

Haaland, D. M., and Easterling, R. G. 1982. Applications of new least-squares methods for the quantitative infrared analysis of multicomponent samples. Appl. Spectrosc. 36:665-673.

Henderson, C. R. 1975. Best linear unbiased estimation and prediction under a selection model. Biometrics 31:423-447.

Hildrum, K. I., Martens, M., and Martens, H. 1983. Research and analysis of food quality. Pages 65-80 in: Control of Food Quality and Food Analysis. G. G. Birch and K. J. Parker, eds. Applied Science Publ., London.

Hildrum, K. I., Martens, H., and Lea, P. 1984. Detection of errors in the NIR-analysis of meat components. Pages 399-400 in: Proc. Eur. Meet. Meat Res. Workers, 30th.

Hoadley, B. 1970. A Bayesian look at inverse linear regression. J. Am. Stat. Assoc. 65:356-369.

Hruschka, W., and Norris, K. 1982. Least squares curve fitting of near infrared spectra predicts protein and moisture content of ground wheat. Appl. Spectrosc. 36:261-265.

Huber, P. J. 1981. Robust Statistics. John Wiley & Sons, New York.

Jensen, S. Å., and Martens, H. 1983. The botanical constituents of wheat and wheat milling fractions. II. Quantification by amino acids. Cereal Chem. 60:172-177.

Jensen, S. Å., Munck, L., and Martens, H. 1982. The botanical constituents of wheat and wheat milling fractions. I. Quantification by autofluorescence. Cereal Chem. 59:477-484.

Joliffe, I. T. 1982. A note on the use of principal components in regression. Appl. Stat. 31:300-303.

Kortum, G. 1969. Reflectance Spectroscopy: Principles, Methods, Applications. Springer Verlag, New York.

Lindberg, W., Persson, J. Å., and Wold, S. 1983. Partial least squares method for spectrofluorimetric analysis of mixtures of humic acid and ligninsulfonate. Anal. Chem. 55:643-648.

Lwin, T., and Maritz, J. A. 1982. An analysis of the linear-calibration controversy from the perspective of compound estimation. Technometrics 24:235-242.

Mandel, J. 1971. A new analysis of variance model for non-additive data. Technometrics 13:1-18.

Mardia, K. V., Kent, J. T., and Bibby, J. M. 1979. Multivariate Analysis. Academic Press, London.

Martens, H. 1981. Qualitative and quantitative morphometry by multivariate image analysis. Chemometric theory, illustrated by a hypothetical example: Near infrared analysis of quality in bacon. Norw. Food Res. Inst. Rep. NR4.

Martens, H. 1985. Multivariate calibration. Quantitative interpretation of non-selective chemical data. Dr. Techn. thesis. NTH.

Martens, H., and Jensen, S. Å. 1983. Partial least squares regression: A new two-stage NIR calibration method. Progress in cereal chemistry and technology. Developments in food science. Pages 607-647 in: Proc. World Cereal Congr., 7th. J. Holas and J. Kratochvil, eds. Elsevier Publ., Amsterdam.

Martens, H., and Martens, M. 2000. Modified jack-knife estimation of parameter uncertainty in bilinear modelling (PLSR). Food Quality Preference 11(1-2):5-16.

Martens, H., and Martens, M. 2001. Multivariate Data Analysis for Understanding Quality. An Introduction. John Wiley & Sons, Ltd., Chichester, United Kingdom.

Martens, H., and Naes, T. 1989. Multivariate Calibration, 2nd ed. John Wiley & Sons, Ltd., Chichester, United Kingdom.

Martens, H., and Stark, E. 1991. Extended Multiplicative Signal Correction and Spectral Interference Subtraction: New Preprocessing Methods for Near Infrared Spectroscopy. J. Pharm. Biomed. Anal. 9(8):625-635.

Martens, H., Jensen, S. Å., and Geladi, P. 1983a. Multivariate linearity transformation for near-infrared reflectance spectrometry. Pages 205-234 in: Proc. Nordic Symp. Appl. Stat. Stokkand Forlag Publ., Stavanger, Norway.

Martens, H., Vangen, O., and Sandberg, E. 1983b. Multivariate calibration of an X-ray tomograph by smoothed PLS regression. Pages 235-268 in: Proc. Nordic Symp. Appl. Stat. Stokkand Forlag Publ., Stavanger, Norway.

Martens, M., Lea, P., and Martens, H. 1983c. Predicting human response to food quality by analytical measurements. The PLS regression method. Pages 185-203 in: Proc. Nordic Symp. Appl. Stat. Stokkand Forlag Publ., Stavanger, Norway.

Martens, H., Westad, F., Foulk, S., and Berntsen, H. 1990. Updating Multivariate Calibration of Process NIR Instruments. Pages 371-381 in: Proc. ISA '90 New Orleans. Instrument Society of America, Research Triangle Park, NC.

Martens, H., Alsberg, B., Foulk, S., and Stark, E. 1991. What Other Spectroscopic Techniques Could Learn from NIR. Pages 221-239 in: Proc. Anal. Appl. Spectrosc. II. A. M. C. Davies and C. S. Creaser, eds. Royal Society of Chemistry, Cambridge, United Kingdom.

Naes, T. 1985a. Comparison of approaches to multivariate linear calibration. Biometrical J. 27:265-275.

Naes, T. 1985b. Multivariate calibration when the error covariance matrix is structured. Technometrics 27:301-311.

Naes, T. 1986. Multivariate calibration by covariance adjustment. Biometrical J. 28:99-107.

Naes, T., and Isaksson, T. 1989. Selection of samples for near-infrared spectroscopy. I. General principle illustrated by examples. Appl. Spectrosc. 43:328-335.

Naes, T., and Martens, H. 1985. Comparison of prediction methods for multicollinear data. Commun. Stat. (Sim. Comp.) 14:545-576.

Naes, T., and Martens, H. 1987. Testing adequacy of linear random models. Report from Norwegian Food Research Institute. Math. Oper. Stat. (ser. Stat.) 78:323-331.

Naes, T., Isaksson, T., Fearn, T., and Davis, T. 2001. A User-Friendly Guide to Multivariate Calibration and Classification. NIR Publications, Chichester, United Kingdom.

Norris, K. 1983. Extracting information from spectrophotometric curves: Predicting chemical composition from visible and near-infrared spectra. Pages 95-113 in: Food Res. Data Anal. H. Martens and H. Russwurm, Jr., eds. Applied Science Publ., New York.

Ott, R. L., and Myers, R. H. 1968. Optimal experimental designs for estimating the independent variable in regression. Technometrics 10:811-823.

Ottestad, P. 1975. Component analysis: An alternative system. Int. Stat. Rev. 43:83-108.

Rabiner, L. R., and Gold, B. 1975. Theory and Application of Digital Signal Processing. Prentice-Hall, Englewood Cliffs, NJ.

Saunderson, J. L. 1942. Calculation of the color of pigmented plastics. J. Opt. Soc. Am. 32:727-736.

Saxberg, B. E. H., and Kowalski, B. R. 1979. Generalized standard addition method. Anal. Chem. 51:1031-1038.

Searle, S. R. 1971. Linear Models. John Wiley & Sons, New York.

Sjöström, M., Wold, S., Lindberg, W., Persson, J. A., and Martens, H. 1983. A multivariate calibration problem in analytical chemistry solved by partial least squares models in latent variables. Anal. Chem. Acta 150:61-70.

Skrede, G., Naes, T., and Martens, M. 1983. Visual color deterioration in blackcurrant syrup predicted by different instrumental variables. J. Food Sci. 48:1745-1749.

Stone, M. 1974. Cross-validatory choice and assessment of statistical prediction. J. Roy. Stat. Soc. B. 36:111-133.

Sundberg, R. 1982. When does it pay to omit explanatory variables in multivariate calibration? Department of Mathematics, Royal Institute of Technology, Stockholm, Sweden.

Vangen, O., and Martens, H. 1982. In vivo estimation of body composition of pigs using computer tomography calibrated by PLS regression. Page 415 in: Food Res. Data Anal. H. Martens and H. Russwurm, Jr., eds. Applied Science Publ., New York.

Westad, F., and Martens, H. 1999. Variable selection in NIR based on significance testing in partial least squares regression. J. Near Infrared Spectrosc. 8:117-124.

Wold, H. 1966. Estimation of principal components and related models by iterative least squares. In: Multivariate Analysis. P. R. Krishnaiah, ed. Academic Press, New York.

Wold, H. 1982. Soft modelling: The basic design and some extensions. Pages 263-271 in: Systems Under Indirect Observations. Causality-Structure-Prediction. Vol. I. K. G. Joreskog and H. Wold, eds. North Holland Publ. Co., Amsterdam.

Wold, S. 1978. Cross validatory estimation of the number of components in factor and principal component models. Technometrics 20:397-406.

Wold, S., Martens, H., and Wold, H. 1983. The multivariate calibration problem in chemistry solved by the PLS method. Pages 286-293 in: Proc. Conf. Matrix Pencils. A. Ruhe and B. Kågström, eds. Lecture Notes in Mathematics. Springer Verlag, Heidelberg, Germany.

Young, F. W. 1981. Quantitative analysis of qualitative data. Psychometrika 46:357-388.

CHAPTER 5

Neural Networks in Near-Infrared Spectroscopy

CLAUS BORGGAARD
Danish Meat Research Institute
Roskilde
Denmark

I. INTRODUCTION

The general purpose in calibrating a near-infrared (NIR) instrument is to replace time-consuming and expensive chemical analyses with a fast and rugged spectroscopic measurement. Traditionally, it is assumed that one can apply Beer's Law, stating that the measured absorbance at specific wavelengths are linearly related to the concentrations of the chemical components of interest within each sample. Accordingly, most of the mathematical techniques applied to the problems in calibrating spectroscopic equipment are based on linear methods. Indeed, methods such as stepwise multiple linear regression (SMLR) and multivariate methods like partial least squares (PLS) and principal component regression (PCR) have proven tremendously successful for creating useful calibrations for NIR spectroscopy, even in cases in which different chemical constituents within the samples are interfering severely (overlapping absorption peaks in the spectra, Chapters 4, 8, and 9).

In most cases, however, the spectra reflect not only the chemical composition of the sample being measured, but also its physical properties such as particle size and temperature. These physical effects give rise to what appears to be deviations from Beer's Law, introducing nonlinear effects through interference. In such cases, one can attempt to remove these nonlinear effects from the spectra through mathematical preprocessing (Chapter 4) before the linear calibration techniques (SMLR, PCR, or PLS) are applied. At the moment, there is no general list of guidelines for which preprocessing method is the appropriate one to use in any specific application for the removal of such nonlinear effects from spectra, as the physics describing scatter phenomena is mathematically and conceptionally a very complex science. For each new application, the NIR spectroscopist will have to find the best preprocessing method through trial and error and, in many cases, a satisfactory method will never be found.

In this chapter, I will discuss an alternative approach to instrumental calibration in the field of NIR analysis, namely artificial neural networks (ANNs).

ANNs are inspired by the way neural systems in living organisms are organized. The human central nervous system, for instance, contains a vast number of neurons (of the order of 10^{11}). Each of these neurons acts as a simple analogue processing unit capable of transforming input signals in a nonlinear way. The neurons are connected to a bunch of other neurons via nerve fibers (axons and dendrites) through which electrical signals can cross from one neuron to another (Fig. 1). This structure is called a neural network. The small gap between axons of the sending neuron and the dendrites of the receiving neuron is called the synapse. The neuron decides whether a received signal (stimulus) is strong enough to be passed on to its neighbors. If the stimulus is too weak, no signal is passed along to its neighbors. If the received signal is sufficiently strong, the neuron will produce a transmitter substance at the tip of its axons that is registered across the synapse by the dendrites of the next neuron. This mechanism of passing along signals between neurons is highly nonlinear. Neurons do not react on any small stimulus that they are subjected to and there is a limit to the amount of transmitter substance that can be produced at the synapses. This kind of response is often referred to as the "all-or-none response."

The number of connections between neurons in the human central nervous system is of the order of 10^{15}. Vast networks like the human brain are, of course, impossible to emulate in a computer program. However, small networks designed to handle specific problems can easily be programmed and have provided solutions for handling simple problems in many fields ranging from artificial sensory systems to systems for predicting trends on the financial markets (Welstead, 1994).

In the following sections, I will discuss two types of ANNs that have been applied with success to problems in the field of NIR spectroscopy—namely feed forward neural networks and Kohonen self-organizing maps (or Kohonen classifiers).

II. FEED FORWARD NEURAL NETWORK TRAINED BY BACK-PROPAGATION OF ERROR

This type of neural network can be used both for creating calibrations for NIR reflectance/transmittance applications that require a high degree of precision and accuracy and for classification purposes.

Feed forward neural networks can be used as an alternative to PLS or PCR. As is the case with PLS or PCR, the neural network model must be trained on a subset of data that is representative of what one can expect to encounter during the life span of the NIR application that is being built. In the case of the feed forward neural network, one starts off with a system of neurons that are interconnected in much the same way as a true neural system in living organisms. Before the training process, the strength of the connections (corresponding to the synapses in the nervous system) are set at small random numbers. The idea is to utilize the adaptiveness of the network in a training process (calibration). One starts by presenting a number of representative examples (spectra) to the network. Hereafter, the strength of the interconnections (called weights) are changed in such a way that the difference between the output from the network and the true answer (given by, for example, chemical analysis) becomes smaller. The same samples are presented to the network over and over again until the error is sufficiently small. After this process, the calibration model is defined by the number of neurons at various levels in the network and by the size of the weight with which data is passed on to neighboring neurons in the network.

III. AN EXAMPLE OF A FEED FORWARD NETWORK

Figure 2 shows a feed forward neural network with three layers of neurons. The top layer is the input layer, and the bottom layer is the output layer. In between these two layers, there is what is known as a hidden layer. In the example shown in Figure 2, the input layer consists of six neurons, the hidden layer contains three neurons, and the output layer contains only one neuron. Figure 2 defines what is called the architecture of the network and could, for example, be useful if there is an NIR instrument that samples an absorbance spectrum at six wavelengths. If measurements are made on, for example, barley, one would most likely be interested in predicting the moisture contents of each of the samples. The trained network in this application would then be expected to give the predicted moisture content of each sample at the output neuron. If one wants to predict both water and protein in the barley samples, one could choose to have two neurons in the output layer, one for each constituent, or have two separately trained networks, one specialized to predict moisture and the other to predict protein (Caruana, 1994).

The ANN shown in Figure 2 contains a number of connections that pass data from the input layer to the output layer. In this case, the number of connections between the input layer and the hidden layer is $6 \times 3 = 18$, and the number of weights from the hidden layer to the output neuron is 3. Additionally, there is the option of implementing connections that pass on information directly from the input layer to the output layer—in this case, six direct connections corresponding to six extra weights have been utilized. Notice in Figure 1 that the network also contains a bias neuron. This neuron always passes on a number equal to 1 and is connected to all neurons in the hidden layer and to the output neuron, thus adding $3 + 1$ additional weights to the network model.

In the training process, all these 31 weights will have to be estimated if the application is to perform well.

In order for the network to describe nonlinear phenomena in the relationship between spectral recordings and the chemical composition of the samples, a nonlinear transformation of data is introduced at the hidden layer. In certain cases, it may be advantageous also to implement this feature at the output layer. But for problems involving regression, this is usually not used. The most commonly used nonlinear transformation in feed forward neural networks is given by the so-called sigmoid function shown in Figure 3.

IV. THE DATA FLOW IN THE FEED FORWARD NETWORK

Initially, all 31 weights in the example network are assigned small random numbers (e.g., between −0.25 and +0.25).

Calibration objects (spectra plus target reference values for moisture) are presented to the untrained network one at a time. If the absorbance readings from the example NIR instrument at the six different wavelengths are called $a_1, a_2, ..., a_6$, one can calculate the predicted output, $\hat{y}$, from the network in Figure 2 as

$$\hat{y} = \sum_{i=1}^{6} a_i u_i + \sum_{j=1}^{3} \sigma\left(\sum_{i=1}^{6} w_{ij} a_i + b_j\right) v_j + B \qquad (1)$$

Here,
- sigma is the sigmoid function (Fig. 3) defined as

$$\sigma(x) = \frac{1}{1 + \exp(-x)}$$

- u_i is the weight of the direct connection from input neuron i to the output neuron;
- w_{ij} is the weight of the connection passing data from neuron i in the input layer to neuron j in the hidden layer;
- v_i is the weight of the connection from neuron i in the hidden layer to the output neuron;
- b_i is the size of the bias weight from the bias neuron (output = 1 at all times) to neuron i in the hidden layer; and
- B is the size of the bias to the output neuron.

Notice that Equation 1 can be easily modified to handle networks with more than a single hidden layer.

The most important feature about Equation 1 is that it is analytical, and the predicted value $\hat{y}$ is written as an explicit function of all the weights. This means that one can calculate the partial derivatives of $\hat{y}$ with respect to any one of these weights—a feature that is utilized for tuning the weights in the training algorithm for ANNs.

V. TRAINING THE NETWORK—TUNING THE WEIGHTS

A number of algorithms exist for adjusting the weights in a neural network for a specific application (Lippmann, 1987; Welstead, 1994). The choice of which of these to use often depends on whether speed of training or prediction accuracy in the final network has priority. Here, the most commonly used method, called back-propagation of error (Lippmann, 1987), will be presented. This method, which is often referred to simply as back-prop, is also the most appropriate for spectroscopic applications in which accuracy is of the greatest importance for regression problems.

The correctness of a calibration model is defined by the term standard error of calibration (SEC) as follows.

$$\text{SEC} = \sqrt{\frac{\sum_{i=1}^{N} (y_i - \hat{y}_i)^2}{N}}$$

Here, the sum is taken over all the objects (training examples) that have been chosen for the calibration data set. N is the number of examples, $\hat{y}_i$ is the predicted reference value for spectra i, and y_i is the true reference value for the corresponding object as determined by, for example, chemical analysis.

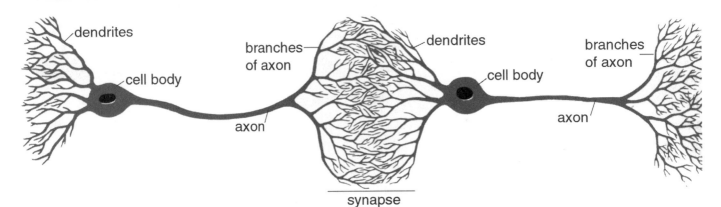

Fig. 1. A neural network.

During the training process, the neural network is repeatedly presented with a number of examples, e.g., NIR spectra and corresponding true values that serve as target values for the prediction. The untrained network, with weights that have initially been assigned random values, will give out prediction results according to Equation 1, leading to very large residuals in $\hat{y}$.

The aim of the training process is thus to find the set of weights in Equation 1 that minimize the SEC. In the back-propagation algorithm, the square of the SEC as a so-called cost function for adjusting the weights is used.

$$\text{COST} = \text{SEC}^2 \times N = \sum_{i=1}^{N} \left(\hat{y}_i - y_i \right)^2$$

With the neural network shown in Figure 2, this cost function defines a 31-dimensional surface in the 32-dimensional space spanned by all the weights in Equation 1 plus the cost itself. This is illustrated in Figure 4, in which the number of weights have been limited to only two.

Through an iterative process, one now wants to change the values of all the weights in Equation 1 so the cost function is reduced. Here, one can utilize the fact that the predicted value $\hat{y}$ has been written as an explicit function of all the weights. This enables us to calculate the partial derivatives of $\hat{y}$ with respect to each of the individual weights w_{ij}, u_i, v_i, b_i, and B (Equation 1) that form the gradient of the cost function.

If the total weights vector is defined as $\mathbf{W} = (w_{11} ...., u_1 ....., v_i ..., b_i ..., B)$, one can write the gradient as

$$d\text{COST}/d\mathbf{W} = \left(\delta\text{COST}/\delta w_{11},...,\delta\text{COST}/\delta u_1,...,\delta\text{COST}/\delta v_1,...,\delta\text{COST}/\delta b_1, \delta\text{COST}/\delta B \right)$$

The gradient of the cost function is a vector that at any given location on the multidimensional surface in the weights space points in the direction of steepest ascent.

The back-propagation training algorithm is then as follows.

1. Initialize all weights w_{ij}, v_i, u_i, b_i, and B to some small random number and introduce a new variable called the sum gradient vector.
2. Set the sum gradient vector to zero.
3. Present n examples (NIR spectra) to the network, yielding a predicted value $\hat{y}$ for each by applying Equation 1. Calculate the new gradient sum vector as

$$\textbf{sum gradient} = \textbf{sum gradient} + d\text{COST}/d\mathbf{W}$$

for each one of the n examples.

4. Modify the weights vector according to

$$\mathbf{W} \rightarrow \mathbf{W} - \text{lambda} \times \textbf{sum gradient}$$

(lambda = learning rate; shown below).

Repeat steps 2–4 for a preselected number of iterations or until the SEC has reached a satisfactorily low value.

By subtracting the **sum gradient** from $\mathbf{W}$ itself in step 4, one is, on average, moving in a "downhill" direction on the cost function surface. However, if one takes too long of strides at a time,

one might miss a useful minimum along the way. Or, if the cost function surface at any given time during the training process is rather flat, one might want to increase the step size. To accommodate for this, the factor lambda is introduced in step 4. Lambda is referred to as the learning rate.

It is important that the n samples chosen in step 3 are randomly selected from the calibration set. The examples should never be presented in order of, for example, increasing analyte concentration, unless the value of n in step 3 is chosen equal to the number of examples, N, in the calibration set. Otherwise the training will result in oscillations around the general direction in which the weights are being changed.

If a value of $n = 1$ is chosen in step 3, one will, in general, move rapidly toward the minimum on the cost surface. But when getting near the minimum, it will often be advantageous to average the gradient over a number of training examples (or maybe the whole calibration set) before modifying the weights in step 4. This is useful if the training data is noisy, e.g., the precision of the reference method is poor or if it is known that one is close to the minimum being looked for on the cost surface.

One can also choose to introduce what is called momentum in the training process (Lippmann, 1987). This is done by not resetting the **sum gradient** to zero after each modification of the weights in step 4. Instead, let the old value of the sum contribute, say, 80% to the new **sum gradient** in the next run-through of the training data. This will have the effect of averaging out effects of noisy data and tend to give a smoother search for the optimal weights and eliminate oscillations in the descent.

After training, the resulting neural network model will have to be validated in the same way as any of the more orthodox methods such as SMLR, PCR, or PLS. If the prediction results for an independent data set prove unsatisfactory, one will have to try a different network architecture (change the number of neurons in

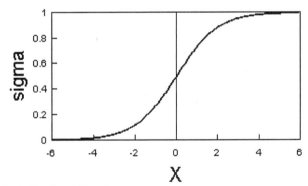

Fig. 3. The sigmoid function.

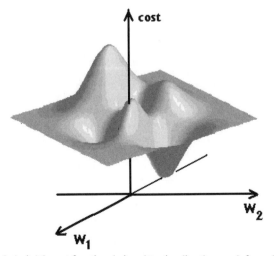

Fig. 4. A simple cost function designed to visualize the search for a global minimum in a neural network using the back-propagation algorithm.

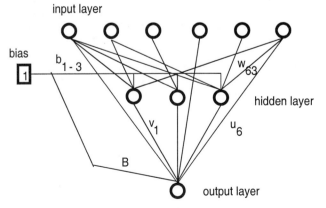

Fig. 2. A feed forward neural network with three layers of neurons.

the hidden layer) or check and see whether the training data is representative of the test set and so on. After this, the network can be trained once more and tested.

VI. HOW TO PRESENT DATA
TO THE NEURAL NETWORK

NIR spectra are often composed of several hundred reflectance or transmittance measurements at different wavelengths on each sample. If one chooses to retain all these variables as inputs for the neural network for training, the number of weights that must be estimated will be near astronomical. This leaves us with two problems, one small and the other rather severe.

The small problem is that the training time, even on a very fast computer, will be quite long (maybe several hours depending on the complexity of the relationships that have to be learned by the network).

The major problem is the curse of dimensionality—meaning that there are too many parameters in the model and too few examples with which to adjust these parameters. With neural networks, this will present a problem if there is a lot of noise in the calibration data. During training, the very adaptive ANN will develop into a model that fits not only the general trends in the relationship between spectra and reference values (these are the values one wants it to learn), but also the random noise that is present in the data. The result of this will be a beautifully low SEC. When this model is tested on independent data, one gets a rather poor standard error of prediction (SEP). This corresponds closely to what is seen in SMLR if too many wavelengths are chosen or in PLS and PCR if too many factors are used in the model.

In most of the applications that have been published, in which neural networks have worked successfully on calibration problems in NIR spectroscopy, the curse of dimensionality has been solved by training the networks on principal component analysis (PCA) or PLS scores (Chapter 4) instead of the raw spectral absorbance values (Gemperline et al, 1991; Long et al, 1990; Borggaard and Thodberg, 1992).

This has the following advantages.
- PLS and PCA retain all relevant information from each NIR spectrum in a few (less than 20) principal components. This means that almost any NIR spectrum (unless it solely consists of random noise) can be perfectly represented by, for example, the scores for the first 10 PCA or PLS components.
- Experience has shown that it is much faster to train a neural network when the input variables are uncorrelated, as is the case with PCA or PLS scores, than to train a network on raw spectral data in which the variables are highly correlated.
- PCA and PLS act as efficient filters against random noise in the spectral data. If the number of factors that are calculated is much less than the number of objects in the calibration set, random noise will only be represented by higher factors in the data compression model.
- A neural network trained on PCA or PLS scores will be quite small, meaning faster training times and a reduction of the chances of overfitting the training data.
- The use of a PLS or PCA data compression step enables one to retain all the advantages of these methods regarding outlier detection.

PLS regression and PCR can be looked upon as data compression followed by a linear regression of the scores on to the reference values. When the PLS or PCA scores are used as input to the network, one only substitutes the linear regression part of these methods with a nonlinear adaptive neural network model.

It is also important to scale the input vectors prior to presenting them to the untrained neural network. This can be done in a number of ways, but typically one can chose to scale each column (PCA or PLS component) in the calibration set to either unit variance or to lie between −1 and +1. In the case of PCA or PLS scores being used, the values will always be centered around zero.

But if raw spectral data is used, one will usually also subtract the average over each variable in the calibration set from each data point in that column. This scaling ensures that all variables before the training influence the network on an equal basis. It is then left to the back-propagation algorithm to find the appropriate weights with which to pass on information from each of the input variables. The reference values should also be scaled in the same way as the inputs. The advantages are that the B weight in Equation 1 will be close to zero (meaning that it will only have to be adjusted slightly during training) and that the training parameters (e.g., learning rates) are nearly the same for almost any data set.

VII. MONITORING THE TRAINING PROCESS

A test set validation can be introduced in the process of training the neural network. This feature serves the same overall purpose as when applied to PLS or PCR in which the performance on the independent test set is used to determine the optimal number of factors to use in the calibration. For training neural networks, an independent test set can be used to monitor the ability of the neural network to generalize from what it has learned, enabling you to stop the training process before the network starts to overfit the training data.

The procedure for this monitoring procedure is as follows.
1. Train the neural network using the back-propagation algorithm on N samples from the training set.
2. Predict the reference value for each object in the test set and calculate the SEP for the test objects. If the value of SEP has been reduced, save all weights in the computer's memory.

Steps 1–2 are to be repeated until the SEP on the test set starts to increase over a large number of iterations of the learning process. The weights of the neural network with the best ability to generalize can be retrieved from memory storage.

As in the case of using test set validation in PLS and PCR, the neural network model will be dependent on the choice of test set objects.

It is, therefore, of utmost importance to select the test set objects with great care. If the PLS or PCA scores are used as input for the network, it is essential that the test set objects have not been part of the PLS or PCA model, because one is testing not only the neural network, but also the data compression model.

Alternative methods of validation in which all available data is utilized in the training process have been developed. A detailed presentation of some of these methods is given by Thodberg (1996).

For a thorough comparison between the performance of feed forward networks and nonlinear PLS methods see the work by Ding et al (1999).

VIII. THE FEED FORWARD NETWORK USED
FOR CLASSIFICATION

The feed forward neural network can also be used for classification purposes. When using the feed forward network as a classifier, one will often choose to have more than one output neuron.

Each of these output neurons is connected to the same hidden layer, thus sharing the same weights from input layer to hidden layer. Ideally, there should be just as many output neurons in the network as there are known classes in the calibration data set. Each of the output neurons is set to react for only one specific class of objects. This type of network is trained on data in which the reference values are presented as vectors with all zeros, except for the component in the vector representing the class to which the input spectra belongs being assigned a value of 1.

This way of presenting data to a feed forward classifier is very similar to the way in which one would present data to a classifier based on PLS-2.

When introducing more output neurons, the cost function that is used will no longer be a scalar function but an array of functions (one function for each output neuron).

Using the feed forward network as a classifier has two disadvantages.

- The network must be trained using supervised learning, which means that one has to know, prior to the training stage, which class each sample belongs to.
- If a new class of objects appears when the trained network is being used routinely as a classifier, the network will not know which output neuron to fire (give a value close to one). Implementing an outlier detection system similar to the one used in standard PCA and PLS models can, of course, solve this problem. Treating the outputs from the hidden layer as standard PCA or PLS scores and comparing these for each new sample with the training set data is an obvious solution.

However, one could also consider using another type of neural network for classification purposes. In the following, a quite different network type, designed to cope with at least the first of the above mentioned disadvantages, will be presented.

IX. KOHONEN SELF-ORGANIZING MAPS

As is the case with PLS and SMLR, feed forward neural networks trained by the back-propagation algorithm are trained using supervised learning. This means that the untrained neural network must have a set of examples together with the correct answers for each of these in order to improve its ability to give useful predictions on new examples. The weights are adjusted during the training process by using the reference values as a teacher. PCA, on the other hand, is trained by unsupervised learning. In PCA, the principal components are deduced solely on the basis of the structure of the input variables. This means that no a priori knowledge of the samples is required for the data compression model.

In the 1980s, Teuvo Kohonen developed a neural network, often referred to as the Kohonen self-organizing map, which, like PCA, is trained by unsupervised learning (Kohonen, 1988, 1989). Recently, a number of papers (Cáceres-Alonso et al, 1995; Remolà et al, 1996) have been published in which this type of neural network is compared with PCA and Soft Independent Modeling of Class Analogy (SIMCA) (Massart et al, 1988) methods for classification purposes in NIR spectroscopy.

X. THE ARCHITECTURE OF THE KOHONEN NETWORK

Usually the Kohonen network consists of a two-dimensional (square of arbitrary side length) lattice of neurons. This lattice is referred to as the Kohonen layer. When applied to NIR spectroscopy, the inputs (spectra) are fed to a one-dimensional array of neurons, one for each wavelength being monitored. Each lattice point (neuron) in the Kohonen layer is associated with a weight vector that has the same dimension as the input array. This is schematically shown in Figure 5.

In Figure 5, there is an *n*-dimensional input vector that could consist of an NIR absorbance spectrum sampled at *n* wavelengths or its first *n* PCA scores. Here, the size of the Kohonen layer is a 4×4 lattice of neurons—and only the input vector $\mathbf{W}_{ij}$ is shown. One must imagine that all of the other 15 neurons in the Kohonen layer are connected to the input array in the same way. When the input vector is presented to the network, the 16 neurons in this Kohonen layer compete among each other for which of them has the weight vector with the closest resemblance to the input vector. In this competition, there is only one winner. However, second prizes may be awarded to the winner's neighbors. Input spectra that are similar to each other will all give a hit at the same neuron. During a training process, the weights of the winning neuron will be adjusted to look even more like these input vectors (spectra). The weights of the neighboring neurons will also have their weights adjusted to match this input pattern, but the adjustment will not be as strong as for the winner. The next time a similar spectrum is presented to the network, the winner and its neighborhood will be even more sensitive to this type of spectra. If the

input data consists of spectra from five different types of samples during the training process, one will see five clusters evolve in the Kohonen layer in such a way that samples that are very different will be placed very far from each other and spectra from closely related samples will be placed in the vicinity of each other. Thus, the Kohonen network maps the input spectra onto a two-dimensional surface (the Kohonen layer), preserving the important topological features of the multidimensional input space.

XI. A TRAINING ALGORITHM FOR KOHONEN NETWORKS

The training algorithm for Kohonen networks is very simple compared with the back-prop algorithm used in the feed forward artificial neural network described previously, and it is computationally very light.

The algorithm is described in the following steps.

1. Often the input vectors are scaled to unit length (optional, but in some cases necessary).
2. Initialize all the weights to the Kohonen layer to some small random number (e.g., between 0 and 0.2).
3. Present an example to the input layer as input vector *a*.
4. Find the weights vector that most resembles the input pattern. This is done by calculating the Euclidean distance between the input vector and the weights vector for each lattice point in the Kohonen layer and then locating the lattice point with the smallest distance. One thus finds the set i',j' for which the following relation is true for all i,j (rows and columns in the Kohonen layer).

$$\left| a - w_{i'j'} \right| \leq \left| a - w_{ij} \right|$$

5. Adapt the weights in the neighborhood of the winning neuron at lattice point i',j' in the following way.

$$w_{ij} \rightarrow w_{ij} + \lambda_{iji'j'} \left(a - w_{ij} \right)$$

in which λ is a parameter (small number) that serves the same purpose as the learning rate in the back-propagation algorithm. It is the product of two separate quantities—the step size ε and a function that decreases as one moves away from the winning neuron in the Kohonen layer.

$$\lambda = \varepsilon \exp \left[-\frac{(i-i')^2 + (j-j')^2}{\alpha} \right]$$

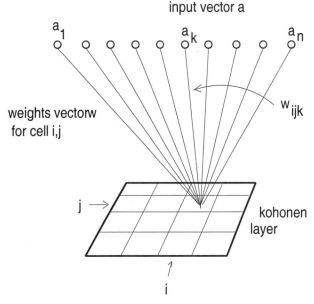

Fig. 5. A schematic of each lattice point (neuron) in the Kohonen layer associated with a weight vector that has the same dimension as the input array.

A

5XS$_2$	5X G$_2$			COD		5X M$_1$ 4X M$_2$ 5X G$_1$
			COD	COD		
				COD	COD	
						M$_2$
		5XS$_1$				
20X Ol						5X Paraf 5X Palm

B

		5XG$_2$			3XM$_1$	2XM$_1$ 5XM$_2$
5XS$_2$						
			5XCOD			5X G$_1$
5X Ol$_3$						
			5XS$_1$			
5X Ol$_4$	5X Ol$_1$ 5X Ol$_2$					5X Paraf 5X Palm

C

		5XG$_2$		5XM$_1$		5XM$_2$
5XS$_2$						
			5XCOD			5X G$_1$
5X Ol$_3$			5XS$_1$			
5X Ol$_2$						5X Palm
5X Ol$_4$	5X Ol$_1$					5X Paraf

Fig. 6. The Kohonen classifier **A,** at 50 iterations; **B,** at 500 iterations; and **C,** at 1,100 iterations.

The Gaussian function in the above equation effectively adapts the weights of the lattice points in the neighborhood of the winning neuron i',j' according to their distance from the winner. The strength of this adaptation falls off as one moves further away from the winning neuron in the Kohonen layer.

Steps 3–5 are repeated with new samples randomly selected from the total data set. The neighborhood size, α, and the step size, ε, are parameters whose values have to be selected at the beginning of the training together with the size of the Kohonen layer. Their values are gradually reduced during the training process, which can be stopped when the Kohonen classifier contains clearly separated clusters.

The weights of this network can be saved and used to classify new objects (input vectors) in order to determine which cluster it belongs to.

In the following, an example will be presented in which the NIR spectra from a number of cooking oils are to be classified. The instrument was a fixed-filter instrument measuring at nine wavelengths in the wavelength range of 800–1,700 nm, and it also monitored the three color coordinates x, y, and z that are used in color video cameras. Spectra were taken in transflectance of four types of olive oil, two types of corn oil, two types of grape seed oil, palm oil, paraffin oil, cod liver oil, and two types of salmon oil (the last three of these can hardly be considered cooking oils).

In order to test the repeatability of the measurements and the sampling presentation technique, five spectra were taken of each oil type over a period of 2 days. The total data set thus consisted of 65 spectra and was used to test a Kohonen self-organizing map like the one shown in Figure 5. Initially, the value of α was set equal to 6 and the value of ε was set to 0.6. The Kohonen network was presented with 1,100 examples, all drawn randomly from the same 65 samples in the data set. During these 1,100 iterations, the value of the neighborhood parameter, α, was reduced linearly to 1 and the value of the step size was likewise reduced to 0.05.

In Figure 6A–C, the winner neurons in the Kohonen classifier for each sample are shown at three different stages of the training process. In Figure 6A–C, the oils are labeled olive Ol$_1$–Ol$_4$, corn M$_1$–M$_2$, grape seed G$_1$–G$_2$, palm, paraf, salmon S$_1$–S$_2$, and cod.

At the early stages of the training process, one sees that the Kohonen classifier cannot differentiate between the four types of olive oil. Neither can it distinguish between the two types of corn oil. Palm oil and paraffin oil are also classified as being one and the same product. However, after 1,100 iterations, there is a complete separation of the 13 different products.

XII. NEURAL NETWORKS— ADVANTAGES AND DISADVANTAGES

Before applying neural networks to a calibration or a classification problem, it is recommended to try standard methods such as SMLR, PLS regression, PLS-2, or PCR. These methods will, in any case, serve as a benchmark for the performance of the trained network on test set objects.

The following gives a short list of both disadvantages and advantages of applying ANNs to calibration problems in spectroscopy.

A. Disadvantages

1. A good deal of intuition/experience is needed in choosing an appropriate architecture for a given calibration or classification problem. A number of parameters must be chosen before the network is trained: network architecture, learning rate, and momentum.

2. The curse of dimensionality exists (Borggaard and Thodberg, 1992). When training a neural network on a given data set, a large number of parameters (weights) have to be optimized, e.g., in the simple neural network in Figure 1, there are 31 weights that have to be estimated. If there is not a large number of objects in the training set (more than 100), the neural network is so adaptive

that it very likely will overfit the training data and develop a network with a poor ability to generalize on new test data.

3. If the networks are very large (in NIR spectroscopy one can easily need several hundred input neurons), the computing time needed for the training process becomes very long. This makes it cumbersome to utilize efficient procedures such as cross-validation that ensure against overfitting the training data.

4. The neural network is a "black box" in the sense that it is difficult to understand how the calibration works by just looking at the size of the weights and the chosen architecture.

5. Neural networks need large data sets for calibration—be careful with neural networks trained on less than $3n$ samples, in which n is the total number of weights in the neural network.

B. Advantages

1. Neural networks are adaptive and possess the ability to model almost any nonlinear relationships.

2. Preprocessing of the data is not nearly as crucial for neural networks as for linear calibration methods, e.g., there does not have to be a linear relationship between spectral response and chemical composition in the samples (it goes without saying that any calibration method will benefit from an intelligent pretreatment of the spectral data).

3. In cases in which one wishes a calibration that is very general, the ANN is certainly a very powerful tool. In, for example, grain analysis, this can be the case if one wishes to have a calibration that can accommodate for regional and year-to-year variability. This can also be the case in the analysis of meat, in which it is practical to have a calibration that works equally well on a wide range of products (Borggaard and Thodberg, 1992; Borggaard, 1994). With linear methods like PLS and PCR, it is often necessary to split a broad calibration into several piecewise linear models that are spliced together. Each of these models is then valid for predicting values of the reference constituent only within a certain concentration range. This results in a calibration containing discontinuities that are very difficult to eliminate. However, the performance of a neural network is much improved if it has been trained on samples representing all the variability in the products to be analyzed as the underlying nonlinearities then become more obvious.

4. In some cases, neural networks simply work wonders!

The work by Zupan et al (1997) has more details on Kohonen and feed forward networks. In the article by Duponchel et al (1999), the very interesting topic of instrument standardization in relation to ANNs is discussed. The neural networks apparently compete very well with more traditional methods of instrument standardization.

XIII. CONCLUSIONS

ANNs are a new tool in NIR spectroscopy and should be placed in the spectroscopists or chemometricians toolbox along side of linear methods such as PLS, PCR, SMLR, and various procedures for classification and discriminant analysis. The neural network approach should only be used in the event that the linear methods do not give satisfactory results (if possible—keep it simple). If neural networks are trained correctly, they will never give worse results that the linear methods and, in some cases, they will work wonders. Examples of the later can be found in various works (Gemperline et al, 1991; Cáceres-Alonso et al, 1995; Borggaard and Thodberg, 1992).

LITERATURE CITED

Borggaard, C. 1994. Modelling non-linear data using neural networks regression in connection with PLS or PCA. Spectrosc. Eur. 6(3):21-27.

Borggaard, C., and Thodberg, H. H. 1992. Optimal minimal neural interpretation of spectra. Anal. Chem. 64(5):545-551.

Cáceres-Alonso, P., and García-Tejedor, A. 1995. Non-supervised neural categorisation of near infrared spectra. Application to pure compounds. J. Near Infrared Spectrosc. 3:97-110.

Caruana, R. 1995. Learning many related tasks at the same time with backpropagation. Pages 657-664 in: Advances in Neural Processing Systems, Proc. Conf. 1994. G. Tesauro, D. S. Touretzky, and T. K. Leen, eds. MIT Press, John Wiley & Sons, London.

Ding, Q., Small, G., and Arnold, M. A. 1999. Evaluation of nonlinear model building strategies for the determination of glucose in biological matrices by near-infrared spectroscopy. Anal. Chim. Acta 384(3):333-343.

Duponchel, L., Ruckebusch, C., and Huvenne, J. P. 1999. Standardization of near-infra-red spectrometers using artificial neural networks. J. Near Infrared Spectrosc. 7(3):155-166.

Gemperline, P. J., Long, J. R., and Gergoriou, V. G. 1991. Nonlinear multivariate calibration using principle components regression and artificial neural networks. Anal. Chem. 63:2313-2323.

Kohonen, T. 1988. An introduction to neural computing. Neural Networks 1:3-16.

Kohonen, T. 1989. Self-Organization and Associative Memory. 3rd ed. Springer, Berlin.

Lippmann, R. P. 1987. An introduction to computing with neural nets. IEEE ASSP Magazine April:4-22.

Long, J. R., Gergoriou, V. G., and Gemperline, P. J. 1990. Analytical chemistry, spectroscopic calibration and quantisation using artificial neural networks. Anal. Chem. 62:1791-1797.

Massart, D. L., Vandeginste, B. G. M., Deming, S. N., Michotte, Y., and Kaufman, L. 1988. Chemometrics: A Textbook. Elsevier, Amsterdam.

Remolà, J. A., Lozano, J., Ruisánchez, I., Larrechi, M. S., and Rius, F. X. 1996. New chemometric tools to study the origin of amphorae produced in the Roman empire. Trends Anal. Chem. 15(3):137-150.

Thodberg, H. H. 1996. A review of Bayesian neural networks with an application to near infrared spectroscopy. IEEE Trans. Neural Networks 7(1):56-72.

Welstead, S. T. 1994. Neural Networks and Fuzzy Logic Applications in C++. Wiley Professional Computing, Wiley, New York.

Zupan, J., Novic, M., and Ruisánchez, I. 1997. Kohonen and counter-propagation artificial neural networks in analytical chemistry. Chemometrics Intelligent Lab. Syst. 38:1-23.

CHAPTER 6

Near-Infrared Instrumentation

W. F. McCLURE
North Carolina State University
Raleigh, NC
U.S.A.

I. INTRODUCTION

Although the theory of infrared (IR) spectroscopy has been known for over 200 years (Herschel, 1800a, b, c), it was not until the 1930s that the first custom-made instruments appeared in industrial laboratories. World War II encouraged the expansion of IR technology as an analytical tool, especially for the analysis of rubber and petroleum products. In the 1940s, the first commercial IR spectrometers became available, and IR spectrometry began significant growth. The first double-beam instrument was sold in 1947 (Archer, 1969).

Over the next 10 years, classical spectropists avoided the near-infrared (NIR). Working with fundamental absorption bands, many of which fell within the mid-IR spectrum of 2.5–15 micrometers (μm), researchers in those days were convinced that overtones and combination absorption bands occurring in the NIR were of little consequence. Several things then changed.

Development of precise manufacturing techniques for fabricating detectors (primarily lead sulfide detectors), the availability of better power supplies, and the advent of the minicomputer revived interest in NIR spectroscopy. Commercialization of NIR spectrometry was joined with ultraviolet/visible (UV/VIS) spectrometry; the Perkin-Elmer model 450 (Perkin-Elmer, Inc., Norwalk, CT) and the Cary 14 UV/VIS/NIR (Varian Instruments, Palo Alto, CA) spectrometers were among the earliest instruments to address NIR spectrometry. NIR did not become a part of IR instrumentation largely because of the prejudice that existed in the IR community at the time and because silicon detectors spanned the visible and part of the NIR regions.

In the 1950s, Karl Norris, an agricultural engineer with the U.S. Department of Agriculture in Beltsville, MD, began to investigate optical properties of dense-light-scattering materials (Norris and Butler, 1961). By 1969, his work led to the development of a computerized NIR spectrometer for analyzing spectra of meat (Ben-Gera and Norris, 1968). Shenk and Hoover (1976) and McClure and Hamid (1980) followed Norris' leadership and, by the mid-1970s, had designed and built their own computerized NIR systems for the analysis of forages and tobacco, respectively.

Performance of NIR instrumentation depends, in large part, on the components that are used in fabrication. The components, unfortunately, are nonlinear in their response to the wavelength of energy involved. Mirrors do not reflect all wavelengths the same way. Combining many components into an NIR spectrometer results in an instrument with very nonlinear characteristics. Fortunately, most of the nonlinearities are removed in the computation of transmittance (T) or reflectance (R).

This chapter deals with many of the nonlinearities associated with the various components that are integrated into a dispersion-type NIR spectrometer. Of course, there are other ways to generate and decode wavelength information (viz, acousto-optical tunable filters [called AOTFs] or diode arrays). Yet, the design principles undergirding the design of a spectrometer are basically the same. Therefore, this discussion is based upon the design and fabrication of the COMP/SPEC I, a custom-built spectrometer. Two COMP/SPECs were built, one for the Bioinstrumentation Laboratory at North Carolina State University, Raleigh, and the other at the Oxford Tobacco Research Station in Oxford, NC. Hardware (the two COMP/SPECs) and software (called CSAS) for the acquisition and analyses of NIR spectra were developed by members of the Bioinstrumentation Laboratory from 1976–1995. These two COMP/SPECs formed a foundation upon which other NIR instruments were developed (Sugiyama et al, 1992; Morimoto et al, 2001; W. F. McClure, *unpublished data*; W. F. McClure and D. L. Stanfield, *unpublished data*; W. F. McClure C. M. Hargrove, M. Zapf, and D. Stanfield, *unpublished data*).

II. COMPONENTS OF NIR SYSTEMS

A. Lenses and Mirrors: Collecting Radiation

Lenses and mirrors used in IR instruments must often be made of special materials, since glasses normally used for the visible region are almost opaque to radiation of wavelengths longer than approximately 2 μm. Figure 1 shows the variation of transmission with wavelengths of various materials along with the reflection of aluminum first-surface material.

Pyrex, although economical, has noticeable absorptions at 1.37 and 2.2 μm, with a diminishing transmission starting at 2.0 m to almost 10% at 2.8 μm. This material is useable, but fused quartz is much more superior. Aluminum first-surface mirrors are acceptable for NIR work, but silver coatings are far superior.

There are two figures of merit that describe optical condensers used with source and dispersion systems. The principal figure of merit is the focal ratio, relative aperture, or, more commonly, the *f*-number, which is defined as

$$f \text{ number} = \frac{f}{\phi} \tag{1}$$

in which f = focal length of the lens and ϕ = clear aperture diameter. The *f*-number is a measure of the light-gathering power (efficiency) of a lens system; the smaller the *f*-number, the greater the light-gathering power of the lens. The theoretical maximum relative aperture of an aplanatic lens is *f*/0.5, but this is practically unobtainable (Melles Griot, 1985).

Figure 2A shows the geometrical relationship between a source L and an *f*/1 condenser lens. Note here that in order to have an *f*/1 condenser

$$F = D = 2R \tag{2}$$

Therefore, γ = arc tan 0.5 = 0.4637 radian. Since, α = 0.5 radian, then

$$R' = R = 0.9274R \tag{3}$$

The area intercepted by one steradian on the surface of a sphere is

$$A_s = (2R)^2 \tag{4}$$

Hence, the area intercepted by the *f*/1 lens on the spherical surface with a diameter R' is

$$A_f = (R')^2 \tag{5}$$

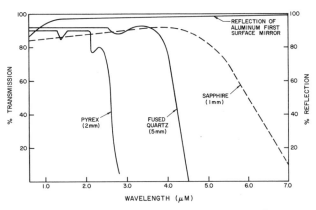

Fig. 1. Transmission/reflection of five optical materials: Pyrex, fused quartz, sapphire, and an aluminum first-surface mirror. Fused quartz and aluminum first-surface mirrors satisfy the requirements of near-infrared spectroscopy.

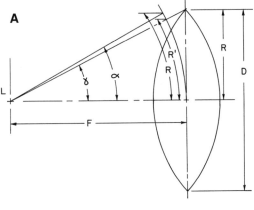

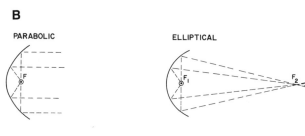

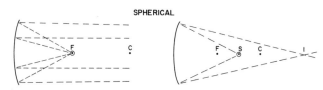

Fig. 2. Optical characteristics of **A,** an *f*/1 lens and **B,** first-surface mirrors. An *f*/1 lens collects only 5.4% of the total energy emitted by a light source. Elliptical and spherical mirrors are recommended when the focal ratio is smaller than *f*/1.

Hence, the total steradians intercepted by an *f*/1 lens is

$$\Omega = \frac{(0.9274R)^2}{4R^2} \tag{6}$$

$$\Omega = 0.6755 \tag{7}$$

The equivalent solid angle is an important aspect of the relative aperture of a condenser lens because lamp irradiance is usually given in watts per steradian. Here, it is seen that an *f*/1 lens would collect only a small portion of the total energy emitted from a point source, or

$$\% \text{ collected} = \frac{1}{12.57 \text{ steradian/sphere}} \times 0.6755 \times 100 \tag{8}$$

$$\% \text{ collected} = 5.4\% \tag{9}$$

This illustrates one means of increasing the illumination available to a monochromator from a given source. However, in order to achieve *f*-numbers less than 1.0, the designer should resort to parabolic or ellipsoidal mirrors (Fig. 2B) (Birth, 1969).

Degree of collimation of a lens system is defined (Birth, 1969) by the size of the source and the focal length of the condenser (Fig. 3A), or

$$D = \frac{\bar{d}}{2f} \tag{10}$$

in which $\bar{d}$ = source size and f = lens focal length. Note that as D becomes smaller, divergence of the beam is smaller (i.e., the collimation is better). Since the filament of sources used in a monochromator are rectangular in shape (to match the shape of the entrance slit of the monochromator), the degree of collimation in the horizontal plane differs from the degree of collimation in the vertical plane.

To increase the power of the source, the dimensions of the source must also increase. Thus, in order to maintain a high degree of collimation, a long-focal-length condenser is necessary (Fig. 3B). Large diameter optics are required to secure a low *f*-number. Therefore, users of NIR systems cannot increase the energy output of a monochromator simply by using a higher-powered source. Extra energy comes only by widening the entrance and exit slits or by replacing the current input optics with optics having a lower *f*-number.

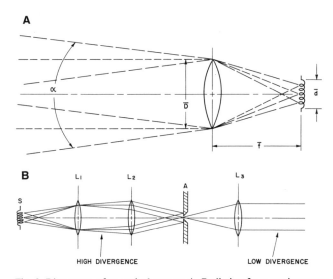

Fig. 3. Divergence of an optical system: **A,** Radiation from a point source located at the focal point of a condenser will yield a parallel beam of radiation, $\overline{D}$. **B,** Since real sources are not point sources but have finite dimensions, the resultant beam has finite dimensions. In instruments in which low divergence is required, an aperture *A* must be used. In dispersion systems, *A* takes the form of a slit that also determines, at the exit, the bandpass of the spectrometer.

B. Radiation Sources

A good radiation source for NIR spectroscopy should have the following characteristics.

1. The beam should emit a continuum of radiation over the range of 0.9–2.6 μm.
2. The intensity of the source should be high enough that extensive signal conditioning (amplification, cross-correlation circuitry, etc.) is not needed.
3. The overall size of the lamp should be small in order to take advantage of low *f*-number optics.
4. The filament of the lamp should be compact, having a projected area equivalent to the slit of the spectrometer.
5. The envelope of the lamp should contain iodine vapor that reduces envelope darkening and extends the useful life.

6. The source should not fluctuate over long periods of time or flicker over short periods of time.

Tungsten-halogen lamps, such as those coded DXM and FAD by the American Standards Association (Fig. 4A), meet the above criteria. Any tungsten lamp with a quartz bulb is satisfactory, but tungsten-halogen lamps provide a longer life and more stability because of the cleaning action of the halogen. Stabilized DC power supplies, rather than AC power supplies, are used extensively to provide radiation that varies little with time. In single-beam systems like COMP/SPEC (McClure and Hamid, 1980), it is extremely important that this power supply be of the highest quality possible, having good regulation and low ripple. Operating the lamps from a 60-Hz source is not recommended, unless the data-recording circuitry is synchronized with the AC source. The irradiance (μW/cm^2 per μm at a distance of 42 cm) of a typical DXM lamp is shown in Figure 4B. The lamp output peaks at 0.9 μm (6.72 μW/cm^2) and tapers off to 1.04 μW/cm^2 at 2.6 μm. The lamps also provide sufficient radiation for operation over the visible spectrum down to 0.4 μm without extensive signal conditioning of the detector signal. Emission (Table I) varies little from lamp to lamp. To date, these lamps are unsurpassed as economical sources of radiation for NIR spectroscopy (Birth, 1969).

Not all the energy of a lamp can be made available at the entrance slit of the monochromator. Assuming *f*/1 entrance optics, the energy available from a 200-W tungsten-halogen lamp is 5.5 μW/cm^2/nm. The area of one steradian at 43-cm radius is (43 cm)^2. Using a nominal bandpass of 10 nm and recognizing that the solid angle of an *f*/1 lens is 0.676, we see that the energy available at the monochromator becomes (Birth, 1971)

$$E_m = \left(5.5 \ \mu W/cm^2/nm\right)\left(43 \ cm\right)^2\left(10 \ nm\right)\left(0.676\right) \quad (11)$$

$$= 68.74 \ mW/10\text{-nm bandpass} \quad (12)$$

TABLE I
Spectral Irradiance
of Three Quartz-Iodine-Tungsten-Filament Lamps[a]

Wavelength (nm)	Lamp QL-2	Lamp QL-5	Lamp QL-10
250	0.0051	0.0052	0.0051
260	0.0093	0.0093	0.0090
270	0.0158	0.0159	0.0155
280	0.0253	0.0252	0.0244
290	0.0380	0.0380	0.0369
300	0.0545	0.0548	0.0532
320	0.104	0.105	0.102
350	0.237	0.242	0.234
370	0.366	0.374	0.363
400	0.643	0.647	0.630
450	1.26	1.26	1.23
500	2.04	2.04	2.02
550	2.93	2.96	2.91
600	3.88	3.94	3.88
650	4.79	4.91	4.80
700	5.54	5.72	5.58
750	6.11	6.32	6.14
800	6.51	6.69	6.49
900	6.72	6.94	6.71
1,000	6.51	6.73	6.53
1,100	6.07	6.25	6.11
1,200	5.53	5.67	5.55
1,300	4.97	5.09	4.98
1,400	4.44	4.52	4.44
1,500	3.93	4.00	3.93
1,600	3.46	3.51	3.45
1,700	3.03	3.06	3.01
1,800	2.63	2.65	2.61
1,900	2.29	2.28	2.26
2,000	1.98	1.97	1.95
2,100	1.73	1.71	1.70
2,200	1.52	1.51	1.50
2,300	1.36	1.34	1.33
2,400	1.22	1.21	1.21
2,500	1.12	1.10	1.11
2,600	1.04	1.03	1.04

[a] Source: Stair et al (1963).

A

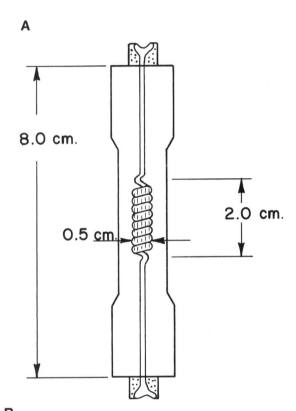

B

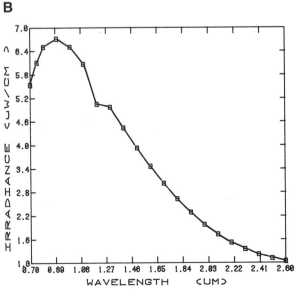

Fig. 4. **A,** Physical size and **B,** spectral irradiance of a DXM tungsten-halogen lamp.

Hence, we can see that only a small fraction of the energy ever reaches the entrance to the monochromator—in this case, only 0.034%.

C. Monochromators

The function of a monochromator is to disperse, or spread out, the radiation according to wavelength. Prisms and gratings are the most popular dispersion elements used in monochromators. Neither type is perfect and the user must take trade-offs in order to achieve an optimum optical system.

There are two types of gratings: the classical ruled-plane grating and the concave holographic grating. The ruled-plane grating has a series of triangular-shaped parallel grooves cut into some solid material (such as glass) with a diamond-shaped tool mount-

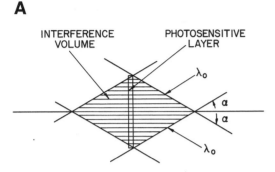

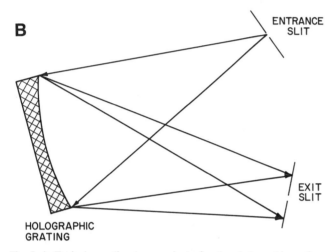

Fig. 5. **A,** Typical recording (or grooving) of a plane holographic grating using laser/etching techniques. **B,** A simplified monochromator using a single concave holographic grating. In this case, the grating, the only inside optical component, provides a very linear dispersion at the exit slit.

TABLE II
Comparison of Ruled and Holographic Gratings[a]

Property	Classically Ruled Gratings	Holographic Gratings
Efficiency	At blaze 60–99%	At maximum 35–99%; efficiency curve can be made flatter than ruled gratings to cover wider spectral domain; in blazed holographic gratings, efficiency curve is identical to ruled gratings
Ghosts	Approximately 10^{-5} of line	No ghosts
Scattered light	At best 10^{-5}–10^{-6} at 5 Å of laser line in the visible	10^{-6}–10^{-8} at 5 Å of laser line in the visible
Size	Generally limited to 8 × 8 in. for groove spacings of more than 600 g/mm	Up to 17-in. diameter in all groove spacings; much larger size, up to 1 m, is possible
Groove density	Maximum of 3,600 g/mm; in general, stray light increases drastically with groove density	Up to 6,000 g/mm; stray light does not increase with groove density
Aberration correction	Not possible	Possible

[a] Source: Hayat (1978).

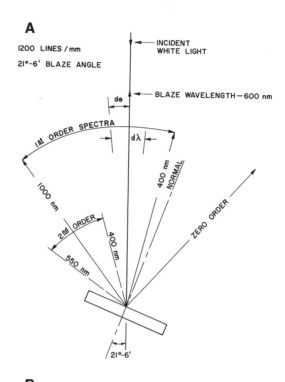

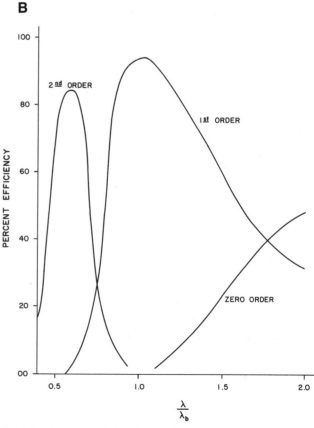

Fig. 6. Grating characteristics: **A,** spectral distribution and **B,** efficiency. The first-order spectrum is the one usually used due to its higher efficiency, broad-wavelength coverage.

ed on a ruling engine. One of the major problems in making a ruled-plane grating is the inaccuracies, or errors, made with the mechanical engine, which results in "ghosts" and "grass" (stray light).

The holographic grating was developed to overcome many of the imperfections in ruled-plane gratings. The term *holographic* refers to the method used for producing this type of grating. Two intersecting laser beams are used to produce interference fringes in a photosensitive material deposited on optically flat glass (Fig. 5A). The interference fringes recorded on the photosensitive material are then developed, thus creating triangular-shaped grooves. Holographic technology has improved to the point in which precision grooves can be made on concave plates, making it possible to produce "flat-field" monochromators without the use of auxiliary optics (Fig. 5B). The simplicity of the optical and mechanical parts of such systems has made possible the manufacturing of simple, high-throughput, narrow-bandpass, low-stray-light spectrophotometers at costs commensurate with filter-based systems. Table II compares the ruled and holographic gratings.

A problem with gratings, which does not exist with prisms, is that light with several wavelengths leaves the grating at the same angle of dispersion (Fig. 6A). This phenomenon is called *overlapping orders*, and the orders vary in their efficiencies (Fig. 6B). Therefore, when using any kind of grating system, filters or prism monochromators must be employed to eliminate the overlapping orders. For example, if a grating is used to scan the NIR region of 1.0–2.5 µm (Fig. 7), the spectrum beginning at 1.0 µm would appear at 2.0 µm as the second-order spectrum. Thus, a high-pass cutoff filter inserted at 1.8 µm would eliminate the second-order spectrum. The third-order spectrum starting at 3.0 µm is of no consequence, since scanning stops at 2.5 µm due to lack of sensitivity of the lead sulfide (PbS) detectors.

Prism monochromators may be used in conjunction with gratings as order sorters. The system is then called a *double-monochromator* system. Figure 8 is an optical diagram of the Cary 17 monochromator. McClure and Hamid (1980) built the COMP/SPEC around a Cary 17 monochromator. Note from Figure 8 that the radiation passes through the prism on its way to the grating. Hence, only a very narrow band of wavelengths reaches the grating (i.e., the orders are sorted out by the prism). The stray-light specification of the Cary 17 is as good as, if not better than, any available (less than 0.1% at 2.65 µm). Unfortunately, Varian Instruments no longer manufacturers this monochromator. However, there are a number of the Cary 14 and 17 instruments still in use and, when these are retired from service, the monochromators would provide the basis for a COMP/SPEC design.

The characteristics of monochromators vary from one design to another. If monochromator characteristics were constant and a single-beam recording of relative lamp energy versus wavelength was made, then the resultant curve would be a function only of the lamp output and detector response curves. Experimentally, this is not found to be the case, as monochromator characteristics contribute significant spectral features to the output.

The reciprocal dispersion (nm/mm) of any monochromator is not constant but varies from one end of the range to the other. For instance, Figure 9 is a plot of the reciprocal dispersion for the

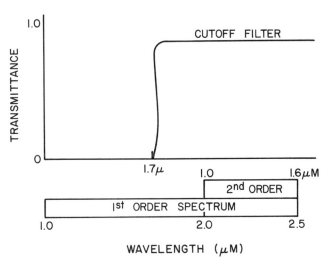

Fig. 7. Overlapping of the orders of a near-infrared grating. A cutoff filter can be used to block the second-order spectrum.

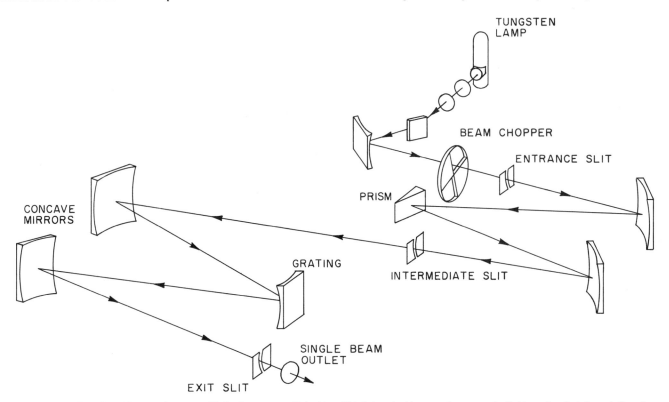

Fig. 8. Diagram of the Cary 17 monochromator (Varian Instruments, Palo Alto, CA). It is a double monochromator, the light passing first through the prism monochromator and then reflected from the grating. Thus, the prism removes the higher-order spectra.

Cary 17. Over the NIR range of 0.9–2.6 μm, the reciprocal dispersion varies from 3.8 down to 2.8 per millimeter of slit width. Therefore, if the monochromator is operated over this range with a fixed 2-mm slit width, the half-intensity bandwidth (HIBW) of the radiation exiting the monochromator will vary from 7.6 nm at 0.9 μm to 5.6 nm at 2.6 μm. The bandpass can be held constant by programming the slit function into the controlling computer.

Observed responses of monochromators are influenced by Wood's anomalies, absorption characteristics of other optical elements (lenses, mirrors, etc.) both inside and outside the monochromator, and water vapor in the radiation path. Figure 10 shows an experimental response curve obtained for a Cary 17 in the single-beam mode. The grating for this system was blazed at 1.6 μm. *Blazing* a grating increases the energy efficiency of the grating over a wide range of angles on both sides of the angle at which the diffracted energy is maximized, which is called the *blaze angle*. The

blaze angle C is the angle between the normal to the broad face of the groove and the grating normal (Fig. 11). A grating can be blazed at an angle specified by the purchaser to maximize efficiency in any part of the spectrum. For example, if the operator intends to work mainly in the region near 2,200 nm, the grating can be blazed at a wavelength in this area. A blazed grating will diffract (break up) up to about 85% of the incident energy at angles close to the blaze angle, and it will maintain high (up to 50%) efficiency at angles of up to 15–20° beyond the blaze angle in the first-order spectrum.

There are only two Wood's anomalies in the NIR, at 0.959 and 1.259 μm. *Wood's anomalies* are spurious absorbances, usually fairly sharp spikes, that are associated with any grating monochromator operating in the single-beam mode and covering a wide wavelength range. They are caused by redistribution of the energy of the primary order among the other orders, including the one leaving the exit slit, and depend partly on the material from which the grating, and particularly its surface, is made. Most gratings are made from glass, with an aluminum or silver coating layer, into which the grooves are cut. More recently, the coating is a photographic film. Silver has a strong absorption band at about 300 nm and is not recommended for gratings intended for measurements in the UV area.

There is to date no way of avoiding Wood's anomalies completely, although the grating can be blazed so that they occur in an area that is of lower importance to the operator. This may cause some loss in efficiency of the grating, but it is important to effect a compromise to the extent that the Wood's anomalies do not interfere with the measurements to be made. For example, measurements made in the area of 1,259 nm may be affected. The sharp dips at 1.38, 2.2, and 2.5 μm are the result of O-H absorption by traces of water vapor in the air path and water in the silica optics. The Cary 14 with Infrasil optics or the Cary 17 with Superasil W optics minimizes the effects of this water absorption. All of these

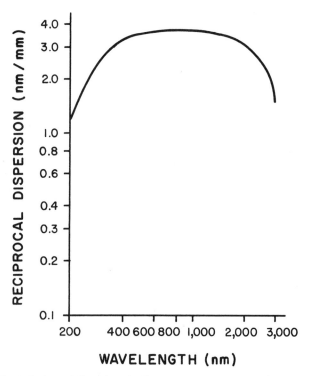

Fig. 9. Reciprocal dispersion of a Cary 17 monochromator (Varian Instruments, Palo Alto, CA). Note that the dispersion is not constant over wavelength but varies from 1.1–3.7 nm/mm.

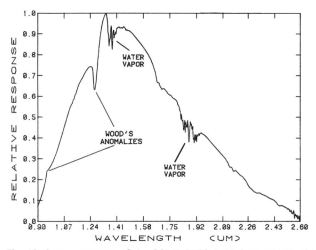

Fig. 10. System response of the COMP/SPEC I (McClure and Hamid, 1980). This response curve includes nonlinearities of the source, detector and monochromator optics. Wood's anomalies and water vapor absorption bands are also present.

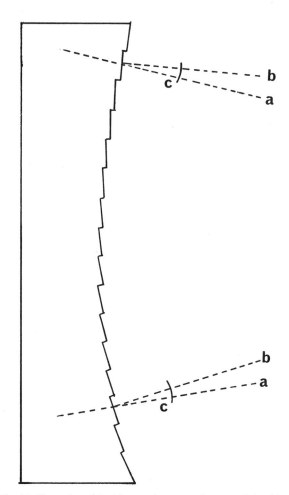

Fig. 11. Illustration of the blaze angle. a = grating normal, b = blaze normal, and c = blaze angle.

anomalies occur only in the single-beam system response curves and are corrected by the computer in the computation of reflectance or log(1/R). If the above anomalies do appear in the reflectance or log(1/R) curves, the system is not performing properly. Gratings, mirrors, detector (PbS), and lamp (tungsten-halogen) all contribute to the nonlinearity of a NIR system (Fig. 10).

D. Filters

The first instrument designed to measure reflected light used filters to define the wavelength range (Taylor, 1919). As the demand for narrow-band interference filters grew, so did the technology of filter manufacturing so that narrow-band filters for almost any region of the optical spectrum can be obtained today.

FIXED FILTERS

The quality of narrow-band interference filters is determined by the percentage of transmission at the peak wavelength, the half-intensity bandpass, and upper (UCO) and lower cutoffs (LCO) (Fig. 12). Filters suitable for NIR instrumentation should have an HIBW of 10–20 nm and a peak transmittance of at least 30%.

Without supplementary filtering, narrow-band interference filters do not have LCO and UCO much beyond one wave band. This phenomenon is similar to grating monochromators (discussed above). Maxima of a given filter with a fundamental maximum at λ_0 (referred to as the first-order bandpass) will also occur at $\lambda_0/2$, $\lambda_0/3$, etc. For example, if the longest wavelength at which maximum intensity occurs at 2.0 μm (λ_0) maxima will also occur at 1.0 μm. 667 μm, etc. The other orders may have the same peak transmission but always will have a narrower or steeper bandpass. LCO filters, used in the same way they are used in grating monochromators, are needed to eliminate these spurious bands.

In some cases, it may be desirable to make use of these narrower and steeper orders. For example, if the second order fell at 1.5 μm, the first order would be at 3.0 μm, and the third order at 1.0 μm. The influence of the first-order bandpass could be ignored since the PbS does not respond in this region, but the third-order band would have to be masked in order to sense only the second-order wavelengths. If the third order is used to isolate the 1.5-μm band, the second order appears at 2.75 μm, and the fourth order at 1.125 μm. This increasing proximity of the side bands acts as a limitation regarding the order of interference that can be made use of for practical purposes.

The literature records many applications of the planar-fixed filter configuration. Birth and Norris (1965) used two filters mounted in a rotating wheel to study the quality of horticultural products; different pigment systems could be studied by changing the filters. McClure (1968, 1969b) designed a two-filter moisture meter for determining the water content of tobacco leaves in the transmission mode; Figure 13 is a simplified diagram of this instrument. Energy from a tungsten lamp passed alternately through two filters (1.82 and 1.93 μm) and through a tobacco leaf. The energy reaching the detector was decoded

with photodiodes, and the difference in the logarithm of the two detector signals was computed for scaling and display on the panel meter.

An obvious disadvantage of the fixed-filter arrangement is that the number of filters that can be mounted on a given filter wheel is limited. However, there is evidence that indicates that the precise positioning of the bandpass of the filters is not necessary to get good estimates of sample chemistry (McClure, 1969a).

VARIABLE FILTERS

Variable interference filters (sometimes called wedge interference filters) can be made by pressing the interference material into the shape of a wedge between two pieces of glass. Norris (1964) constructed a spectroradiometer using a linear wedge interference filter for the range of 0.4–1.2 μm. This instrument was later used (Downs et al, 1964) for measuring irradiance in plant growth chambers.

Circular variable interference filters are available for the NIR range (Birth, 1971; Corion Corp., 1983). These are made in circular segments, each segment covering only a portion of the NIR region. However, with the advent of the holographic grating, both linear and circular variable filters became too expensive for most applications. Much better resolution with lower stray light is obtainable with the gratings and at lower costs.

TILTING FILTERS

Filter parameters are generally expressed with the incident beam collimated and normal to the surface of the filter. When the filter is tilted from the perpendicular, or the beam is other than normal to the surface of the filter, the bandpass broadens, its center wavelength shifts toward shorter wavelengths, and its peak transmission is reduced (Fig. 14). The extent of these changes is ultimately

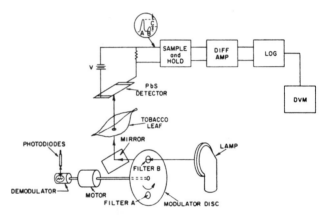

Fig. 13. Schematic/block diagram of a near-infrared moisture meter for tobacco, a dual interference filter design.

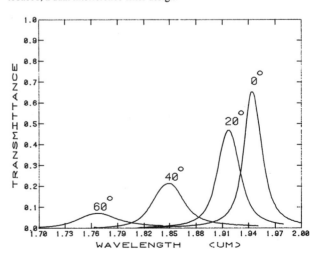

Fig. 14. Effect of tilting a narrow-band interference filter from normal to the light beam. Note that the peak wavelength shifts toward shorter wavelengths (regardless of the directions of tilt from normal) and that the half-intensity bandwidth broadens.

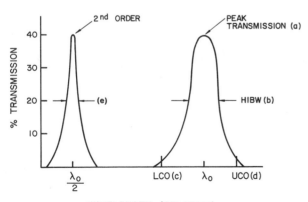

Fig. 12. Characteristics of narrow-band interference filters: (a) peak transmission, (b) half-intensity bandwidth (HIBW), (c) lower cutoff (LCO), (d) upper cutoff (UCO), and (e) second-order bandpass.

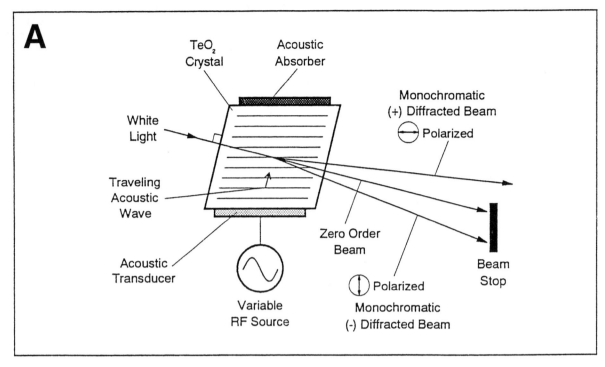

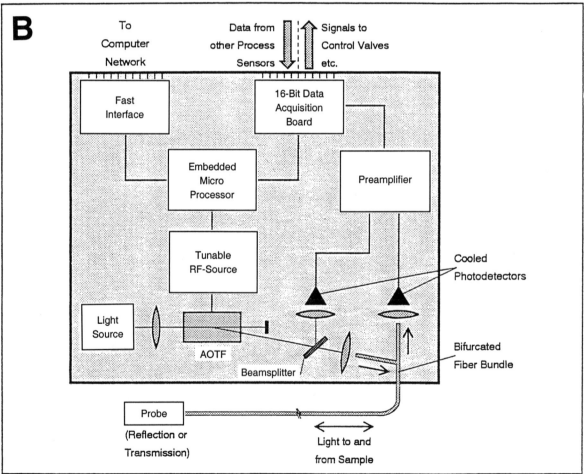

Fig. 15. Simplified schematic of **A,** an acousto-optical tunable filter (AOTF) and **B,** an AOTF-based spectrometer for measuring the reflectance or transmittance of solid samples. **B,** The AOTF crystal is physically small compared with grating monochromators. The wavelength output is selected by changing the radio frequency applied across the crystal. Hence, wavelengths may be selected quickly and in any order desired. (Source: Brimrose Corp., Baltimore, MD.)

determined by the number of degrees of tilt from the normal angle of incidence and/or the cone angle of the beam. The formula for determining the shift of the center wavelength with the angle (≤30°) of incidence is

$$\lambda_\phi = \lambda_0 \sqrt{\frac{n^2 - \sin^2 \theta}{n}} \qquad (13)$$

in which λ_ϕ = center wavelength at the angle, λ_0 = center wavelength at normal incidence, θ = incident angle in degrees, and n = effective index of refraction of the total filter.

Of course, the narrower the bandwidth of the filter, the more noticeable are the effects of the angle of tilt. Although the wavelength shift and bandwidth change can generally be theoretically predicted by use of the formula above, exact wavelength and bandwidth changes for a given angle of incidence and/or cone angle are best determined empirically for each individual filter, as in Figure 14.

In electro-optical systems, the most common off-normal situation encountered is one in which the radiation source is incident to the filter surface in either a convergent or divergent configuration. The effect encountered in such an arrangement is substantially less than that produced by collimated off-normal rays of equivalent angles.

AOTFs. Technology for generating monochromatic energy for spectrometry was enhanced considerably by the introduction, in the 1990s, of the AOTF—a no-moving-parts monochromator (Brimrose Corp., Baltimore, MD). The crystal of choice for the NIR appears to be the tellurium dioxide (TeO_2) birefringent crystal (Fig. 15A) used in a noncollinear configuration in which the acoustic and optical waves move through the crystal at different angles. The crystal is carefully cut in order to make it acoustically resonate in response to radio frequency (RF) vibrations. Acoustic waves, generated by an RF source driving a piezoelectric transducer attached to one side of the crystal, change the refractive index of the material. If energy from a tungsten-halogen source (white light) is focused onto one side of the crystal, the resulting output from the opposite side are two monochromatic beams, one vertically polarized and the other horizontally polarized, diverted out of the zero-order beam. Both diverted beams have the same wavelength that is mathematically related to the velocity and frequency of the acoustic wave, the birefringence of the crystal, and a parameter computed from physical dimensions of the crystal. A beam stop blocks the unused beams as indicated in Figure 15B.

Brimrose Corp. indicates that the bandpass of the diverted beams can be as narrow as one nanometer full-width-half-maximum (FWHM) transmission and transmission efficiencies can be as high as 98% with the input intensity divided equally between the two polarized beams. Another feature of the AOTF is the ease of adjusting intensity by simply varying the power of the RF energy applied to the piezoelectric crystal. Published specifications for the TeO_2 crystal used as an NIR spectrometer include (i) having wavelength purity (out-of-band transmissions) as low as 1×10^{-5}, (ii) having guaranteed wavelength repeatability error less than ±0.05 nm for TeO_2, (iii) being rugged with no moving parts, (iv) having high-speed scanning or "snap-to" wavelength selection, and (v) being easily interfaced to and controlled by a computer. The "snap-to" feature of an AOTF crystal is particularly useful in an NIR spectrometer. Once calibrations are developed, it becomes sufficient to read only a few wavelengths (perhaps ≤5) in order to compute composition.

In spite of the apparent advantages, AOTF does have limitations, some very severe. For example, the design of the crystal forces an angular input with a narrow field of view (low *f*-number). This makes the AOTF crystals less attractive for applications in which the optical information (or signal) is weak, particularly for imaging spectroscopy. Increasing the bandpass of the crystals beyond 10–20 µm is difficult, if not impossible, thus excluding

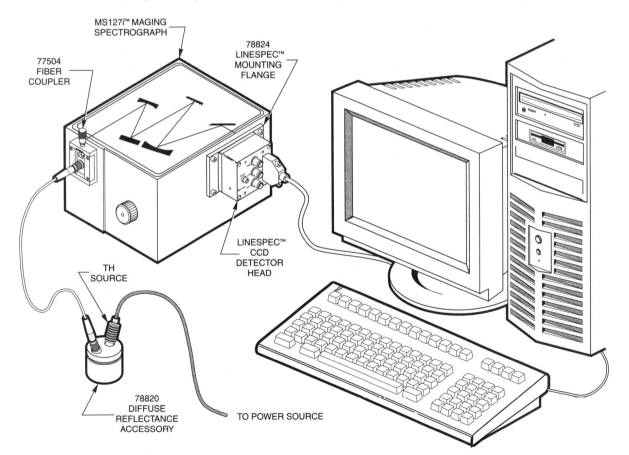

Fig. 16. Typical diode-array spectrometer utilizing fiber optics to collect reflectance spectra of solids. With a fixed grating, the dispersed spectrum is captured by a diode array. The resolution of the spectrum is fixed by the number of elements in the diode array. (Used with permission of Thermo Oriel, a Thermo Electron business, Stratford, CT.)

many broad-band applications. Research with multichannel Bragg cells is an attempt to overcome these obstacles.

Diode Arrays. Fabrication of silicon-based sensors in linear arrays resulted in the development of the diode array spectrometer (DAS). A DAS is a no-moving-parts spectrometer (Fig. 16). Once it is calibrated for wavelength, both the grating and the diode array are fixed with respect to each other. The system can be packaged in small boxes and spectrometers can be made portable.

The multichannel feature of diode arrays is probably its most important advantage over single detectors, like the photomultiplier. Arrays come with hundreds of detectors making it possible to acquire an equal number of data points (or wavelength data points) in a single readout of the array. In a DAS, one readout of the array constitutes a spectrum. Furthermore, numerous spectra (readouts of the array) can be made in fraction of a second. Consequently, the DAS is excellent for monitoring dynamic spectral changes of a sample entity. Two good practical examples of this is monitoring the cross-linking of proteins in surimi during heating or changes in fermentation systems.

Diode arrays differ dramatically from single detectors in the way they measure optical energy. Photomultipliers, for example, measure radiant flux (watts), while diode arrays measure radiant energy (joules). Each diode element accumulates charge in a "well" developed by the impinging photons. If the energy is constant, the accumulated charge is proportional to the radiant flux times the exposure time. This single features makes the diode array applicable to low energy measurements by simply exposing the array for long periods of time. Basically, the diode array acts as an electronic sensing device requiring that the exposure time be adjusted to obtain the best signal.

The most popular diode arrays (and least expensive) are silicon-based devices. Their spectral sensitivity is specified to be between 180–1,180 nm, but their useful range is more like 200–1,050 nm, where the sensitivity drops to 20% of the peak response at 700 nm.

Dark current is a major problem with the diode array. Once triggered to take a reading, each element will accumulate charge even when in the "full dark" and will eventually saturate if the accumulation (integration) time is long enough. This becomes a critical problem if the diode array is required to make measurements of low levels of radiant energy. In practice, at room temperature, the dark current charge is always subtracted from the charge developed. The dark current effect can be minimized by thermoelectrically cooling the array. If the arrays can be cooled to about 5°C, the integration time can be as long as 4 min.

In NIR spectrometry, the wavelength resolution is determined by the spatial resolution (or physical separation) of the individual elements in the array. Niquist sampling theory dictates that the spatial resolution of an array is about twice the width of a single element. Hence, a 1,024-element array would have a spatial resolution of 50 μm.

Diode array spectrometers are input-only devices. That is, energy from the sample (either in transmittance or reflectance mode) enters the spectrometer, where it is dispersed by a grating to fill the diode array. The output of each element of the array constitutes a measure of the radiant energy at a given wavelength; reading all the elements constitutes a spectrum. All the energy entering the DAS is absorbed internally; there is no exit slit, as is the case for the monochromator.

The fact that the energy entering the diode array is the measured beam is a distinct disadvantage. For example, it is not easy to collect and focus diffuse reflectance energy into the DAS. Use of first-surface mirrors to collect diffusely reflected energy is, at best, cumbersome and inefficient. However, diode array spectrometers are readily interfaced to fiber optics for sensing of both reflectance and transmittance energy. Furthermore, the ends of fiber optics can be formed to the shape of the entrance slit on the DAS, greatly improving throughput efficiency.

E. Detectors

IR detectors may be characterized by three basic parameters: their spectral response, their speed of response, and the minimum amount of radiant power they can detect. Other parameters, such as the temperature of operation, magnitude and type of signal produced, and the recommended circuit that best couples the detector to a readout device, are of less importance.

IR detectors may be classified into two broad categories according to principle of operation: thermal detectors and photon detectors. Thermal detectors have been available since the early 1800s. Their principle of operation involves the absorption of IR energy by a temperature-sensitive surface. Generally, the response time of thermal detectors is too long (approximately 10 msec) to be considered for NIR applications. One unique advantage of thermal detectors is their uniform sensitivity at all wavelengths.

Although thermal detectors have been around since the early 1800s, photon detectors are relatively new, having been developed during the years 1960–1995. The operation of photon detectors requires that the incident radiation be strong enough to

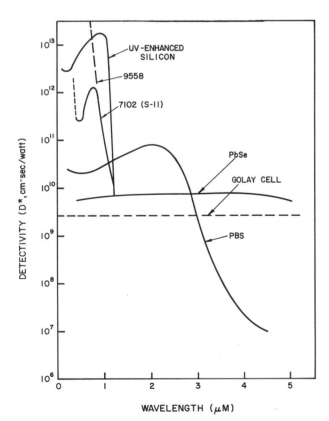

Fig. 17. Detectivity (*D**) of the more popular near-infrared (NIR) detectors at room temperature. These curves were obtained at a chopping frequency of 90 Hz with an amplifier bandwidth of 1 Hz. The lead sulfide detector has adequate detectivity for NIR work.

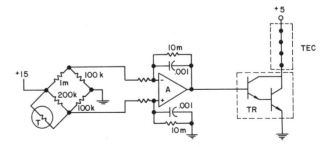

Fig. 18. Thermoelectric cooling control circuit for the lead sulfide detectors in the COMP/SPEC I (T Fenwal GA51P32, 100 K 25°C [Newark Electronics, Chicago, IL]; A Signetics 5556 [Newark Electronics]; TR Motorola HEP-S9102 [Newark Electronics]; and TEC four Borg-Warner 930-17 thermoelectric element [Borg-Warner Thermoelectrics, Chicago, IL])

liberate charge carriers from either the crystal lattice (intrinsic detectors) or from impurities intentionally added to the host crystal (extrinsic detectors).

The minimum detectable power, called *noise equivalent power* (NEP), is usually measured by placing a black body of known temperature (usually 500°K) at a convenient distance from the detector. Its radiation is interrupted periodically (1,000 Hz) by means of a rotating sector disk, such that a detector signal of constant frequency is obtained. Calculating the radiant power received at the detector, the detector noise is measured, usually with a narrow-band amplifier, when the detector is shielded from the radiation source. The NEP is then calculated from the relationship

$$\text{NEP} = \frac{P}{S/N} \tag{14}$$

in which P = power received by the detector, S = signal voltage, and N = noise voltage.

The more commonly used term for characterizing the sensitivity of IR detectors is detectivity, D, in which

$$D = \frac{S/N}{P} \tag{15}$$

or, in comparison of detectors, a normalized quantity, $D*$, which is the detectivity of a detector of a 1-cm^2 area whose noise is reduced to that obtained with an amplifier of 1-Hz bandwidth, or

$$D* = \frac{S/N}{P_D}\left(\frac{f}{A}\right)^{1/2} \tag{16}$$

in which f = amplifier bandwidth, A = detector area, and P_D = the power density at the detector. The spectral curves for the more popular NIR detectors are shown in Figure 17.

Both the 7102 and the 9558 photomultiplier tubes (PMTs) (Hamamatsu Corp., Middlesex, NJ) have higher $D*$ than PbS, but both have limited operating ranges. In the COMP/SPEC, for example, the operating range of the 9558 is 0.4–0.85 μm. UV enhancement of silicon (Longerich and Ramaley, 1974) also improves its performance in the NIR, but its range of operation in the COMP/SPEC is limited to 0.55–1.15 μm. PbS can be operated over the entire region from 0.5–4.0 μm with a $D*$ that is only slightly larger than lead selenide (PbSe), but the characteristics of other elements in the system (i.e., lamps and monochromator) limits its usefulness to 0.9–2.6 μm.

Silicon, unlike PbS, is not especially sensitive to temperature change and need not be thermoelectrically cooled, especially when operated in an air-conditioned laboratory. From 0.4–0.95 μm, the deviation of absolute spectral responsivity with temperature is less than 0.5% per degree centigrade.

Cooling can be accomplished using thermoelectric elements. The PbS detector arrangement, described later, is cooled with four Borg-Warner 930-17 thermoelectric cooling elements (Borg-Warner Thermoelectrics, Chicago, IL). The circuitry for controlling temperature is given in Figure 18. This circuit controls the detector temperature to within ±0.01°C for a laboratory temperature variation of ±5°C.

The thermistor is mounted in the cooled block of aluminum (Fig. 19A, circular piece) within 1 mm of a detector. The hot block (Fig. 19B, square piece of aluminum) is mounted against the surface of the sample compartment to provide additional heat conduction to the outside.

III. COMPUTERIZED SPECTROPHOTOMETRY: THE COMP/SPEC

An instrument that reaches the user today without incorporating some type of digital computer technology is probably obsolete before it is pressed into service. NIR spectrophotometry, with the exception of Fourier transorm infrared reflectance (FTIR), is probably more dependent on computers than any other phase of spectrophotometry.

The literature reveals several instruments that were interfaced to a computer in order to enhance their function and improve ease of operation. Longerich and Ramaley (1974) described the circuitry needed to connect a Cary 14 spectrophotometer to a model 2114B Hewlett-Packard computer (Hewlett-Packard USA, Palo Alto, CA). An optical tachometer was attached to the wavelength drive mechanism of the spectrophotometer for wavelength encoding.

Mattson and Smith (1977) interfaced a Nova 1220 computer (Data General Corp., A Division of EMC Corp., Hopkinton, MA) to a Perkin-Elmer 180 dual-beam spectrometer. The abscissa (wave number) carousel was stepping-motor driven; an electromechanical encoder sent signals to the wave-number readout as the carousel rotated. Ordinate values were converted to ratiometer counts between 1 and 12,000. Both abscissa and ordinate values were transmitted in serial ASCII format from the spectrometer to the computer. A 10-bit digital-to-analog converter was used to drive an X-Y recorder for displaying the spectra.

Norris and Barnes (1976) briefly described a computerized spectrophotometer incorporating a Cary 14 monochromator and a Nova 2/10 computer. Shenk and Hoover (1976) discussed a computer-controlled spectrophotometric system that utilized a Jobin-Yvon HRS-2 monochromator (Jobin-Yvon, Inc., Edison, NJ). Bran+Luebbe (Bran+Luebbe GmbH, Norderstedt, Germany), Foss/NIRSystems (Foss/NIRSystems, Inc., Silver Spring, MD),

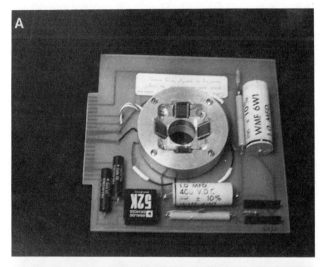

Fig. 19. Mounting arrangement for the lead sulfide (PbS) reflectance detectors in the COMP/SPEC I (McClure and Hamid, 1980): **A,** the circular cold block to which the PbS cells are bonded with a thermally conductive glue and **B,** the square heat sink that conducts the heat to the instrument cabinet.

and LT Industries (LT Industries, Inc., Rockville, MD) all offer computer-controlled instruments, but detailed designs are proprietary. McClure and Hamid (1980) described in detail the design of the COMP/SPEC. Its design will serve as a means of drawing together the components discussed above into one system.

A. General Design

A block diagram of the computerized spectrophotometer is shown in Figure 20 and a pictorial view is shown in Figure 21. While many of the components used in this system have been upgraded since its construction, the design principles are the same regardless of the components used. The system is computer controlled and mounted on rollers so that it can be conveniently positioned within the laboratory.

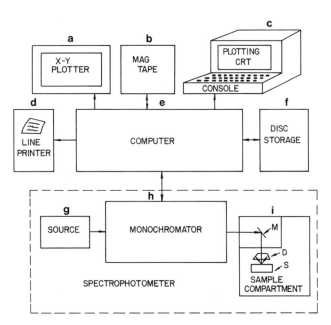

Fig. 20. Block diagram of the COMP/SPEC I (McClure and Hamid, 1980), a custom-built computerized spectrophotometer: **a,** Tektronix 4662 plotter [Tektronix, Inc., Irvine, CA]; **b,** Data General (DG) 6027 magnetic tape drive [Data General Corp., Hopkinton, MA]; **c,** Tektronix 4010 plotting CRT; **d,** OKIDATA 2350 printer [OkiData, Mt. Laurel, NJ]; **e,** DG NOVA 2/10 computer; **f,** two disk drives: DG 6030 and DG 6045; **g,** DXM tungsten-halogen lamp; **h,** Cary 17 monochromator [Varian Instruments, Palo Alto, CA]; and **i,** sample compartment with lead sulfide detector.

Fig. 21. The author prepares a sample for scanning on COMP/SPEC I (McClure and Hamid, 1980) located in the Bioinstrumentation Laboratory at North Carolina State University. Another COMP/SPEC (COMP/SPEC II) was constructed for the U.S. Department of Agriculture scientists at the Oxford Tobacco Research Station, Oxford, NC.

B. Optomechanical

The Cary 17 is a double monochromator consisting of a prism and a grating monochromator in series, permitting measurements in excess of six optical density units (OD) before stray light becomes a problem. High energy output in the NIR region of the spectrum was achieved by specifying the blaze wavelength of the grating to be 1.6 µm. Even though scanning range of the Cary 17 is 0.18–3.2 µm, the use of PbSe detectors for NIR measurements restricts the useful scanning range from 0.9–2.6 µm with a nominal bandpass of 7 nm.

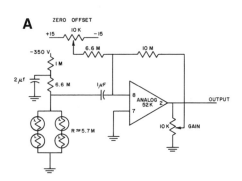

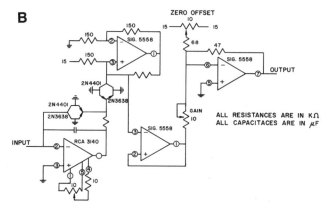

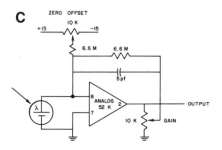

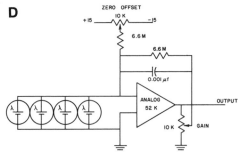

Fig. 22. Schematics of four detector circuits designed for the COMP/SPEC I (McClure and Hamid, 1980): **A,** lead sulfide reflectance; **B,** photomultiplier tube (PMT) log transmittance; **C,** silicon transmittance; and **D,** silicon reflectance.

The Cary 17, unlike the Cary 14, is driven by a stepping motor with a stepping resolution of 200 steps per revolution. The gear train coupling the motor to the cam screw of the monochromator provides further reduction, giving a wavelength resolution of 0.01 nm per step. The advantages of the stepping-motor drive over the conventional synchronous motor include its simplicity, resettability, reversibility, and superior flexibility for operating under computer software control. Both the scanning rate and data interval can be controlled by software. Furthermore, wavelengths can be encoded by a software counter without additional hardware. The work of the author thus far has been at a scanning rate of 10.8 nm/sec with a return speed of 18 nm/sec. Controls on the front of the monochromator allow operator intervention for setting the starting wavelength before going to computer-controlled scanning.

The light source is a Sylvania DXM tungsten-halogen lamp (Bulb Direct, Pittsford, NY) rated at 250 W at 30-V input. The light is chopped mechanically at 60 Hz before entering the monochromator. Slits of the monochromator are manually adjustable to give a bandpass of up to 11.0 nm for a maximum width of 3 mm.

Light from the monochromator vertically enters the sample compartment (32 × 23 × 23 cm) by being reflected at a right angle by an ellipsoidal first-surface mirror. The mirror has a short radius of 14.98 cm and a long radius of 24.38 cm. When the mirror is positioned with the short axis in the horizontal plane, the slit of the monochromator is focused near the top of the sample compartment for reflectance measurements. Turning the mirror 90° focuses the beam near the bottom of the sample compartment for transmittance measurements. A permanently mounted card holder at the top and a removable card holder at the bottom of the sample compartment make it easy to convert from transmittance to reflectance measurements, respectively, and vice versa.

The system currently operates in the single-beam mode. As compared with the double-beam mode, the single-beam mode has the advantage of high energy throughput, which is important for optically dense light-scattering materials. Furthermore, signal-to-noise ratio is higher in this mode because of the absence of cross-talk noise encountered in double-beam systems. Cross-talk noise, a characteristic of double-beam instruments, is produced when the light source is divided to illuminate the sample and the internal standard. A signal from one beam can reach the other and interfere with the measurement producing the cross-talk noise.

C. Optoelectronic

Separate detectors are required for the visible and NIR regions of the spectra. An EMI 9558 PMT will cover the region of 0.3–0.90 μm for high OD applications in the transmittance mode. Silicon solar cells work well over the region of 0.4–1.2 μm in both the reflectance and transmittance modes, and PbS photoresistive cells cover the region of 0.9–3.0 μm. Because of the dynamic range of the energy throughput of the monochromator and the response of the detectors, the practical scanning range of the system is limited in the single-beam mode to 0.4–0.9 μm for the visible and 0.9–2.6 μm for the NIR (using PbS detectors). Four detector circuits are shown in Figure 22. With the exception of circuit B, each circuit is constructed on an 18 × 18-cm, custom-etched, printed circuit card that plugs directly into prewired 44-pin, pc-edge connectors mounted in the sample compartment. Circuit B was mounted directly on the bottom of the 9558 PMT housing. An AC synchronous 1,800-rpm motor drives a disk with two open sectors for modulating the light beam entering the monochromator. As shown in Figure 23, the detector preamplifier output is a square waveform, the peak-to-peak magnitude of which is proportional to full light (A) and no light (B) and is in phase with the AC line voltage.

Demodulation of this signal is accomplished by the circuit in Figure 24. The circuit consists of two sample-and-hold modules (Analog Devices SHA-5 [Analog Devices, Inc., Norwood, MA]) and a difference amplifier (Analog Devices 52 K). The inputs of both SHA-5s are connected to the output of the detector preamp. If the sample-and-hold pulses given to the SHA-5s are properly timed, the output of the difference amplifier will be the voltage difference between points A and B of the detector output signal.

The sample-and-hold pulses, TIM-1 and TIM-2, are generated by the circuit in Figure 25. It is essentially a phase-synchronized rectifier that gives two 30-Hz, 20-sec pulse strings at 180° apart that are used to trigger the SHA-5. TIM-1 drives SH-1 and TIM-2 drives SH-2 (Fig. 24). The output of the difference amplifier is a signal proportional to the intensity of the light reaching the detector.

D. Digital Interface

The digital interface was designed using the Data General Corporation's general-purpose interface board with 4040 basic interface and 4041 input/output data registers (Archer, 1969). The 4040 basic interface contains the prewired circuitry that connects the interface to the data and control lines on the bus, the networks for passing the interrupt and data channel priority signals along the bus, and a device selection network that allows a choice of any device code by putting in appropriate jumpers. The 4041 data registers act as buffers to control the data flow between custom logic and the internal data bus. All the above circuitry is mounted on a standard 38 × 38-cm board that is divided in the middle by two rows of approximately 200 wire-wrap pins, of which 48 are available for custom interfacing. Half the board is reserved for customer logic and is configured for mounting 65 14- or 16-pin integrated circuits (ICs).

Commercially available components were mounted on the general-purpose board for control of the monochromator and the analog-to-digital converter. The device selection network was jumpered for a device code of 21 for both the analog-to-digital converter and the monochromator control. Scanning direction of the monochromator was provided via the output buffer according to the status of DATA 14 and DATA 15 (Fig. 26).

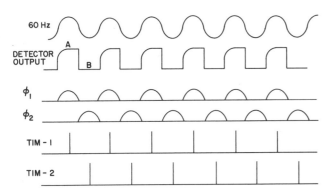

Fig. 23. Modulation/demodulation timing pulses for detector signals. The timing pulses are generated from the two phases (ϕ_1 and ϕ_2) of the 60-Hz power. A = full light; and B = no light.

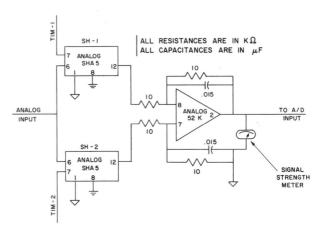

Fig. 24. Schematic/block diagram of the demodulation circuit consisting of two sample-and-hold amplifiers and a difference amplifier.

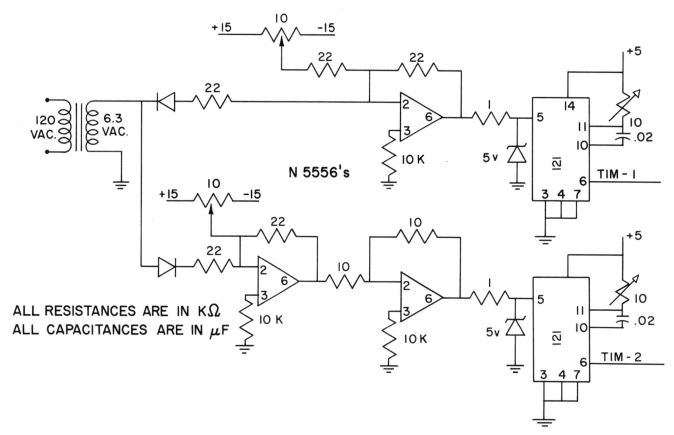

Fig. 25. The circuit used to generate the sample-and-hold trigger pulses, TIM-1 and TIM-2. Monostable chips (74121 [Newark Electronics, Chicago, IL]) are used to shape the 60-Hz phases into pulses.

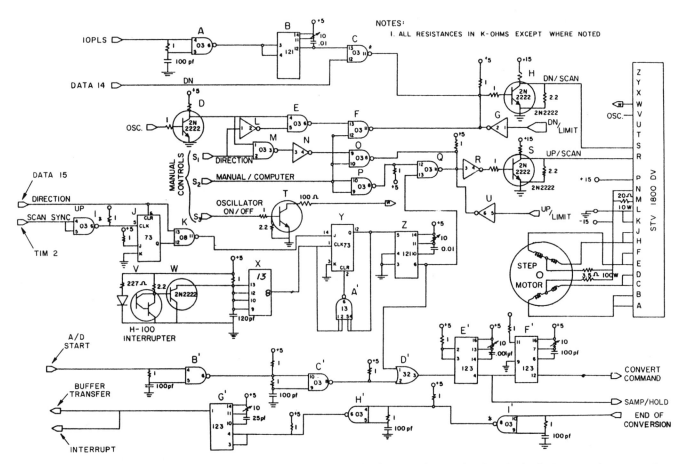

Fig. 26. Circuit diagram of the scan-control interface and A/D converter control circuit. Data-taking was timed to take data between stepping-motor pulses.

As diagrammed in Figure 26, a mechanical chopper synchronously rotating at 1,800 rpm in conjunction with the H-100 interrupter (V) (Newark Electronics, Chicago, IL) gives a forward scanning pulse rate of 1,080 pulses per sec. The output of the interrupter is buffered with a transistor (W) and shaped by a Schmidt trigger (X) (Newark Electronics) before going to the synchronizer (Y). The computer initiates forward scanning by raising DATA 15 to logic 1 and lowering DATA 14 to logic 0. This releases the clear of the flip-flop J so that it can be set by the next TIM-2 pulse. Since the flip-flop J is negative-edge triggered, the presence of TIM-2 when DATA 15 goes high will not set the flip-flop. The synchronizer (Y) shapes the output of the interrupter, which is timed with the one shot (Z). The output of the one shot is passed through gates Q and R to the driving transistor (S) for the translator module. Pulses at the input of S move the stepping motor in the forward (UP/SCAN) direction.

The output of Z also triggers the one shot E′, which triggers the one shot F′, which furnishes the convert command to the A/D converters. The OR-gate (D′) allows additional START pulses from the computer to generate additional conversions as required by the data-taking scheme.

The end of convert signal from the A/D converter is double-buffered through gates I′ and H′ and taken inside the computer. The output of the one shot G′ transfers the binary conversion of the A/D converter to the output of the data input buffer and, at the same time, sets the DONE flip-flop and raises the interrupt flag. At this time, the data are ready to be brought into the computer with the data-input-authorization (DIA) instruction.

Reverse scanning of the monochromator is initiated by setting DATA 14 to 1 and DATA 15 to 0. There is no requirement for synchronization in the reverse mode because no data are taken. Pulses generated by IOPLS from the computer can be varied as the system requires. For example, reverse scanning begins at a rate of 1,000 pulses per sec for 1,000 steps and then changes to 1,800 steps per sec until the scan reaches within 1,000 steps of the starting wavelength, whereupon the speed is reduced to 1,000 steps per sec for the last 1,000 steps.

The stepping motor of the monochromator is bidirectionally controlled by a SLO-SYN STM 1800 DV translator module with logic inputs that can be easily interfaced to a computer. A single 10- to 15-V negative change of voltage with a minimum pulse width of 50 µm (fall and rise time not to exceed 2 µsec) at pins R or S of the translator module will cause the motor to step down and up the wavelength scale, respectively. Two transistors pulled up to +15 V are adequate to interface the translator device to the transistor-transistor logic (TTL) of the computer. The module is capable of driving the stepping motor in a four- or eight-step switching sequence. The eight-step sequence was employed in the COMP/SPEC, giving a scanning resolution of 0.01 nm.

The manual control circuitry is shown in detail in Figure 26. In the manual mode, DATA 14 and DATA 15 are a logic 0. The output of NAND gates (C) and (Q) is high, allowing outputs of gates (F) and (P) to activate the driving transistors (H) and (S), respectively. The oscillator OSC is on the translator pc-board and is activated by switch S_3. Switch S_2 changes the operation from computer to manual, and S_1 determines the direction of manual scanning. Manual scanning is asynchronous because no data are taken.

IV. PERFORMANCE OF THE COMP/SPEC

A summary of the specifications and certain performance indexes for the COMP/SPEC are given in Table III. Data for computing performance indexes were taken at a scanning rate of 10.8 nm/sec and a return rate of 18 nm/sec. Thus, the forward-scanning (data-taking) time was 2.62 min and the return (reset) time was 1.57 min, for a total time per spectra of 4.19 min. Each spectrum scanned contained 1,700 data points, with each point being the average of 1,126 analog/digital conversions.

A. Photometric Noise

Photometric noise in a computerized spectrophotometer is a measure of total system noise inherent in the detectors, amplifiers, and A/D converter. Photometric noise for the COMP/SPEC was computed from two reflectance spectra, 900–2,600 nm, of a Halon reflectance standard (Acton Research Corp., Acton, MA) run as a sample with a 20% attenuation (0.7 log[1/R] units). A log(1/R) spectrum was computed for each reflectance spectrum, and the two log(1/R) spectra were subtracted so that the noise data are clustered about zero. This noise spectrum, S_n, was used to compute the photometric noise. No smoothing was used in any computation.

Peak-to-peak noise was computed by subtracting the minimum and maximum data points in S_n. Bias b was obtained by averaging all the points in S_n, or

$$b = \frac{\sum_{n=1}^{N} d}{n} \tag{17}$$

TABLE III
Specifications and Performance Indexes of COMP/SPEC

Specifications	Performance Indexes
General	
Light source	Sylvania DXM (250 W at 30 VDC)
Slits	Adjustable 0–3 mm
Bandpass	0–11 nm for range 900–2,600 nm
Wavelength drive	Slo-Syn TYPE MO91-FD06
Scan speed	10.8 nm/sec
Return speed	18 nm/sec
Optical path	Single beam
Range	0.4–2.6 µm
A/D conversion time	10 µsec maximum
A/D word size	15 bits plus sign bit
Performance	
Absorbance limits	7 OD with 9558 photomultiplier tube
Silicon	5 OD
Lead sulfide	4 OD
Stray light	0.0001% for range 0.24–0.5 µm
	0.001% for range 0.21–1.8 µm
	0.1% at 2.65 µm
Wavelength precision	0.005 nm at 1682.4 nm
Wavelength resolution	0.01 nm
Photometric noise	162 µOD

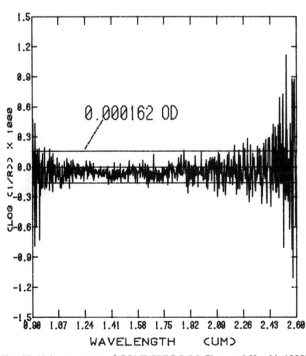

Fig. 27. Noise spectrum of COMP/SPEC I (McClure and Hamid, 1980). The standard deviation of the noise is 162 µOD.

in which b = bias, d = value of the nth data point, and N = number of data points in S_n.

The root mean square (RMS) noise was the computed square root of the average of the square of the data points in S_n, or

$$S_{RMS} = \sqrt{\frac{(d)^2}{n}} \qquad (18)$$

in which S_{RMS} = RMS noise.

Plotting S_n as a function of wavelength (Fig. 27) shows the RMS noise to be 162 μOD with a peak-to-peak noise of +1,150 and –1,112 μOD at 2.536 and 2.576 μm, respectively. The average offset was 44 μOD. The noise was higher on each end of the spectrum, where the signal-to-noise ratio is lower.

B. Wavelength Precision

Wavelength precision is extremely important in NIR spectroscopy, much more important than accuracy. Correlograms, showing the relationship between wavelength information (log[1/R], etc.) and chemistry of all samples in a calibration set, often exhibit sharp peaks. A shift in the wavelength mechanism from one

spectrum to another would cause excessive errors for both calibration and prediction. Wavelength accuracy, on the other hand, can be off by several nanometers without causing any problems, since the calibration and prediction sets are usually run on the same machine and wavelength inaccuracies are compensated for in the collection of the data. Moving of calibration equations from one instrument ("master") to another ("slave") requires that the slave be allowed to deviate from prescribed wavelengths, picking its own optimum wavelengths. The latter procedure is necessary in recognition that the wavelength mechanism may not register exactly with that of the master.

Wavelength precision (Table IV) for the COMP/SPEC varied slightly with wavelength. The standard deviation was determined to be 0.055, 0.005, 0.008, and 0.008 for the polystyrene wavelengths of 1,143.9, 1,682.4, 2,163.4, and 2,306.5 nm, respectively. These values were determined by calculating (by interpolation) the zero-crossing of a quadratic polynomial first derivative (through nine points) for each of the unsmoothed log(1/R) 1,700-point spectra.

This procedure is extremely sensitive to treatment of the data. Smoothing the data prior to computation of the derivatives will give different zero-crossings. However, the precision figures should vary very little from one treatment to another.

TABLE IV
Wavelength Precision of the COMP/SPEC

Reading Number	Wavelength Number			
	1	2	3	4
1	1,143.9643	1,682.4187	2,163.4509	2,306.5693
2	1,143.9640	1,682.4129	2,163.4722	2,306.5598
3	1,143.9634	1,682.4129	2,163.4722	2,306.5600
4	1,143.9779	1,682.4103	2,163.4635	2,306.5657
5	1,144.1007	1,682.4078	2,163.4540	2,306.5807
6	1,144.0965	1,682.4026	2,163.4593	2,306.5578
7	1,143.9675	1,682.4074	2,163.4599	2,306.5594
8	1,143.9642	1,682.4061	2,163.4507	2,306.5661
9	1,143.9671	1,682.4057	2,163.4495	2,306.5527
10	1,143.9643	1,682.4035	2,163.4566	2,306.5548
Average	1,143.9929	1,682.4087	2,163.4559	2,306.5626
Std. dev.	0.0548	0.00462	0.00797	0.00775
Coefficient of variation	4.8×10^{-3}	2.7×10^{-4}	3.7×10^{-4}	3.4×10^{-4}

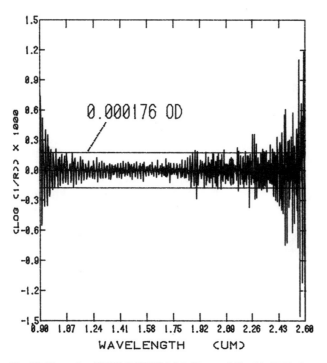

Fig. 28. The noise of COMP/SPEC I (McClure and Hamid, 1980) determined by Fourier analysis is 172 μOD.

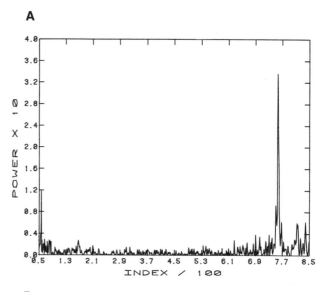

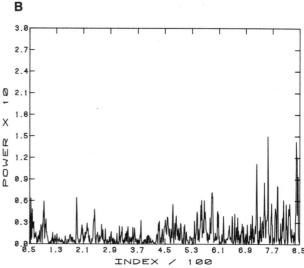

Fig. 29. Power spectrum from **A,** COMP/SPEC I (McClure and Hamid, 1980) and **B,** COMP/SPEC II providing information relative to an instrument anomaly in COMP/SPEC I, a 4.8-Hz spike at point 759. This peak never appears in the power spectra of COMP/SPEC II.

C. Fourier Analysis of Instrument Performance

Fourier analysis of NIR spectra obtained with dispersion (monochromator) systems is a quick, reliable way to monitor instrument anomalies. Generation of smoothed spectra (for determining noise), derivative spectra (for determining wavelength precision), and power spectra (for monitoring scanning anomalies) can be accomplished from the Fourier domain with much less distortion than that encountered in wavelength space.

Figure 28 is a noise spectrum generated with the aid of Fourier analysis. It was obtained by transforming the $\log(1/R)$ spectrum, $S1$, to Fourier space retaining only the first 200 Fourier coefficient pairs. Another $\log(1/R)$ spectrum, $S2$, was computed from these coefficients. The difference spectrum $(S1 - S2)$ is the noise data in Figure 28. It is obvious that generation of $S2$ was a smoothing operation invoked by dropping Fourier pairs 201 and higher. The RMS noise in Figure 27 is 176 µOD with peak-to-peak noise of 1,190 µOD and −1,450 µOD at 2.597 and 2.566 µm, respectively. Offset for the noise spectrum in Figure 27 was less than 1.0 µOD.

Figure 29 shows the power spectrum for two different COMP/SPEC systems. Both power spectra were computed from a single tobacco spectrum. Note that, in Figure 29A, there is a very powerful spike at index 757, one that never appeared in the power spectrum of COMP/SPEC II (Fig. 29B). This spike appeared in all spectra recorded on COMP/SPEC I, regardless of the material being scanned. The index represents a frequency of 4.8 Hz and is thought to be associated with a bad gear in the drive train of the monochromator.

TABLE V
Typical Data Directory for Spectral Files Recorded with COMP/SPEC

Displacement	Definition	Value
1	Disc number	1
2	Section	0
3	Format	4
4	Initial wavelength	910
5	Wavelength interval	200
6	Points in each spectrum	840
7	Number of spectra	60
8	Initial sample number	1
9	Sample number increment	1
10	Data type	1
11	Spectrum size in records	27
12	Wavelength offset	10
13	Y size in records	3
14	Location of first Y	1,621
15	Number of Y data	3
16	Type of smooth	1
17	Smooth interval	21
18	Type of derivative	0
19	Derivative interval	0
20	Divide by R?	0
21	Delta OD?	0
22	Delta for OD	0
23	Antilog?	0
24	Log?	1
25	Shrink factor	2
26	Scale	0
27	Shift	0
28	Pshift	0
29	WINC (wavelength incr.)	2
30	Plotter speed	10
31	Cursor speed	10
32	OD offset	0
33	Number of scans	0
34	Last sample number on file	60
35	Pulses per point	100
36	Record number	2,188
37	Spectrum number	60
38	APNP1	1,700
39	Fourier transformation	0
40	Number of Fourier pairs	0

[a] Directory information, June 11, 1984, File: LR1.

V. SOFTWARE FOR COMP/SPEC

Results obtained with any computerized spectrophotometric system depends on the software packages developed for its support. COMP/SPEC software was developed in house, independent of all other packages that may be available. A brief discussion of COMP/SPEC software packages is included to give the reader an appreciation of the type of programs needed in order to take advantage of the NIR spectrophotometric hardware.

COMP/SPEC software comes in three packages: Scanning/Analysis (SCANL), Analytical Software Programs (ASP), and

TABLE VI
Command Options in SCAN Software Package for COMP/SPEC

Command Displacement	Program	Description
0	DATAQ	Data acquisition (scanning)
1	SANLS	Spectrum analyzer
2	PNCH	Print spectral data
3	UCALB	Write data directory (DD) to diskette
4	RDWTD	Read/write spectrum to/from diskette
5	UDDLR	Update DD
6	DDLR	Display DD
7	PCAL	Calibrate analog plotter
8	DSPLY	Display spectra on scope
9	SMDR	Smooth spectrum/display
10	CDERV	Compute derivative spectrum/display
11	DFSP	Compute difference of two spectra
12	CALB	Set up DD
13	TRSF	Transform entire spectra file
14	MNUL	Releases for manual control
15	EODS	Return to CLI

TABLE VII
Programs in Analytical Software Package for COMP/SPEC

Program Type/Name	Description
1. Transformations	
A. FTRSF	Conventional transformations (smoothing, derivative, shrinking, etc.)
B. FOTRSF	Fourier transformation
C. FOPWR	Power spectra transformation
2. Spectrum handling	
A. FDFSD	Difference spectroscopy
B. SHUFFLE	Shuffle any spectrum of spectra from one file to another
C. SPLIT	Split a file into two or more parts
D. SUBADD	Corrects a set of spectra from one standard (e.g., ceramic) to Halon reference standard
E. STAT	Prints spectrum statistics (maximum/minimum, std. dev., etc.)
F. AVER	Compute average and std. dev. spectra of any number of spectra
G. CWAVE	Change wavelength range of spectrum
H. SHIFT	Shift spectrum along wavelength axis for plotting, difference spectroscopy, etc.
3. Calibration and prediction	
A. SMLR	Stepwise multiple linear regression program
B. LRRD	For any chemical constituent attached to the spectra file, computes gap-2nd derivatives and determines best ratio of derivatives for the range of wavelengths specified
C. PRED	Take calibration equations from equation files established by SMLR and predict unknowns, or knowns; IF known, pred. statistics are provided
D. MCANS	Multiple constituent analyses; take any number of established equations (set up by program EQUATION) and predict a constituent from a set of spectra
E. EQUATION	Set up established equations for a product; MCANS uses these equations for predicting unknowns

TABLE VIII
Computerized Spectrophotometric Analytical System
Command Summary

Command Code	Function
Handling spectrum and chemistry arrays	
RS #3	Read a spectrum numbered 3
RY #3	Read Y data set number 3
TY #–# ...	Type all (TY) or a range of Y data values
TW #–# ...	Type range of spectrum wavelength values
TI #–# ...	Type range of spectrum index values
TXDS [/X/Y] "FILENAME"	Type contents of external data set
LY #–# ...	List (type) range of Y data set names
LYP	Print all Y data set names
PY #–# ...	Print all (PY) or range of Y data values
PW #–# ...	Print range of spectrum wavelength values
PI #–# ...	Print range of spectrum index values
EDITS [I#/W#]	Edit current spectrum (at I-index W-WL)
EDITY [#]	Edit current Y data set (at index #)
DIF WL1 WL2 ...	Difference of two wavelengths (WL1-WL2)
RAT WL1 WL2 ...	Ratio of two wavelengths (WL1/WL2)
PV[P] #–# ...	Peaks and valleys report (P-print)
PVX[/P] "FILENAME"	Peaks and valleys report for external file
YST [P/L]	Y data statistics (P-print, L-list all)
VARY [/P] #–# ...	Pooled variance of Y data sets
CORW [P] WL WL ...	Correlation statistics between WL pairs
CORWC [P] WL/C#	Wavelength/constituent correlation
CORWRC [/P] WL/C#	Wave ratio/constituent correlation
POSS [P]	List/print all positive spectra in file
NEGS [P]	List/print all negative spectra in file
MM #–# ALL	List max. and min. VALUES from spectra
SCALE?	List the current spectrum scale factor
SET SCALE #	Set the scale factor to the number #
SCALE [/M/D/A/S] #–#	Scale the specified spectra
FRESET	Reset the current file (remove the scale)
CS	List current spectrum number
CY	List current Y data set name
DIR [P/M]	Directory list (print/maintenance)
Plotting spectra and external (X-Y) files	
PLOT [#–#,#]	Plot current spec. or range from in file
PLOTF "FILENAME"	Plot a plot file named "FILENAME"
PLOTY X# Y# ...	Plot two Y data sets as X and Y coordinates
REGLF "FILENAME"	Draw regression/error/std. dev. line
SYMF "FILENAME"	Draw symbols at plot file points
CHART	Create a grid pattern on the plotter
BOX	Create a grid pattern without tick marks
LABEL	Label the plotting grid
TIC# WL ...	Tick mark with rotating # at wavelength WL
LINE X Y X Y ...	Draw a line (XY to XY, etc.) on plotter
HL # ...	Draw a horizontal line on plotting grid
VL # ...	Draw a vertical line on plotting card
NOTE# "MESSAGE"	Print "MESSAGE" with rotation on plotter
PRINTF# "FILE"	Print contents of ASCII file on plotter
PDIR [M]	List/maintenance plotting parameters
File commands	
GET FILE XX	Get input file named XX
FILE	List current input file name
INITIALIZE[/F] "DIR" ...	Initialize a directory(/F-full)
RELEASE "DIR" ...	Release a directory
FLOAT "FILE" INTO "FILE"	Convert a file to floating point
COPY "FILE" INTO "FILE"	Copy a file into another file
CXYPF X# Y# ...	Create an external plot file from Y data
YINPUT "FILENAME"	Attach a Y data set to the current file
NEWY "Y-LABEL"	Attach a dummy Y data set to current file
? [/P/S]	List (P = print, S = sort) available commands
STOP	Exist CSAS; return to CLI

Computerized Spectrophotometric Analytical System (CSAS). Before discussing these packages, it is helpful to take a look at the organization of the spectra data files for COMP/SPEC.

A. COMP/SPEC File Structure

All spectra files are organized into three parts: data directory (DD), spectra, and chemistry (sometimes called Y data). The DD and space for spectra are established on a diskette before the scanning of a set of samples begins. Spectra files are organized into blocks of 64 16-bit words each. The first block in the file is devoted to the DD; only 40 locations are currently used. The first two words of each spectrum array is the sample number. Position of the chemistry arrays is computed when a spectra file is established.

A typical DD is given in Table V. Much of the information in DD is self-explanatory. The DD is updated when samples are added to the file (D7), when Y data (chemistry arrays) are added to the file (D13–D15) and when transformations (smoothing, derivative, Fourier) are performed (D16–D25 and D39–D40). SCANL also interfaces to an analog X-Y plotter that uses special DD information (D27–D32); this feature is used in displaying spectra on the Tektronix model 611 storage display. D33 controls the number of scans to be averaged for one sample.

B. Scanning/Analysis

The SCANL software package contains 15 programs that are accessible from the console. The programs (Table VI) permit scanning or collection of spectral data, making changes to the DD, making data transformations (smoothing, derivatives, differences, etc.), printing of spectral data, and display of spectra on the storage scope. The programs are intended for qualitative spectral analyses during the scanning process; hence, it is not necessary to exit the scanning process in order to determine trends in the data.

C. Analytical Software Package

The analytical software package for COMP/SPEC is a set of programs, written in Fortran, that provide the user with more powerful tools for making detailed spectral analyses. Table VII lists the major programs in this package. A detailed discussion is beyond the scope of this chapter on hardware, but a brief description of each program is given in Table VII.

D. Computerized Spectrophotometric Analytical System

Software for COMP/SPEC (called CSAS) has evolved over 30 years, from 1960 to 1990. During the course of software development, it became apparent that many spectra "housekeeping" functions should be brought under the roof of one program, and CSAS was developed to meet this need. It is a program written in 132 overlays so that it will run on a 64-kilobyte computer. Table VIII is a list of the functions that can be performed with CSAS. Only the major functions will be discussed.

Using CSAS, not only can chemical data arrays be edited and attached to spectral data, but also external X-Y files can be established. Spectra and external X-Y files may be plotted. Statistical data with respect to spectra and chemistry can be obtained. Pooled variance of chemistry data (multiple analyses) is easily obtained. Correlations between single wavelength and wavelength ratios versus chemistry can be generated. Finally, the two commands NOTE and PRINTF are useful for labeling graphs and making slides.

ACKNOWLEDGMENTS

I would like to thank these people for their contributions: Brenda Mason for typing the manuscript, M. L. McLester for his technical assistance and proofreading, and my many other colleagues for their patient listening as I talked about ideas. Especially, I thank my wife, Judy, for her understanding during all those Saturdays when I was away from home.

LITERATURE CITED

Archer, E. D. 1969. Infrared analysis-I. Lubrication 55:13-24.

Ben-Gera, I., and Norris, K. H. 1968. Direct spectrophotometric determination of fat and moisture in meat products. J. Food Sci. 33:64-67.

Birth, G. S. 1969. Development of a high intensity optical system for non-destructively evaluating the interior quality of agricultural products: Engineering analyses of a high energy spectrophotometer. Dep. Agric. Eng., Purdue University, Lafayette, IN.

Birth, G. S. 1971. Spectrophotometry of biological materials. Ph.D. thesis. Purdue University, Lafayette, IN.

Birth, G. S., and Norris, K. H. 1965. The difference meter. U.S. Dep. Agric. Agric. Res. Serv. Tech. Bull. 1341. Washington, DC.

Corion Corp. 1983. Optical Components. Corion Corp., Holliston, MA.

Downs, R. J., Norris, K. H., Bailey, W. A., and Klueter, H. H. 1964. Measurement of irradiance for plant growth and development. Am. Soc. Hortic. Sci. 85:663-671.

Hayat, G. S. 1978. Holographic grating update. Electro-Opt. Syst. Des. 10(April):50-54.

Herschel, W. 1800a. Investigation of the powers of the prismatic colours to heat and illuminate objects; with remarks, that prove the different refrangibility of radiant heat. To which is added, and inquiry into the method of viewing the sun advantageously, with telescopes of large apertures and high magnifying powers. Phil. Trans. Roy. Soc. Lond. 90(13):255-283.

Herschel, W. 1800b. Experiments on the refrangibility of the rays of the sun. Phil. Trans. Roy. Soc. Lond. 90(14):284-292.

Herschel, W. 1800c. Experiments on the solar, and on the terrestrial rays that occasion heat; with a comparative view of the laws to which light and heat, or rather the rays which occasion them, are subject, in order to determine whether they are the same, or different. Phil. Trans. Roy. Soc. Lond. 90(15):293-331.

Longerich, H., and Ramaley, L. 1974. Digital interface for a Cary 14 spectrophotometer. Anal. Chem. 46:2067-2071.

Mattson, J. S., and Smith, C. A. 1977. An on-line minicomputer system for infrared spectrometry. Chapter 2 in: Computers in Chemistry and Instrumentation. Vol. 6. J. S. Mattaon, H. B. Mark, Jr., and H. C. McDonald, Jr., eds. Marcel Dekker, New York.

McClure, W. F. 1968. A rapid method for measuring moisture content of cured tobacco: Research report. Dep. Biol. Agric. Eng., North Carolina State University, Raleigh.

McClure, W. F. 1969a. Affect of shrinking spectral data on estimating chemical composition. Dep. Biol. Agric. Eng., North Carolina State Univ., Raleigh.

McClure, W. F. 1969b. Design of a near infrared moisture meter for tobacco. Int. Res. Rep., North Carolina State University, Raleigh.

McClure, W. F., and Hamid, A. 1980. Rapid NIR measurement of the chemical composition of foods and food products. Part I: Hardware. Am. Lab. 12:57-69.

Melles Griot. 1985. Optics Guide III. Melles Griot, Irvine, CA.

Morimoto, S., McClure, W. F., and Stanfield, D. L. 2001. Hand-held NIR spectrometry: Part I. An instrument based upon gap-second derivative theory. Appl. Spectrosc. 55:182.

Norris, K. H. 1964. Simple spectroradiometer for the 0.4 to 1.2 micron range. Trans. ASAE 7:240-242.

Norris, K. H., and Barnes, R. F. 1976. Infrared reflectance spectrocomputer design and application. In: Proc. Technicon Int. NIRA Conf., 7th. Bran+Luebbe, Chicago, IL.

Norris, K. H., and Butler, W. L. 1961. Techniques for obtaining absorption spectra on intact biological samples. IRE Trans. Bio-Med. Electron. (BME) 8:153-157.

Shenk, J. S., and Hoover, M. R. 1976. Infrared reflectance spectrocomputer design and application. Pages 122-125 in: Advances in Automated Analysis. Technicon, Tarrytown, NY.

Stair, R., Schneider, W. E., and Jackson, J. K. 1963. A new standard of spectral irradiance. Appl. Opt. 2:1151-1154.

Sugiyama, J., McClure, W. F., and McLester, M. 1992. Building your own NIR spectrometer system. Pages 1:61-66 in: Making Light Work: Adances in Near-Infrared Spectroscopy. I. Murray and I. Cowe, eds. VCH, Aberdeen, Scotland.

Taylor, A. H. 1919. Measurement of reflectance using filters. J. Opt. Soc. Am. 4:9-10.

CHAPTER 7

Contemporary Near-Infrared Instrumentation

DAVID L. WETZEL
Kansas State University
Shellenberger Hall
Manhattan, KS
U.S.A.

I. INTRODUCTION

At the time of previous reviews by this author (Wetzel, 1983, 1986) and when the first edition of this book appeared, there were less than a dozen instruments from which to choose to do what in the past was primarily quantitative analysis using the near-infrared (NIR) region of the spectrum. That is no longer the case (Wetzel, 1998). The contemporary NIR instruments discussed in this chapter include those that use electronic wavelength switching (diode array or acousto-optic tunable filter instruments), Fourier transform (FT), grating monochromator, interference filter, discrete light-emitting diode (LED) source, special purpose, and imaging instruments. In the 1980s, several interference filter-based instruments were on the market. These included both discrete filter and tilting filter varieties. The former provided rugged, stable, wavelength reproduction and allowed quantitative analysis to be performed with limited discrete wavelengths. In the discrete filter instruments, data collection at each filter was completed before indexing to the next turret mounted filter position. Tilting filters provided a sequence of wavelengths, depending on the angle of incidence. This feature allowed whole segments of the spectrum to be scanned, and, from these segments, groups of wavelengths were used as the basis of quantitative analysis.

Both slow-scan, grating monochromator instruments and rapid-scan, vibrating grating instruments were introduced that allowed collection of adjacent wavelengths and the use of various plotting or quantitative functions. The grating monochromator instruments also provided a classical spectroscopic look at the resulting spectral features rather than a purely statistical approach. In the early days of scanning NIR instruments, the use of interferometer (i.e., FT-NIR) instruments was not routine and, in fact, infrequent. These Fourier transform instruments were converted to NIR application by changing the beam splitter, the source, and the detector on a classical mid-infrared (mid-IR) interferometer instrument. Formerly, the purchase of such an instrument would have been justified for classical mid-IR spectroscopic use and then possibly adapted for NIR use as a secondary purpose.

Much of the previous filter and grating monochromator instrumentation was documented in the first edition of this book. Not all of it will now be reviewed, because much of it has been relegated to historical rather than contemporary interest. Filter instruments were introduced initially by DICKEY-john, Neotec, and Technicon. Most of these models still endure, although not necessarily under the same company or model name. The tilting filter instrument was discontinued by NIRSystems (the successor to Neotec) and replaced with a rapid-scan, grating monochromator instrument. Other NIR filter instruments have emerged, however, for day-to-day quantitative analysis in the field, particularly for routine monitoring. It is estimated that 85% are still filter instruments.

Among the scanning NIR instruments, great changes have been made, particularly within the 1990s. Random wavelength access is readily available from an instrument that has scanning capability. Very rapid response photodiode detectors, such as those composed of indium gallium arsenide (InGaAs), have been introduced to enhance the speed of data acquisition. Electronic wavelength switching as a means of scanning the spectrum or for random wavelength access provides opportunities for speed in the scanning or monitoring process. Two areas of electronic wavelength switching discussed in this chapter include the grating polychromator diode array and the acousto-optic tunable filter spectrometers. These scanning electronic wavelength switching instruments do not require moving parts, which may be an advantage for industrial use.

In the grating polychromator, the rays exiting the postsample dispersive device fall simultaneously upon segments of a detector array. Data from each element in the array are polled with what could be described as electronic wavelength switching. This is in contrast to the classical grating monochromator, in which rotation of the grating is necessary to aim a particular ray through the monochromator exit slit before transversing the sample and hitting the detector. In the very near infrared region, the silicon photodiode array is used. The engineering for silicon photodiodes and instrumentation in this region is advanced, and silicon arrays are produced mostly at a low cost. At longer wavelengths within the NIR region, however, the arrays are considerably more expensive because of the solid-state devices required. One has the choice of germanium to extend the wavelength at the expense of having a higher-than-desirable noise or InGaAs that works at ambient temperature quite well up to 1,700 nm. An extended-range phosphorous-doped InGaAs detector that reaches 2,400 nm requires cooling to achieve maximum "D*", and such a detector array is still expensive at the present time and thus limited to small dimensions.

The monochromator of an acousto-optic tunable filter spectrometer (TFS) involves a solid-state device (usually a TeO_2 crystal) that is tuned to pass selected optical frequencies by imputing select ultrasonic frequencies. This device, an acousto-optic tunable filter (AOTF), operates without moving parts, and the optical tuning is controlled completely by the software that directs the ultrasonic input to provide random wavelength access, continuous scanning, scans of short regions of the spectrum, and even programmed variable duration of data collection time at one point. Both types of systems that utilize electronic wavelength switching can accomplish the wavelength switching very rapidly and can benefit immediately by the rapid response of the more recently available NIR photovoltaic detectors.

In the economics of instrument manufacturing, the development of FT-NIR instrumentation came about as a result of the upsurge of Raman spectroscopy. In Raman spectroscopy, the results of molecular vibrations are observed as shifts in the frequency of scattered rays from the frequency of incident excitation radiation. To avoid fluorescence, NIR lasers were employed for Raman excitation, and FT–mid-IR instruments were converted for NIR use to analyze the Raman spectra resulting from NIR excitation. Thus, in a short period of time, there was a reason for producing NIR-optimized, FT spectrometers. Several of the well-established mid-IR classical spectroscopic instrument companies now provide conversion of their classical mid-IR instruments or have produced dedicated NIR FT instruments. In addition, two interferometers that are exclusively FT-NIR scanning instruments, produced by companies new to FT spectroscopy, have emerged on the market. These, based on polarization interferometry, have been produced and dedicated primarily to rapid qualitative analysis in a manufacturing setting or to use as an inspection tool.

Another form of NIR instrumentation is that involving discrete sources. Discrete sources are produced with a group of sequentially fired, NIR LEDs. These solid-state devices are chosen for their particular emission wavelengths. However, the wavelength limitation is imposed by a very small interference filter incorporated with individual LEDs. These discrete source instruments were introduced first by the Trebor Inc. (Rossenthal, 1980). Rays coming from discrete sources through a sample then are collected with a lens and fall onto a silicon photodiode. These instruments are used almost exclusively for NIR transmission measurements, and their use is limited to the very near infrared region established by the upper end of the silicon photodiode range at 1,050–1,100 nm. Two instrument companies produce tabletop or hand-held instruments of this variety. Another company produces on-line fiber optic monitoring based on this same technology.

Instruments with specialized configurations designed for select application are listed in Table I. These are discussed not as separate topics, but within the context of the type of instrument under consideration. Table II includes a partial list of instrument developers by type of instrument. The newest types of NIR instrumentation are presented first and the earlier types at the end of this chapter, followed by a brief reference to imaging instruments of the near future.

II. ELECTRONIC WAVELENGTH SWITCHING: DIODE ARRAY INSTRUMENTS

This type of instrumentation is described as scanning with electronic wavelength switching. Under this general heading, the author will discuss particularly the use of photodiode array polychromators. These devices have become commonplace in the 1980s and 1990s in the ultraviolet (UV) and visible ranges, where the technology of silicon photodiode arrays has been highly developed. Ever since Hewlett Packard (Wilmington, DE) introduced a diode array UV detector for high-performance liquid chromatography (HPLC) in the late 1970s and another diode array, high-market-end, laboratory benchtop UV spectrometer, a drastic change has come about in UV, and subsequently in fluorescence, instrumentation.

The concept of using a different detector dedicated to each particular wavelength of interest is not new. In classical arc/spark atomic emission spectrographic analysis in the steel industry, the time required to develop and read a photographic plate was costly, because molten steel had to be held in that state to obtain elemental spectrographic analysis prior to its pouring. To speed up the analytical process, the direct-reader spectrograph was produced. It employed a different dedicated photomultiplier at each location on the Roland circle of the spectrograph corresponding to a particular atomic emission line. The multiple detectors accumulated separate signals simultaneously and integrated them over a number of seconds. A capacitor was charged to reflect the photocurrent over time, and the stored information was dumped by discharge of each capacitor into an information channel dedicated to that atomic emission line. The instrument subsequently solved appropriate analytical simultaneous equations to provide a direct readout for each element of interest. Because no two independent detectors had the same sensitivity, each required a separate calibration and voltage control on each photomultiplier, and the amplifier gain had to be maintained very tightly.

Two technical advances have made possible the modern diode array for spectroscopic instruments: (i) the development of excellent photovoltaic solid-state devices, and (ii) the use of a memory chip to store correction coefficients and to compensate for the difference in sensitivities of each different element in the diode array. Also, solid-state engineering, particularly for production of silicon photodiodes, has become advanced, so that the yield of hot pixels in an array is high and the cost of their manufacture is lowered. The color-matching devices in paint stores for a number of years have relied on a multichannel, silicon photodiode array coupled with a replica grating for use in the visible region of the spectrum.

In this relatively low-cost device, radiation (white light) reflected off the paint was conducted to a slit, a grating, and subsequently to carefully positioned photodiodes. Typically, a dozen or fewer detectors were required to perform the paint-matching exercise. Typical users of this device (employees of a paint store) probably had no idea of exactly what was going on inside of their color-matching optical device. An early attempt to convert one of these simple devices for use in the NIR region was described by David Honigs at the Boston Federation of Analytical Chemistry and Spectroscopy Societies (FACSS) meeting in 1988 (Honigs and DeThomas, 1988). In this case, the photodiode array simply was moved over to a position where the NIR rays would fall on various elements, and data actually were reported based on responses at several wavelengths in the NIR. The author is familiar with attempts by other experimenters to utilize UV/visible diode array instruments and to obtain NIR data by moving the grating. This failed in several instances, because certain manufacturers put the order-sorting filter, required with gratings, on top of the diode array.

One notable success in adapting a silicon photodiode array to NIR was the work done at the Center for Process Analytical Chemistry (CPAC) in the Department of Chemistry, University of Washington, Seattle, by Professor Callis and coworkers. The motivation for this particular work was to produce a portable, lightweight octane analyzer. A diode array instrument and a laptop computer were put together to produce such an instrument operating in the very near infrared region (880–1,050 nm) that was described previously (Lysaght et al, 1990).

TABLE I
Specialized Configuration or Application

Fiber optic sampling probe	Double transmittance (transflectance)
Whole-seed or pellet optimized (dedicated)	High speed detector (real time quantity)
Noncontact monitor	High speed scanning (real time quality)
Moisture (dedicated)	Electronic wavelength switching
Octane (dedicated)	No moving part operation
Clear fluid monitoring (dedicated)	NEMA-enclosed industrial instrument
Diffuse reflectance (dedicated)	Automatic wavelength referencing to standard
Diffuse transmittance	Single wavelength light scatter probe
Polarization polymer test dichroic ratio before and after elongation	
Oscillatory time resolved data points along perturbation waveform	
Imaging functional group mapping	

One prominent, commercial, silicon photodiode array instrument, PIONIR 1024, was produced initially and introduced by the Perkin-Elmer Corp. (Norwalk, CT) in the Process Analytical Instruments Division. The PIONIR, now a product of Orbital Sciences Corp. (Pomona, CA), operates with a fixed holographic grating and a photodiode array detector with a spectral range of 800–1,100 nm. It has a spectral resolution of less than 3 nm and a dynamic range of 25,000:1 at 850 nm, based on a 15-sec measurement cycle. Wavelength repeatability is rated as ±0.002 nm scan to scan, with a long-term wavelength accuracy of ±0.01 nm. The manufacturer reports that the 15-sec measurement for data could give information on as many as 32 different properties simultaneously (Perkin-Elmer Corp., 1992). The PIONIR comes equipped with a 200-μm-diameter fiber composed of a low OH silica core in cladding. The optical fiber is designed to withstand the high pressures and reasonably high temperatures that would be anticipated in an industrial setting. It is designed for remote and automatic usage and provides various control outputs, so that a process control computer could be driven from the information obtained with this monitor. It reportedly has a 500-hr source lifetime. It is equipped with self-diagnostic characteristics to operate as a remote system and operates in a cabinet that is pressurized with particle-free, oil-free, vapor-free, and dry compressed air. A water-cooled system also is available.

In the broader NIR region, a highly efficient and very rapid-response, highly sensitive, InGaAs photovoltaic detector is now on the scene. This was known to be available in the past, but the prices when it was first introduced were prohibitive. However, now in the array form, it has become a reality.

In the early 1990s, LT Industries, Inc. (Rockville, MD), introduced a commercial photodiode array instrument in which the source of illumination directs radiation toward a sample and from the sample by way of fiber optics and a multiplexer, to a spectrograph operated as a polychromator (Fig. 1).

The polychromator has an entrance slit of approximately 100 μm and a concave grating rated at f/3. Located on the Roland circle is an array of 256 InGaAs diodes. In fact, an array of 512 could occupy the same position. The diodes are 100 μm high × 30 μm wide and are spaced on 50-μm centers. The optical range covered by this array is 800–1,750 nm. A limit of 1,750 nm is specified,

TABLE II
NIR Instrument Types and Their Developers

Type	Developer
Scanning instruments	
Electronic wavelength switching	
Diode array grating polychromator (Section II)	University of Washington (Seattle) 1990 silicon range photodiode
	Orbital Sciences Corp. (Pomona, CA) silicon range octane instrument
	LT Industries, Inc. (Rockville, MD) indium gallium arsenide range (Powerscan)
	KES Analysis (New York) indium gallium arsenide
	Perten Instruments North America, Inc. (Reno, NV) indium gallium arsenide parallel array
	Ocean Optics (Dunedin, FL)
	Bühler, Inc. (Uzwil, Switzerland) dedicated plastics sorting device
	Carl Zeiss GmbH (Jena, Germany)
	On-Line Technologies (East Hartford, CT) diode array/linear variable filter (no grating)
Acousto-optic tunable filter random access spectrometer (Section III)	Kansas State University (Manhattan) 1986 research model
	Brimrose Corp. of America (Baltimore, MD)
	Fibertech (Rockville, MD)
	Rosemount Analytical, Inc. (Orrville, OH) industrial lead sulfide system
	Bran and Luebbe (Nordestadt, Germany) octane analyzer
	Various one-of-a-kind homemade systems
Interferometer	
Fourier transform spectrometers (Section IV)	Bomem/Hartmann & Braun (Quebec, Canada)
	Bruker Optics (Billerica, MA)
	Nicolet Instrument Co. (Madison, WI) (currently Thermo-Nicolet)
	Bio-Rad Spectroscopy Division (Cambridge, MA)
	ATI (Madison, WI)
	MIDAC Instrument Co. (Costa Mesa, CA)
	Mattson Instrument Co. (Madison, WI) (currently Thermo-Mattson)
	On-Line Technologies (East Hartford, CT)
Polarization interferometers (Section IV)	Bühler, Inc. (Uzwil, Switzerland)
	Bran and Luebbe (Nordestadt, Germany)
Grating monochromator (Section V)	U.S. Department of Agriculture (Beltsville, MD) research model
	Pacific Scientific (Silver Springs, MD) (currently Foss NIRSystems)
	LT Industries, Inc. (Rockville, MD)
	Technicon Industrial Systems (Buffalo Grove, IL) (currently Bran and Luebbe)
	Perstorp Instruments, Tecator Div. (Höganäs, Sweden) (currently Foss Tecator)
	Guided Wave Process Analytical Systems (El Dorado Hills, CA)
	Analytical Spectral Devices (Boulder, CO)
Discrete wavelength instruments	
Interference filters (Section VI)	Dickey-John (Auburn, IL)
	Infrared Engineering Ltd. (Maldon, Essex, United Kingdom); NDC Infrared Engineering (Irwindale, CA)
	Perten Instruments North America, Inc. (Reno, NV)
	Oxford Instruments (Oxen, United Kingdom); LECO, Inc. (St. Joseph, MI)
	Technicon Industrial Systems (Buffalo Grove, IL) (currently Bran and Luebbe)
	Kett Co. (Tokyo, Japan); Kett U.S. (Villa Park, CA)
	Zeltex, Inc. (Hagerstown, PA)
Discrete-source light-emitting diodes + filter (Section VII)	Zeltex, Inc. (Hagerstown, PA) (formerly Trebor Co.)
	Futrex, Inc. (Hagerstown, PA)
	Katrina, Inc. (Hopkinton, MA) (currently Moisture Systems Corp.)

but we recognize a limited sensitivity beyond 1,700 nm in this particular array, even though it is operated with two-stage thermoelectric cooling. Each pixel covers approximately 3.8 nm at the particular positioning with the grating currently used. Another fact concerning this particular 256-diode element (or 512) InGaAs diode array is that it is operated in a parallel readout mode (LT Industries, Inc., 1992). The particular commercial version of this instrument that is described, known as the Powerscan, is produced and offered on the market by LT Industries, Inc.

One feature of the Powerscan is that it has a built-in, optical multiplexing unit, which enables it to have seven optical fiber arms to acquire data from seven remote-sampling measurement sites. As a multiplexer, it does not read simultaneously, but reading can be done sequentially or operated in a random-site access. In this mode of operation, it does a complete sampling of all the wavelengths at the one site, the multiplexer switches to another position and gets the incoming radiations of the optical fiber, and then the spectrograph subsequently takes a multiwavelength snapshot at that site, and so on. The typical source that comes with this is a high-intensity, quartz-tungsten-halogen lamp. The source lamp is followed by an appropriate lens that sends the energy through an optical fiber bundle element. Various sample probes are available. Illumination of the sample from one direction will give rise to energy coming away from the sample, which can be 180° (in a transmittance format) or anything in between that and an offset of a few degrees from the input (sometimes referred to as interactance).

Optically, this instrument operates very rapidly, partly because of the availability of high-speed detectors in this range. It collects 100 scans per second and has computational speed that is not limited to that acquisition rate. The built-in digital signal processor is able to accumulate the data at 100 scans per second and carry it through appropriate analytical transformations to perform log functions, derivative functions, derivative spectra, and calculated analyses all within the collection time. Another item of the optics is a built-in polystyrene disk, which serves as a reference. The purpose of this standard is to assure wavelength accuracy. The built-in standard can be inserted on call from the computer to apply to any of the seven channels from which the multiplexing is done. The purpose of the polystyrene is not just for a wavelength check but also to serve as a sample of known spectral features up and down the spectrum, so that diagnostic functions can be performed on any one of these information channels.

Referencing is accomplished by using the same properly cleaned probe either on an appropriate solvent or against air. This is done on a periodic basis. Alternately continuous intermittent referencing is possible using one of the seven samples on the multiplexer and another probe dipped into an appropriate reference. Obviously, differential spectra could be produced in this manner. To perform a match with any other diode array system, a dual-wavelength mode of referencing can be used. In this case, a readable wavelength that has no reason to change, based on the composition of the material being measured or the sample matrix, serves as a pseudo baseline-intensity reference source to apply to a classical Beer's Law expression or statistically produced empirical expression.

The Powerscan is supported with the usual type of software in terms of its function, and the current software, although updated, is compatible with previous software developed for use on other products from the same company. This software includes partial least squares (PLS), multiple linear regression (MLR), principal

component regression (PCR), and various transformations, such as derivatives to go along with the conventional $\log(1/R)$ or absorbance expression. The Powerscan includes control software and routine operational software but also is equipped with method development capabilities. The market for this equipment is definitely industrial, and it is housed in a level-4 NEMA cabinet. Because it is equipped for heavy-duty industrial use, but also can be used for method development, it is no longer necessary to download a laboratory method to some remote implant system. Both analytical functions can be accomplished with the same system.

Another diode array system patented by Stark (Stark, 1991) of KES Analysis (New York, NY) was initially commercialized by Perten Instruments North America, Inc. (Reno, NV). A very rapid, parallel readout InGaAs array (DA-7000) instrument engineered at the Chatham, IL, facility was introduced that samples a 5-in.-diameter area to perform quantitative analysis in an industrial setting. The parallel array has sensing elements aligned at uniform wavelength spacing throughout the spectrum, as dispersed by the stationary grating. This instrument has a silicon array for 400–900 nm functioning in parallel with an InGaAs array for the range of 950–1,700 nm. Parallel simultaneous sensing uses all the energy at each spectral point and performs its amplification and analog-signal processing with a separate signal channel for each sensing element. This differs from other diode array spectrometers that do not utilize parallel processing but sample the respective energy of each sensing element in sequence to produce a sequential series of measurements.

In parallel operation, optimal signal conditioning can be performed without time-sharing among individual diodes in the array. This reportedly provides an enhanced dynamic range and the full multiplex advantage (Perten Instruments North America, Inc., 1996). The system provides oversampling of each observed point in the spectrum, so there is negligible time-skew of recorded spectral data, and dual-beam processing provides continuous referencing of background effects. Chopper-synchronization provides automatic rejection of ambient light throughout the spectrum and permits noncontact measurements. The parallel channel diode arrays with their sampling rate of 600 spectra per second and parallel processing challenge the speed and sensitivity of other diode array or monochromator-based instruments. A spectral freeze-frame view of dynamic processes and rapidly moving samples results when the sample is placed on a transparent surface a few inches above the instrument. Light hitting the sample is reflected back to the polychromator to accomplish the optical measurement. A fiber optic-coupled version of this diode array instrument is available for other reflectance or transmission sampling tasks. Communication via RS-485 at 2-MHz clock speeds allows multiplex polling of four different sampling systems.

A grating polychromator is embedded into a specialized plastic sorting device (the "Niriks" system of Bühler, Inc. [Uzwil, Switzerland]) that is dedicated to single use. In this instrument that is now a product of Büchi, Inc., of Flawail, Switzerland, output from various pixels of the germanium photodiode array are reportedly weighted appropriately to emulate the principal-component eigenvector at each of these wavelengths. The configuration (weighting of terms and choice of wavelengths) is designed specifically for each application. Typically, this procedure is followed for four "factors" in parallel operation, and multiplexing is accomplished via divided fiber optics coming from a common fiber optic bundle exposed to white light (quartz-tungsten-halogen) transmitted through or reflected by the unknown plastic specimen subjected to sorting. From the typical response to the four factors, discriminate analysis is performed from a fixed menu such as polyethylene, polyethylene terephthalate, polypropylene, or polyvinyl chloride.

Optical Solutions, Inc. (Folsom, CA) produces a fiber optic-equipped portable instrument with a stabilized light source and diode array detection and chemometrics package for use in the NIR region.

Carl Zeiss GmbH (Jena, Germany), a company of long standing in the optics field, initially introduced a diode array polychromator in 1997 at Pittcon. This spectrograph is available with

Fig. 1. Block diagram of grating polychromatic diode array spectrometer showing the **A,** source; **B,** sample cell; **C,** slit; **D,** stationary holographic diffraction grating; **E,** diode array (mounted on the Roland circle with respect to the grating); and **F,** signal processing electronics.

silicon or InGaAs diode arrays with 128 or 256 pixels as an original equipment manufacturer (OEM) item, complete with software to control the acquisition of the optical data. Incorporation of the spectrograph with a custom NIR instrument was available initially. Currently, an off-the-shelf MCS511 analytical instrument is produced. Three wavelength ranges are available within 800–2,400 nm.

Recently, a special circuit board with the optical fiber and spectrograph attached has become available to plug into an ordinary personal computer (PC). Such a device is available from Ocean Optics, Inc. in Dunedin, FL. The Ocean Optics system employs a grating polychromator for dispersion like the other instruments discussed above. However, it features a thermoelectrically cooled 256-element lead sulfide (PbS), linear charge-coupled device (CCD) array with a range of 1,000–3,000 nm (Ocean Optics, Inc., 1993). The pixel resolution is approximately 1 nm in the general use configuration. Unlike the photodiodes that produce a voltage buildup upon exposure, the CCD elements are charged prior to exposure, and subsequent change of this charge (photon counting) is monitored for each pixel with an integrator clock. Photo current discharged through an analog-to-digital chip produces counts for each of the 256 pixels. Ocean Optics, Inc. also produces an instrument with a 1,024-pixel CCD camera in the silicon-sensitive wavelength range.

A linear, continuously variable interference filter in series with a photodiode array of either silicon or doped PbS is the basis of the instrument from Textron Systems (Wilmington, MA). In collaboration with Case/New Holland Corporation (Burr Ridge, IL), Textron Systems produced an in-the-field, continuous flow grain analyzer (Anonymous, 1999). These analyzers measure protein and oil levels in whole-seed corn, wheat, and soybeans in real time. This solid-state system designed for harvest combines is applicable to grain elevators, feed systems, and processing plants.

III. ELECTRONIC WAVELENGTH SWITCHING: ACOUSTO-OPTIC TUNABLE FILTER SPECTROMETER

Acousto-optic tunable filter spectrometers (TFSs) allow the taking of random wavelength accessed data without the necessity to sweep through all data points. However, scanning by acquisition of data at sequential points is an option. The main feature of this approach is that no moving parts are required, which made development of an industrial on-line process monitor based on this technology an attractive and optimistic goal of the author and coworkers beginning in the mid-1980s.

Acousto-optic tunable filters (AOTFs) were proposed by Harris and Wallace (1969). The use of acousto-optic modulators is common in everyday devices such as laser printers. Ultrasonic energy fed into an acousto-optic crystal, through which the light must pass, alternately transmits or blocks that light. In AOTF devices, a piezoelectric transducer bonded to an acousto-optic crystal, usually composed of tellurium dioxide, is used to insert ultrasonic radio frequency (RF) energy into the crystal. A particular tuned optical frequency passes through the filter corresponding to the RF (ultrasonic) tuning frequency applied. (The crystal is designed to be tuned within a particular wavelength range.) An ultrasonic absorber is placed at the opposite side of the crystal. The solid-state engineering of this device is important. Optimization for anisotropic forms of such a device was contributed nearly 2 decades ago by Chang (1974). In construction, once the axis of the raw crystal has been established by X-ray diffraction, the cleavage of the crystal and architecture of the substrate from which a device is produced will determine the performance characteristics of the filter. The wavelength range, the size of the aperture, and the resolution are of concern in writing specifications for production of a custom crystal to perform a particular wavelength switching task. Acousto-optic tunable filter theory, equations, and a more detailed discussion are found in the recent five-volume spectroscopic treatise edited by Chalmers and Griffiths (Wetzel et al, 2001).

AOTFs received considerable early usage in laser laboratories for angular deflection of the beam, which was performed as a rapid, electronically switchable function. Electronically switchable dual-wavelength devices based on AOTF also were used early in the fiber optics communication field to locate poor connections. Demonstrating the use of an AOTF as a means of scanning the spectrum has been relatively easy from the beginning of the production of such devices (Gottlieb et al, 1980), but producing spectroscopic data worthy of quantitative analysis in the NIR region was a completely different matter. We undertook that task at Kansas State University (KSU), Manhattan, KS, in the mid-1980s because we could find no public record of such a quantitative instrument for use in the NIR region. Based on early work conducted in Manhattan, KS, a patent application was filed, which was subsequently granted for a quantitative instrument based on this technology (Kemeny and Wetzel, 1989). Quantitative data on corn oil in freon as well as numerous spectra were made public at the Denver 1987 American Chemical Society meeting (Wetzel et al, 1987). A pulsed, mid-IR stack monitor based on a ternary mixture, IR-transmitting crystal was introduced by Westinghouse (Pittsburgh, PA) and displayed for sale in the late 1980s.

The original 1986 continuous-wave version of the KSU acousto-optic tunable filter spectrometer (TFS) instrument was built around a custom-designed AOTF crystal using mostly components scavenged from other NIR instruments, including the source and detector. Nicol polarizers from a polarimeter were used in the first instrument as well as a mechanical chopper. The RF energy was inserted into the crystal from an oscillator that had been ordered specifically to deliver appropriate frequencies to drive this crystal. Software was written locally to drive the oscillator by way of a digital-to-analog converter through an appropriate voltage ramp to produce the desired ultrasonic frequency. That particular crystal had an optical response corresponding to the ultrasonic input. Figure 2 shows the block diagram of the 1986 acousto-optic TFS.

NIR radiation from the source was focused on the entrance to the crystal, by way of a polarizer that was tuned, and subsequently assumed an angular deflection as it passed through a second polarizer, which was rotated 90° to the initial polarizer, before striking the sample and subsequently a lead sulfide detector. A phase-sensitive detector (lock-in amplifier) received the chopped signal that went to an expansion-board-mounted A/D converter in a PC. The designation of scanning the spectrum with electronic wavelength switching is readily apparent, because the scanning function is driven from the central processing unit (microcomputer), and the detector signal from the optical process also is accepted by that same computer. Those early experiments lacked optical efficiency, and some electronic deficiencies also occurred, but the painstakingly obtained data led the way to future development.

Since that time, what has become known as the KSU research model acousto-optic TFS was developed over a period of years. Every single optical component of the original working instru-

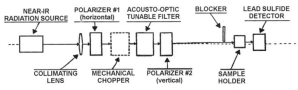

Fig. 2. Block diagram containing optical arrangement of the 1986 quantitative near-infrared Acousto-Optic Tunable Filter Spectrometer. Note that this instrument included a mechanical chopper that was eliminated soon after be using frequency modulation. Source (quartz-tungsten-halogen), lens, horizontal polarizer, chopper, acousto-optic tunable filter (7-mm aperture custom-made tellurium dioxide), vertical polarizer, blocker (razor blade), sample holder (for film or 1-mm quartz cuvette), and lead sulfide detector (initially 10 × 10 mm). The first model had radio frequency ultrasonic input to the crystal from an oscillator controlled through a D/A converter attached to an expansion board of a personal computer (PC). Using the lead sulfide detector required a sequential preamplifier, lock-in amplifier (Ithaco model 393 [Ithaca, NY]), and A/D converter attached to an expansion board of a PC.

ment was changed, and three other AOTF devices were substituted. As our learning curve developed, the progress achieved was reported in a series of Eastern Analytical Symposium, Pittcon, and FACSS presentations, culminating with the Hirshfeld student award to physicist A. J. Eilert (Eilert and Wetzel, 1991) and a subsequent Pittcon invited presentation by the author (Wetzel, 1992). The final version was equipped with an AOTF device from Crystal Technology, Inc. (Palo Alto, CA). In the PbS detector version of the research instrument, chopping was produced by sequentially tuning and detuning the AOTF, and a dedicated phase-sensitive detector from Evans, Inc. (Berkeley, CA) was incorporated for routine usage (Wetzel and Eilert, 1990). In an advanced intermediate version, a thermoelectrically cooled PbS detector from New England Photoconductor (Taunton, MA) was used to give excellent reproducible data.

An effort to take advantage of this speed of the electronic wavelength switching was made by incorporating an Epitaxx, Inc. (Los Angeles, CA) phosphorous-doped InGaAs detector. This thermoelectrically cooled, photovoltaic detector has a rapid operating speed, and the phosphorous-doped version extends the wavelength range to 2,400 nm, considerably beyond the 1,700-nm cutoff of the conventional InGaAs detector. An operational amplifier was used in place of a phase-sensitive one, and a frequency synthesizer provided RF to the AOTF. The research model still incorporates polarizers; however, the Glan Thompson polarizers selected for the final version gave a high efficiency of polarization. They transmit in the region of interest and provide good rejection when crossed. A blocker was added to give geometric restriction and not rely completely on the combination of the tuning and polarizer efficiencies to eliminate untuned radiation and the tuned ray not being used.

Figure 3 shows the advanced configuration of the KSU research model. This was referred to as the "fastest gun in the West." It was tested as a flow-through monitor by using HPLC pumps and a gradient programmer to produce calibrations on binary mixtures (Wetzel, 1992; Eilert, 1995). Spectral subtraction was used to prove wavelength reproduction and linearity of response (Eilert et al, 2000). Random wavelength access provided a relatively high duty cycle compared to other scanning instruments, and this was partially responsible for the quantitative success. Speed of the instrument for use as a transient monitor was established by mechanically introducing different thicknesses of plastic sheets in the beam. Using a two-wavelength expression, 480 analyses were possible in 3 sec, which amounts to an analysis time of 8.3 msec. Isophotonic data accumulation was demonstrated to enhance the signal-to-noise ratio at wavelengths where radiation intensity was low. A longer accumulation time was programmed to enhance the signal-to-noise ratio. Other wavelengths where the signal intensity was adequate (either because the instrument was transmissive and sensitive in that region or because no strong absorbers existed) could be sampled for a much shorter period of time. This software-controlled interactive data acquisition enhanced the performance of the instrument in some cases for quantitative purposes.

While the KSU research model instrument was being improved, the builders, under contract, provided technology transfer to a team of eight engineers at the Bran and Luebbe, Elmsford, NY, facility to

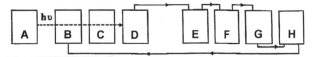

Fig. 3. Block diagram of the final version of the Kansas State University research model Acousto-Optic Tunable Filter Spectrometer. Shown are the **A,** source (reflectively enhanced quartz-tungsten-halogen); **B,** monochromator consisting of near-infrared polarizers (Glan Thompson) and acousto-optic tunable filter driven by **H,** a radio frequency synthesizer; **C,** sample cell; and **D,** indium gallium arsenide(P) detector (thermoelectrically cooled) with **E,** an operational amplifier leading to **F,** the A/D converter and **G,** personal computer.

produce a commercial instrument. Robert Rachlis was the optical engineer on the project directed by G. J. Kemeny, the coauthor of the earlier patent (Kemeny and Wetzel, 1989). Three TFS units of the range of 1,100–1,700 nm were built at Elmsford, NY, and a commercial instrument was introduced under the name Infra-Alyzer AOTS at the New York meeting of the Pittsburgh Conference on Analytical Chemistry and Applied Spectroscopy in 1990.

After the Elmsford, NY, group that designed and built the commercial TFS instrument was disbanded and the Bran and Luebbe operation moved to the parent company facility in Nordestadt, Germany, essentially the same instrument was reintroduced after considerable delay under the name InfraPrime as a high-cost, process monitoring device for measuring octane number in the petroleum industry and for selected industrial monitoring in the European chemical industry. The 1990 commercial instrument and its successor use the ordinary ray and extraordinary rays for monitoring and for referencing in a double-beam mode.

After the previously discussed custom research model acousto-optic TFS instruments were refined to their highly developed state and early commercialization starts had been made, Brimrose Corp. of America (Baltimore, MD) aggressively entered the field of manufacturing NIR instruments. This company, a well-established OEM supplier of optical devices, first produced AOTF devices and RF drivers as an OEM supplier and subsequently built a commercial TFS-based process monitor (Luminar 2000) as well as a tabletop instrument. Brimrose Corp. uses a different scheme for referencing than that previously described, and its initial instruments were exclusively fiber optic-coupled with probes for transmission measurement of liquids or for diffuse reflectance measurement of liquids. For diffuse reflectance measurement of solids (e.g., powders, polymer pellets, and food products), a remote fiberless Luminar 2030 "Free Space" system is used in a process-hardened enclosure. This vertically integrated American company headed by a Ph.D. physicist progressed from producing the solid-state AOTF device and its RF driver, to manufacturing complete NIR instruments, and finally to developing customized sampling, control software, and a chemometrics package capable of sending real-time results to a process controller.

Since its initial entry into the NIR field, Brimrose Corp. has grown rapidly as an instrument company and expanded its product line to include specialized systems. Its latest product is the Luminar 3000 "Sentinel," which is a rapid pharmaceutical analyzer designed for 100% inspection of tablets, capsules, and vials on-line. This manufacturer also markets the Luminar 2035 "Seed Meister" product for measuring oilseeds quantitatively by rapidly analyzing single seeds nondestructively and then sorting them into high-oil, low-oil, and reject receptacles. To accomplish this, the developer produced chemometric analytical software in addition to instrument control and sampling automation software. Brimrose Corp. has gone worldwide with its AOTF-based NIR instrument and opened offices in London (United Kingdom), Mainz (Germany), and Sao Paulo (Brazil). Its industrial version is distributed widely, and a miniaturized version, the size of a man's wallet, with minimal power requirements has been produced by Brimrose Corp. for the National Aeronautics and Space Administration for space flights scheduled early in the 21st century. Brimrose also has a thin-film acousto-optic TFS and AOTF-based imaging microscopes for use with NIR-sensitive cameras.

Another acousto-optic TFS is produced by Rosemount Analytical, Inc. (Orrville, OH). This strictly industrial quality control monitor operates with a PbS detector. Fibertech of Rockville, MD, produces custom AOTF instruments. Various other companies have offered custom instrument production of AOTF-based equipment but have not offered off-the-shelf models.

Various one-of-a-kind NIR acousto-optic TFS instruments were built in government or academic laboratories, including Professor Siesler's at the University of Essen, Department of Physical Chemistry, Essen, Germany, which was used to monitor the stretching of polymers for dichroism studies (Huehne et al, 1992). In comparison to any grating monochromator scanning device, acousto-optic TFS instruments are extremely wavelength reproducible. With the KSU research model, only a single scan was required in a few milliseconds to produce a spectrum that needed no smoothing. In contrast, all grating monochromator spectra shown involve coadding several spectra or taking one at a very deliberate pace. Even FT-NIR instruments cannot accumulate data of this quality as rapidly. Complete software control of the data acquisition is the key to tailoring a flexible acquisition system that maximizes the signal-to-noise ratio and enhances quantitative analytical performance.

IV. FT-NIR INSTRUMENTS

An interferometer provides certain advantages that apply to spectroscopy in the mid-IR region. These include the multiplex advantage and the throughput advantage. Throughput of an interferometer instrument is greater because, unlike a grating monochromator, it has no entrance or exit slits. In an interferometer, the entering beam of radiation from the source encounters a beam splitter. This optical device reflects approximately half of the incident radiation and transmits the other half. In one pathway, a second mirror is encountered, and radiation coming off the second mirror rejoins the rays coming straight through the beam splitter to proceed onward to the sample target. Because the second mirror is moving, the pathways to and from the movable mirror are variable as a function of time. At different mirror positions, a difference in pathlength produces interference. Data accumulated during the time of oscillation are subjected to fast Fourier transformation. The optical frequency resulting from interferometry of a broad-range infrared beam is a cosine function of the difference between the two mirror paths. In the mid-IR region, the challenge always is conserving a low amount of energy; thus, the throughput is very important. In the NIR region, bright sources are available, and throughput is less essential. Also, with all of the wavelengths simultaneously hitting the detector, the ability to discriminate between small differences in intensity has been regarded as an instrumental challenge. Throughput thus is no longer the issue in NIR, but the linear dynamic range of the A/D converter overtakes throughput when striving to increase the signal in an NIR interferometer type of instrument useful for application to quantitative analysis.

Although wavelength resolution never has been regarded as an issue in NIR, wavelength reproduction certainly has been. Actually, an interferometer offers both excellent resolution in comparison to a grating monochromator and excellent wavelength reproduction. The latter is made possible by using interference fringes from a laser in a path parallel to the IR path to allow verification of the mirror position. Figure 4 shows a typical path of a Michelson interferometer. Note the alternate pathways of the light emerging from the beam splitter and recombining at the secondary surface of the beam splitter either in or out of phase. Before dedicated FT-NIR spectrometers were manufactured, a number of isolated laboratories used NIR accessory capability for conversion of a stock mid-IR instrument to NIR use. These will not be discussed in this report.

An industrial monitor based on an FT-NIR instrument is produced by Applied Automation/Hartmann & Braun (Bartlesville, OK). This employs the Bomem MB155 or 160 interferometer bench produced by Bomem/Hartmann & Braun (Quebec, Canada). The particular interferometer employed in the Bomem Michelson series is regarded as rugged with reference to vibration. Optically, it is described as a cube-corner retroreflector design. This reduces the actual length without reducing the optical pathlength. The motion of the mirror is supported on a steel hinge device. The source is a quartz tungsten halogen lamp, and the detector is lead selenide (PbSe). This is used because its speed of response is greater than that of PbS. The author (Baudais et al, 1990; Wetzel and Eilert, 1992) had an early opportunity to test the Bomem MB155 for quantitative analysis before it was incorporated into the Applied Automation Systems instrument.

For this work, chemicals were chosen that had different characteristic spectral features. When scattering samples were avoided, the original instrument performed for quantitative purposes in a way comparable to dedicated-design, quantitative NIR instruments of the past. The placement of scattering samples in the beam of a standard sample compartment did not provide for a sufficiently large angle of collection. The modern version now dedicated to NIR use differs in two ways from the original benchtop MB155. It uses fiber optic coupling, and the probe is designed in such a way that light scattering does not reduce the collection of radiation. In addition, an effectively higher solid angle of collection is preserved through the use of various probes attached with optical fibers to the interferometer and the detector. Also, the file structure for collection of scans was changed. Formerly, all FT instruments had one file for a single spectrum resulting from coadded scans of one specimen. For performing various chemometric procedures, whether MLR or global statistical techniques, in which the whole spectrum was used, a different file structure is required, and this has now been accommodated with appropriate data-acquisition software.

The Nicolet Instrument Co. of Madison, WI (currently Thermo-Nicolet), makes a fiber optic probe that is provided as an accessory for its Magna series interferometer bench. The Magna bench is equipped with a readily changeable beam splitter, so it is possible to convert from mid-IR to NIR in just a few minutes. Also, when it is outfitted for both NIR and mid-IR, a mirror can be switched from one source to the other. In the Nicolet FT-NIR, the moving mirror is supported by a glass bearing for the mirror sliding motion. Nicolet Instrument Co. employs a calcium fluoride beam splitter for the NIR instrument but also can provide a quartz beam splitter. Unlike the other FT-NIR instruments previously mentioned, this one utilizes a special PbS detector that was developed to have a high speed and a high D*. This particular PbS detector eliminates the need to deal with a shift in the frequency of maximum response, and, at the same temperature, the D* of the PbS is greater than the D* of PbSe.

Before 1996, this long-standing FT-IR instrument company did not specifically market an FT-NIR product, but owners of its instrument could readily make a conversion and have fiber optic sampling available by purchasing an accessory. The author has had the opportunity to test such a system for quantitative analysis and report the results (Wetzel, 1995; Wetzel and Sweat, 1997). All in all, it received good marks, with the only limitation being that the sampling area of the probe required averaging multiple probings when dealing with heterogeneous samples. Data file organization and chemometric analytical software was bundled with a Nicolet standard-model interferometer bench configured with appropriate source, beam splitter, and detector for NIR interum use in both transmission and reflectance sampling configurations. In 2000, a new spectrometer designed from the ground up specifically for FT-NIR was introduced by Nicolet Instruments that has five detectors, an integrating sphere for diffuse reflectance, transmission capability, and a fiber optic probe with its own detector.

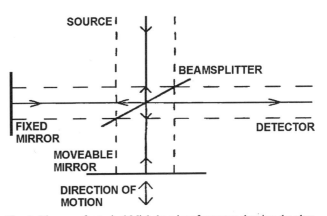

Fig. 4. Diagram of a typical Michelson interferometer showing the alternate light paths from the beam splitter. Rays reflected to the movable mirror and travel back to the beam splitter to be united with those from the fixed mirror before entering the sample and detector.

Both standard (1,100–1,700 μm) and extended (1,100–2,400 μm) range In-Ga-As detetors are used.

Bio-Rad Spectroscopy Division (Cambridge, MA), another long-standing manufacturer of FT-IR instruments, markets an FT-NIR version of their standard interferometer bench that employs a fiber optic probe provided by a third party. It uses the same source as all NIR instruments, a quartz beam splitter, and a PbSe detector. Third-party suppliers of optical fibers and probes for coupling to instruments of this type are Galileo Electro Optics Corp. (Sturbridge, MA) and Graseby Controls, Inc. (Greensboro, NC).

The Mattson Instrument Co. of Madison, WI (currently Thermo-Mattson), a manufacturer of FT-IR spectrometers for many years, has a dedicated FT-NIR instrument, the Genesis NIR System. This instrument can be configured with a variety of NIR fiber optic probes for remote sample analysis. This system uses third-party chemometric software.

The MIDAC Instrument Co. of Costa Mesa, CA, produces an instrument that performs oscillations on a mechanical bearing. One version of a MIDAC instrument designed for field use at long paths (kilometers) has been used for some time in both mid-IR and NIR for air monitoring. MIDAC NIR field instruments have been used extensively in environmental work for volatile organic compounds by Dr. Fately's spectroscopy research group in the Chemistry Department at KSU (Chaffin et al, 1995).

Bruker instruments made in Germany and sold in North America out of Billerica, MA, by Bruker Optics recently have entered the field of FT-NIR. This company traditionally has been known for its laboratory instruments, including the high-end, mid-IR instrumentation. Bruker Optics entered the NIR field with the Vector 22/N, a dedicated FT-NIR spectrometer, described as an instrument for quality assurance/quality control and on-line process monitoring in industrial environments. It employs a variety of fiber optic accessories, as do the other instruments, and as many as four accessories can be multiplexed optically to the spectrometer for multisite sampling. A 0.8-nm spectral resolution is quoted at 2,000 nm, and a 0.004-nm wavelength accuracy is quoted. The interferometer is equipped with what is described as a frictionless piston. It employs a PbSe detector. Data files have been constructed in the typical NIR chemometrics style to allow statistical method development, and the software provided (OS/2) is written by Bruker. Sample spectra are matched to those of knowns by Euclidean distance at each stored data point in the spectrum. Discriminate analysis is performed by use of principal component analysis (PCA) and subsequently applying Mahalanobis distances to the PCA factors.

All of the preceding interferometers described are versions of the classical Michelson interferometer. The Analect well-known "Trancept" interferometer (now a product of Orbital Sciences, Pamona, CA) employed another type of interferometer that uses a refractive wedge, which is oscillated back and forth in the beam to produce the difference in pathlength based on the thickness and refractive index. This system is not vibration sensitive. It also uses a cube-corner retroreflector mirror design and a CaF$_2$ beam split-

ter. Instruments made by Analect have, in the past, been adapted industrially to monitor heated polymers under extrusion in the mid-IR and NIR regions by enclosure in NEMA cabinets. This system comes with custom chemometric software and can be equipped with third-party software and a third-party fiber optic probe.

Another type of refracting optical wedge interferometer, a polarization interferometer, is produced in Switzerland and is licensed to Bühler (Uzwil, Switzerland) and to Bran and Luebbe (Nordestadt, Germany). The Bühler instrument, now a product of Buchi (Flawail, Switzerland), uses polarizers on either side of dual refracting prisms. The moving prism has a large amplitude of oscillation (Bühler, Inc., 1993). Therefore, the oscillation takes place rather slowly. The speed of operation is dictated by the slow oscillation; however, the large amplitude allows precise repositioning. In a typical operation, multiple scans are coadded to produce a spectrum. Although quantitative analysis is not ruled out with this instrument, it was produced primarily as an inspection tool to establish and verify qualitatively the identity of chemical stock in an industrial setting.

Any scanning NIR instrument typically uses spectral matching, in which a cosine function at each point in the spectrum is compared to a cosine function for the spectra of reference materials. However, other forms of discriminant analysis also are employed. In particular, the software furnished with the Bühler qualitative probe employs factor analysis. The factors produced by this chemometric procedure are then utilized in a discriminant analysis scheme. Typically, the best "hits" from a library of reference spectra will be given with figures of merit on each of the best matches chosen. The Bran and Luebbe polarization interferometer with the fiber optic coupling works similarly. The Bühler instrument is called the NIRVIS, and the discriminant software is NIRCAL. The Bran and Luebbe instrument is called the InfraProver.

A common feature of FT-NIR instruments is that the wavelength reproduction is excellent. In fact, it is so good that use of spectral subtraction should be possible. The author has demonstrated that, and Figure 5 shows the result of subtracting the spectrum of a solid sample of a light form of coffee whitener from one obtained from coffee whitener of the regular form to produce the spectrum of lipid (Wetzel, 1995). As previously mentioned, in the NIR region, some of the traditional advantages that are obtained with FT-IR are not necessarily an issue. However, with FT-IR, the wavelength precision is excellent, as is the precision in terms of the intensity readings for a number of reasons not discussed.

On-Line Technologies (East Hartford, CT) has produced an FT-NIR spectrometer with an accurate, highly stable, monolithic interferometer (PLS, Dear Park, NY). This interferometer (Blaier et al, 1999) was assembled from fused quartz plates in which the two mirrors and beam splitter (also made from quartz) are mounted, and axial movement of a hollow retroreflector separates and recombines the beam. The rigid alignment possible with the all-quartz system locks in the optical elements without distorting the exiting wavefront of the system. In addition to optical stability, the design, with no metal chassis, is not affected by changes in ambient temperature.

V. GRATING MONOCHROMATOR INSTRUMENTS

The NIR region of the spectrum has long been attainable for qualitative purposes, as the extended range of some standard grating monochromator visible instruments or an alternate range of a few IR instruments. The classical NIR experimentation of Wilbur Kaye (Kaye, 1955) and Kermit Whetsel (Whetsel, 1957) on various chemicals at Tennessee Eastman, Kingsport, TN, and of J. D. S. Goulden at the University of Redding (Redding, United Kingdom) (Goulden, 1968), were done with such instruments. Work of Kaye and Whetsel involved the use of baseline-corrected peak height (Beer's Law) in quantitative monitoring of liquids. Goulden obtained diffuse reflectance spectra of dairy products. The later era of direct, solid-sample, quantitative NIR, in which chemometric methods and diffuse reflectance were used, involved instruments

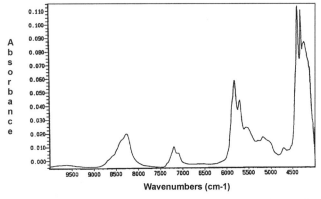

Fig. 5. Spectrum of lipid resulting from spectral subtraction with Fourier transform-near-infrared of coffee creamer lite from regular.

that required a much higher signal-to-noise ratio. Conserving the signal in a grating monochromator instrument involves compromise in the choice of the slit width and requires that the diffraction grating and all other optical components have a very low "*f*" number compared to most standard monochromators.

The chemometric quantitative NIR work reported in the laboratory of Karl Norris (U.S. Department of Agriculture [USDA], Beltsville, MD) used the modification by David Massey of a Varian grating monochromator (Ben-Gera and Norris, 1968). The hardware modification involved replacing the grating with one blazed for the NIR region and also replacing the source and the detector. The resulting customized device was driven by locally produced software. That instrument, copied for use in various academic laboratories, has been described previously (Massey and Norris, 1965), so it will not be discussed further here. NIR instrument companies also have produced various prototypes of grating monochromators for in-house use. Selections of filters to be incorporated into interference filter instruments were based on work with some of these custom-made scanning NIR instruments.

The author had the opportunity to use one of these early "one-of-a-kind" grating monochromator instruments to scan the spectrum of isolated components of agricultural products as well as the products themselves. The concept of quality control determination of bran in flour milling streams was an outgrowth of 1976 scanning work with spectra of individual portions of wheat gluten, wheat starch, cellulose, and bran compared to mid-IR spectra obtained with the KBr pellet technique from the same samples. In this work, the overtone and combination bands were related to fundamental vibrations (Wetzel and Mark, 1978).

Grating monochromator instruments equipped with very large grating ($f = 1.8-2.0$) for quantitative analysis became available to the public soon after and initially were of two types. The fast-scan type involved a vibrating grating that initially was driven mechanically. This instrument was produced by Pacific Scientific (the successor to Neotec and the predecessor of Foss NIRSystems [Silver Springs, MD]). The results of numerous rapid scans were coadded to average out the signal over many scans. This averaging not only contributed by averaging the intensity at each wavelength, but also served to compensate for any slight shift in wavelength between scans. The early, mechanically driven, rapid-scan, grating monochromators have been superseded with a much better system, so they are not discussed here.

In contrast to the rapid scan performed by oscillation of a grating is the stepping motor-driven, slow-scan, grating monochromator. A version of this type was produced by Technicon Industrial Systems (Tarrytown, NY) and known as the InfraAlyzer 500. It is currently a product of Bran and Luebbe (Buffalo Grove, IL). This instrument was built in the classical optical instrumentation style mounted on a very heavy base plate. The very large grating had an *f* value of 1.8 and was operated with a sine bar drive that was driven by a stepping motor. It used an enhanced source, in that the radiation from the lamp not directed toward the monochromator was focused by a spherical mirror back onto the filament. This instrument is worthy of mentioning many years after its initial production because the concept involved having a top-of-the-line scanning instrument that was matched to a series of filter instruments. The objective was to have a grating monochromator instrument that was capable of collecting data and developing methods that subsequently could be downloaded to either the model 400/450 or the model 300/350 filter instrument or the current model 2000.

In order to accomplish this, the grating monochromator instrument also was equipped with an integrating sphere to maintain optical compatibility (solid angle of collection, etc.) with corresponding filter instruments. With programmed pulses to the stepping motor, the grating was rotated to a different angle to produce a new corresponding wavelength at the exit slit. At this wavelength, the signal intensity was obtained from the sample. Next, a signal was read from a reference reflector, followed by a signal read again from the sample. Subsequent to that, a dark current

reading was taken. After the stepping motor received the next pulse, the grating moved to the next frequency, and the cycle of sample, reference, sample, and dark was taken again.

This grating monochromator instrument was designed to operate as a stand-alone device to run analyses; however, perhaps only a few of these were ever used in that manner. An auxiliary PC was used to drive the scanning function for data accumulation and to accept the data and archive it for statistical manipulation, method development, plotting, and various displays.

Currently, such an instrument represents an older design, but the concept of downloading quantitative methods from a grating monochromator instrument to a filter instrument that had the same basic sample optics is worthy of note. In particular, data received from the grating monochromator instrument reflected the wavelength transmission distribution profile of the slit. A mathematical function was provided to take the data from the profile of a slit and convert it to the typical profile of an interference filter, whose maximum transmittance was centered at the same point as that selected from the monochromator instrument. Use of the monochromator "mother instrument" involved making all data points in the spectrum accessible, whether or not a filter had ever been manufactured for that wavelength.

Offspring methods resulted from manipulating the data to make them appear as filter data and performing regressions to produce expressions capable of achieving quantitative analysis. Once these expressions were produced with wavelengths selected from all wavelengths in the spectrum, specialized filters could then be ordered or manufactured to produce a dedicated filter instrument for a particular purpose. In fact, this was done not only by the instrument manufacturer but by users of this family of instruments. Some of the filter instruments produced subsequently had specialized new filters replacing existing filters or being added in a vacant spot on the filter wheel. Single-purpose filter instruments thus were tailored to meet certain analytical needs, including on-line monitoring for flour mill control (Wetzel, 1986).

In the field, a large number of modern rapid-scan, grating monochromator instruments are currently in use. Most of these have been produced by Foss NIRSystems (the successor to NIRSystems/Pacific Scientific/Neotec) and by LT Industries. The Foss NIRSystems model 6500 and the LT Industries Quantum 1200 are used at numerous industrial locations. Both of these NIR instrument companies are committed exclusively to grating monochromator operation and do not produce a filter instrument. Since the early days of the first mechanically cam-driven, oscillating grating, these instruments have been very much improved. In this discussion, specific information available to the author is used to describe the Foss NIRSystems equipment and, in the absence of detail, the workings of the LT Industries Quantum series are inferred. The first improvement over the mechanically driven, rapid-scan, grating device was avoiding the moving contact involved with the cam drive. In the model 6500, something similar to the micropositioning technology (digitally driven) that repositions a computer disk drive to a particular address was used to reposition the grating. Thus, the position of the grating could be stepped rapidly and digitally through a series of points. With the emerging low cost of computer memory, it was no longer necessary to preselect the points so that each angle would correspond to a particular wavelength. Instead, this was done in a sweep motion through all points available with postoscillation selection of the data channels of interest corresponding to the wavelengths of interest.

This was a somewhat revolutionary concept in spectroscopic instrumentation. In the mechanical sine bar-driven era, it was quite a feat to preselect the mechanical steps to correspond to appropriate intervals of wavelength or frequency. In the modern era of low-cost computer memory, it was simpler to have an uncomplicated, digitally driven sweep and then to throw away all of the unwanted data that were accumulated between the points of interest. In the LT Industries Quantum series, a spring operates in a clocklike oscillating motion that eliminates the sensitivity to shock and vibration. The frequency is constant, as in a tuning fork. A bit of make-up energy compensates for the normal friction. The instrument also accumulates data at all angles during the forward oscillation of the grating.

An early failure of rapid-scan, grating monochromator systems was in wavelength reproduction. This problem was solved by periodically introducing a wavelength standard into routine operation of the instrument. A standard such as polystyrene is introduced automatically to allow correction in the wavelength and achieve reproducibility. The equation for the grating assures compliance of the other frequencies, once a home frequency has been established with a calibration standard. While the grating is oscillating in one plane, an order-sorting filter revolves in a perpendicular plane. The filter changing has to be timed properly to perform its function. Actually, a series of order-sorting filters is incorporated into a rotating wheel so that the proper order-sorting filter is present at the right part of the grating oscillation.

As far as scanning NIR instruments are concerned, currently, these much-improved, rapid-scan, grating monochromator instruments enjoy the most widespread usage throughout the industry. Foss NIRSystems currently has the largest share of the market. They have a wide variety of sampling devices, including fiber optically coupled probes. Industrially hardened versions are available, and operation at high temperature and high pressure is possible with these instruments. Once the improvements were made to the rapid-scan, grating monochromator instruments of both of these companies, considerable effort went into sampling accessories, probes, and housing these systems in NEMA cabinets for industrial usage.

The software for grating monochromator instruments becomes considerably more complex. With the collection of data at a series of points along the spectrum, global methods of data treatment are used readily by having three points close together on the spectrum, with the center point representing the wavelength of interest and points on either side serving as local baseline corrections. The method of second difference (sometimes called second derivative) requires selection not only of the central frequency but also of a smoothing to apply to the data obtained from scanning and the increment, referred to as the gap, between the central point and the baseline points. Some users of grating monochromator scanning instruments are committed to using every single point on the spectra that they scan. Another approach is to disregard regions of the spectrum that have little analytical information or have a spectral interference and treat the data from miniscans in certain regions where useful information is to be had. Other global methods, such as PCA, PLS, and neural networks, may be applied.

Much effort has gone into method development in all instruments, but the multiplicity of methods is greater for those in which the whole spectrum is scanned. Usually, the instrument package includes method development software and perhaps some predeveloped calibrations. Foss NIRSystems has its own industrial software. It also has third-party software available for models used for agricultural purposes. LT Industries has software licensed from 3M Corp. (St. Paul, MN). A standardization of the data format to allow transfer of data obtained from one instrument to the computer of another instrument or that of a third party has been possible through the American Society for Testing Materials (ASTM) E-13 Committee on Molecular Spectroscopy. E-13 has a subcommittee on IR and a subsubcommittee on NIR. Instrument manufacturers thus provide a way to export data obtained on their instrument into a JCAMP format, which produces an ASCII file. Although standardized, this file is unfortunately quite voluminous. However, this ASCII format then can be imported, manipulated, and dealt with for comparison purposes.

Another grating monochromator instrument is that produced by Guided Wave Process Analytical Systems, now a part of Ocean Optics (Dunedin, FL). Guided Wave has, since its origin, produced grating monochromator NIR instruments specifically to send an optical ray through a single optical fiber for use as an industrial probe. Incident radiation attenuated by the sample of a transmitting fluid is returned to the detector of the instrument for intensity measurements. The use of this instrument is dedicated to the analysis of clear liquids. Emphasis is placed on multiplexing the instrument to a number of fiber optic probes at different parts of the process (UOP, 1991). Dedicating this instrument to a single optical fiber requires certain limited optical requirements. Sampling methods all are done at the end of the fiber and in a transmission mode. The grating used has an *f* number of presumably 4 or less. The time of one scan is 2–3 sec, including referencing, and the nominal value for resolution is 6 nm.

Bruins Instruments (Puchhein, Germany), a well-established European OEM monochromator manufacturer, produces the Omega 20 and Omega 40 grating monochromator instruments that include the NIR range. They use gratings with an *f* number of 4.0 or less. The modular design allows various solid or liquid sampling devices to be attached, and fiber optic probes also are available with these instruments.

Another grating monochromator instrument was originally developed for relatively specialized use in Sweden by Tecator, a subsidiary of Perstorp Instrument Co. The Tecator instrument, now a product of Foss, was designed to illuminate and analyze, in a transmission mode, whole seeds or large granules of solid materials, such as plastic pellets or pharmaceutical tablets. It operates in the range of 800–1,100 nm and it takes data at 200 points in that range. Each data point is used to produce a PLS analytical expression. What is peculiar about this particular grating monochromator instrument is that it functions as a black box. It is a smart system that cannot be accessed by the user, so there is no way of exporting data to be exchanged with other spectroscopists. It must be said that this instrument has enjoyed a degree of success, having been accepted as a whole-seed analyzer in various parts of the world. The Foss instrument described in the section on filter instruments is aimed at the same whole-seed market. This market also is the target of Perten Instruments's rapidly rotating discrete filter instrument.

A specialized grating instrument was produced previously at the University of Washington by Professor Callis and coworkers for octane determination in gasoline. That particular instrument operated with the very near infrared region from 800–1,100 nm and used a silicon diode array polychromator rather than a monochromator. It was described under the section on diode arrays.

Another grating monochromator instrument is produced by Analytical Spectral Devices (Boulder, CO). It is called the Field-Spec Chem NIR analyzer. This is a portable analyzer for production environments. It uses a combination of a diode array for short wavelength and a rapid-scan, grating spectrometer with a single moving InGaAs photodiode detector to provide NIR spectra with a sampling interval of 1 nm and spectral resolution of 10 nm (Analytical Spectral Devices, Inc., 1995). The lower range corresponds to that of the InGaAs detector (1,000–1,800 nm), and the higher range extends to 2,500 nm. The scanning time is 100 msec. This device is battery operated, comes in a suitcase, and is equipped with a laptop computer.

VI. INTERFERENCE FILTER INSTRUMENTS

Interference filters have been used in a large percentage of NIR quantitative analysis instruments for industry. In the early 1970s, the first few years in the renaissance of NIR, exclusively filter instruments were available on the market. Interference filters typically have a transmittance of up to 70%, with a nominal bandwidth of 10 nm at half height. Specialized filters are available with a narrower bandwidth, but the 10-nm filter typically is used.

Wavelength selection with filter instruments usually involves some mechanical means of interposing a particular filter into the beam in order to obtain data at that corresponding select wavelength. The filter-moving assemblies that have been produced operate either in a carousel (Knepler, 1974), with continuous motion of a filter wheel (the original DICKEY-john design), or another configuration. Formerly, a paddle wheel system was employed for the Neotec tilting filter configuration (Webster, 1975). Most of the contemporary wavelength-selection systems employ a turret mechanism, in which the filter wheel is indexed to the position for one filter and remains static in that position until sufficient data are accumulated before it cycles to the next position of the turret to continue the process (Funk, 1980; Judge and Lipshutz, 1980). Also, a continuously moving, discrete filter system reminiscent of the original DICKEY-john carousel mount has been brought back into use.

The mechanism of referencing is particularly important in reflectance instruments. In one system, reference data at each filter from a cycle of filter changes are stored in the computer prior to introduction of the sample. For a reflectance reference, a standard reflecting gold mirror or ceramic surface is used. Once the sample is inserted into the filter instrument, the filter cycle is repeated, and the intensity of radiation reflecting diffusely off the sample is measured and ratioed to the intensity of radiation at that same filter coming off a standard reflector. By definition, the ratio of these intensities is called the reflectance (R), and the log of $1/R$ is the reflectance equivalent of the classical spectroscopic absorbance value.

Another mechanism for collecting reference information is to do this at each filter position at the same time the data are being collected from the sample. In such a system, the movement of a mirror placed in the incident beam coming from the filter will alternately direct its radiation to impinge on the sample or on a standard reflector. In this case, the standard reflector, the gold-plated surface of an integrating sphere, is part of the optical collection system. The ratio of intensities produces a value of reflectance, $R = I_{sample}/I_{ref}$. This is repeated for every different filter position that the turret mount indexes. Two filter instruments currently on the market employ this type of referencing. In the original Technicon InfraAlyzer (I/A) 400 instrument (currently Bran and Luebbe), a mirror placed before the integrating sphere tilts the beam to direct it to the sample or alternately to divert it to the inside surface of the sphere. The Percon instrument, which also uses an integrating sphere, temporarily interposes a mirror between the integrating sphere and the sample to obtain reference intensity against which the intensity off the sample is ratioed. In all cases, movements of both the mirror and the filter turret are controlled by stepping motors. The new version of a continuously moving, discrete filter instrument will be discussed with reference to that particular instrument.

Besides the mechanism of filter motion, the means of referencing the optics of the diffuse reflectance radiation collection also distinguish one filter instrument from another. It is well known that increasing the area of a detector contributes to the noise of the system. However, a large detector area also allows collection of a greater solid angle. The geometry of these detectors is also quite important in retaining a large solid angle of collection. Theoretically, an integrating sphere placed very close to a diffusely reflecting surface would collect radiation from a solid angle of nearly 2π steradians. The holes at the entrance and the sample end of the integrating sphere subtract part of that solid angle away, as does the blackened rim surrounding the entrance. Also, the detectors themselves occupy a certain amount of surface space, lessening the theoretical 2π steradian solid angle of collection.

The purpose of an integrating sphere is to collect at all angles of azimuth between the object's reflecting surface and the zenith (normal to the sample surface). Also, the spherical mirror provides 360° collection. Averaging of directionally preferential reflection based on the sample texture is accomplished in this way. The purposes of an integrating sphere are to collect radiation in all directions at all angles continuously and to direct it onto one or more detectors mounted in the sphere. This process requires multiple bounces, because radiation does not necessarily hit the detector directly. Another mechanism to collect radiation at all sides is to have separate detectors mounted at a 45° angle to the normal on four sides of the sample.

An alternative to this has been to put two detectors in a tent configuration on two sides of the sample and in very close proximity. Still another arrangement uses a single detector mounted at a 45° angle from the incident radiation. In this design, continuous rotation of the sample cup in time averages any differences based on direction. Time-averaging of directional differences also is accomplished by still another system, in which, instead of rotating the sample cup, the mount of the detectors is oscillated. Undoubtedly, through the years of operation of filter-based NIR instruments, abundant excellent data have been produced that were not directionally dependent. Certain samples, however, may have a directional preference texture resulting from a slicing or shearing mechanism in their production. In homogeneous powders or on smooth polymer films, it probably is not an issue at all.

Historically, the first commercial filter instrument patented was the "Grain Analysis Computer" (GAC II) produced and marketed by DICKEY-john of Auburn, IL (Knepler, 1974). The first instrument was of a carousel design. It was mass-marketed subsequently by Technicon Industrial Systems (Tarrytown, NY) as the InfraAlyzer model 2.5. A tilting filter instrument was patented (Webster, 1975) soon after the DICKEY-john instrument and was introduced onto the market by Neotec Instruments (Silver Springs, MD). Descriptions of these two instruments appeared in the first edition of this book and will not be repeated. The current era of filter instruments began in 1977, when DICKEY-john (Funk, 1980) and Technicon Industrial Systems (Judge and Lipshutz, 1980) independently produced turret-design instruments. The DICKEY-john instrument used a single, thermoelectrically cooled, PbS detector mounted at a 45° angle. A rotating sample cup was used, and a filter wheel allowing for eight different filters was produced. Subsequent generations of this basic design have endured, and many are currently in use.

These instruments are provided with printers to print out the results, computer attachments for acquisition of data, and software to develop calibration equations on site. Instruments of the Technicon InfraAlyzer (I/A) 400 and 300 series employed an integrating sphere and had filter wheels to accommodate either 10 or 19 different interference filters. These systems employed data collection at a single wavelength, with referencing at that wavelength through the tilting mirror mechanism previously described, and no rotating of the sample cup was used initially. The model I/A 300 was the workhorse for routine use in industry where a limited menu of filters was available. The model I/A 400 was supplied with a menu of 19 standard filters for use in the grain and food industries. A somewhat altered menu of filters was required for the textile industry.

This basic unit also was adapted for handling liquid samples. In particular, the model I/A 400D, for use with dairy products, included a homogenizer and a sophisticated, thermostated liquid drawer that employed a double transmission (also referred to as transflectance) mechanism. The integrated sphere-based optical system was adapted for use with liquid samples by having radiation go through one thickness of the sample, and then scatter back off a diffusely reflecting surface beneath the sample and return through the liquid, before being collected in the integrating sphere. Thus, the liquid attachment for the integrating sphere instrument allowed it to be used for either scattering liquid samples, such as milk, or clear liquids. A variety of sampling accessories was developed to accommodate granular solids, pastes, or viscous liquids. This basic design was incorporated into the more economically engineered model 350 and model 450 InfraAlyzers subsequently produced and marketed by Bran and Luebbe. The model 2000 is the current filter instrument.

The Percon line of Perten Instruments (Huddinge, Sweden) also uses a turret mechanism for filter changing. The horizontal beam is in-line, with an integrating sphere that has the sample window mounted perpendicular to the bench top. Opposite the integrating sphere is a quartz window with a refillable sample chamber for handling powders. The powders are compressed with a spring-loaded tamping tool. Referencing is done at each individual filter position for the turret by interposing a planar, diffusely reflecting, gold surface between the window leading to the sample from the integrating sphere. After analysis, the solid sample is removed by opening the sample compartment door and brushing out the sample; the compartment is then closed and refilled with the next sample. These instruments are controlled with a built-in computer, but data acquisition can be controlled by, and data exported to, an external computer used for calibration purposes. Custom software is provided by this company for method development.

Another tabletop interference filter-based instrument that is produced by Oxford Instruments (Oxen, United Kingdom) is currently marketed in the United States by LECO, Inc. (St. Joseph, MI). In this instrument, the filter-changing mechanism is described as a turret that indexes into position and collects data. Six, ten, or twenty interference filters are used depending on the model. Averaging of the collection of the reflected beam at different directions is done by oscillation of the pair of detectors in a plane parallel to the sample. The collection angle is approximately 45°. These instruments have their own software package for performing calibrations (Oxford Instruments, 1992).

Another area of use for interference filter NIR instruments is that of analyzing whole seeds. Analysis of whole seeds requires a considerable amount of averaging. Two filter instruments are available that operate rapidly to allow the averaging required for whole seeds. One of these is the Foss filter instrument. In this case, several filters are used in alternating angles of tilt to produce a larger number of data points than there are filters. It was designed as a lower cost alternative to compete with the previously cited Tecator whole-seed instrument that is based on a grating monochromator that employs PLS statistical chemometrics.

Another interference filter instrument designed to operate rapidly and accommodate whole-seed analysis is produced by Perten Instruments. This instrument configuration is the product of the their engineering group in Europe. In this case, a return has been made to the GAC II 1974 DICKEY-john continuously rotating filter wheel (carousel mechanism), where the metal between filters in which they are mounted acts as a chopper, so that it goes from dark, filter 1, dark, filter 2, dark, filter 3, etc., until it recycles to dark, filter 1, dark, filter 2, dark, filter 3. The rapid rotation of a filter wheel in which a large number of filters is mounted produces one multiwavelength scan on each rotation. The corresponding optical responses of successive scans are coadded to produce an averaging at each filter. The averaging serves to account for differences occurring in time or scanning a different part of the sample at a different rotation of the wheel. It will time-average the heterogeneity out of the sample. This instrument is typically equipped with 12 filters. Typically 30 rotations are coadded for a measurement duration of 3 sec. PbS detectors are used. Filter instruments of this type are used primarily in the range of 1,100–1,400 nm. Sampling with this instrument typically is done in a reflectance mode with a PbS detector. The system measures 120 subsamples of whole grain to produce a single analysis.

Another type of interference filter NIR instrument is the noncontact meter. In all of the instruments previously described, the specimen has been in direct contact with a window of the NIR filter instrument. There are many cases in everyday use when it is not convenient to have a granular material, in particular, reproducibly compressed against a cell window. Materials moving on a conveyer belt or transferred by a web are other examples. Traditionally, the most commonly used noncontact interference filter NIR instrument was the moisture meter. More recently, the noncontact approach has been employed for constituents other than moisture. One such line of instruments is manufactured in Japan by the Kett Co. Those models sold by Kett U.S. (Villa Park, CA) and Zeltex Inc. (Hagerstown, PA) are described in Section VII on discrete source instruments.

A laboratory tabletop model of the Kett noncontact instrument has a base plate in which the open cup sample is placed at a measured distance (several inches) below the optical sensing head. The changing of filters in the beam is accomplished by mechanical means similar to that described for the contact interference filter instruments. The solid angle of collection as the reflected radiation goes to the detector is constant, so long as the height of the surface of the sample is constant. In the tabletop laboratory model, this probably is not a great problem. For a sample moving along a conveyer belt, different elevations of the sample would produce different solid angles of collection for a sensing head mounted in a particular geometry with respect to the conveyer belt. Various attempts have been made to optically scramble the returning rays and thus average out solid angle of collection differences as part of the empirically derived expression. Any uncertainty resulting from inconsistency of the solid angle of collection will then be added to the signal obtained.

In the case of moisture analysis, where there is a prominent OH band that is highly absorbing, the relative uncertainty produced by slight differences in geometry may not be important. However, when it is necessary to make measurements of an NIR band that has only subtle changes throughout the concentration range of the analyte in question, the relative analytical uncertainty could be affected severely by the optical geometry changes. This same company markets noncontact instruments (Kett U.S., 1993) primarily in the moisture analysis area. However, it does supply corresponding filters and capability for oil, protein, and other major components.

One hand-held or tripod-mounted noncontact inspection device produced by Kett (model KJT 100) that is of particular interest resembles a video camera in its size and shape. The incident beam of NIR radiation is projected from a port in this instrument. A concentric cone of red visible light also is projected. This cone acts as a range finder to allow the operator to adjust the testing device closer to or further from the material being tested and assure a constant solid angle of collection. This reproducibility is necessary between a series of measurements or between samples and standards.

Infrared Engineering Ltd. (Maldon, Essex, United Kingdom) produces a family of filter-based monitors for a variety of industrial applications from brewing to plastics to tobacco inspection that are sold in the United States as products of NDC Infrared Engineering (Irwindale, CA). One noncontact monitoring system has received considerable acceptance in the tobacco industry. In the earlier years of using NIR spectroscopy as an analytical technique, the question often was asked, "Can you look at a leaf of tobacco with NIR and give an answer regarding the nicotine, reducing sugars, and moisture contents?" Formerly, when ground tobacco samples were analyzed for nicotine and moisture, the nicotine calibration necessitated taking into the account the reducing sugars present, because their variation caused a variation in the grinding. An on-line model TM55 plus Infrared Engineering noncontact meter for use in the tobacco industry produces analytical results with a few filters for nicotine, total reducing sugars, and moisture. Typically, the optical sensing head is suspended above a conveyer line containing the tobacco. The TM55 plus can be operated remotely via a two-way serial link, and 16 sensors can be controlled by a computer over a single RS 485/422 loop (Infrared Engineering, 1988). The use of this device is apparently compatible with the needs of this particular industry. For the brewing industry, in-line alcohol and original gravity are measured by transmission across a 1-mm gap in an optical fiber equipped stainless steel probe. Signals are transmitted up to 100 m away (Infrared Engineering, 1993).

Over the years, a number of noncontact moisture meters have been produced and sold. It is probably not possible to take into account all of those that have appeared at one time or another. One prominent company of long standing in this area is Moisture Systems Corp. (Hopkinton, MA). Its analyzer consists of an optical head mounted above a conveyer. As many as four filters are arranged to be mechanically placed into the beam. Incident radiation is projected, and reflected radiation is collected at a typical distance of a few inches. Movement of the sample while filters are being changed is a consideration with all on-line systems, but a considerable amount of averaging is necessary, particularly when applying filter instruments to on-line systems. It is the experience of the author that when a filter instrument is used in a contact mode for granular material (e.g., intermediate product streams in a flour mill), taking a running mean of perhaps five rapid (13-sec) analyses is essential in order to smooth out any transients produced as artifacts of sampling or nonreproducible contact with the window (Wetzel, 1986). Using a running mean prevents inappropriate corrective action based on a single spurious analysis, but upward or downward trends are spotted readily with this type of device.

VII. DISCRETE SOURCE INSTRUMENTS: LEDs PLUS FILTERS

For use in very near infrared (850–1,100 nm) transmission work, the source array instrument was introduced years ago by Trebor Inc. Successors of the Trebor company are Zeltex, Inc. (Hagerstown, MD) and Futrex, Inc. (Gaithersburg, MD). Another independent company located in Maryland, Katrina, Inc. (Hagerstown, MD), also used the source array technology. The concept of discrete sources is not new in spectroscopic analysis. In particular, a discrete source for each element was required for atomic absorption (AA) spectroscopy. Unfortunately, in typical AA instrumental configurations, that meant running each element separately. Perhaps the simplest example is flameless AA, in which the mercury atom line emitted from a mercury source (germicidal lamp) was attenuated by passage through a tube containing vapor of mercury atoms circulated from a reducing medium containing the sample.

A discrete source NIR instrument has a number of NIR-emitting diodes. Each of these diodes may be fired in sequence, and, thus, the same detector with appropriate synchronized timing can be used for each of the sources. The actual wavelength selection is contributed partially by the emission characteristics of the chosen LED. However, the actual wavelength limitation or narrowness of the source is dependent on the attached miniature interference filter. Typically, an instrument of this type would be equipped with anywhere from 2 to 13 LEDs and filters. Because multiple sources cannot come from exactly the same point in space, they are directed toward a sample through a Fresnel lens. With this arrangement, any one of the discrete sources can equally and evenly illuminate the entire sample.

One of the first uses of the source array instrument was for determining moisture, oil, or protein in whole seeds of various commodities. The optical arrangement involves radiation from the source through a Fresnel lens into a typically 2-cm-thick transmission cavity containing the sample. On the other side of the sample cell, a lens collects the radiation transmitted by the sample and directs it to a silicon photodiode detector. One tremendous advantage in using an LED as a source is that it has a projected lifetime of 20 years. One disadvantage of LED sources is that probably no two LED/filter combinations are identical, and, thus, each instrument is calibrated independently once it has been configured with an appropriate number of filters and LEDs. The Zeltex instruments are the industrial versions and the successors to Trebor.

The Zeltex ZX800 is a tabletop instrument that will accommodate whole seeds. It measures protein, moisture, oil, fat, starch, sugar, or alcohol. One feature of the patent upon which this instrument is based is that a reading related to temperature is incorporated into the scheme of the calibration. Thus, not only are intensity values given for each different wavelength impinging on the photodiode, but the voltage produced on the temperature-sensitive element is incorporated into the calibration. The model 880 is similar to the earlier Trebor instrument in configuration and is intended primarily for granular materials, whole grains, flakes, powders, and pastes. In the Zeltex 880, the sample actually is moved, so that data are taken from different parts of the presumably heterogeneous sample. Several measurements are averaged in order to produce a representative analysis.

A portable version is the Zeltex 100 F/C (Zeltex, Inc., 1991). This hand-held device can operate on batteries or on AC. In the Zeltex 100, a sample in liquid form is placed in a glass jar and would be interrogated through the glass. The portable 100 series has some specific configurations, such as for octane. The portable octane meter is referred to as the Zeltex 101. The ZX 50 is designated specifically for food and agriculture (including whole grains, such as wheat, barley, soybeans, or rice) to measure constituents,

including protein, moisture, and oil. Calibration software is obtainable with any of these instruments.

The Futrex instrument line is similar to the Zeltex as far as the source array; however, the Futrex line is designed and calibrated for health and medical uses such as measurements of body fat or, potentially, glucose by noninvasive means. A tabletop version of the Futrex body-fat meter designed for health club or athletic team use is a wavelength-specific instrument (Futrex, Inc., 1990). To determine body fat, measurements must be made at the same corresponding part of the arm each time and calibrations run for this measurement in the same way. In general, for the determination of fat, a relatively constant body moisture is assumed, and the ratio of a CH absorption band of the lipid to an OH band of the water is determined. Another wavelength or two may be used for baseline correction for varying amounts of scattering or overall opacity. A very low-cost, mass-marketed product uses the response of broadband radiation entering the tissue and coming back as a simplified way of analyzing a person's body fat. This is not a wavelength-specific spectroscopic measurement, but rather is dependent on the overall optical characteristics of the tissue.

Katrina, Inc. also is located in Hagerstown, MD. Although it is not connected with Zeltex, Inc. or Futrex, Inc., it operates on a license of the same patent on which they are based. Katrina, Inc. deals exclusively with on-line monitoring systems, including the 12-channel Protronics P-112 (Katrina, Inc., 1994). The sources are still NIR-emitting diodes, and the detector is still a silicon photodiode. However, 8-ft long, rather large bundles of optical fibers (approximately the size of a vacuum cleaner hose) are used to transfer sufficient radiation from one place to another so that it can go through various products. These might include chocolate or all sorts of foods. A product such as a thin section of a doughy material can be monitored on a conveyer web. Although the Katrina process monitoring systems are made up of various optical modules, the installation and calibration are engineering feats. The calibration is specific for the particular location and the particular product at that location. It is emphasized by the people dealing with such a system that the longevity of the source justifies using a reasonable amount of time for setup and calibration, because once it is functional, it should operate for a considerable length of time. This instrument and technology are now the property of Moisture Systems (Hopkinton, MA).

VIII. SPECIAL PURPOSE INSTRUMENTS

SINGLE-KERNEL OR GRANULE CONSTITUENT DETERMINATION

Brimrose Corp. introduced in 1996 a commercial instrument referred to in an earlier section that analyzes and automatically sorts seeds. One commercial plant-breeding organization alone has six of these running continuously 24 hr a day for nondestructive genetic selection for oil and protein. Perten Instruments has a single kernel instrument as a product of its European group, and the USDA laboratory at Manhattan, KS, has a working one-of-a-kind instrument assembled on location.

STATIC POLYMER TESTING FOR WHEAT GLUTEN STRENGTH

An automated, dedicated, polarized NIR instrument for molecular testing of the strength of wheat gluten films was produced at KSU (Wetzel and Sweat, 1999). With this system, gluten strength is inversely related to the "delta dichroic ratio" determined from combination bands derived from the amide I and amide II fundamentals with the N-H fundamental stretching vibration. Dichroic ratios are calculated before and after strain in a gluten film is imposed by a standard work input (stretch), in which the delta dichroic ratio is the difference. The magnitude of the delta infers the degree of molecular orientation. Programmed film rotation in the polarized beam of a transmission acousto-optic TFS and its electronic wavelength switching are combined with stepping motor screw-driven, bilateral stretching of the film to a predetermined multiple of the original length. A single microcomputer controls the pneumatically actuated film rotation and tunable wavelength switching. It calculates and stores the dichroic ratio obtained, controls the metered elongation, and plots the delta dichroic ratio versus elongation.

DYNAMIC POLYMER TESTING

Dynamic testing of a polymer during oscillatory stretching is the purpose of the dedicated NIR instrument in which the mechanical perturbation, wavelength switching, and time-resolved data acquisition are all controlled by the same microprocessor (Sweat, 1999; Sweat and Wetzel, 1999, 2001). The polarized beam of an AOTF monochromator is used to determine the changes in molecular orientation of functional groups in a polymer subjected to a repetitive, small-amplitude, mechanical stretching cycle. The polymer film in the sample position is held between metal jaws and is perturbed in a sine wave cycle by a piezoelectric actuator that drives the moving jaw. The instrument allows the collection of spectral data in the 250–2,450-nm range at a time resolution as short as 10 μsec. With this instrument, two specimens of the same chemical composition but a different degree of molecular orientation can be readily distinguished by differing relaxation times, and copolymers of different monomer ratios are characterized by their in-phase and quadrature dynamic spectra. The NIR instrument is a potential rugged (for industrial environment use), simpler, lower-cost version of previous sophisticated step-scan FT or dispersive mid-IR oscillatory instruments previously used in the polymer field. It is also applicable to thicker specimens that can be tested "as is" without casting a film thin enough for transmission in the mid-IR.

MINIATURE INSTRUMENT

An NIR instrument based on acousto-optic TFS, the size of a man's wallet, has been engineered and produced at Brimrose Corp. as an on-board device for 2004 and 2005 space explorations by NASA (Wang et al, 1994). In addition to miniaturization, it was necessary to conserve power to allow maximum use on location (Mars) after the long trip.

IX. IMAGING

NIR imaging can be accomplished in one of two ways. Obtaining spectroscopic data at several positions in a sample in a grid pattern enables the reconstruction of monochromatic functional group images by extracting absorbance peak area data at one wavelength from all points in the grid and plotting it. An idealized way to produce an image is to have a focal plane array detector in which the response at a particular wavelength is recorded simultaneously at all points. With a focal plane array, it is necessary to send the beam of radiation through a filtering device, and, for each wavelength that comes through the filtering device, a different snapshot is taken of the x,y field. Recently, in the area of vibrational spectroscopy, including both the NIR and the mid-IR, the production of monochromatic images has been accomplished. It is relatively easy working in the silicon photodiode region to produce an NIR image, since detector cameras have been available in this region for some time. Unfortunately, one is limited to a wavelength range of approximately 850–1,050 nm.

Use of a PbS CCD camera produces NIR images at somewhat longer wavelengths. The author has attempted to use such a camera where the visible radiation was cut off by a 1,050-nm filter. Operation of the PbS camera is rather slow. In fact, the camera has an image memory that shows up with quick motion. Such a device was used and demonstrated more than 20 years ago at the First International Diffuse Reflection Conference (Chambersberg, PA) by a physician who was using it on women patients for breast cancer detection. In this case, light from a tungsten source was conducted by way of a Lucite rod to illuminate the breast of a patient from the underside. NIR radiation coming through the breast was photographed by a PbS camera, and the physician was able then to view on the video monitor the very near infrared image to look for abnormalities within the breast caused by calcification or

by some greater amount of blood present in a particular region of the breast. That physician examined 200 patients by both mammography and the NIR imaging technique and reported that three cases of tumors were found by each technique. Of those three cases, one was found by mammography that was not found by the NIR technique, and one was found by the NIR technique that had been missed by mammography. The other two were found by both methods.

NIR imaging on a macro scale in reflection with an InSb focal plane array (IR camera) using filters and white light (tungsten) illumination has been used routinely during surgery and diagnostic procedures by R. A. Lodder and coworkers (Dempsey et al, 1997) at the Pharmacy Department of the University of Kentucky, Lexington. More recently, similar activity has been reported at the Institute of Biodiagnostics of the National Research Council at Winnipeg, Manitoba, Canada (Hewko et al, 1999). A platinum silicilide camera has also been used. R. A. Lodder and coworkers (Dempsey et al, 1996) have also employed an InSb array with fiber optic catheters to examine arteries in stroke patients. Medical diagnostic fiber optic catheters have subsequently been commercially produced by InfraReDx (Lexington, KY). An NIR laser has been used with a remote NIR camera for two-dimensional imaging deposits from aerosols on whole tobacco leaves (Dausman et al, 1999).

Microscopic imaging has been achieved by way of an AOTF used to filter NIR radiation going into a microscope. A focal plane array camera mounted on the microscope produces the monochromatic image from the frequency of radiation transmitted by the filter. The AOTF does not significantly distort the image and has a relatively large aperture. This was reported in *Applied Spectroscopy* by the team of Treado, Levine, and Lewis (Treado et al, 1992a). The wavelength limitations of the camera used images restricted to those that highlighted either an O-H stretch combination or overtone or a C-H or N-H stretch combination or overtone. Thus, it was possible to demonstrate the oil/water differences rather dramatically. The same group at the National Institutes of Health (Bethesda, MD) also produced microscopic Raman imaging. A focal plane array camera was also used, but in this case, the AOTF filter position was in the microscope after the incident radiation had struck the sample. The secondary radiation emitted from the sample to produce the Ramam Stokes lines was filtered from all parts of the image simultaneously. Chemically dependent contrast was produced from the Raman microscopic image (Treado et al, 1992b).

Point-to-point scanning by mid-IR using a microspectrometer has given rise to a matrix of data points. Selected absorption bands from each of these data points have been used. Baseline-corrected (absorbance) peak areas calculated at each data point have been interpolated to produce two-dimensional contour maps or three dimensional maps showing the absorbance peak area at a particular wavelength at a particular x,y address in the specimen. Preliminary experiments have been done in the extended range of one such commercial instrument, the IRμs (Spectra-Tech, Inc., Shelton, CT), that overlaps with NIR. Unfortunately, the coating of the beam splitter of the mid-IR FT-IR instrument does not allow transmission of NIR at frequencies greater than 5,800 cm^{-1}. In 1995, in an attempt to go to shorter wavelengths, the author, in connection with personnel at Spectra-Tech, Inc., configured an FT-IR as an FT-NIR in connection with an IR microscope to use with various thick nylon and polyethylene terphthalate fibers as well as polymer films on an experimental basis. FT-NIR microspectroscopy has since been used on a variety of specimens too thick for mid-IR. Data taken with this type of instrument has produced spectra at different points of the specimen, and these can be constructed into NIR functional group microspectroscopic images as commonly done in the past for mid-IR.

In 1994, an InSb focal plane array camera was first coupled to a step-scan FT-IR spectrometer by Neil Lewis and Curtis Marcott in Cincinnati, OH, at the Proctor and Gamble laboratory. Subsequently, locally configured instruments of this type were in use in Cincinnati and at the instrumentation research facility of the National Institutes of Health (Bethesda, MD). Using a fast frame-grabber with an on-board processor, it was possible to collect a very large number of spectra (one for each pixel) simultaneously in only a few minutes (Lewis et al, 1995). As a result, experimenters are swamped with data in a relatively short time. Besides the NIR wavelength range below 2,500 nm, the InSb is sensitive in the O-H, O-D, and C-H stretching vibration region of the mid-IR. The first experimenters were primarily interested in this region. Although rapid NIR imaging is readily available by this sophisticated and high-cost system, that application has subsequently been overshadowed by imaging in the fingerprint region of the IR with a 64 × 64-pixel, cooled MCT focal plane array detector (Marcott et al, 1999). This type of equipment has also been commercialized by Bio-Rad Spectroscopy Division (Cambridge, MA), Spectral Images (Olney, MD), and Bruker Optics (Billerica, MA). Spectral Images also produces filter-based imaging systems.

X. SUMMARY

There is now a wealth of instruments to choose from, as NIR analytical spectroscopy has emerged from its status in 1983 as a "sleeper" among spectroscopic techniques (Wetzel, 1983). Now that FT-NIR instruments have become cost effective, they are available from a variety of vendors. Solid-state technologies that produce fast, sensitive, low-noise photodiodes and AOTFs have made possible high-speed electronic wavelength switching instruments with no moving parts via diode array polychromators or AOTF spectrometers. Interference filters and multiple-source instruments continued to perform routine tasks. New instruments are available from known instrument companies, and new companies have sprung up. Industry's analytical needs are being served and at last, as predicted (Wetzel, 1983), the legitimacy of NIR as an analytical discipline has been recognized in the classical spectroscopic community.

ACKNOWLEDGMENTS

This is Contribution No. 96-325B of the Kansas Agricultural Experiment Station, Manhattan.

LITERATURE CITED

Anonymous. 1999. Case Textron System. Spectroscopy 14(10):54.

Analytical Spectral Devices, Inc. 1995. Field spec—Spectrometer for transmittance, reflectance, and absorbance measurements. Tech. Bull. Analytical Devices, Inc., Boulder, CO.

Baudais, F. L., Eilert, A. J., Buijs, H., and Wetzel, D. L. 1990. FT-NIR correlation transform quantitation of industrial organic fluids. Paper 551 in: Annu, Meeting Fed. Anal. Chem. Spectrosc. Soc., 17th.

Ben-Gera, I., and Norris, K. H. 1968. Determination of moisture content in soybeans by direct spectrophotometry. Isr. J. Agric. Res. 33(1):64-67.

Blaier, Z., Vrouillette, C., and Carangelo, R. 1999. A monolithic interferometer. Spectroscopy 14(10):46-49.

Bühler, Inc. 1993. Bühler FT-NIR universal spectrometer NIRVIS system. Tech. Bull. Bühler, Inc., Uzwil, Switzerland.

Chaffin, C. T., Marshall, T. L., Fateley, W. G., and Hammaker, R. M. 1995. Infrared analysis of volcanic plumes: A case study in the application of open-path FT-IR monitoring techniques. Spectrosc. Eur. 7(3): 18-24.

Chang, I. C. 1974. Noncollinear acousto-optic filter with large angular aperture. Appl. Phys. Lett. 25(7):370-372.

Dausman, E., Budevska, B., and Lodder, R. A. 1999. Imaging areas of deposition by 2-D tunable near-IR laser spectrometry. Paper 322 in: Annu. Pittsburgh Conf. Anal. Chem. Appl. Spectrosc., 50th.

Dempsey, R. J., David, D. G., Buice, R. G., Jr., and Lodder, R. A. 1996. Biological and medical applications of near-infrared spectrometry. Appl. Spectrosc. 50(2):18A-34A.

Dempsey, R. J., Cassis, L. A., Davis, D. G., and Lodder, R. A. 1997. Near infrared imaging and spectroscopy in stroke research: Lipoprotein distributions and disease. Ann. N.Y. Acad. Sci. 820:149-169.

Eilert, A. J. 1995. Acousto-optic tunable filter spectroscopic instrumentation for quantitative near-IR analysis or organic materials. Ph.D. dissertation. Kansas State University, Manhattan.

Eilert, A. J., and Wetzel, D. L. 1991. Acousto-optic tunable filter spectroscopy in the near-infrared. Paper 217 in: Annu. Pittsburgh Conf. Anal. Chem. Appl. Spectrosc., 42nd.

Eilert, A. J., Sweat, J. A., and Wetzel, D. L. 2000. Parabolic concentration of diffusely transmitted near-IR radiation in an acousto-optic tunable filter spectrometer. J. Near Infrared Spectrosc. 8:239-250.

Funk, D. 1980. Analysis instrument. U.S. patent 4,193,116.

Futrex, Inc. 1990. Futrex-5000 body fat and fitness computer. Tech. Bull. Futrex, Inc., Hagerstown, MD.

Gottlieb, M., Feichtner, J. D., and Conroy, J. 1980. Programmable acousto-optic filter—A device for multispectral optical processing. Proc. SPIE-Int. Soc. Opt. Eng. 232:33-41.

Goulden, J. D. S. 1957. Diffuse reflexion spectra of dairy products in the near infra-red region. J. Dairy Sci. 24:242-251.

Harris, S. E., and Wallace, R. W. 1969. Acousto-optic tunable filter. J. Opt. Soc. Am. 59:744.

Hewko, M. D., Sowa, M. G., Payette, M. D., Mansfield, J. R., and Mantsch, H. H. 1999. Medical applications on in-vivo near infrared spectroscopic imaging. Paper 321 in: Annu. Pittsburgh Conf. Anal. Chem. Appl. Spectrosc., 50th.

Honigs, D. E., and DeThomas, F. 1988. Matching instrumental and spectral information bandwidths in NIR correlations. Paper I 28 in: Annu. Meeting Fed. Anal. Chem. Spectrosc. Soc., 15th.

Huehne, M., Eschenaure, U., and Siesler, H. W. 1992. Acousto-optic tunable filter NIR-Spectrometer for rapid process-control. Proc. SPIE-Int. Soc. Opt. Eng. 1575:214-215.

Infrared Engineering. 1988. TM55plus. Tech. Bull. Infrared Engineering, Maldon, Essex, United Kingdom.

Infrared Engineering. 1993. Liquidata—In Line Measurement of Beer. Tech. Bull. Infrared Engineering, Maldon, Essex, United Kingdom.

Judge, J. F. X., and Lipshutz, V. G. 1980. Infrared analyzer. U.S. patent 4,236,076.

Katrina, Inc. 1994. The I-1000 INSPECTOR fact sheet. Tech. Bull. Katrina, Inc., Hagerstown, MD.

Kaye, W. 1955. Near-infrared spectroscopy (a review). II. Instrumentation and technique. Spectrochim. Acta 7:181-204.

Kemeny, G. J., and Wetzel, D. L. 1989. Optical analysis method and apparatus having programmable rapid random wavelength access. U.S. patent 4,883,963.

Kett U.S. 1993. Infrared moisture meters. Tech. Bull. Kett U.S., Villa Park, CA.

Knepler, J. T. 1974. Grain analysis computer circuit. U.S. patent 3,828,173.

Lewis, E. N., Treado, P. J., Reeder, R. C., Story, G. M., Dowrey, A. E., Marcott, C., and Levin, I. W. 1995. Fourier transform spectroscopic imaging using an infrared focal plane array detector. Anal. Chem. 67: 3377-3381.

LT Industries, Inc. 1992. Analyzer-rapid scanner for constituent and physical properties. Tech. Bull. LT Industries, Inc., Rockville, MD.

Lysaght, M. J., Kelly, J. J, and Callis, J. B. 1990. Field testing and quality evaluation of aviation fuels by near-infrared spectroscopy and multivariate statistics. Paper 1113 in: Annu. Pittsburgh Conf. Anal. Chem. Appl. Spectrosc., 41st.

Marcott, C. A., Reeder, R. C., Sweat, J. A., Panzer, D. D., and Wetzel, D. L. 1999. FT-IR spectroscopic imaging microscopy of wheat kernels using a mercury-cadmium-telluride focal-plane array detector. Vibrat. Spectrosc. 19:123-129.

Massey, D. R., and Norris, K. H. 1965. Spectral reflectance and transmittance properties of grain in the visible and near infrared. Trans. Am. Soc. Eng. 8(4):589-600.

Ocean Optics, Inc. 1993. Fiber optic spectrometers—UV, VIS and NIR and accessories. Tech. Bull. Ocean Optics, Inc., Dunedin, FL.

Oxford Instruments. 1992. Near infrared analysis with the Oxford QN. Tech. Bull. Oxford Instruments, Oxon, England.

Perkin-Elmer Corp. 1992. PIONIR 1024. Tech. Bull. Perkin-Elmer Corp., Norwalk, CT.

Perten Instruments North America, Inc. 1996. The new near infrared analytical alternative DA-7000. Tech. Bull. Perten Instruments North America, Inc., Reno, NV.

Rossenthal, R. D. 1980. Apparatus for optically analyzing a sample. U.S. patent 4,540,282.

Stark, E. 1991. Grating spectrometer. U.S. patent 4,997,281.

Sweat, J. A. 1999. Time resolved dynamic and static dichroic measurements of polymers with acousto-optic tunable filter instrumentation. Ph.D. dissertation. Kansas State University, Manhattan.

Sweat, J. A., and Wetzel, D. L. 1999. Dynamic oscillatory polymer testing with an AOTF spectrometer. Paper 425 in: Annu. Conf. Fed. Anal. Chem. Spectrosc. Soc., 26th.

Sweat, J. A., and Wetzel, D. L. 2001. Near infrared acousto-optic tunable filter based instrumentation for the measurement of dynamic spectra of polymers. Rev. Sci. Instrum. 72(4):2153-2158.

Treado, P. J., Levin, I. W., and Lewis, E. N. 1992a. Near-infrared acousto-optic filtered spectroscopic microscopy: A solid-state approach to chemical imaging. Appl. Spectrosc. 46:553-559.

Treado, P. J., Levin, I. W., and Lewis, E. N. 1992b. High-fidelity Raman imaging spectrometry: A rapid method using an acousto-optic tunable filter. Appl. Spectrosc. 46:1211-1216.

UOP. 1991. Guided wave process analytical systems. Tech. Bull. UOP, El Dorado Hills, CA.

Wang, X., Zhang, H., and Soos, J. 1994. A wallet-size NIR spectrometer based on a light emitting diode array and a miniature acousto-optic tunable filter. Paper 1245 in: Annu. Pittsburgh Conf. Anal. Chem. Appl. Spectrosc., 45th.

Webster, D. R., 1975. Optical analyzer for agricultural products. U.S. patent 4,404,642.

Wetzel, D. L. 1983. Near-infrared reflectance analysis—Sleeper among spectroscopic techniques. Anal. Chem. 55:1165A-1171A.

Wetzel, D. L. 1986. Chemical sensing using near-IR reflectance analysis. Pages 271-296 in: Fundamentals and Applications of Chemical Sensors. D. Schuetzle and R. Hammerle, eds. American Chemical Society, Washington, DC.

Wetzel, D. L. 1992. Fastest gun in the West! Paper 1135 in: Annu. Pittsburgh Conf. Anal. Chem. Appl. Spectrosc., 43rd.

Wetzel, D. L. 1995. FT-NIR: Its place among other quantitative NIR instruments and its utility for imaging and qualitative analysis. In: Int. Conf. Near Infrared, 7th.

Wetzel, D. L. B. 1998. Analytical near infrared spectroscopy. Pages 141-194 in: Instrumental Methods of Food and Beverage Analysis. D. L. B. Wetzel and G. Charalamnous, eds. Elsevier Science, Amsterdam.

Wetzel, D. L., and Eilert, A. J. 1990. Quantitative analysis with a high duty cycle solid state random wavelength access near-infrared system. Paper 1195 in: Annu. Pittsburgh Conf. Anal. Chem. Appl. Spectrosc., 41st.

Wetzel, D. L., and Eilert, A. J. 1992. Quantitative FT-NIR of thermochemical properties of industrial organic fluids. Proc. SPIE-Int. Soc. Opt. Eng. 1575:523-524.

Wetzel, D. L., and Mark, H. 1978. Scanning NIR of grains, oilseeds, and their components. Paper S2.1 in: ICC Symp.: Use of Near-Infrared Techniques. World Bread Congress, 6th.

Wetzel, D. L., and Sweat, J. A. 1997. FT-NIR comparative quantitative performance for complex samples. Mikrochim. Acta 14(suppl.):325-327.

Wetzel, D. L., and Sweat, J. A. 1999. Dedicated polymer orientation polarization response gauge. Paper 423 in: Annu. Conf. Fed. Anal. Chem. Spectrosc. Soc., 26th.

Wetzel, D. L., Kemeny, G. J., and Eilert, A. J. 1987. Using an acousto-optic tunable filter in near-infrared spectroscopy. In: Natl. Am. Chem. Soc. Meeting, 193rd.

Wetzel, D. L., Eilert, A. J., and Sweat, J. A. Tunable filter and discrete filter near-IR spectrometers. In: Handbook of Vibrational Spectroscopy. J. M. Chalmers and P. R. Griffiths, eds. Wiley, Chichester, United Kingdom. In press.

Whetsel, D. B. 1968. Near-infrared spectrophotometry. Appl. Spectrosc. Rev. 2(1):67.

Zeltex, Inc. 1991. The next generation of near-IR analyzers. Tech. Bull. Zeltex, Inc., Hagerstown, MD.

CHAPTER 8

Implementation of Near-Infrared Technology

P. C. WILLIAMS
Grain Research Laboratory
Canadian Grain Commission
Winnipeg, Manitoba
Canada

I. INTRODUCTION

This chapter is for newcomers to the field of near-infrared (NIR) applications. Most of the recent advances in NIR technology have been in chemometrics, including artificial neural network (ANN) and genetic algorithm software. These advances have already begun to unfurl their valuable contributions to future generations of NIR users. This chapter addresses basic principles, which are unlikely to change. But first some philosophy. (i) When users have assembled comprehensive sample sets and have recorded good quality spectra with reliable reference data, practically any software will enable them to develop excellent calibrations. If they have not, even the most advanced software will not be able to generate excellence from mediocrity efficiently. (ii) Quantitative NIR technology can be used in two ways. The first is the classical approach, in which the technology is used to provide accurate analytical data for individual constituents. The second type of application is that of screening.

Large populations, or single samples at the time of arrival/ delivery, can be rapidly tested by NIR to identify samples that require further testing by more time-consuming and expensive methods. An example is the falling-number test for sprouting damage to wheat. The test takes about 10 min, including sample preparation, which is too slow for use at time of delivery or unload. The standard error of prediction (SEP) is about 40 sec, and samples can be screened in less than a minute to identify whether the wheat is safely high, obviously low, or in between. These in between samples would require testing by the reference method, but about 70% of the reference testing would be eliminated. There are many applications of this type in grain handling and plant breeding.

The first commercial NIR reflectance instruments were reflectance devices, designed for analysis of protein content in soybeans and later wheat (Rosenthal, 1971). In 1980, NIR transmittance and, still more recently, diode array and acousto-optical tunable filter (AOTF) instruments joined the field, which is growing very quickly. The scope of NIR reflectance/transmittance technology, originally limited to grain analysis, has grown to include applications in candy, cosmetics, dairy products, food, feed, forage, fossil and other fuels, manure, meat, pharmaceuticals, soils, textiles, tobacco, and in many other industries. The most recent applications have extended the scope of NIR reflectance technology to the fields of diagnostic medicine and environmental studies (e.g., Malley and Williams, 1993; Tong et al, 1995). Both quantitative and qualitative applications are widely used in many fields.

One inconvenience that has persisted since the dawn of NIR reflectance application is the need for separate calibration of any instrument for every commodity, constituent, or parameter. This chapter describes the implementation of a new NIR reflectance testing operation, including calibration, and the monitoring of accuracy and repeatability/reproducibility. It includes a commentary on the interpretation of the results of calibration. In this text, "composition" refers to content of discreet *constituents*, such as moisture, oil, or carbon, and "functionality" to some attribute, or *parameter*, of the material that contributes to, or detracts from, its usefulness, such as kernel hardness in wheat, malting potential in barley, and digestibility in forages.

The procedure described is proposed mainly for quantitative applications to agricultural commodities of plant and animal origin and their derived products (including textiles), forages, foods, feeds, and their ingredients. These complex materials are made up of carbohydrate, protein, fat or oil, fibrous components, and water, in an array of combinations. Calibrations for application to simpler materials, such as those encountered in the pharmaceutical and petrochemical industries, can be achieved with fewer samples, and operating ranges in composition can be prepared with known amounts of the material to be determined. These applications, as well as qualitative applications, are discussed elsewhere in this monograph.

Important advances in NIR reflectance software and computer hardware have improved the speed and versatility of calibration. For large-scale, day-to-day analytical NIR reflectance operations, the preferred method of the future will likely become the ANN or "nearest neighbor" or "local" type of calibration, offered by Infra-Soft International Inc. (Port Matilda, PA). Both of these calibration approaches depends on the accumulation of very large databases, with each item possessing full spectral and analytical data. In the local method, each "new" sample is predicted using the spectra from the database (neighbors) that are the nearest match to that of the unknown sample.

The ultimate system will be one in which an instrument company or a "host" laboratory under contract from the instrument company will develop calibrations based on samples supplied by diverse clients. Software would enable automatic prediction of results upon receipt of spectra, using the local or ANN concept. The instrument company (or host laboratory) will accept spectra from clients by e-mail. Having acquired an appropriate instrument, their clients would need only to scan the samples and would never need to develop their own calibrations. The spectra would automatically be e-mailed to the host, or "master," instrument, the data processed, and clients would receive their results by return e-mail. Instrument performance can already be monitored over long distances using modem control or e-mail.

Many of the precepts of this chapter are also applicable to this type of calibration. But even this "Utopian" concept calls for the host laboratory to develop calibrations. For the majority of new applications, and for the benefit of newcomers to the field, this chapter provides a set of guidelines for development of a typical "classical" NIR reflectance calibration and for its evaluation.

The efficiency of an NIR reflectance calibration is usually evaluated by means of applied statistics. In most common use are the coefficients of correlation (r), determination (r^2), and regression (b); the intercept (a) and bias; and the SEP, which is the standard deviation (SD) of the differences between NIR reflectance and reference data. The higher the value of r^2 and the lower the SEP, the more effective is the calibration. The importance of the bias should not be overlooked. This is discussed later in this chapter. Finally, the standard error of the NIR test itself must be evaluated during calibration, since this is part of the overall error. This is determined by scanning a check sample periodically during the scanning of the calibration samples. The SD of the sought-after parameter is the standard error of the test (SET).

Calibrations for qualitative applications are dealt with in Chapter 13. These are rather different in development than calibrations for quantitative analysis, since the above statistics are not required. The efficiency of a qualitative NIR reflectance calibration lies in the degree to which it identifies the materials for which it is developed and (ideally) categorizes the rest.

II. CALIBRATION DEVELOPMENT

Most of the references and examples refer to grain, but the commentary applies equally to any other type of quantitative application. The steps are essentially the same, whether the instrument is

a relatively simple bench-type filter instrument used to determine moisture and protein in grains or the more powerful scanning monochromator. The main differences among applications lie in the reference methods used to calibrate and monitor the testing and in sample preparation and presentation to the instrument.

Computerized scanning spectrophotometers allow more flexibility than filter instruments in wavelength range and mathematical treatment of the optical signals. They can serve as "workhorses" in routine analysis and process control. In addition, they may be used for research in the development of calibrations for parameters that are more demanding than constituents such as protein, moisture, and oil. Recent explorations are uncovering a role for NIR reflectance technology in investigative studies on functionality and on changes that take place during processes such as dough-mixing, baking, malting and brewery fermentation, and product development in many fields.

Many of the filter and monochromator instruments used in day-to-day analysis employ the optical signals in the $\log(1/R)$ (apparent reflectance) or $\log(1/T)$ (apparent transmittance) form and may be factory calibrated. Methods for calibration include multiple linear regression (MLR) and partial least-squares (PLS) regression. These still greatly outnumber the research instruments (which also use the $\log[1/R]$ or $\log[1/T]$ form, either directly or in derivatized format). The most up-to-date instruments enable calibration transfer among instruments, using the "master/slave" instrument concept and network or modem control of accuracy. These advances include development of neural network calibrations, which enable analysis of, for example, several different commodities, using a single calibration for each constituent.

Computerized spectrophotometers can be calibrated with MLR or PLS regression. PLS regression is based on factors derived from principal component analysis of the optical data and uses both optical and reference data. Ideally, the calibration (sometime called k) constants (i.e., regression coefficients) determined by MLR should be as few as possible for a given instrument or mathematical treatment. Similarly, the number of factors used in a PLS calibration should be as low as practicable. Calibrations involving whole seeds or grains require more wavelength points or PLS factors than calibrations developed for powdered materials, such as flour or ground grain. As a general rule, the more highly correlated the optical and reference data, the fewer the regression constants required to effect a reliable calibration.

Table I outlines the steps necessary for calibration of an NIR reflectance or transmittance instrument. These are discussed in the next section.

TABLE I
Steps in Calibration of a Near-Infrared (NIR) Reflectance/Transmittance Instrument

No.	Description
1	Identify variance sources
2	Identify or develop reference test methods
3	Verify accuracy and reproducibility of reference methods
4	Identify or develop the sample preparation technique
5	Identify the NIR reflectance or transmittance instrument
6	Assemble samples and document
7	Prepare samples
a	Remove foreign material (unless future samples are to be tested in an "as-received" condition)
b	Blend and subdivide as necessary
c	Grind, homogenize, or otherwise prepare
d	Blend thoroughly after grinding
e	Carry out any other necessary preparation steps
f	Store carefully and document
8	Read optical signals
9	View the spectra where possible
10	Select samples on basis of spectral characteristics
11	Perform reference analyses on all samples or on samples selected on the basis of spectral characteristics
12	Add reference data to optical data
13	Select samples on basis of composition
a	Sort samples on the basis of reference data
b	Eliminate excess samples with similar characteristics
c	Partition samples into calibration and verification sets
14	Perform mathematical transformations (computerized spectrophotometers only)
a	Log($1/R$) (or log[$1/T$]) smoothing
b	First derivative
c	Second derivative
15	Compute calibration constants
a	Step-up regression
b	Forward stepwise regression
c	Backward stepwise regression
d	Principal component analysis/partial least-squares regression (PCA/PLS)
e	Neural networks
f	Other
16	Enter calibration constants to instrument
a	Simple calibration
b	"Universal" constants
17	Verify accuracy of calibration by analysis of validation sample set
18	Perform slope/bias correction as necessary
19	Transfer calibration to other instruments
20	Ideally, repeat calibration with fresh sample sets

A. Implementation Steps

IDENTIFY VARIANCE SOURCES

Sources of variance include chemical composition, physical texture, bulk density after grinding, sample type (including variety), seed size before grinding, variability in seed size distribution, color, texture, the presence of features such as surface down (e.g., in cottonseed), awns, seed shape, growing location and season, storage conditions, formula changes, sources of ingredients, ingredient changes, temperature, barometric pressure, humidity, processing conditions, viscosity, stage of maturity of material being analyzed, presence of foreign materials, soil depth and texture, and other features. Other types of variables affect applications in areas such as environmental or clinical studies.

For cereal grains and other plant material, chemical composition and texture can both be strongly affected by growing location and growing season, and these are important variables. Growing location includes soil conditions, farmers' husbandry (including fertilizer use, irrigation, tillage, weed control, etc.), local weather conditions year-round and during the harvest period, altitude, local fungal diseases, and insect pests and other agents. The stage of maturity is a significant variable in forages. It affects both composition and texture, and, as a result, it affects the spectral form. Moisture, cellulose, lignin, and soluble sugars are among the constituents most strongly affected by stage of maturity. The stage of maturity also influences sample preparation technique. For foods, variables include sources of raw materials and variability in processing conditions, canning and other forms of preservative storage, and other factors.

The location of the NIR reflectance instrument is another important source of variance. Grain elevators and flour or feed mills may be subject to large fluctuations in voltage, temperature, and dust conditions. All three can influence the performance of NIR reflectance and transmittance machines. Factory conditions may include extremes of temperature and relative humidity and various types of dust and other material that may be carried by the air. Very dry conditions may result in the generation of static electricity, which affects the handling of some ground (powdered) materials and even small seeds.

Analysis of whole seeds is more prone to the influence of temperature than is analysis of ground seeds. For example, grinding frozen samples, part of NIR reflectance sample preparation, tends to equalize sample temperature due to the heat developed during grinding (Williams et al, 1982). No grinding is necessary for the NIR reflectance analysis of intact materials (grains and other materials, such as fresh forages or manures). This makes the technique more susceptible to the fairly wide ranges in moisture content and sample temperature that can occur from day to day and even within a day. Temperature variation must be built into the calibration by including samples at temperatures that cover the range in temperature anticipated under normal analytical conditions. This correction technique appears to be more successful with monochromator-based than with filter-based instruments.

Temperature of processed materials may vary from very high immediately following processing, through room temperature, to very low temperatures during storage or transportation. For example, flour streams immediately from the rolls may be at a temperature of up to 50°C. If composition is to be monitored by NIR reflectance in-line, as well as on the finished blended and cooler products, these temperature fluctuations should be incorporated into the calibrations. Other variations in ambient conditions (e.g., in manufacturing plants) should also be included. Differences in relative humidity can introduce variable amounts of error between and during days. Changes in the relative humidity have been shown to affect noise levels in a computerized spectrophotometer (Davies and Grant, 1987). Differences in barometric pressure caused by altitude can influence the accuracy of reference tests, which may in turn affect the accuracy of NIR reflectance or transmittance testing by causing biases in calibration or monitoring. This is particularly important when the reference testing calls for the use of boiling liquids. Even the normally reliable Kjeldahl test is subject to the influence of altitude.

Dust and various types of dirt are ever-present factors in NIR reflectance testing. They accumulate on the surface of the ceramic standards of NIR reflectance machines, on the cell covers (particularly on quartz), and on any internal face that is not protected. The periodic cleaning of these surfaces is an integral part of preventive maintenance in NIR reflectance technique. Check samples should be analyzed immediately following cleaning of an internal surface to detect any changes in bias.

Chemical composition is important because of the interactions among constituents during sample preparation and wavelength selection. The most important constituents in agricultural materials and any materials with a biological source are water, oil (fat), cellulose, and protein. Water, oil, and cellulosic components affect the way in which a grain or seed (or forage plant, etc.) breaks down during grinding. They affect particle size and shape, bulk density, and the way in which the ground material packs when a sample cell is loaded.

High moisture levels (above 50%) influence the wavelength range over which the instrument can provide reliable analysis. At very high moisture levels, such as in fresh crops, meats, and fish, lead sulfide detectors are not always reliable above 1,800 nm when NIR reflectance/transmittance instruments are to be used. Figure 1 illustrates the log(1/*R*) and second-derivative spectra of fresh meat. The fresh meat contained over 70% water. Bands

above 1,900 nm are much reduced. The wavelength range of 700 (more recently 650)–1,800 nm has been found effective for the analysis of many commodities (Williams and Sobering, 1993). The influence of high moisture content also depends on the commodity. Moisture contents of even 15–20% have a very significant influence on the spectra of oilseeds, such as canola seed.

Water is the most important constituent of any material in the agricultural and food (and probably other) industries, since it features prominently in both sample preparation and reference analysis. When calibrations are to be developed for water or any other constituent, it is essential that either all of the water be removed before analysis or that the water content be accurately determined. Water content usually varies independently of all other constituents, and it affects the concentration of the other constituents proportionally. The instrument scans the sample "as is" and interprets the signal according to the reference data. For some applications, it is important to convert all constituents to the original moisture basis before adding the reference information to the optical data in preparation for calibration or monitoring. For example, in the case of cooked materials such as canned peas (moisture content 65–70%), the protein content of the material analyzed (and eaten!) is only about 7%, whereas that of the unprocessed peas is closer to 25%.

Modern instruments can be calibrated to provide results on a constant-moisture basis, although this is not truly applicable in commodities in which the range in moisture content is very high, such as corn. (The moisture content of corn can vary from 12% to more than 40%.) Experience with wheat and corn has shown that over a range of up to 3–4% moisture, instruments can be calibrated accurately to determine protein, oil, and other constituents on a constant-moisture, as well as on an as-is, moisture basis. When the moisture range is higher, accuracy is often improved by calibration to constituents other than moisture on an as-is basis. As a separate operation, the moisture content should be determined and the composition *computed* to the constant-moisture basis by suitable programming.

Accuracy of analysis for both moisture and the other constituents is essential to effective calibration. This is particularly evident when NIR reflectance technology is to be applied to fresh material, such as plant material, or animal (including human) tissues or organs. Accurate determination or removal of water may be difficult due to the possibility of driving off volatile constituents other than water.

A further word about moisture. If the instrument is to be calibrated to report results for the analysis of any constituent other than moisture itself, directly on a moisture-free basis, it is essential to ensure that the material *is* moisture free when the reference analysis is carried out. A small amount of residual moisture can lead to variable errors if the instrument is to be calibrated for a constituent that is present in high concentration. For example,

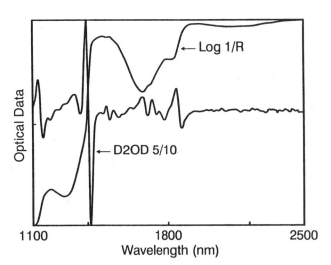

Fig. 1. Log(1/*R*) and second-derivative spectra of fresh meat.

residual moisture of up to 3% in soybeans, which contain over 40% protein, can lead to errors of up to 1.5% in protein content as reported, if the residual moisture is undetected. The instrument will "see" the sample as is and will report low "moisture-free" results for (in this case) protein content.

Oil also affects particle behavior. It tends to cause "pasting," which can give trouble in grinding, even at levels as low as 5%. For example, in grinding oats or corn on a cyclone grinder, it is necessary to clean the screen between samples, whereas with grains such as wheat, barley, peas, or lentils, the grinder is self-cleaning. Oil can cause particles of high-oil-bearing seeds, such as sunflower or canola (oilseed rape), to agglomerate after grinding so that, after a few days, the results of a repeat test may not agree with the original due to changes in the particle characteristics.

Cellulose and cellulosic materials change particle shape and bulk density. Cellulose is very light, and, the more cellulose present, the lighter the ground sample and the lower the bulk density. Particles of highly cellulosic materials are usually longer, relative to their width, than particles of less fibrous substances. As plants mature, their structural components become lignified. Lignin is a complex material that greatly strengthens plant stems, such as straw. This enables the stems to continue to hold the developing grains and seeds above the ground as their weight increases due to the formation of starch and other seed food reserves. Lignin also affects the way in which straws and other stem material, including wood, is reduced in grinding.

Materials high in cellulosic fiber components often generate static electricity, and ground samples may "fly" from the spatula, cling to the sides of containers, and orient themselves on the surface of sample cells, etc.

Table II gives the bulk density of some ground materials of plant origin.

Proteins can also affect particle behavior, since they vary in their reaction to static electricity. This is noticeable with materials such as oats and some pulses (food legume seeds). When dealing with materials of animal origin, proteins are often present in the form of ligaments and cartilages, which introduce a further peculiarity to sample preparation. Fish scales are particularly difficult to incorporate evenly into ground, dried fishmeal. Protein can also influence the texture of commodities such as grains. Some types of wheat become harder as protein content increases from, for example, 10 to 18%. On the other hand, hard red spring and hard red winter wheats, and other wheat types of similar texture, become significantly softer as the protein content increases (Pomeranz and Williams, 1990).

In some cases, other constituents can influence particle behavior. For example, materials such as the leaves of *Atriplex* species ("salt-bush") may contain very high levels of minerals; levels of above 20% have been reported. The mineral content can substantially change particle bulk density and behavior in grinding. High mineral content can also affect spectral form.

Color is yet another factor that may influence the shape of spectra. Although many measurements are made at near-infrared wavelengths where color should not be a factor, in practice, very dark samples can cause bias at lower wavelengths and should be included in any sample sets in which dark-colored material is likely to occur. Figure 2 illustrates the effect of color on the spectra of canola samples. The higher spectra are those of the darkest seeds.

IDENTIFY AND DEVELOP REFERENCE TEST METHODS

The NIR reflectance instrument reports data in terms of the reference data fed into it during calibration. The reference test methods are usually those that have been used by the operator before the introduction of NIR reflectance. It is a good idea to review the accuracy and precision of reference methods before calibrating an NIR reflectance machine. Suitable reference methods for analysis of agricultural and many other materials are described in the *Approved Methods of Analysis* issued by the American Association of Cereal Chemists (American Association of Cereal Chemists, 2000) and by the *Official Methods of Analysis* of Association of Official Analytical Chemists International (1995).

In specialized cases, it is necessary to develop reference methods specific to the application. Methods of this type should be very carefully evaluated from the aspects of their relation to reproducibility by other laboratories and to the functionality of the material. Some industrial applications, in which the instrument is used to monitor composition with respect to one or more ingredients, or to pure chemical constituents, can be calibrated directly to specific composition. Operators can prepare their calibration and validation samples by weighing out the required amounts and then carefully blending the prepared samples. This type of calibration can be affected using only a few samples (e.g., 12–20).

The reference test of choice for protein content is still the Kjeldahl test in most laboratories, although the Dumas (combustion analysis) procedure is rapidly gaining in popularity due to its superiority in precision and its slight but consistently greater efficiency in measuring all the nitrogen present in the sample. Over 30 sources of error have been identified in the Kjeldahl test (Williams, 1974), and, later in this monograph, an even larger number of factors are described that affect the NIR reflectance procedure. The NIR reflectance and Kjeldahl test procedures both include sampling and preparing the sample, but the Kjeldahl test for total nitrogen or protein also involves reagent preparation, weighing, digestion, and distillation.

The Dumas test is not free from error, although the error sources are fewer than those of the Kjeldahl test. The chief sources of error have been identified and discussed (Williams et al, 1998). These are summarized in Table III.

VERIFY ACCURACY AND REPRODUCIBILITY OF REFERENCE METHODS

Reproducibility or precision can be determined by periodic testing of check samples. These should be carefully prepared and thoroughly blended before use. The samples should be stored at a

TABLE II
Bulk Density of Some Common Commodities

Material	Bulk Density, g/cm³
Faba bean	0.639
Field pea	0.588
Lentil	0.542
Wheat flour (hard red spring)	0.514
Corn (maize)	0.506
Wheat (hard red spring)	0.495
Soybean hulls	0.489
Wheat (soft white spring)	0.450
Soybean	0.415
Wheat flour (soft white spring)	0.395
Barley	0.390
Wheat bran	0.349
Oats	0.345
Defatted canola seed	0.338
Dried alfalfa	0.240
Barley straw (mature)	0.142

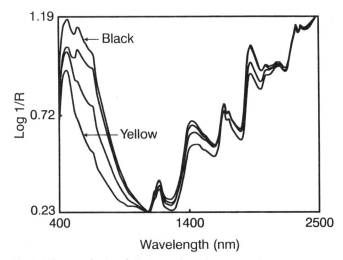

Fig. 2. Influence of color of whole canola seeds on spectra.

reduced temperature and protected from moisture loss (or gain) and possible deterioration due to light or infestation. Each test should involve fresh sample preparation, since it should represent the entire test procedure. Reproducibility can also be determined by introducing "blind" duplicates to the system, ideally four to five times each week. These are samples that have been previously tested by the laboratory and are reintroduced under another identification number. *Repeatability* can be determined by periodic analysis of the sample *after* preparation—this evaluates the operators' efficiency in presenting the sample to the instrument.

Accuracy can often be evaluated in industrial commodities only in terms of check samples of "known" composition, prepared by the operator. The determination of "true" accuracy in materials of biological origin is complicated by the fact that the constituents that are being determined are often not well defined, so that "spiking" samples to evaluate recovery (e.g., with pure protein) is not possible. Subscription to authorized check sample programs, such as those of the AACC and Smalley Check Samples (available, respectively, from the American Association of Cereal Chemists, St. Paul, MN, and the American Oil Chemists Society, Champaign/Urbana, IL) enables operators to check on the accuracy of their laboratory or instrument and provides information as to how their laboratory/instrument compares with others. The actual operation may also serve as a check on accuracy. In the absence of the use of such a check sample system, consistent yields of results that are too high or low will eventually be reflected in terms of inferior end-product production and customer complaints.

IDENTIFY AND DEVELOP
A SAMPLE PREPARATION TECHNIQUE

Sample preparation is extremely important to both reference and NIR reflectance testing and can account for 60–70% of the overall error of the testing (Hildebrand and Koehn, 1944). Sample preparation is defined in this monograph as "transformation of the sample into the form in which it will be analyzed, without causing any changes in functionality or composition (other than in moisture content)." It includes documentation, blending, subsampling, removal of foreign material, grinding or some other form of size reduction, reblending, and storage.

Blending and subsampling before size reduction entail careful study of the original sample to ensure that it is thoroughly blended and that the subsample abstracted for analysis is truly representative of the population from which the original sample was drawn. Foreign material is usually removed from grains and seeds by equipment that combines the use of air aspiration and sieves. Removal of small seeds, such as canola seed from wheat or barley, is very important in NIR transmittance analysis, since the presence of seeds can have a strong and variable influence on the accuracy and reproducibility (Williams et al, 1985). Blending is important in all substances, including liquids, which tend to stratify due to changes in temperature. It is particularly important in milk, in which the fat globules move fairly quickly. Milk analyzed in transmittance mode should be agitated continuously during scanning, since the upward movement of fat globules may cause a significant "particle size" effect. This is the case whether NIR transmittance or transflectance is used—in transflectance mode the composition of the surface may change constantly due to the accumulation of fat.

Sampling of soil and sediments (under lakes, rivers, and oceans) involves taking samples at different depths, as well as from different locations. Different depths reflect different ages in sedi-

ments, and, in soils, they represent not only different also the effects of cultivation and levels. Sampling and sa moval, are critical and manures. ence anal than

state for size reduction. This subsampling. It is important to select the optimum for the material to be analyzed. For some pulses, such as faba or "horse" beans, are very hard may be too large to enter the grinding chamber of a grinder such as the Cyclone Sample Mill grinder (Udy Corp., Fort Collins, CO). This can be overcome by prereduction of the beans in a hammer mill set at a coarse setting. This produces a coarse meal, too coarse for accurate and precise analysis but that can be fed into another grinder to give the final laboratory sample. Other "hard" seeds include field pea, chickpea, and palm kernel.

Oilseeds tend to paste during grinding and can clog the screen of any grinder fitted with a screen. This can be largely avoided by grinding soft seeds, such as canola and sunflower, in a high-speed impeller mill in short "bursts" of about 15 sec. The partially ground, caked material is stirred with a spatula between grinding periods. The Retsch Centrifugal mill (Brinkman Instruments, Westbury, NY) has been found very useful in grinding seeds of high oil content, such as canola, sunflower, and flax seeds (as well as other types of grains). Soybeans, and even corn and oats, contain enough oil to clog grinder screens. Table IV suggests some grinders suitable for preparation of various grains and seeds. The Braun, Krups, and Moulinex grinders are high-speed impeller grinders. They are all kitchen appliances that are available from department stores.

Blending of the ground sample after grinding is an essential part of sample preparation. No matter what type of material is to be reduced, it will always stratify during the grinding and must be thoroughly remixed before analysis. This is particularly important in the case of fine materials such as wheat flour and in cases in which a large sample has been ground (e.g., 300 g in a Falling Number model KT 3100 grinder [Perten Instruments, Inc., Springfield, IL]). Liquids, including milk and fruit juices, and slurries also stratify due to differences in composition and temperature, and they should be well stirred before and during analysis.

Laboratories that have become accustomed to the use of a particular grinder over many years may find it unacceptable to change to a new grinder. The actual grinder used in sample prepa-

TABLE III
Sources of Error in Dumas Testing

Sampling	Subsampling
Sample preparation	Blending after grinding
Moisture correction of results	Poor instrument maintenance
Impure gases, especially oxygen	Impure EDTA
Moisture in EDTA	Residual nitrogen (air) in weighed sample
Sample size	Access of sample to instrument (liquids only)
Sporadic use of instrument	Insufficient EDTA and/or blank tests
Weighing error	Instrument malfunction

TABLE IV
Some Grinders for Laboratory Testing[a]

Workload/Material	Grinder	Action
High		
Cereals, pulses	KT 3303[b]	Burr
	Cyclone[c], Cyclotec[d]	Impeller
Medium		
Cereals, pulses	KT 3303[b]	Burr
	Cyclone[c], Cyclotec[d]	Impeller
	KT 3100[b]	Hammer
	Christy/Norris[e]	Hammer
	Microhammer Cutter Mill[f]	Hammer
	Retsch[g]	Centrifugal
Oilseeds, including soybean	Retsch[g]	Centrifugal
	Braun[h], Krups 75[h]	Impeller
	Moulinex[h]	Impeller
Forages	Wiley no. 4[d]	Cutting
	Christy/Norris[e]	Hammer

[a] Other grinders are described by Williams (1984).
[b] Available from Perten Instruments, Inc., Springfield, IL.
[c] Available from Udy Corp., Fort Collins, CO.
[d] Available from Fisher Scientific, Inc., Chicago, IL.
[e] Available from Christy, Ltd., Scunthorpe, England.
[f] Available from Glen Mills, Clifton, NJ.
[g] Available from Seedburo, Inc., Chicago, IL.
[h] Available from many department stores.

ration is not important, provided the samples used in future analysis by NIR reflectance will continue to be prepared using the same grinder. However, it is important for the laboratory to determine that its grinder of choice will continue to be available into the far distant future; otherwise, the time to change grinders is when the laboratory decides to employ NIR reflectance technology.

The advent of whole-seed NIR transmittance analyzers has resulted in significant improvements in the accuracy and reproducibility of grain analysis, due mainly to the elimination of the errors inherent in sample preparation and sample cell loading. But this has introduced other sources of error. First, the moisture content of whole grain is usually significantly higher than that of the ground grain upon which the reference analysis is based. For some purposes, it is important to report the composition of the grain as received rather than on a constant-moisture basis. This calls for accurate determination of moisture in the whole grain and in the ground grain. The composition based on the moisture content of the ground grain used in reference analysis must be recomputed on the basis of the original moisture content of the whole grains (or other materials).

Temperature is another source of error. Most NIR reflectance instruments are sensitive to temperature. In some cases, temperature sensitivity can be compensated for by incorporating samples at different temperatures into the calibration and validation sample sets. In reflectance instruments that require ground sample, the action of grinding warms samples of grain even at subfreezing temperatures to room temperature (Williams et al, 1982). In the case of instruments that do not require the sample to be ground, cold samples enter the instrument the way they are. Not only may these samples be below freezing, but the analysis of a series of them cools the instrument, which can lead to further significant errors. The same applies in reverse to the analysis of samples of very high temperatures, such as freshly harvested grain.

When calibrating the instrument to compensate for temperature, the chilled, or heated, samples used in calibration should be scanned in batches of at least six, all of which should be at about the same temperature. This exposes the instrument to the effect of cooling, or warming, of the instrument itself. Under operational conditions, truckloads or railway carloads of grain will arrive at the testing location at more or less the same temperatures, depending upon the time of year. Samples of industrial materials, such as flour, that are analyzed by NIR reflectance in-line (during or immediately following processing) may be warm to hot at the time of scanning, and the NIR reflectance instruments should be calibrated with samples in the anticipated temperature range.

Sampling and sample preparation are dealt with in more detail by Williams (1992).

IDENTIFY NIR REFLECTANCE OR TRANSMITTANCE INSTRUMENT

You don't need an ax to cut flowers! For day-to-day determinations of protein, moisture, etc., and possible extensions to oil and fiber components, discrete filter instruments are suitable and less expensive. If it is necessary to report the moisture content of whole seeds, either NIR reflectance or transmittance instruments are available, and the analytical process is nondestructive, whereas the sample preparation necessary for NIR reflectance testing of ground grains or seeds causes variable moisture loss. If analysis for functionality parameters or applied research is anticipated, a computerized spectrophotometer is recommended, or at least an instrument with a large number of discrete filters and the capability of using comprehensive software to exploit the wavelength or filter range to the fullest extent. Many applications involve in-line analysis. In this monograph, "in-line" implies that the sample is continuously being scanned and tested during the manufacturing or handling process. "On-line" means that samples are withdrawn during the process or after the processing is complete, but with the instrument adjacent or close to the operation.

Transferability of calibrations among instruments is a prerequisite for any operation that operates more than one instrument, as is also the capability of networking. Identification of the NIR reflectance instrument is particularly important if the laboratory is contemplating any long-term (3 years or more) research, since the instrument may become obsolete before the research is complete if a new instrument that represents a significant advance in technology is introduced.

Sample access, or presentation, is an important aspect of NIR reflectance instrument design. Ideally, the instrument should be able to accommodate a large sample (at least 100 g), to minimize sampling error. On the other hand, for applications in plant breeding, it may be necessary to analyze samples of only a few grams. Some NIR reflectance scanning spectrophotometers, such as the Foss/NIRSystems models 5000 and 6500 (Foss/NIRSystems, Inc., Silver Spring, MD), have sample access systems capable of handling samples from 120 g or more to a few milligrams. Their software enables the instrument to record the same spectra at several levels of loading, so that, for example, a calibration developed on a completely filled cell can be used to predict parameters in cells only one-quarter-filled, and vice versa.

Another aspect of sample access is the degree to which gravity is involved in the movement of sample into and out of the instrument. High-moisture grains do not flow easily, and an instrument that relies on the sample flowing smoothly into and out of the instrument may not be suitable for the day-to-day analysis of high-moisture grains, such as maize, the moisture content of which can exceed 50%. In the case of instruments that rely on a sample cell, it may be difficult to fill the cell consistently with high-moisture grain. Packing the grain by tamping onto a firm surface may assist in filling the cell, but it may be very difficult to empty the cell after analysis. Cells such as the coarse-sample cell offered by Foss/NIRSystems, Inc. and the Natural Products cell developed by Infra-Soft International Inc. overcome this fairly satisfactorily by enabling simple removal of the back of the cell. Instrument manufacturers have designed an extensive series of sample presentation cells, as well as custom-designed systems for in-line and other specific sample presentation.

Gravity also affects in-line instruments, since the withdrawal for scanning of samples of materials of many particle sizes and shapes, from fine material such as cement or flour up to larger particles, often depends on gravity. The incorporation of some form of agitating device to the sampling system may be required to improve the uniformity of sample withdrawal and disposal.

For some time, NIR reflectance instruments have been employed for in-line analysis for purposes of segregation, process control, quality assurance, and other purposes. Speed of analysis is very important in in-line NIR reflectance analysis, particularly where this involves analysis of materials on moving belts, or materials that themselves form the "belt," such as dough sheets in biscuit and cracker factories, paper in paper mills, etc. The new, ultrarapid generation of diode array and AOTF instruments (Chapter 7) will likely find many applications in these areas.

Application of NIR reflectance technology to the analysis of materials such as sewage and fresh manure is becoming important in monitoring of the impacts of operations such as hog and chicken production on the environment. These materials present new exigencies, such as their offensive odor. This can be overcome by presentation of the samples to the instrument in plastic bags. Unwanted absorbers contributed by the plastic can be excluded by using a piece of the same plastic as reference in reflectance mode. Liquid materials such as manures can also be scanned in scintillation vials, provided provision is made for agitation to prevent settling of the solid components.

ASSEMBLE SAMPLES AND DOCUMENT

The assembly of samples can be the most difficult and time-consuming operation in setting up an NIR reflectance or transmittance calibration. The samples should include all of the anticipated variance in terms of growing location, season, etc., as outlined in the first step, identify variance sources. This may represent a large number of samples. Unless samples have been accumulated before the calibration is developed, assembly of samples from several growing seasons may take several years, during which the calibration should be monitored and, where necessary, updated each year. Similarly, in the case of industrial processing, samples should be accumulated from batch to batch, etc. In all such cases, the samples should be scanned at the time of receipt (since that is the way they will be scanned in real time), and stored for subsequent checking of the instrument. Storage of samples becomes important in long-term studies. Samples stored at 5°C in heavy-duty plastic bags, further protected by tightly lidded plastic pails, can be preserved without significant change for several years.

Failure to include all sources of variance may lead to outliers in subsequent analyses. An outlier is defined (in this monograph) as a sample of the predicted NIR reflectance result that differs from the true result by three or more times the SEP. Outliers are discussed more thoroughly in Chapter 9. When a laboratory is considering purchase of an NIR reflectance or transmittance instrument, or if the laboratory is likely to extend its NIR reflectance testing to another constituent or parameter, it should start accumulating samples of all types in readiness. Samples can be preserved in cold storage as described above until the laboratory is ready to develop the calibration. If the laboratory already possesses the instrument and seeks to develop a new calibration or calibrations for a new commodity, spectra should be recorded at the time of receipt of the samples, since that is the time at which they will be scanned when the calibration has been developed. Spectra retain their properties forever, whereas samples may not.

Documentation should provide the operator with all of the relevant information concerning each sample. The correct identification must be attached to each sample in a manner that will not cause subsequent confusion. *About 4% of all errors in NIR reflectance, and particularly in reference testing, can be attributed to mistaken identification.* In the case of reference testing, this can result in the wrong analytical procedures being applied to the samples. The information should include the date and, where necessary, the time of day, in the case of industrial process development and production. The year of accession should be included in the date. Growing season is an important source of variance (Williams and Cordeiro, 1985), and different seasons should be identified. The stage of maturity of the material such as forage is an important factor.

PREPARE SAMPLES

For materials such as grains and seeds, comments further to those in the fourth and sixth steps (identify and develop sample preparation technique and assemble samples and document, respectively) are unnecessary, beyond the note that it is a good idea to retain samples of foreign material removed from samples. If a sample emerges as an outlier, the foreign material removed may provide useful background information as to the source of the sample. For many materials, sample preparation may be more time-consuming and demanding. For example, the preparation of fresh forages, and particularly silages, requires specialized treatment to ensure that the highly variable moisture content has been addressed, without affecting the composition with respect to volatile constituents, such as fatty acids and esters.

For many environmental applications, a further prerequisite is the identification of a suitable medium, the analysis of which will provide a reliable indication of the state of the environmental condition under investigation. This may mean the collection of living organisms of plant or animal origin, which are representative of the location or region. In the case of clinical investigations, analysis of body liquids, such as blood and urine, is often sufficient to enable diagnosis of a disease and monitoring of its progress or cure. Storage and preparation of these biological materials for reference analysis may involve more specialized procedures, such as freeze-drying, but the speed of NIR reflectance analysis, besides being noninvasive, may be able to by-pass some of the laborious sample preparation necessary to reference analysis.

READ OPTICAL SIGNALS

Computerized spectrophotometers all record the optical signals on their hard disk. Most modern stand-alone instruments can re-tain the signals of sufficient samples to enable monitoring, and they can be interfaced to a personal computer (PC). Some stand-alones such as the Foss/Tecator InfraTec (Foss/Tecator AB, Höganäs, Sweden) are capable of both storing and processing the optical signals (up to 100 scans in the case of the latest Foss/Tecator InfraTec models). Otherwise, the optical signals must be recorded manually and later transferred to a computer for calculating the regression (calibration) constants.

The spectra *are* the samples. They contain all of the information that makes up the sample, including chemical composition and the interactions among constituents that cause the sample to possess its unique physical properties and functionality. The spectrum will be the same, no matter what instrument is used to record it. Differences in resolution and in the quality of the spectra will certainly occur, but the actual spectral information is based on what is in the sample and cannot be changed.

Variance occurs among successive scans of the same sample depending on the instrument and the sample presentation method, and, to get the best practicable calibrations and analytical data, the operator should take time to determine the optimum number of scans that are needed. Normally, results improve with duplicate, and up to quadruplicate, scans, which can subsequently be averaged. If the operator finds that triplicate scans are needed to secure the required accuracy and precision, three NIR tests are still very much faster than a single test by the reference method.

VIEW THE SPECTRA

If possible, the spectra should be examined before any further steps are taken. Noisy spectra and other abnormalities can be detected and the spectra removed or rescanned. The spectra can often be viewed during recording and abnormalities noted at that time. In the case of many benchtop filter instruments, viewing the spectra is not possible; instrument performance is monitored by check samples and the instruments' built-in diagnostics.

Computerized spectrophotometers have diagnostic software to enable monitoring of instrument performance. This should be used regularly, at least twice every week, and should be run at different times of the day, to monitor instrument response to ambient conditions. Some computerized spectrophotometers are sensitive to relative humidity, and noise "spikes" can occur if the relative humidity changes even momentarily. If the diagnostic software is run first thing in the morning every day, the results may indicate that the performance is abnormal. This may be misleading, since the instrument should be allowed to warm up for at least 15 min before use. The most up-to-date benchtop instruments carry software options that indicate whether optical signals appear to the instrument as outliers. The appearance of outliers should not be confused with spectral aberrations, unless the frequency of their appearance suddenly begins to increase.

SELECT SAMPLES ON BASIS
OF SPECTRAL CHARACTERISTICS

For an organization such as a company that has no laboratory, laboratory analysis can become expensive. For example, a licensed laboratory may charge the company $5.00, $10.00, or $50.00 per test (with possible volume reductions) for relatively simple tests such as moisture (oven), protein (Kjeldahl or Dumas combustion analysis), or oil (extraction or nuclear magnetic resonance). Costs per test for additional constituents or parameters that call for more demanding analyses may be considerably higher. The initial cost of testing 100 or more samples can add substantially to the purchase price of the instrument, particularly if the instrument is to be calibrated for several constituents or parameters. To alleviate this, instrument and software companies have developed software to enable selection of samples for calibration and validation on the basis of spectral characteristics alone. The operator can select perhaps 20–30 samples from a large population with sufficient spectral diversity to provide a reliable calibration. Reference testing can then be limited to these samples.

Spectral selection of samples is a useful technique, and it is relatively simple to perform with modern software. The system is effective in reducing the preponderance of samples that cluster about the population mean, and up to two-thirds of the samples can thus be eliminated. It works very well in situations in which the wavelengths for making the measurements are not highly subject to interferences from other constituents. Protein content in cereals is a good example. Many applications involving discrete substances, such as pharmaceuticals, are well suited to spectral sample selection.

The most advanced software for spectral selection employs statistics derived from the Mahalanobis distances of the optical data. No reference data are involved. The number of samples selected is in turn based on an arbitrary value by which data for individual spectra differ from one another. The principle is that when two samples furnish spectra that are essentially similar across the wavelength range, there is need for only one of them. The selected spectra are stored as the calibration set, and the remaining samples can be used in validation. An effective way of achieving calibration and validation by this method is to set up the calibration based on spectral selection and validate it by cross-validation (preferably using single-sample iteration). The calibration can then be further validated by predicting the composition (or functionality) of the remaining samples, preferably by analysis of a second validation set that has not been involved in the selection.

The software usually recommends exclusion of certain samples that are identified as spectral outliers. These may be excluded or included at the discretion of the operator. In practice, these samples should be included as far as practicable, since they represent variance in the population that may actually assist in stabilizing the calibration in the long term. As well, samples that arrive for analysis may be identified as spectral outliers but may predict within the boundaries of the SEP. There are some points to note with the spectral method of sample selection.

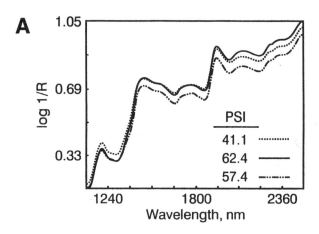

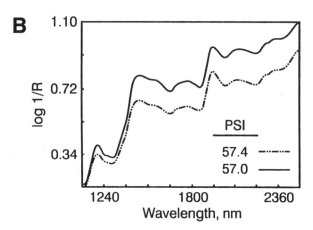

Fig. 3. Spectra of wheat **A,** with different texture but similar spectra and **B,** with similar texture but different spectra.

First, in addition to composition, spectral characteristics can be strongly influenced by variance in particle size, shape, and distribution. Similar spectra can be identified for samples that differ in composition and functionality, and vice versa. Variations in spectral bands caused by the particle effect may be affected by factors other than chemical constitution, but they can influence sample selection. As a result, a sample set selected solely on the basis of spectral variance may have satisfactory distribution of composition with respect to one or more constituents, particularly if the constituents are intercorrelated, but it may not have good distribution of composition of others. This is particularly patent in the case of multiple calibrations in which moisture content is one of the constituents, since moisture content is not correlated with any other constituent (except dry matter).

Second, as a corollary to the above, assuming the system is successful in the selection of calibration samples for one constituent, the same sample set, selected on the basis of spectral characteristics only, may not be equally effective for calibration for a second constituent, *but the selection process can only be carried out once*. This difficulty has been overcome by software that can employ PLS, which uses reference as well as spectral data to compute the "scores" used in the development of calibrations.

Third, differences between spectra of samples that differ considerably in composition may be subtle, and spectra that are rejected due to their similarity to other spectra may not be similar in composition to those spectra that are retained. This is illustrated by Figure 3A, which shows spectra of two wheat samples that differ widely in texture (particle size index) yet have similar spectra. Figure 3B shows spectra of two wheat samples of similar texture, but with very different spectra.

Fourth, interferences from other constituents may bias the selection. For example, oil and cellulose both have strong absorbances between 2,300 and 2,350 nm. Commodities such as canola, sunflower, and sesame are rich in cellulosic components, and some of the samples selected for an oil calibration on the basis of their spectra may be higher in cellulose than in oil, and vice versa.

Fifth, spectral sample selection may be a useful approach, but it still requires the accumulation, preparation, and scanning of a large

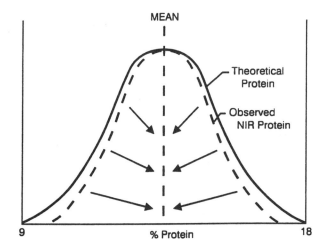

Fig. 4. The Dunne effect.

TABLE V
Computing the Dunne Effect

Source of Data	r[a]	Mean	Low	High
Original reference result		13.50	8.90	15.90
Computed NIR reflectance result[b]	0.85	13.50	9.59	15.54
	0.95	13.50	9.13	15.78
	0.98	13.50	8.99	15.85

[a] Coefficient of correlation.
[b] Formula for near-infrared (NIR) reflectance calculation: Computed result = mean + r × (original reference result − mean).

number of samples in order to include all of the anticipated variance from which to select the best sets for calibration and validation, particularly when calibrations are to be developed for the determination or prediction of several constituents and parameters.

PERFORM REFERENCE ANALYSIS

Reference analysis is crucial, since the accuracy of the calibration, and particularly its subsequent monitoring, depends largely on that of the reference testing. Modern NIR reflectance/transmittance instruments are very precise in their operation, often far more so than reference methods. No matter how precise the instrument, it is impossible to assess its performance properly if the reference results to which it is being matched are unreliable.

The argument has been advanced that absolute accuracy may not be essential to the development of a calibration equation, provided the errors are unbiased. The instrument will interpret the optical signals according to its own precision, will continue to make measurements with the same precision, and will consequently furnish results after calibration. This is a fair assumption. But the fact remains that, if the accuracy of subsequent analysis by the instrument is to be compared with that of the same source of reference analysis, it is essential that the reference analysis used in the evaluation and monitoring of the instrument be both very accurate and precise. Inaccuracies in the results of monitoring are usually attributed to the NIR reflectance instrument.

ADD REFERENCE DATA TO OPTICAL DATA

Reference data must be attached to the correct optical data. Although this sounds like a simple prerequisite, one of the most consistent reasons for outliers is the attempt to marry the reference information to the wrong partner!

SELECT SAMPLES ON THE BASIS OF COMPOSITION

If sufficient samples are available with reliable reference analysis, it is recommended that calibrations be developed using both spectral and reference data (rather than selection using spectral data alone). There are three substeps to follow: first, list samples from lowest to highest reference data; second, remove excess samples with similar composition; and third, prepare a calibration set and (ideally) two validation sets.

List Samples from Lowest to Highest Reference Data. This step is self-explanatory, and is optional. If all samples are ranked

before compiling calibration and validation sample sets and the calibration and validation sets are selected at regular intervals from the ranked samples, this guarantees distribution of samples representing the range in composition in all sample sets. Otherwise, either the calibration or the validation sets may have an overabundance of samples at some composition levels and be bereft of others. If the calibration is to be evaluated using cross-validation this step is unnecessary.

Remove Excess Samples. Most populations are characterized by the Gaussian or "normal" distribution of samples about the mean, the familiar bell-shaped distribution. Depending on the type of material, and the constituent or parameter to be determined, the distribution about the mean can vary in width. Narrow distributions have relatively fewer samples at extreme levels than broad distributions.

When the assembled samples show a marked "peak" in composition or functionality about the mean, the samples up to one SD above and below the mean should be reduced in numbers. The numbers of samples in that range will then be not excessively greater than those at the more extreme reference data levels. If possible, Gaussian distribution of samples with respect to composition should be avoided in the development of NIR reflectance (or any other) calibrations. Calibrations developed with sample sets having Gaussian distribution of composition with respect to a particular constituent may cause the results of future analyses to regress toward the mean (the "Dunne" effect) (Fig. 4).

Results of predictions at high concentrations will appear to be lower, while those at the low end will appear higher than are the actual (reference) results. The size of the error is related to the frequency distribution, the error of both reference and NIR reflectance testing, and the coefficient of correlation. It will be most pronounced if most of the samples are closely distributed about the mean, with relatively few at 2.5 or more SDs away from the mean, and if the standard error per test of either or both reference and NIR reflectance techniques is high.

For a given sample, the degree of error can be approximately computed using the coefficient of correlation and the difference between the reference result of the sample and the mean result for the population. Examples are given in Table V. The higher the correlation between NIR reflectance and reference data, i.e., the more accurate the NIR reflectance analysis, the lesser will be the Dunne effect.

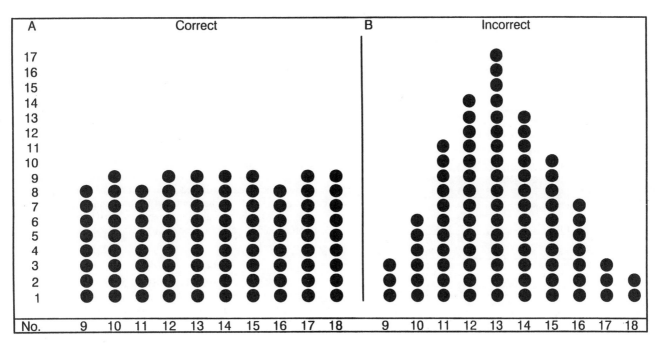

Fig. 5. Distribution of samples for calibration sets.

Ideally, sample sets for calibration should be assembled with uniform distribution of composition across the anticipated range. It is good practice to limit calibration and validation sample sets to approximately the same number of samples at regular intervals of reference data. For example, a population of 120 samples with a range in composition of 9–18% should ideally contain about 12 samples at each percentage "interval" from 9–18%. Samples over and above the selected number at levels close to the mean may be discarded from the calibration exercise. Figure 5 illustrates good and poor sample distribution (on the basis of composition).

Some sample selection software based on spectral characteristics, such as ISI and WINISI software (InfraSoft International, Inc., Fort Matilda, PA), is effective in eliminating excessive samples with similar spectral characteristics. This is because samples closest to the mean in a Gaussian distribution are most likely to be similar to each other in spectral characteristics. The number of these actually selected by ISI software will be proportionately fewer than the number of samples at the extremes. This is an important advantage of sample selection by spectral methods for calibration development.

Assembling sample sets for calibration for more than one constituent or parameter can be more complicated, because a sample

TABLE VI
Setup Table for Compiling Calibration and Validation Files

1	51	101	151	201	251	301	351	401	451
2	52	102	152	202	252	302	352	402	452
3	53	103	153	203	253	303	353	403	453
4	54	104	154	204	254	304	354	404	454
5	55	105	155	205	255	305	355	405	455
6	56	106	156	206	256	306	356	406	456
7	57	107	157	207	257	307	357	407	457
8	58	108	158	208	258	308	358	408	458
9	59	109	159	209	259	309	359	409	459
10	60	110	160	210	260	310	360	410	460
11	61	111	161	211	261	311	361	411	461
12	62	112	162	212	262	312	362	412	462
13	63	113	163	213	263	313	364	413	463
14	64	114	164	214	264	314	364	414	464
15	65	115	165	215	265	315	365	415	465
16	66	116	166	216	266	316	366	416	466
17	67	117	167	217	267	317	367	417	467
18	68	118	168	218	268	318	368	418	468
19	69	119	169	219	269	319	369	419	469
20	70	120	170	220	270	320	370	420	470
21	71	121	171	221	271	321	371	421	471
22	72	122	172	222	272	322	372	422	472
23	73	123	173	223	273	323	373	423	473
24	74	124	174	224	274	324	374	424	474
25	75	125	175	225	275	325	375	425	475
26	76	126	176	226	276	326	376	426	476
27	77	127	177	227	277	327	377	427	477
28	78	128	178	228	278	328	378	428	478
29	79	129	179	229	279	329	379	429	479
30	80	130	180	230	280	330	380	430	480
31	81	131	181	231	281	331	381	431	481
32	82	132	182	232	282	332	382	432	482
33	83	133	183	233	283	333	383	433	483
34	84	134	184	234	284	334	384	434	484
35	85	135	185	235	285	335	385	435	485
36	86	136	186	236	286	336	386	436	486
37	87	137	187	237	287	337	387	437	487
38	88	138	188	238	288	338	388	438	488
39	89	139	189	239	289	339	389	439	489
40	90	140	190	240	290	340	390	440	490
41	91	141	191	241	291	341	391	441	491
42	92	142	192	242	292	342	392	442	492
43	93	143	193	243	293	343	393	443	493
44	94	144	194	244	294	344	394	444	494
45	95	145	195	245	295	345	395	445	495
46	96	146	196	246	296	346	396	446	496
47	97	147	197	247	297	347	397	447	497
48	98	148	198	248	298	348	398	448	498
49	99	149	199	249	299	349	399	449	499
50	100	150	200	250	300	350	400	450	500

set may have uniform distribution in one constituent but not in the others. In some cases, there is a negative relationship between two constituents. Distribution of both will then be relatively uniform, but in others, particularly in the case of constituents such as fiber components and moisture, which are not highly correlated to any other component (and usually not to each other), the pattern of distribution of composition may be poor. Under such circumstances, more samples should be accumulated and tested to extend the range and balance the distribution, and the process of assembling calibration and validation sets must be repeated.

Prepare Calibration and Validation Sample Sets. Prepare sample sets, one for calibration and (ideally) two for validation of the effectiveness of the calibration. Sample sets should be set up so that the ratio of calibration to validation samples is 3:1. If calibration sets are too small, the calibration may be sample sensitive, and analysis of fresh batches of samples may show significant differences in accuracy.

A useful procedure is as follows. Using the original sample file sorted by reference data and, with excess samples removed, transfer the first three samples into the calibration set; leave the next two samples, and then take samples 6–8 into the calibration set; leave the next two, and then take samples 11–13; and so on, taking groups of three and leaving the intermediate two samples until you reach the end of the original set. Then set up the first validation set using samples 4, 9, 14, 19, etc., and the second validation set using samples 5, 10, 15, 20, etc. This will provide a calibration and two validation sets, and, provided that excess samples around the mean have been discarded, it will assure a uniform distribution of reference data in all three sets.

Table VI illustrates a method of setting up samples for original sample sets from populations of up to 500. Spectra of samples with numbers 4, 9, 14, 19, up to 494 and 499 make up the first validation set. Numbers 5, 10, up to 495 and 500 make up the second validation set.

This system is applicable to populations of 100 or more spectra. For smaller sample sets (between 60 and 100), it is preferable to sort the samples, as above, and then divide them into two sets (odd and even) and use them in preparing reciprocal calibration/validation sets. For populations of up to 60 spectra, cross-validation is recommended. Cross-validation is discussed later in this chapter.

When samples are to be accumulated to represent two or more growing seasons, or industrial processing conditions, ideally, samples with a range of composition should be accumulated for all seasons, processing conditions, and other factors. Otherwise all of the "low" or "high" samples may be associated only with one set of conditions, and future samples from different conditions may become identified as outliers.

In a perfect world, the calibration and validation sample sets should not be related to one another, but both should embrace the same variance dimensions. The samples used in validation should, as far as practicable, be assembled separately from those used in calibration, but should include all of the identified sources of variance. The method described above will not achieve this, and, ideally, a further set of samples should be assembled to act as a true set of "unknown" samples for validation. If the validation samples have been extracted from the original total population of spectra, the SEP can be identified as the standard error of validation (SEV). If the validation samples derive from a different population, the data obtained on prediction should be described as the true SEP.

PERFORM MATHEMATICAL TRANSFORMATIONS (COMPUTERIZED SPECTROPHOTOMETERS ONLY)

Modern software for mathematical treatment permits a wide range of options for smoothing (noise reduction) and derivatization of the optical $\log(1/R)$ signal. The theory and practice of smoothing, derivatization, Fourier transform, and PLS regression are all described in detail by Hruschka (1987) and Martens and Naes (1987). Two actual data points are involved in a first derivative and three data points in a second derivative. The method for computing derivatives may differ among the compilers of NIR reflectance software. Most instruments record optical data at intervals of 2 nm. Software such as WINISI, Vision (Foss/NIRSystems, Inc.), and NSAS (Foss/NIRSystems, Inc.) use 2 nm per segment as the basis for the development of derivatives.

Variance introduced as a result of particle characteristics can be reduced by transformation of the $\log(1/R)$ signal into the first or second derivative. Derivatization of the $\log(1/R)$ signal also reduces baseline drift. Calibrations generated by dividing the derivatized signal at one wavelength point by that at another (the "quotient version") have been reported to reduce potential error still further (Norris and Williams, 1984). Derivatization of the signal before sample selection on the basis of spectral characteristics can facilitate spectral sample selection by reducing the amplitude of spurious spectral variance caused by particle size and shape and moisture content. Scatter correction may also be applied to achieve further reduction of system noise.

Derivatization, the use of the quotient algorithm, PLS, neural networks, and scatter correction all represent efforts to improve the accuracy of calibrations developed from the raw $\log(1/R)$ optical data and, particularly, to improve accuracy of those calibrations that involve nonlinear data. Care should be exercised in the application of scatter correction, since the removal of scatter may result in the removal of spectral variance components that are necessary to the development of stable calibrations.

The objective of calibration development is to provide the means for future reliable analyses rather than to provide a set of attractive statistics (achievable by the elimination of spectral outliers). The accuracy of prediction of reference data in samples identified as outliers may actually be superior to that of other samples that have not been identified as such.

There is no "magic" mathematical treatment of optical data that will give the best predictions for all constituents in all materials, and as many treatments as are available should be tested to optimize the treatment for a particular application. Personal microcomputers with large memory capacity and very fast math coprocessors allow the operator to generate calibrations in a few seconds. These may be based on several versions of derivatized signal, Fourier transform, and PLS regression, and the validation of all calibrations by analysis of prediction samples can be complete in an hour. This has revolutionized wavelength selection and mathematical optimization, which were hitherto relatively time-consuming.

COMPUTE CALIBRATION CONSTANTS

This is done by regressing the optical signals at all wavelength points for all samples against the reference data. Forward or backward stepwise, step-up MLR, and PLS regression are the methods most frequently used.

Forward step-up regression selects the first wavelength point on the basis of the highest coefficient of correlation between optical and reference data. This is referred to as the primary wavelength. The second, and subsequent, wavelength points are then computed. The primary and subsequent wavelengths remain in place in the order in which they were selected, so that the selection of all subsequent wavelengths is influenced by the first wavelength.

Forward stepwise regression software usually calculates the first wavelength point in the same way as forward step-up regression, but, for subsequent points, the first point selected may be eliminated in favor of a wavelength point with which the second point is more compatible. The first point may or may not become "reinstated" when further points are calculated. The system compares all wavelength points during the calculation to ensure that each additional wavelength is truly optimum to accompany those already selected. The first wavelength point selected by stepwise regression should theoretically be the same as the first wavelength point selected by step-up regression. In some cases it is, and, in others, the first wavelength points selected by stepwise or step-up MLR usually appear among wavelength points selected subsequently. Stepwise is usually more effective than step-up regression. Table VII illustrates differences in the effectiveness of prediction of moisture content in whole wheat using step-up and stepwise MLR.

Both methods identified a similar wavelength point for the primary wavelength. The stepwise system replaced the original selection with different wavelengths, which progressively improved the statistics, whereas the step-up method was forced to employ the same original wavelength throughout.

With backward stepwise regression, the first computation includes all of the available wavelength points (e.g., all 20 of a 20-filter instrument). The second and subsequent computations remove selected wavelength points from the equation. The ensuing predictions of the validation samples identify the optimum equation/wavelengths to be used.

The most common source of error in the MLR procedure is the "overfitting" of data. The SD of differences between NIR reflectance and reference data for the calibration sample set is referred to as the standard error of calibration (SEC). As one adds data points to the equation, the SEC and the multiple coefficient of correlation progressively improve. This is a mathematical phenomenon and does not imply a commensurate improvement in the accuracy of the calibration.

The most practical way to establish the optimum number of wavelength points is to develop calibration equations with as many wavelength points as possible (e.g., nine), and then use all of the equations to predict the results in the validation set. The coefficient of determination (r^2) and SD of differences between NIR reflectance and reference data for the validation sample set, referred to as the SEV or SEP, may continue to improve throughout. This implies that all of these wavelengths may be necessary for effective calibration for that commodity and that constituent or parameter. On the other hand, the values of r^2 and SEV may improve up to, for example, five wavelength points, and then remain constant or deteriorate. In this case, the optimum number of wavelength points is five, using that mathematical treatment and wavelength range. This is illustrated in Table VIII, obtained from a calibration for the prediction of wet gluten content in wheat.

The use of two validation sets is recommended. The operator may optimize and "fine-tune" the calibration (by testing different

TABLE VII
Prediction of Moisture Content in Whole Wheat by Step-Up and Stepwise Regression

Regression Method	Wavelength, nm				r[a]	SEP[b]	RPD[c]
	1	**2**	**3**	**4**			
Step-up	2,288				0.371	2.82	1.08
	2,288	1,932			0.989	0.444	6.83
	2,288	1,932	2,284		0.989	0.450	7.74
	2,288	1,932	2,284	2,306	0.836	1.66	1.83
Stepwise	2,280				0.370	2.82	1.08
	1,960	2,060			0.991	0.406	7.47
	1,400	1,200	1,680		0.994	0.332	9.14
	1,540	1,340	1,400	1,360	0.996	0.266	11.40

[a] Coefficient of correlation.
[b] Standard error of prediction.
[c] Ratio of the SEP to the standard deviation of the reference data.

mathematical treatments of the log[1/R] data and changing the wavelength range) using one validation set. The calibration can then be tested further by predicting results in the second validation set, which has not been involved in the optimization.

Additional wavelengths past the primary wavelength are selected (by MLR) to compensate for an area of variance that the first wavelength is not able to accommodate. Most of the variance inherent in a population of ground, or otherwise powdered, commodities can be accounted for by two or three points (sometimes even a single point is adequate with derivatized data). Subsequent wavelengths may be associated only with system noise and may not contribute to the integrity of the calibration. This becomes apparent when the calibrations are used to predict results in the validation set. Experience has indicated that more wavelength points are required with whole-grain than with ground-grain applications.

In either forward or backward stepwise MLR, wavelength points can be stipulated by the operator. If the operator has the opportunity to view the multiple correlation plot after the computation of the first wavelength point, he or she may find areas of high correlation that appear to be more suitable than the point selected, due to having wider correlation "bands." These provide the most stable calibrations.

If a correlation plot indicates that a wavelength that has not been selected appears to be more suitable for use in a calibration, the operator can stipulate the wavelength and recompute the calibration by step-up or stepwise regression. In practice, this rarely results in a better calibration (the computer knows more than we do!). Either forward or backward stepwise regression may be used in selection of wavelengths (filters) to be used in discrete-filter instruments, such as the DICKEY-john (DICKEY-john Corp., Auburn, IL), Bran+Luebbe (Bran+Luebbe, North America, Inc., Buffalo Grove, IL), and Perten instruments. The number of filters ranges up to 40 in stand-alone machines, but calibrations rarely use more than four. Use of more than this number increases the risk of overfitting the data. All modern discrete-filter reflectance machines operate with the log(1/R) signal in its raw form. Smoothing is achieved by averaging up to 128 readings for each filter.

PLS regression uses all of the available wavelengths, which theoretically not only eliminates the possibility of overfitting data, but also enables more efficient use of the optical data, since a wavelength, even though it may not be selected in a calibration, may contribute toward the stability of the calibration by reducing some sources of outliers. Up to 15 factors are usually researched in the computation of PLS calibrations. Final calibrations commonly employ up to 10–12 factors.

In practice, PLS regression is also predisposed to a type of overfitting. During the computation of the equations, a certain proportion of the spectral variance represented by the calibration sample set is accounted for by each principal component. Usually most (over 90%) of the variance has been accounted for by the first four or five components, and subsequent factors may appear to improve or worsen the accuracy of prediction slightly. This phenomenon represents a form of noise, since it results from the interaction of the instrument and sample in the same way as the selection of multiple wavelength points by MLR. Small improvements or reductions in accuracy (r^2, SEV, and bias) that occur after the initial improvements in accuracy statistics have "leveled off" should be viewed with caution. Later PLS factors are often associated with system noise. They may indicate that the calibration/validation package has become sample sensitive, and the addition or removal of a single sample from either the calibration or the prediction set may influence the system to an inflated degree.

Modern high-speed computers enable the computation of calibration equations in seconds and their evaluation in minutes. The most effective method of appraisal is to generate calibration equations for up to six to eight filters for discrete-filter instruments (wavelength points for computerized spectrophotometers) and up to four or more wavelength points for instruments that use tilting filters and derivatized signals or for monochromators. Up to 15 or more PLS factors can be computed for computerized spectrophotometers. Each calibration equation is then used to predict the constituent or parameter in the validation sample set. In contrast to MLR calibrations, in which the r^2 and SEP values often worsen after the optimum number of wavelength points has been determined, in the case of PLS calibrations, the statistics remain essentially constant after the optimum number of factors has been determined.

The wavelength range available to computerized spectrophotometers should also be optimized. In many cases, the higher wavelengths do not contribute to the efficiency of the calibration. Several areas of wavelength contain absorbers that enable the prediction of common constituents. The wavelength range of 600–1,800 nm has been found particularly useful in NIR reflectance analysis of several agricultural commodities.

A useful exercise is to optimize the calibration as described, using the calibration and one validation set, and then to use the best calibration equation to analyze the second validation set. The two validation sets can then be combined and used to develop a calibration, using the same parameters as those identified during the development of the original calibration. The calibration is then used to analyze the original calibration sample set. The evaluation statistics, wavelengths, number of factors, etc., should be essentially the same for all three prediction exercises.

ENTER CALIBRATION CONSTANTS TO INSTRUMENT

Inserting the constants into modern NIR reflectance instruments is achieved simply by pressing the appropriate keys. There are several types of calibration in use. The first and most common is the simple calibration, which entails the assembly of samples to provide calibration and validation sets. Many examples of this type of calibration used in industry are "one time" calibrations, developed for the determination of a specific constituent in a specific commodity. Most stand-alone filter instruments require access to an auxiliary computer with MLR capability to compute the calibration constants.

Calibrations based on the quotient version of mathematical treatment of the log(1/R) signal could be used in some earlier (approximately 1978) instruments that used the derivatized form of the log(1/R) value. Most of the variance due to particle characteristics was eliminated by use of this algorithm. Only one wavelength point was necessary, and the calibration constants could be calculated using a hand calculator with simple regression programming. The Pacific Scientific (now Foss/NIRSystems) models 101, 102, and 4250 employed this system.

Universal constants (factory calibrations) are becoming widely used in stand-alone instruments. These are constants based on many sets of calibration constants generated over (usually) several years, using spectral data from several instruments of the same model. For some instruments, such as the DICKEY-john GAC III or Instalab series, the Perten Inframatic, the Bran+Luebbe Infra-Alyzer, or, more recently, the Foss/Tecator InfraTec series, many sets of constants have been generated. These can be used to develop "universal" calibrations. The instruments can then be factory calibrated by installing the universal constants and then

TABLE VIII
Comparison of SEC and SEP[a]

No. of Wavelengths	r	SEC	r	SEP
1	0.965	1.32	0.967	1.36
2	0.969	1.26	0.968	1.34
3	0.971	1.32	0.969	1.32
4	0.984	0.92	0.975	1.17
5	0.984	0.90	<u>0.977</u>[b]	<u>1.14</u>
6	0.985	0.88	0.977	1.14
7	0.985	0.87	0.976	1.15
8	0.986	0.84	0.976	1.15
9	0.987	0.83	0.976	1.16

[a] r = Coefficient of correlation; SEC = standard error of calibration; and SEP = standard error of prediction.
[b] Underlined values are optimum (i.e., highest r, lowest SEP, and fewest wavelengths).

checking performance with check samples having a range of concentration. Slope/bias adjustments are often necessary to obtain consistent results among instruments, but this has greatly simplified the calibration of this type of instrument, which is still the "bread and butter" instrument of the NIR reflectance industry in many areas.

The concept of universal factory-generated calibrations has been facilitated, in the case of the Foss/Tecator InfraTec, by software that enables standardization of the instruments so that all instruments of the same make and model provide essentially the same spectra from a set of samples. Calibrations for stand-alone instruments, such as those in the Foss/Tecator InfraTec series, which employ a monochromator as wavelength source, can be installed either by PC and modem or by means of a diskette. The calibration is developed on a single instrument, the master instrument, and transferred to the desired number of slave instruments. In the case of the Foss/Tecator InfraTec series, PLS regression is used to develop the calibration. The samples are scanned only in the master instrument. A small sample set is used to verify the accuracy of the calibration after transfer to other instruments. The subsequent accuracy of the Foss/Tecator series of NIR transmittance instruments can be monitored and controlled via modem. This process is becoming standard for interlaboratory reproducibility in operations such as grain analysis.

ANN calibrations are proving to be very successful in the application of some models of NIR reflectance instruments to the large-scale analysis of grains and seeds. A single calibration can be used to predict constituents such as protein and moisture contents in several widely different types of grains (e.g., wheat of different classes, barley, and maize) using a single calibration. Development of ANN calibrations involves combining scans recorded on many instruments of the same make and model, together with the reference data with which the scans are associated.

VERIFY ACCURACY (AND REPRODUCIBILITY) OF CALIBRATION BY ANALYSIS OF VALIDATION SAMPLE SET (OPTIMIZE WAVELENGTH RANGE AND MATHEMATICAL TREATMENT OF SPECTRAL DATA)

The efficiency of a calibration is evaluated by using the calibration to analyze the validation sample set and computing the SD of differences (SEV or SEP), bias, and coefficient of correlation between NIR reflectance (or transmittance) and the reference data. Provided there are enough samples in the validation set or sets, this is the recommended approach. The mathematical treatment can be optimized by developing calibrations for each mathematical treatment using a range of segment and gap sizes. This is illustrated in Table IX. When the optimum combination has been determined, the calibration can be fine-tuned by changing the wavelength range. Calibrations can often be improved by the elimination of wavelength areas at the extremes of the wavelength range.

A useful method for evaluation of a calibration involves the ratio of the SEP to the SD (RPD) statistic. This is the RPD of the reference data of the validation sample set; it is discussed more fully later in this chapter. The calibration that gives the highest RPD, together with the highest coefficient of correlation, is the best.

Beware of high coefficients of correlation based on a "dumbbell" distribution of samples. These are sample sets in which there are a number of samples of, for example, low protein content, few or no samples at intermediate levels, and another set of samples at levels of high concentration. This makes the correlation appear to be high, whereas, within the relatively small range at either the high or the low level, the correlation may be very poor. This may mean that the calibration is useful simply for sorting out the two levels roughly but is of little value within either level. If such a situation is encountered, the next step is to assemble further samples at each level and develop separate calibrations.

This interpretation of dumbbell correlations may be misconstrued, since the implication is that the coefficient of correlation is inflated, due to variance at extremes of composition, with no information about samples of intermediate composition. Actually,

the relationship between the NIR reflectance signals and the reference data is usually the same at intermediate levels of composition as it is at the extremes, and if a calibration provides satisfactory results at extremes of composition, it should also do so at intermediate levels (Anonymous, 1996). This is illustrated in Figure 6.

The deviations from the regression line are of the same order of magnitude at any point along the range of composition, and highly statistically significant coefficients of correlation will persist at either extreme of the population if the overall correlation is high. Improvement in the accuracy of predictions can often be achieved by improving the accuracy of sample preparation and reference procedures.

PERFORM SLOPE/BIAS CORRECTION

The above method of evaluation of the efficiency of a calibration generates a coefficient of correlation and enables the calculation of the regression of NIR reflectance on reference data. The coefficient of correlation should be high—at least 0.96 for prediction of composition in most agricultural and food commodities. If it is not, there is little that can be done with that set of data, beyond researching the cause of the poor correlation. If it is caused by a few outliers, the authenticity of the outliers should be verified by retesting by both NIR reflectance and reference methods. If the poor correlation is caused by a large proportion of outliers, then either the method for sample preparation or for developing reference data may require further attention, or NIR reflectance may not be applicable in that situation.

If the correlation is high, results at one or both ends of the range of reference data may indicate that some of the differences between NIR reflectance and the reference data can be corrected by a slope adjustment. When comparing reference and NIR reflectance results, the slope (regression coefficient) should be 1.0 or close to 1.0. If it is, the results will not be greatly improved by a slope adjustment. If the slope is greater than 1.05 or less than 0.95, a slope adjustment will improve the accuracy (but not the precision) of predictability. The more the slope differs from 1.0, the more the accuracy of the analysis can be improved by slope correction. But caution is needed if the slope deviates too much from 1.0, because the system may be very sample sensitive.

A slope/bias adjustment is applied by multiplying all of the constants (including the intercept) by the associated regression coefficient and then adding the intercept gained from the data comparison to the corrected calibration intercept. The validation sample set should then be reanalyzed and the statistical comparison repeated, to verify to what extent the results have been improved. This will change the magnitude of errors at the extremes of the reference data, but it will not change the values of r or the SEP, since the relative deviations of each individual result from the theoretical will be maintained.

TABLE IX
Influence of Optimization of Segment and Gap for Prediction of Protein Content in Whole-Wheat Kernels Using the First Derivative of the Log($1/R$) Near-Infrared Reflectance Signal

Seg-ment	Statistic[a]	Gap				
		2	4	10	20	40
2	r	0.988	0.987	0.988	0.986	0.984
	SEP (N)	0.288 (6)	0.298 (9)	0.291 (8)	0.316 (6)	0.335 (8)
4	r	0.987	0.986	0.988	0.988	0.984
	SEP (N)	0.298 (8)	0.313 (6)	0.289 (8)	0.296 (8)	0.333 (7)
10	r	0.988	0.987	0.985	0.986	0.983
	SEP (N)	0.291 (6)	0.306 (8)	0.325 (6)	0.309 (6)	0.344 (9)
20	r	0.986	0.986	0.987	0.985	0.989
	SEP (N)	0.316 (6)	0.318 (8)	0.303 (9)	0.326 (7)	0.283 (9)
40	r	0.983	0.984	0.984	0.984	0.985
	SEP (N)	0.346 (9)	0.338 (7)	0.340 (9)	0.331 (9)	0.322 (9)

[a] r = Coefficient of correlation; SEP = standard error of prediction; and N = number of wavelength points used.
[b] Underlined values are optimum.

A certain amount of stigma has been applied to the use of slope/bias adjustment in NIR reflectance technology. For practical purposes, there is nothing wrong with applying a slope/bias adjustment provided that (i) the slope does not differ from 1.00 by too great a margin (e.g., below 0.8 or above 1.15), and (ii) the calibration remains stable after the adjustment. This can be verified only by testing check samples over a period of at least several days. Modern instruments retain stable calibrations from year to year, with only minimal adjustment required from time to time.

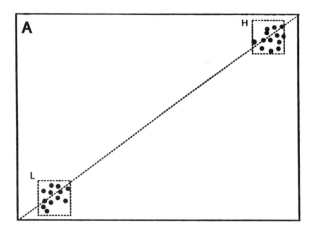

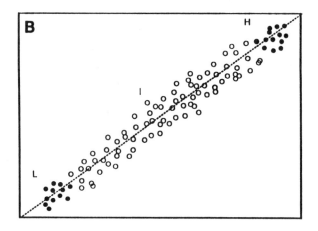

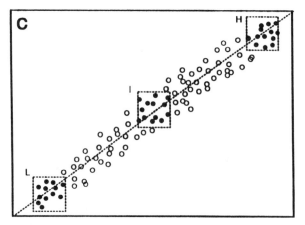

Fig. 6. Illustrations around a correlation plot. L = low, I = intermediate, and H= high. **A,** "Dumbbell" distribution; **B,** full correlations plot; and **C,** distribution of data points about the line of best fit is essentially the same at low, intermediate, and high areas of the full range. (Reprinted from Anonymous [1996] with permission of NIR Publications, Ltd., Chichester, United Kingdom. Copyright NIR Publications 1996.)

TRANSFER CALIBRATION TO OTHER INSTRUMENTS

This assumes that the other instruments are of the same type and model. Calibration transfer is an important aspect of the suitability of an NIR reflectance instrument for a particular organization. A company may wish to employ several, or a large number, of machines and not have the time to calibrate all of them individually. Calibrations can be transferred manually, by floppy disk, by modem transfer, or by e-mail over long distances. After transfer, the accuracy of the calibration in the new instrument can be verified by the analysis of check samples. Reference samples that can be exchanged among laboratories are an aspect of NIR reflectance technology likely to become very important as its luster spreads.

An important adjunct to calibration transfer is the concept of normalization or standardization of instruments of the same type. The concept was introduced by Foss/NIRSystems (then Neotec Corp.) in 1978, but it has been improved by the development of software by InfraSoft International Inc. The concept involves establishment of one machine as the master instrument. A set of samples is scanned on each instrument (of the same make and model). The optical signals are recorded from the "satellite" or slave instruments, and the InfraSoft software enables the spectra to be corrected to conform to that of the master. This improves the efficiency of calibration transfer.

Modem control of instruments is another option. Instruments are connected to a master instrument, which may be located at a distance from the instrument itself. Calibrations are entered into remotely located instruments from a PC via the modem, and the instrument performance is monitored and controlled via modem. Check samples are analyzed at the site, the results transferred to the control center, and adjustments made via the modem. There is essentially no limit to the distance between the master and satellite instruments. Modem control does demand that an independent telephone line be installed to connect with the modem. Modem response can also be fairly slow and is best operated in a closed system.

This brings up the concept of networking instruments. This would appear to be the ultimate method for the control of instrument accuracy, as well as for the detection of differences in instrument precision. Apart from the modem approach, other systems for networking include circulation of optical and analytical data via diskettes (from remotely located instruments to central control laboratories), and e-mail. *All three methods for instrument monitoring and control depend on the distribution of standard samples of commodities or finished products.*

The diskette and e-mail methods do not call for a dedicated telephone line. Transfer of data via diskette, using courier service, enables regular monitoring of the performance of instruments, including remotely located instruments. The e-mail method has particular appeal, in that it is much faster. Spectra of standard samples can be e-mailed to the central laboratory, together with the NIR reflectance results, as recorded on the remote (or slave) instrument. The spectra can be used to predict the result, using central laboratory calibration to detect changes that may have occurred in the remote instrument due to the instrument itself or the operator.

IDEALLY, REPEAT CALIBRATION OPERATION, USING FRESH SAMPLE SETS

This is particularly important in applications in which the wavelengths selected do not apparently relate to any wavelength normally associated with the constituent or parameter to be measured (e.g., in the prediction of functionality parameters, as distinct from composition). When this occurs, the calibration may be abnormally sample sensitive. When the calibration exercise is repeated using different sample sets of the same type of material, the wavelengths or filters selected should be the same as, or close to, those of the original calibration, and the equation constants should be of the same dimensions. Similarly, if PLS is used to develop the calibration equations, the same number of factors should be used, and display of the weights should show that the distribution of loadings and weights generated during the development of PLS equations is comparable. This gives credibility to the calibration—and peace of mind to the operator!

B. Monitoring Instrument Performance

Monitoring of accuracy and precision is an important aspect of NIR reflectance testing. An NIR reflectance instrument should be monitored daily, using a "known" reference-tested sample and up to six randomized "unknown" samples. Check samples should be stored in plastic bags in tightly lidded plastic containers (tin pails may rust in cold storage) at 5°C. Periodically, a set of 20 samples should be tested on the instrument. The first, third, fifth, etc., samples can be tested singly and the even-numbered samples in duplicate. The last half of the samples should include randomized blind duplicates. Typical results are detailed in Table X.

In this series, the SD (of duplicates) was determined on the basis of blind duplicates, which included errors incurred during sampling and sample preparation, as well as the error of the test itself.

III. SIMPLIFIED APPROACH TO THE INTERPRETATION OF CALIBRATION EFFICIENCY

The preceding section has outlined the steps to be taken in the establishment of a reliable calibration. The following section is intended to explain the statistics employed.

A. Accuracy and Precision

These two terms are often confused. In this monograph, *accuracy* is identified as "closeness to the actual result." It is truly applicable only to the determination of composition, since only elemental constituents such as carbon and nitrogen can be determined with exactness. Anything other than this type of constituent can only be estimated. The authentic amount present of a constituent such as protein cannot be positively determined beyond doubt, since the exactness of the nitrogen-to-protein conversion factors are not impeccable and may vary slightly even from sample to sample. Even water content can only be estimated in biological materials, since minute amounts of water persist after prolonged drying or Karl Fischer extraction. On the other hand, in many industrial applications, the determination of some pharmaceutical compounds, or even organic substances such as glucose or urea, which can be obtained in a very high degree of purity, can be achieved with a very high degree of accuracy.

When samples with sufficiently well-defined reference values are available, the degree of *accuracy* can be defined as the mean of the NIR reflectance-predicted results for the samples minus the mean of the check sample reference values. Such samples are very difficult to obtain for biological materials.

Precision is defined as the degree of *reproducibility* of the result, regardless of accuracy. It indicates consistency in the analytical data and should be determined at the same time as accuracy is evaluated. If a procedure is not capable of giving good precision, long-term accuracy is impossible to achieve.

Precision is evaluated by repeated analysis of a well-blended check sample, using all of the steps necessary to the final analytical procedure, including sample preparation. It is expressed as the standard error of a single test (SET). The SET is the SD of the results of the analysis of the check sample.

Another source of confusion is that between *repeatability* and *reproducibility*. The International Standards Organization (ISO) differentiates between repeatability and reproducibility as S_r and S_R, respectively. The following reflects the ISO usage. *Repeatability* does not include all sources of error. It is the SD of repeated tests carried out on the same sample by a single operator (or team of operators if a team is involved) in the same laboratory, by the same instrument, on the same day. It may or may not in-

clude the sample preparation steps, but, in the case of NIR reflectance analysis, it enables operators to monitor their own efficiency in presenting the sample to the instrument.

True *reproducibility* includes all sources of error. It should include sampling, sample preparation, and cell loading or presentation to the instrument, as well as the actual test. It is the SD of repeated analyses of the same sample by different operators from the same laboratory or from different laboratories or instruments on different days. The dimensions of the error terms normally increase from repeatability to between-operator reproducibility to between-laboratory reproducibility. In most NIR reflectance applications, reproducibility within a laboratory is adequately determined by the analysis, including sample preparation, of check samples on different days, ideally by the different operators who normally use the instrument for the application. For most practical purposes, this is sufficient.

Provided a laboratory is satisfying the needs of its organization and clients, there is really no need for it to concern itself with among-laboratory reproducibility (although pure statisticians would likely not agree with this!). Should the laboratory take part in a collaborative study ("ring test"), the degree to which it agrees with the other collaborators will become apparent and may call for some changes in operation.

The establishment of precision enables the identification of the factors influencing the procedure and the development of methods for avoiding or correcting for them. Storage of the check samples is also an important factor, since the results of check sample analysis cannot be expected to reflect reproducibility if the sample itself is changing.

To evaluate a calibration fully, a precision or reproducibility file should be created during the development of the calibration. This includes check samples, duplicated cell loading, readings taken at different positions of the cell, and readings taken at the

TABLE X
Suitable Monitoring Sequence
for Near-Infrared Reflectance Protein Testing[a–c]

Container Number	Sample Identification	Protein %
<u>1</u>	<u>1</u>	<u>14.22</u>
2	2	12.72
3	2	12.79
<u>4</u>	<u>3</u>	<u>14.07</u>
5	4	14.66
6	4	14.62
<u>7</u>	<u>5</u>	<u>12.96</u>
8	6	14.73
9	6	14.66
<u>10</u>	<u>7</u>	<u>16.31</u>
11	8	12.61
12	8	12.70
<u>13</u>	<u>9</u>	<u>12.83</u>
14	10	15.13
15	10	15.10
<u>16</u>	<u>11</u>	<u>15.75</u>
17	4R	14.74
18	4R	14.70
<u>19</u>	<u>7R</u>	<u>16.40</u>
20	5R	13.04
21	5R	13.03
<u>22</u>	<u>13</u>	<u>11.85</u>
23	6R	14.61
24	6R	14.75
<u>25</u>	<u>15</u>	<u>13.44</u>
26	14	12.47
27	14	12.50
<u>28</u>	<u>17</u>	<u>13.87</u>
29	11R	15.75
30	11R	15.78

[a] Standard deviation (SD) (duplicate test; includes sampling error) = 0.065. Coefficient of variability (CV), % = 0.46.

[b] SD (duplicate grinds) = 0.05. CV, % = 0.33.

[c] Odd-numbered samples (underlined) are tested once; even-numbered samples are tested in duplicate. R = randomized blind duplicates.

same position of the cell with no reloading. Table XI illustrates a typical sequence of readings taken during the development of a calibration for a scanning spectrophotometer or any other type of NIR reflectance spectrophotometer. The analysis for Table XI was performed on ground grain.

The check samples with numbers 1, 10, and 25 (and others to follow in that sequence) would be combined in a *precision* file. The precision file is used to predict composition, using calibrations developed for eventual prediction of the required parameters. Precision is determined by computing the SD of the predicted results of the check samples (e.g., Table IX, Nos. 1, 10, 25, etc.). The SD of differences between the various types of duplicates (reloading, turning, etc.) can be determined by computing the SD of differences between the various types of duplicates (reloading, turning the cell through 180°, etc.).

The reason for the check sample readings is to determine the SET, which encompasses all sources of error, and to preserve continuity in the determination of precision from file to file, day to day, etc. The relative contributions of each of these error sources to the overall error can be determined by analysis of variance. In computing these latter aspects of precision, the errors of both testing methods (NIR reflectance and reference) are involved, so that division by the root of 2 is necessary before these values are combined with the SD of the check samples (to get the SET). This enables the determination of the significance of the individual error sources incurred in repeated analyses of the same sample.

Numbers 11–14, 26–29, and others of that type are combined in a separate precision file. Analysis of the SD between the results obtained for numbers 11–12, 26–27, and others of that type enables the determination of the errors incurred by reloading different samples. Numbers 12–13, 27–28, and others of the same sequence enable the determination of the errors incurred by turning the sample cell through a segment of a circle (where applicable). Numbers 13–14, 28–29, and others in this sequence enable the determination of the error of the instrument in rereading a sample that is in the same position as the previous sample. This is the only

"visible" source of error attributable solely to the instrument itself. All other samples are used in the creation of the calibration and validation sample sets.

The error factor ascertained by turning the loaded cell through a segment of a circle is caused by distribution of the ground or powdered sample at the face of the cell (usually the bottom of the cell), and it is only partly a function of the way the cell is loaded by the operator. It is affected by static electricity, among other things, and differs for different materials. Ground oats and many forages are more susceptible than less-cellulosic materials, and quartz windows are more susceptible to this source of error than glass windows.

When a center is operating more than one instrument, the overall precision can be estimated by pooling the data obtained from check sample analysis.

The extent to which precision is itself affected by sampling, sample preparation, and test procedure can be identified by testing under different conditions. For example, the precision of the NIR reflectance test has been determined (i) for wheat that was ground separately for each test, (ii) for the same ground wheat but with separate sample cell loading, and (iii) by using a sealed sample cell ("permacell") placed in the sample cell holder in the same position for every reading. This series was used to determine precision of both NIR reflectance protein and moisture testing (Table XII).

The permacell values represent the actual repeatability of the NIR reflectance instrument. The aspects of error contributed by all loading and sample preparation can be extracted from the total error. This is done by squaring the SET and subtracting the squared value for sample loading from the figure for separate grinds to determine the error due to grinding, and then subtracting the squared value for permacell or instrument error from the value for separate loading to determine the proportion of total error due to sample loading. The square root is then extracted from the residual figure. In this way, the proportions of error for the different operations can be identified. This procedure is illustrated in Table XIII, using the wheat data of Table XII.

These values suggest that the proportion of the SET from different causes varies between constituents. For example, cell loading was a bigger source of error than grinding in the case of protein, but grinding was by far the biggest factor in NIR reflectance testing for moisture. This is logical, since the grinding action affects both the moisture content and the moisture $\log(1/R)$ signal. The effect of grinding increases directly as does the original moisture content of the sample. As the grinder is used for more samples, it heats up and causes more moisture loss. The moisture $\log(1/R)$ values at the respective wavelengths in the NIR reflectance region are stronger than those of protein over the same concentration range. This contributes to the observation that the precision of the moisture signal is likely to vary more from sample to sample than that of the protein signal.

TABLE XI
Sequence of Samples
for Calibration Development and Determination of Precision

No.	Sample[a]	Constant 1	Constant 2
1	Check 01	1.00	1.00
2	L-1228	13.2	405
3	L-1236	10.6	350
4	L-1321	14.1	385
5	L-1277	13.7	390
6	L-1302	12.9	400
7	L-1275	12.2	410
8	L-1290	14.6	400
9	L-1247	13.4	370
10	Check 02	1.00	1.00
11	L-1284	15.7	405
12	L-1284 RL	1.00	1.00
13	L-1284 RLT	1.00	1.00
14	L-1284 RLNT	1.00	1.00
15	L-1278	16.6	366
16	L-1164	11.9	340
17	L-1250	13.7	398
18	L-1267	14.1	380
19	L-1253	13.9	372
20	L-1281	13.4	410
21	L-1292	12.8	400
22	L-1299	14.1	390
23	L-1275	13.5	385
24	L-1264	14.4	378
25	Check 03	1.00	1.00
26	L-1287	11.4	320
27	L-1287 RL	1.00	1.00
28	L-1287 RLT	1.00	1.00
29	L-1287 RLNT	1.00	1.00
30	L-1262	13.9	400

[a] RL = cell reloaded; RLT = cell reloaded and turned through 45, 60, 90, or 180 degrees; and RLNT = cell reloaded and reread without turning.

TABLE XII
Precision (Standard Error of the Test [SET])
of Near-Infrared Reflectance Protein and Moisture Testing of Wheat[a]

Procedure	SET for Protein	Coefficient of Variability, %	SET for Moisture	Coefficient of Variability, %
Separate grinds	0.168	1.18	0.371	3.28
Separate loading	0.135	0.95	0.117	1.04
Permacell	0.065	0.46	0.051	0.45

[a] $N = 12$.

TABLE XIII
Distribution of Error for Near-Infrared Spectroscopy Testing
of Wheat for Protein and Moisture

Error Source	Protein Proportion of Error, %	Moisture Proportion of Error, %
Grinding and sampling	35.4	90.1
Cell loading	50.4	8.0
Residual (instrument error)	15.0	1.9

In a similar way, the absolute precision of the reference tests for protein, moisture, and oil have been determined on wheat (protein and moisture) and canola seed (oil). The reference tests used were the Kjeldahl procedure, the single-stage air-oven procedure, and Goldfisch extraction with anhydrous diethyl ether (American Association of Cereal Chemists, 2000; methods 46-12, 44-15A, and 30-20, respectively). The results and the proportion of error are summarized in Table XIV.

All tests for Table XIV were carried out by experienced technicians. The testing error in reference analysis includes mixing and weighing of the ground sample. The sample size for reference testing for moisture and oil was larger than for protein. Digestion conditions and mixing and weighing of the ground sample are the most important sources of error in the actual testing by the Kjeldahl method. The contributions of larger sampling error and more sources of error in the actual testing result in a proportionally higher basic testing error for protein than for moisture and oil. The influence of sampling on the error of Kjeldahl testing can also be isolated from the total error and has been reported to be as high as 34% of the total error or precision of testing wheat for protein content (Hildebrand and Koehn, 1944). The best way of improving overall analytical precision is to improve sampling, subsampling, and grinding techniques and to develop a consistent weighing technique, in the case of reference analysis, and cell loading procedure, in the case of NIR reflectance testing.

The accuracy of an NIR instrument for determination of a particular parameter is a function of the reference analysis, the calibration, and the actual instrument error. Once the calibration is developed and validated, accuracy is no longer strictly a function of the operator, since it can be changed to conform to what is considered to be the *true* result by changing the intercept. The reliability with which this true result can be achieved is precision. Precision is partly a function of the calibration, but mainly a function of the operator, in sample preparation and accessing the sample to the instrument. Accuracy is established only once (at the time of calibration). Precision affects every result, since it derives from every test. For the most part, the instrument cannot be blamed for poor precision.

While the importance of a high degree of precision is emphasized, the accuracy cannot be overlooked—otherwise all results will be reported precisely wrong! Bias can become costly, and every effort should be made to minimize (preferably to eliminate) it.

B. Statistical Terms Necessary to the Evaluation of Accuracy and Precision

First, a word of explanation about dependent and independent variables. The x axis data are usually referred to as the *independent* variables and the y axis data the *dependent* variables, since all values of y depend on the sampling and analytical efficiency of x. In NIR reflectance technology, when computing a set of calibration constants, the reference values (Dumas or Kjeldahl protein, oven moisture, extracted oil, etc.) are the dependent variables (y) and the optical data ($\log[1/R]$ or derivatized $\log[1/R]$) are the independent variables (x). When evaluating accuracy by computing the SEP (for calibration validation), the NIR reflectance results are predicted reference data and depend on the integrity of the samples and the calibrated NIR reflectance instrument. Here, the reference data become the independent variables. For routine NIR reflectance analysis, the NIR reflectance optical data again become the independent variables, and the predicted values (NIR reflectance protein, oil, in vitro dry matter digestibility, etc.) are the dependent variables.

The efficiency of an NIR reflectance calibration is usually evaluated by means of applied statistics. Several terms are needed for the correct interpretation of statistical analysis of the results of NIR reflectance testing. Unless all of these are correctly appraised, the operator may draw conclusions that are incorrect and can lead to frustrating and often costly discrepancies.

EXAMPLE OF WHEAT STATISTICS
Some useful statistics are summarized in Tables XV–XIX. The data were culled from analysis of railway carloads of wheat for protein content and include the following.
1. The mean result of testing the samples by the reference method (chemical or physical) of analysis, x. (The mean of the NIR reflectance values is y.)
2. The SD of both the reference and NIR reflectance series of samples used in calibration and validation or monitoring, SD_x and SD_y.
3. The mean difference, or bias, d, between reference and NIR reflectance data.
4. The coefficient of correlation between reference and NIR reflectance results, r.
5. The coefficient of determination, r^2.

TABLE XIV
Distribution of Error in Standard Testing for Protein, Moisture, and Oil[a]

Sample Source	Standard error			Proportion of Error, %			Error Source
	Protein[b]	Moisture[b]	Oil[c]	Protein	Moisture	Oil	
Separate grind	0.178	0.256	0.870	70.0	92.8	92.4	Sampling and grinding
Single grind	0.098	0.069	0.240	30.0	7.2	7.6	Testing, instrumental

[a] N = 24.
[b] Results from testing wheat.
[c] Results from testing canola seed.

TABLE XV
Example: Wheat Protein Content

Reference Protein, % (x)	NIR[a] Reflectance Protein Series A, %	Difference (yA − x)	NIR Reflectance Protein Series B, %	Difference (yB − x)	NIR Reflectance Protein Series C, %	Difference (yC − x)
13.1	13.2	+0.1	13.4	+0.3	12.5	−0.6
12.6	12.6	±0	12.8	+0.2	12.8	+0.2
10.7	11.0	+0.3	10.8	+0.1	11.0	+0.3
15.9	15.6	−0.3	16.3	+0.4	15.5	−0.4
9.4	9.9	+0.5	9.3	−0.1	9.3	−0.1
12.1	12.2	+0.1	12.2	+0.1	12.3	+0.2
14.3	14.1	−0.2	14.6	+0.3	14.2	−0.1
17.8	17.2	−0.6	18.4	+0.6	17.9	+0.1
16.5	15.8	−0.7	17.0	+0.5	16.2	−0.3
10.2	10.6	+0.4	10.2	±0	10.3	+0.1
14.7	14.3	−0.4	15.0	+0.3	15.0	+0.3
11.4	11.7	+0.3	11.5	+0.1	11.6	+0.2

[a] NIR = near-infrared.

6. The regression coefficient, b, and intercept, a, for the same values.
7. The pattern, or distribution of differences, between NIR reflectance and reference results.
8. The standard error of a single test, SET (or precision).
9. The standard error of performance/prediction, SEP.
10. The coefficient of variability between SEP and x, CV.
11. The root mean square deviation, RMSD.
12. The ratio of the SEP to the SD_y, RPD.

Formulas for calculating all of these parameters are given in Table XIX.

EXPLANATION OF TERMS

Mean. The mean is the arithmetic mean of the reference or NIR reflectance data. It is calculated by summing all values of x (or y) and dividing by n, the number of observations.

Standard Deviation. The SD is a means of expressing the variability, or variance, in the data. The terms SD_x or SD_y are used variously in the text to denote the SD of the reference data, depending upon whether the SD is used in calibration or validation.

Bias. When calculated from the NIR reflectance predictions of data in the validation sample set, the bias, or mean difference between reference and NIR reflectance data, is a measure of the overall accuracy of the calibration. In the real world, the bias is one of the most important statistics. When payments are being made on the basis of composition, such as premiums or discounts for protein or oil content, biases on individual samples mean money, and an offset of as little as 0.1% can incur considerable profits or losses when large volumes of commodities are involved. Biases can occur even when the coefficient of correlation and SEP statistics indicate that an excellent calibration has been developed. The main causes of bias include changes in the source of raw materials, samples (grains or other commodities) from different locations or seasons, changes in processing conditions, changes in ambient temperature and humidity, and others. *Bias cannot be ignored.*

Coefficient of Correlation. The coefficient of correlation (r) shows the degree to which two sets of data (x and y data, which are usually reference and NIR reflectance results) agree with each other. Perfect agreement, with no differences at all between the two data sets, results in an r value of 1.000. In practice, this is impossible, since a certain amount of error in either x or y data (or both) is unavoidable. The x and y data may be either positively or negatively correlated. Table XX provides guidelines for the interpretation of r.

Coefficient of Determination. The coefficient of determination (r^2) shows the proportion of the variance in y data attributable to variance in the x data. For example, an r value of 0.97 will give an r^2 of 0.941. This means that 94.1% of the variance in y can be accounted for by variance in x. It follows that 5.9% of the variance in y is attributable to other factors, such as sample preparation, reference testing, etc. Values of r^2 will always be positive, regardless of the sign of r.

Regression Coefficient and Intercept. The regression coefficient, b, and the intercept, a, show the degree to which values of y can be predicted from those of x. In a perfect relationship between x and y (in which both x and y refer to the same types of data, e.g., protein content by NIR reflectance and reference method), the values of r and b will both be 1.000 and a will be 0.000. Again, a certain amount of error is unavoidable, so that b will be more or less than 1.000 and a will differ from zero. In NIR reflectance technology, the error level of the optical data is usually lower than that of the reference data. This is an asset to the use of MLR in the development of calibrations, in which the assumption is made that there is no error in the x values (the optical data). As in the case of r, regression coefficients and intercepts can also be positive or negative.

Distribution of Differences. The pattern, or the distribution of differences between NIR reflectance and reference results, should be studied for any new calibration involving materials that have not been tested before.

In Table XVII, the series B SEP data show that the results at the lower protein level agree closely with the reference data and that the overall bias is 0.23%. If the detailed results are examined, it is apparent that the NIR reflectance results became progressively higher relative to the reference results at higher protein levels. Eight different patterns may characterize the relationship between NIR reflectance and reference data. These are itemized in Table XXI.

A slope change may be accompanied by a positive or negative overall bias in which a slope/bias correction is necessary. Biases

TABLE XVI
Statistical Treatment of Data from Table XV

| Statistic | Abbreviation | Statistics of Raw Data | | | |
		Reference	Near-Infrared Reflectance (Series A)	Near-Infrared Reflectance (Series B)	Near-Infrared Reflectance (Series C)
Mean	x or y	13.23	13.18	13.46	13.22
Standard deviation	SD ($\Sigma_{x\,or\,y}$)	2.65	2.27	2.85	2.58
Number of observations	N	12	12	12	12
Sum of x^2 (or y^2)	Σx^2 (or Σy^2)	2,176.11	2,142.24	2,263.07	2,169.26
Sum of x (or y)	Σx (or Σy)	158.7	158.2	161.5	158.6
Sum of xy	Σxy	…[a]	2,158.30	2,219.03	2,169.26

[a] … = No data available.

TABLE XVII
Statistics of Comparison for Data in Table XV

Statistic	Abbreviation	Series A (x:y_a)	Series B (x:y_b)	Series C (x:y_c)
Mean difference between x and y (bias)	d	−0.042	+ 0.233	+ 0.008
Standard deviation of differences	SDD (SEP[a])	0.396	0.206	0.294
Root mean square of the differences	RMSD	0.399	0.319	0.294
Sum of differences squared	Σd^2, $(\Sigma x - y)^2$	1.75	1.12	0.95
Sum of differences	Σd, $(\Sigma x - y)$	−0.50	2.80	−0.10
Coefficient of variability	CV	2.99	1.56	2.22
Coefficient of correlation	R	0.999	0.999	0.994
Coefficient of determination	R^2	0.998	0.998	0.988
Regression coefficient	b	0.85515	1.07619	0.9667
Intercept	a	+1.874	−0.7744	+0.433
RPD[b]		6.7	12.9	9.0

[a] Standard error of prediction.
[b] Ratio of SEP to SD.

can also occur with no slope difference. Slope-difference types 7 and 8 can be corrected with an intercept adjustment. All of the other types of bias require a slope/bias adjustment if the slope differs from 1.0 by more than ±0.05 or if the differences between NIR reflectance and reference results at either extreme are greater than 1.5% of the mean.

Slope/bias correction is carried out by multiplying all of the calibration constants (k or f values), including the intercept, by the regression coefficient, b. The intercept, a, is then added to (or subtracted from) the original calibration intercept. This changes the magnitude of errors at the extremes of the reference data, but it does not change the value of r.

The values of Table XV–XVII are typical of the type of values likely to be obtained in the NIR reflectance analysis of wheat for protein content, but they are applicable to any application. The examples given in Table XVIII illustrate the application of the respective regression equations to predict the values of y from the observed values of x. The RPD statistic in Table XVII relates the SEP (or SEV) to the SD of the reference data in the sample set used for validation.

Tables XV–XVII illustrate several features of the application of NIR reflectance technology. Series A illustrates a high-low bias; series B data had a progressively high bias throughout the series; and both series showed that, despite individual biases of up to 0.7, it was possible to achieve a very high coefficient of correlation. The data from Tables XV–XVII also illustrate that a high coefficient of correlation may be misleading unless all of the associated statistics are studied. Application of the regression equations showed that the results of series A became biased upward at the low end and downward at the high end, whereas the results of series B were high at the high end. The series C results also showed a high correlation, despite large biases (0.5 and 0.6) in two samples, while application of the regression equation gave acceptable results over the full range.

Standard Error of a Single Test. The SET was described in detail in the section that defined "precision."

Standard Error of Prediction. The SEP is the SD of differences between NIR reflectance and reference values. The formula for its calculation is given in Table XIX and incorporates a correction for bias. The SEP should be computed from the results of the prediction of a set of samples that have not been used in the development of the calibration. This sample set is the *prediction*, or the *validation*, sample set. Ideally, the sample set used in validation of a calibration should consist of samples that are completely unrelated to the calibration sample set. More often, the validation samples are part of a single population from which both calibration and validation samples sets are compiled. In these situations, there is a good case for using the term SEV, as distinct from the SEP.

Computation of the results of the prediction of the samples used in the actual development of the calibration gives the SEC, which is the SD of differences between NIR reflectance and reference samples in the calibration sample set. Using MLR, the r and SEC statistics progressively improve as more terms are added. If

TABLE XVIII
Application of Regression Statistics,[a] $x = a + by$

	Predicted x		
Observed y	NIR[b] Reflectance Series A	NIR Reflectance Series B	NIR Reflectance Series C
9	9.57 (9.6)	8.91 (8.9)	9.13 (9.1)
11	11.29 (11.3)	11.06 (11.1)	11.07 (11.1)
13	12.99 (13.0)	13.22 (13.2)	13.00 (13.0)
15	14.70 (14.7)	15.37 (15.4)	14.93 (14.9)
17	16.41 (16.4)	17.52 (17.5)	16.87 (16.9)

[a] Based on the data of Table XV.
[b] NIR = near-infrared.

TABLE XX
Guidelines for Interpretation of r

Value of r	r^2	Interpretation
Up to ±0.5	Up to 0.25	Not usable in near-infrared reflectance calibration
±0.51–0.70	0.26–0.49	Poor correlation: reasons should be researched
±0.71–0.80	0.50–0.64[a]	OK for rough screening; more than 50% of variance in y accounted for by x
±0.81–0.90	0.66–0.81	OK for screening and some other "approximate" calibrations
±0.91–0.95	0.83–0.90	Usable with caution for most applications, including research
±0.96–0.98	0.92–0.96	Usable in most applications, including quality assurance
±0.99+	0.98+	Usable in any application

[a] Due to rounding off, there are no values of 0.65, 0.82, etc. in this table.

TABLE XIX
Equations for Computation of Validation Statistics

Standard error of performance	SEP	$\left\langle \Sigma(x-y)^2 - \left\{[\Sigma(x-y)]^2/N\right\} \middle/ N-1 \right\rangle^{1/2}$
Root mean square difference	RMSD	$\left\{\left[\Sigma(x-y)^2\right]/N - 1\right\}^{1/2}$
Mean difference	d (bias)	$\Sigma(x-y)/N$
Mean x	x	$\Sigma x/N$
Mean y	y	$\Sigma y/N$
Standard deviation of x	SD$_x$	$\left\{\Sigma x^2 - \left[(\Sigma x)^2/N\right]/(N-1)\right\}^{1/2}$
Standard deviation of y	SD$_y$	$\left\{\Sigma y^2 - \left[(\Sigma y)^2/N\right]/(N-1)\right\}^{1/2}$
Coefficient of variability	CV	$(\text{SD}\times100)/x$
Coefficient of correlation	r	$\dfrac{\Sigma(x\times y) - \left[(\Sigma x \times \Sigma y)/N\right]}{\left\langle\left\{\Sigma x^2 - \left[(\Sigma x)^2/N\right]\right\}\times\left\{\Sigma y^2 - \left[(\Sigma y)^2/N\right]\right\}\right\rangle^{1/2}}$
Coefficient of determination	r^2	
Coefficient of regression	b_{yx}	$\dfrac{(\Sigma x\times y) - \left[(\Sigma x\times\Sigma y)/N\right]}{\left\{\Sigma y^2 - \left[(\Sigma y)^2/N\right]\right\}} = r\times\dfrac{\text{SD}_y}{\text{SD}_x}$
Intercept of regression	a	$y - (bx)$
Ratio of SEP to SD$_x$ (ref. data)	RPD	SD_x/SEP

the validation exercise indicates that r is low and the SEP is unacceptably high, the calibration set can be predicted. The SEC may indicate one or more gross outliers, the correction of which may bring about a significant improvement in the actual r and SEP values.

Coefficient of Variability (CV). The CV is the population SD × 100 divided by the population mean. It is expressed as a percentage. It relates the SD to the mean and provides a more realistic evaluation of the importance of the SD.

The SD should be compared with the mean reference result to evaluate the significance of the values. This is achieved by the CV. Tables XV–XVII illustrated three different types of data. The comparison between NIR reflectance series A and the reference data showed that, although the overall mean of the NIR reflectance data agreed closely with that of the reference data, there was a SEP of nearly 0.4, indicating that the NIR reflectance determinations were rather variable. The CV was 3.0%, but should ideally be between 1 and 1.5% of the mean value for the reference data. The SEP of the NIR reflectance series B was about one half of that of the series A, and the CV of 1.6% was acceptable.

The size and interpretation of the CV depends partly on the source of the data. The CV of reference testing for composition constituents such as protein or moisture contents should be close to 1.0%. For quality assurance applications, the CV should be about 1.0–1.5%. Values of 2–3% are acceptable for the CV. Values higher than 3% for reference testing for composition should be investigated to determine the reasons. Values of up to 5% may accrue from the determination of the reproducibility of functionality parameters. On the other hand, a CV of 5% may be satisfactory when NIR reflectance is used to predict parameters such as texture differences, wheat "strength" parameters, or digestibility. Table XXII gives some guidelines for the interpretation of the CV.

The mean and CV are adjuncts to the SD and SEP and are very useful in their correct interpretation.

Root Mean Square of the Differences. The RMSD between NIR reflectance and reference results gives another measure of the accuracy of a calibration equation. It is not corrected for bias, and the RMSD should be accompanied by the bias. If no bias exists, the SEP and RMSD will be identical.

Use of the RMSD statistic, rather than the SEP, has been suggested. The data in Table XVII show how the RMSD is related to the SEP. Where there was little or no bias, the SEP and RMSD were about equal. In the case of the series B data, in which there was a positive bias of 0.233, the RMSD was substantially higher than the SEP. The separate SEP and bias values are preferred for reporting accuracy since the RMSD by itself incorporates the bias but does not indicate its magnitude or sign. This is immaterial if both mean x and mean y are reported.

The SEP and the RMSD are amalgamations of the combined errors of both reference and NIR reflectance tests and the sampling, sample preparation, and variability factors inherent in testing an array of randomly distributed unknown samples. The SEP can be reduced to terms that relate to the efficiency of the NIR reflectance instrument itself and its calibration. The SET of both reference and NIR reflectance procedures can be extracted from the SEP to give the true test error (TTE). This is illustrated in Table XXIII, which uses data from a comparative study of NIR reflectance instruments (Williams et al, 1983).

The TTE was derived by squaring the SET of both the Kjeldahl and the respective NIR reflectance instruments, subtracting both from the square of the SEP to obtain the variance of the test error, and extracting the square root, as follows.

$$\text{TTE} = [(\text{SEP})^2 - (\text{SET}_{\text{ref}})^2 - (\text{SET}_{\text{NIR}})^2]^{1/2}$$

The very low TTE of instrument 7 is an indication of the potential of the NIR reflectance technique when a very high-resolution monochromator and a powerful computer with comprehensive interfacing and software are used.

Often a high SEP is associated with a high SET, and the TTE may be much lower. This indicates that there is scope for improvement in sampling, sample preparation, and cell loading before attempting real-time NIR reflectance analysis. It is theoretically impossible for the SEP to be lower than the combined SET for the reference and NIR reflectance techniques. According to this theory, the minimum SEP likely for NIR reflectance determination of protein in wheat is about 0.175 by a commercial bench-type instrument when the Kjeldahl method is used as the reference. This can be reduced to about 0.12 when the Dumas method provides the reference data.

By the same token, the SEP should theoretically always be higher than the SEC. For calibrations developed using MLR, the SEC is also affected by the number of wavelength points used, since the mathematics of MLR results in a progressive increase in r and decrease in the SEC.

In practice, the SEP (or SEV) may not always be higher than the SEC. For some applications, the precision of the NIR reflectance instrument may be superior to that of the reference method. Sample selection for calibration and validation sets may result in some of the calibration samples having a higher reference test error than any of those in the validation set, and the SEV may be slightly lower than the SEC. If the SEP is very much higher than the SEC (e.g., SEP = about 3 × SEC), either there has been a high degree of overfitting or there is a major error in at least one of the samples used in validation.

When using PLS in developing calibrations, although the use of all available wavelengths may lead the operator to believe that overfitting is impossible, very attractive values for r and SEC for up to 15 factors do not necessarily mean that those values will be achieved for the SEP. A type of overfitting commonly occurs, and the r and SEP values often indicate that the optimum number of factors is different from the optimum indicated by the software. This is because the PLS equations have been developed from the optical data of the calibration samples, whereas the SEV is determined from a different sample set (the validation samples).

Ratio of SEP (or SEV) to SD. The RPD (Williams and Sobering, 1993; Williams, 1987) is a simple statistic that enables the evaluation of an SEP in terms of the SD of the reference data. It is calculated by dividing the SD of the reference values used in the validation, or prediction, SD_x, by the SEP. The efficiency of NIR reflectance analysis is determined by the size and consistency of deviations from reference analyses (SEP). If the SEP value is similar to the SD_x (it may be even higher!), it means that the in-

TABLE XXI
Different Types of Slope Difference
in Near-Infrared Reflectance Analysis

No.	Low Reference Result	High Reference Result
1	Accurate	High
2	Accurate	Low
3	High	Low
4	Low	High
5	High	Accurate
6	Low	Accurate
7	High	High
8	Low	Low

TABLE XXII
Guidelines for Interpretation
of the Coefficient of Variability (CV) Statistic

| CV Value (%) | Situation/Interpretation | | |
	Reference Tests, Protein, etc.	NIR[a] Reflectance Constituents	NIR Reflectance Functionality, etc.
Up to 0.5	Exceptional	...[b]	...
0.6–1.0	Excellent	Exceptional	...
1.1–2.0	Very good	Excellent	Exceptional
2.1–3.0	Good	Very good	Excellent
3.1–4.0	Fair	Good	Very good
4.1–5.0	Poor	Fair	Good
5.1+	Needs investigation	Poor	Fair

[a] NIR = near-infrared.
[b] Unlikely to be attained.

strument is not predicting the reference values at all, and the operator could equally well report the mean of the original data. The SEP should be much lower than the SD_x, and ideally the ratio of the SD_y to the SEP should be 5 or higher.

In the case of NIR reflectance series A in Table XVI, the RPD was 2.65/0.396 = 6.7, and, for series B, the RPD was 12.9. For a series of fairly uniform samples in which the SD is not high, the RPD may not be able to be as high as 5, but it should indicate that the SEP is appreciably lower than the SD. For example, RPD values of 3 or 4 will verify accurate analysis if the SD is only 0.4–0.5. From a practical point of view, if the SD is very low for a population of reasonable size (60 or more), this may indicate that the variance is so low that analysis is not necessary, except for quality control. This type of situation may appear, for example, in a flour mill and in many industrial applications in which the objective is to maintain uniformity in the product and minimize variance. Despite the low range in reference values, with its consequent low SD, there is still a need for regular frequent analyses to ensure that specifications are being met.

Another method of standardizing the SEP is to relate the SEP to the range of the reference data, as proposed by Starr et al (1981). The ratio error range (RER, i.e., the ratio of the SEP to the range) is computed by dividing the range in reference values of the validation samples by the SEP. It should also be as high as possible, but it can be inflated by a single sample at an extreme of the concentration, whereas the SD is not so markedly affected by a single observation. In a breeding program, when screening of the material is inhibited by very large numbers of samples, time per test, and expense, an RER of 10 or more indicates that the NIR reflectance instrument is predicting the data at better than 10% of the mean. This is usually good enough for screening, particularly in early generations.

Table XXIV illustrates the significance of some values of the RPD and RER.

INTERPRETATION SUMMARY

The SEP indicates the variability in deviations of x from y, and the bias shows the average amount by which the results differ. The SEP and the bias together pinpoint the overall accuracy of the test procedure. The RPD relates the SEP to the SD and simplifies the interpretation of the SEP. High values for the RPD (ideally 5 or more, but at least 3) indicate efficient NIR reflectance predictions.

The coefficient of correlation, r, indicates the closeness of fit between the NIR reflectance and reference data over the range of composition. A high r value with a low SEP and bias, together with a slope close to 1.0, means that the NIR reflectance results are accurate over the anticipated range and likely to remain so, provided that these statistics were based on a sufficient number of observations.

A high correlation coefficient (0.95+) with a bias at one or both extremes of the composition means that these discrepancies are also practically certain to persist. This is illustrated by NIR reflectance series B in Table XVII.

The slope indicates the degree to which NIR reflectance predicted values change relative to reference values. A slope of 1.0 is

excellent, since it shows that the rate of change in both sets of data is identical. Deviations from 1.0 of greater than 0.05 are likely to require slope (and possibly bias) correction. Deviations greater than 0.1 are more significant and require investigation as to the cause. The regression statistics b and a can be used in the equation $y = a + bx$ to correct the slope and bias.

Slope/bias adjustment improves the accuracy of prediction at the extremes of the reference data. The coefficient of correlation will not change. If the coefficient of correlation is low (0.8 or less), it is usually not possible to obtain consistently high accuracy by NIR reflectance analysis. A low coefficient of correlation between NIR reflectance and reference data means that the NIR reflectance analysis has not been successful. If all sources of error for poor NIR reflectance results have been carefully studied, including reference analysis, and no improvement can be achieved, the sad truth may be that NIR reflectance is not applicable to the analysis!

The coefficient of correlation, r, (preferably r^2), the bias, and the RPD are the most meaningful statistics for "instant" appraisal of analytical efficiency by NIR reflectance.

C. The Calibration (k) Constants

The calibration constants are regression coefficients, together with an intercept. The size of the calibration, or k, constants depends on both the optical and reference data. They will be larger for derivatized signals than for $\log(1/R)$ data and will also be large where reference data values are high (e.g., hundreds of ppm, etc.). The k constants of some stand-alone instruments are software specific and dependent upon the format used by the manufacturer to translate the optical signals from the instrument into regression coefficients for the purpose of computing calibration equations. For example, the DICKEY-john constants ranged from ±0 to about 1.5 for prediction of protein and moisture contents in wheat, whereas those of the Technicon (now Bran+Luebbe) InfraAlyzers were much higher. Both instruments employed discrete filters and the $\log(1/R)$ algorithm.

The size of the constants can be reduced in cases in which the reference data consist of large numbers by dividing the reference data by 100 or 1,000 before computing the calibration constants. Very large k constants (10,000 or higher) may be accompanied by sample sensitivity, and a small number of samples (in extreme cases even a single sample) can have a significant influence on the SEP and the long-term reliability of the calibration. Recent studies involving whole grains and seeds have shown that this may not necessarily be true for those applications.

As the SD of the reference data increases, the coefficient of correlation can also be expected to increase, provided the precision of testing by both reference and NIR reflectance methods is satisfactory. Simultaneously, the SEP may be expected to increase, commensurate with the SD. This is illustrated in Table XXV.

D. NIR Reflectance Software

Software is what drives NIR reflectance spectrophotometers. The first edition of this monograph contained only a small section on software, the software used in those days to operate the then Pacific Scientific (now NIRSystems) model 6350 scanning spec-

TABLE XXIII
True Test Error (TTE) of Some Near-Infrared Reflectance Instruments for Testing Wheat for Protein[a]

Statistic[b]	Method/Instrument[c]						
	1	2	3	4	5	6	7
SEP	0.171	0.186	0.275	0.218	0.244	0.241	0.145
SET	0.089	0.123	0.138	0.116	0.118	0.105	0.110
TTE	0.146	0.107	0.221	0.162	0.194	0.198	0.032

[a] Data by Williams et al (1983).
[b] SEP = standard error of prediction, and SET = standard error of a single test.
[c] 1 = Kjeldahl; 2 = InfraAlyzer model 400; 3 = InfraAlyzer model 300; 4 = DICKEY-john GAC III; 5 = Neotec GQA 31EL; 6 = Neotec model 101; and 7 = Beltsville universal computerized spectrophotometer.

TABLE XXIV
The RPD[a] and RER[b] Statistics

RPD Value	RER Value	Classification	Application
0.0–2.3	Up to 6	Very poor	Not recommended
2.4–3.0	7–12	Poor	Very rough screening
3.1–4.9	13–20	Fair	Screening
5.0–6.4	21–30	Good	Quality control
6.5–8.0	31–40	Very good	Process control
8.1+	41+	Excellent	Any application

[a] Ratio of the standard error of performance to the standard deviation of the reference data.
[b] Ratio error range.

trophotometer. This was the only software commercially available at that time for application to scanning spectrophotometers. The software used to develop calibrations for the benchtop instruments of the day was forward step-up and stepwise and backward stepwise MLR, available in several statistical packages.

Principal component analysis, PLS regression, ANN, and genetic algorithms have joined the original three forms of regression. There are now several types and versions of software for use with NIR reflectance instruments. These are summarized in Table XXVI.

Software can be grouped into two main classes, dedicated and generic. The first type is dedicated to NIR reflectance instruments. It enables the recording of spectra and the operation of the instrument in regular analytical mode, and it features instrument diagnostics, as well as spectral manipulation, statistical analysis, and graphics. Generic software does not provide the means for recording spectra or for routine analysis. It is usually not limited to use in the NIR reflectance wavelength range and may be adaptable to processing other than spectral data.

The purposes of NIR reflectance software include the following.

1. Recording of spectra;
2. Regular (routine) analysis;
3. Spectral manipulation, i.e., (i) viewing of spectra; (ii) addition and editing of sample identification; (iii) addition and editing of reference data; (iv) file set-up and organization (with spectral selection or selection on the basis of composition); and (v) spectral addition or subtraction;
4. Mathematical treatment of optical data (smoothing and derivatization);
5. Application of scatter correction;
6. Development of calibration equations;
7. Evaluation of calibration equations (statistical analysis);
8. Slope/bias correction;
9. Outlier detection;
10. Discriminant analysis;
11. Instrument diagnostics; and
12. Graphics, to illustrate publications, etc.

Provided that the operator applies the same mathematical treatment to the optical data in exactly the same format, it should be possible to obtain the same results, in terms of *r* and SEP, with any software assemblage. Different software packages present features such as file manipulation, scatter correction, instrument diagnostics, and graphics in their own unique manner. Some software systems employ the optical data in different ways to generate "segments" and "gaps." Instruments may take optical data at different intervals, and instruments are available that take readings at intervals from 0.5 nm to as high as 5 nm. The smoothing needed to produce the same derivative in terms of nanometers per wavelength point varies accordingly.

The change to Windows-driven software, while it was an inevitable step, has the disadvantage that every action requires the use of the "mouse." DOS-driven software is operated directly through the keyboard and is considerably faster than mouse-operated software. Not all advances are beneficial from every aspect!

It is permissible to employ more than one software system in the development of a calibration. The fastest software can be used to optimize mathematical treatment and wavelength range. The calibration can then be fine-tuned by transposing the optical data to the format used in the more comprehensive software, but a great deal of time can be saved by going directly to the best mathematical treatment, as revealed by the faster software.

A good example is that offered by Foss/NIRSystems. The old DOS-driven NSAS is very fast, and in its AutoCal mode allows simultaneous generation of up to nine equations for several mathematical treatments in a few (2 or 3) seconds. The operator is able to set up calibration and validation sets and then identify rapidly whether, for example, it is preferable to pursue the log($1/R$) signal in its "raw" form or to optimize the segment and gap for first or

TABLE XXV
Influence of SD on *r* and SEP for Prediction of Protein and Moisture Contents of Wheat[a]

Constituent	N	SD	12–14		11–15		9–17	
			r	SEP	*r*	SEP	*r*	SEP
Protein, %								
12–14	76	0.57	0.911[b]	0.236	0.961	0.272	0.977	0.344
11–15	137	1.08	0.924	0.218	0.968	0.271	0.982	0.307
9–17	176	1.62	0.918	0.226	0.967	0.272	0.982	0.303
Moisture, %								
12–14	37	0.58	0.956	0.170	0.987	0.226	0.994	0.235
11–15+	72	1.38	0.946	0.188	0.987	0.224	0.994	0.243
9–17+	109	2.20	0.959	0.165	0.986	0.233	0.994	0.238

[a] SD = standard deviation; *r* = coefficient of correlation; SEP = standard error of prediction.
[b] Underlined values are optimum for data sets.

TABLE XXVI
Software for Use with Near-Infrared Reflectance Instruments

Type	Company	Address	Contact Information
Dedicated			
WINISI	InfraSoft International	Port Matilda, PA, U.S.A.	814/237-0867 (fax)
NSAS	Foss/NIRSystems	Silver Spring, MD, U.S.A.	301/236-0134 or 301/989-1485 (fax)
Vision	Foss/NIRSystems	Silver Spring, MD, U.S.A.	301/236-0134 or 301/989-1485 (fax)
Sesame	Bran+Luebbe	Buffalo Grove, IL, U.S.A.	847/520-0855 (fax)
SpectroMetrix	LT Industries	Rockville, MD, U.S.A.	301/468-2230 (fax)
AnaTec	Buhler Corporation	Uzwil, Switzerland	+41 71 955 3356 (fax)
Delight	Buhler Corporation	Minneapolis, MN, U.S.A.	612/540-9246 (fax); www.buhlerusa.com
	D² Development	LaGrande, OR, U.S.A.	mayesd@dsquared-dev.com
Generic			
GRAMS 386	Galactic Industries	Salem, NH, U.S.A.	603/898-6228 (fax)
Pirouette	InfoMetrix	Woodinville, WA, U.S.A.	206/402-1040 (fax)
Unscrambler	CAMO	Corvallis, OR, U.S.A.	541/757-1402 (phone); www.camo.com
Unscrambler	CAMO AB	Trondheim, Norway	+47 73 514 257 (fax)
MatLab	The MatWorks, Inc.	Matick, MA, U.S.A.	508/647-7001 (fax)

second derivative. NSAS does not enable any form of scatter correction. Having identified what is likely to be the most suitable mathematical treatment, the files can be imported to another software system, such as the more comprehensive Windows-driven WINISI. This (and other) software allows several options for scatter correction and also provides systems for population structuring to improve the distribution of samples with respect to their optical identities.

Graphics probably offer the most scope for different systems of presentation. The most usual application of graphics is for purposes of publication. Three-dimensional graphics allow the operator to observe whether certain samples "belong" in the population and whether groups of samples within the population are concentrated in one "area" of the population. This is of interest when outliers appear to occur in groups. Such samples either must be accommodated in an overall calibration (which may restrict its efficiency) or a separate calibration must be developed for these interlopers. Three-dimensional graphics are also useful to illustrate how trends occur as a result of the influence of more than two variables.

Another option is that of spectral addition or subtraction. For example, subtraction of the spectrum of water from that of an as-received sample can illustrate the influence of the water on the spectrum. Conversely, water can be added to very high levels (e.g., 30%) to observe the influence of high levels of moisture. Another application is the scanning of samples of, for example, high and low kernel texture in barley. Subtraction of the low- from the high-texture spectrum indicates wavelength areas where texture has the most influence on the spectra. These wavelength areas can be stipulated in the development of calibrations for parameters such as hot water extract. Another method of subtraction is to use a sample with low reference data as a reference and then scan a sample with a high reference value. The instrument will subtract the low (reference) from the high sample, and the "spectrum" obtained will illustrate the areas where the two samples differ most in the reference parameter.

E. Cross-Validation

If the sample set is small (up to 60), the calibration is best evaluated by using cross-validation. Cross-validation can be carried out by dividing the population of samples into equal "blocks" and eliminating samples one block at a time, but the most effective method is to eliminate observations one at a time ("one-out" cross-validation). By this technique, all samples are used in development of the equation, and all are predicted, yet none of the samples is actually used in the calibration used in the prediction.

The operation is carried out by eliminating the first sample (or block of samples) from the equation development, developing the calibration without it, and then reinstating it and predicting it. All residuals are recorded. The same steps are followed, eliminating the samples one at a time, or one block at a time, until all have been predicted, but none of them has actually been used in the development of the calibrations used in its prediction. The SD of the residuals is the SEP, and the predicted NIR reflectance/transmittance results are statistically compared with the reference data to give the coefficient of correlation. The final equation is based on the combination of lowest r^2 and highest SEC (or SEV) (i.e., SD of [NIR reflectance cross-validation results] − [reference results]).

The procedure suffers from the criticism that the samples used for validation (in this case cross-validation) must inevitably be selected from the original sample set, whereas, ideally, samples used in evaluation should be garnered from a different source. Cross-validation is also applicable to large populations, but the computation of "one-out" cross-validation may be prohibitively more time-consuming.

A preliminary exercise may be carried out by the "classical" calibration method, using calibration/validation samples in the ratio of about 2.5:1, to establish the most suitable mathematical format and wavelength range. Provided that the statistical data

look reasonable, the cross-validation can be performed using optimal mathematical treatments selected by this preliminary exercise. The process is time-consuming unless the operator has access to software that performs cross-validation automatically.

Table XXVII illustrates the use of cross-validation to determine the feasibility of using NIR reflectance to predict the concentration of seven ingredients in a baking formula.

In this case, cross-validation indicated that the instrument was satisfactory for monitoring concentrations of constituents 3–6, but it would have been questionable for constituents 1 and 2 and was not acceptable for constituent 7. The sample set was small, and the preliminary MLR study showed that the efficiency of prediction of constituent 7 was promising. The implication of the cross-validation exercise was that the provision of more samples with a greater diversity of variance in constituent 7 could lead to a satisfactory calibration for the monitoring of constituents 1, 2, and particularly 7. This would lead to a more effective evaluation of the possibility for developing a satisfactory calibration for the monitoring of these constituents.

Spectral selection of samples, as described under the tenth step (select samples on basis of spectral characteristics) in the calibration development section, is an effective application of cross-validation. The calibration is developed using selected samples and the best combination of mathematical treatment and wavelength range, as determined by cross-validation. The calibration is then further tested by the prediction of the remaining samples, which have been stored as the validation set.

F. Interpretation of PLS Calibrations for Functionality

The development of calibrations by PLS regression involves computing loadings and weights, or "scores," for each factor of variance (Chapter 4). Normally, up to 15–20 factors are computed in NIR reflectance analysis using computerized scanning spectrophotometers. The loadings and weights can be displayed across the spectrum. *Loadings* indicate areas where differences in variance coincide with wavelengths known to be associated with molecular groupings and vibrations, such as C-H stretching. *Weights* indicate the degree to which the variance has actually been used in the computation of the factors across the wavelength range, and, as a result, they are rather more useful than loadings in the interpretation of the process of the development of the calibration.

When the weights are displayed across the wavelength range, "peaks" and "valleys" indicate areas of positive and negative influence of molecular species on the course of the development of the calibration. A prerequisite of the use of PLS weights is a knowledge of the molecular groupings responsible for absorbances in different areas of the NIR reflectance spectrum. Table XXVIII summarizes the chief absorbers for the main constituents of materials of plant and animal origin.

Displaying the weights is a particularly useful indication of what has occurred during the development of calibrations for the prediction of functionality (as distinct from composition). Attention has been drawn to the use of weights in the interpretation of

TABLE XXVII
Comparison of Multiple Linear Regression (MLR)
and Cross-Validation, Using a Small Sample Set

Statistic[a]	Constituent						
	1	2	3	4	5	6	7
MLR							
r^2	0.92	0.96	0.99+	0.95	0.98	0.99	0.79
SEP	0.01	0.01	0.11	0.03	0.11	0.19	0.02
RPD	5.4	8.6	19.1	7.8	10.1	13.1	2.7
Cross-validation							
r^2	0.76	0.78	0.99	0.91	0.95	0.98	0.12
SEP	0.02	0.03	0.16	0.05	0.17	0.24	0.04
RDP	2.7	3.0	13.4	4.5	6.6	10.2	1.2

[a] r^2 = Coefficient of determination, SEP = standard error of prediction, and RPD = ratio of standard deviation of reference data to SEP.

TABLE XXVIII
Wavelengths of Principal Absorption Bands
for Common Constituents of Biological Material

Constituent	Strong	Fair	Weak
Protein			
			708[b]
			808
			868
		908[b]	982[b]
		1,018[b]	1,140
	1,692	1,186	1,276
	1,734	1,428	1,360
	1,930	1,498	1,454
	1,978	2,202 sh	1,578
	2,054	2,308	1,628 vw
	2,172	2,346	1,798
	2,274		1,824
	2,466		2,108
			2,380
			2,418
Cellulose (dry)			
			678[b]
		758[b]	816[b]
		982[b]	914[b]
		1,216	1,004[b]
	1,364	1,488	1,156
	1,428 vs	1,588	1,272
	2,054 sh	1,702	1,636 vw
	2,076	1,828	1,768
	2,104	1,918[c]	2,188 vw
	2,270		
	2,332 vs		
	2,480 vs		
Starch			
			614[b]
			758[b]
		914[b]	1,160
		986	1,268
		1,200	1,360
	1,432	1,584	1,750
	1,928[c]	1,700	1,826
	2,094	1,780	2,188
	2,282	2,370	
	2,318	2,474	
Oil			
			660[b]
		758[b]	816[b]
		928	1,210
		1,042[b]	2,144
	1,724	1,162	
	1,762	1,390	
	2,306	1,410	
	2,346	1,896	
		1,932	
		2,008	
		2,120	
		2,270	
		2,384	
Water			
		718[b]	
		758	810[b]
		964[b]	894[b]
	1,410		1,116
	1,460		1,154
	1,906		1,778
			2,208
Starch (dry)			
		610[b]	
		754[b]	820[b]
		914[b]	1,062[b]
		980[b]	1,272
		1,204	1,574
		1,360	1,750
	1,432	1,700	1,784
	2,282	2,044	1,830
	2,322	2,094	1,910
	2,446	2,478	2,180

[a] sh = Shoulder, vs = very strong, and vw = very weak.

[b] These bands are third and fourth overtones and are relatively strong absorbers.

[c] Probably water.

NIR reflectance determination of the composition of oat bran (Williams et al, 1991), and, more recently, in interpretation of several aspects of wheat composition and functionality (Williams and Sobering, 1996). In the case of oat bran, strong "bands" of variance in the traditional oil and protein wavelength areas indicated that these two constituents were exerting a strong influence on the predictability of dietary fiber in oat bran.

In the latter article, the first three factors accounted for over 96% of the total variance in the prediction of kernel hardness in ground-wheat meal, whereas, in the case of whole-wheat kernels, all 15 factors computed (using PLS regression) did not account for more than 95% of the total variance. Figure 7A and B illustrates the distribution of weights for the prediction of kernel texture in ground and whole-wheat kernels, using PLS regression.

Reference data were particle size index (PSI) data (American Association of Cereal Chemists, 1995). The PSI test involves grinding and sieving. Soft wheat kernels release more flour upon grinding and have PSI values of about 70–75 (%), whereas very hard wheat kernels, such as those of durum wheat, have PSI values of 35–40. The distribution of variance for the first factor in the ground-wheat meal calibration had the appearance of an "upside down" spectrum. This indicated that particle size was the most important factor influencing the ability of NIR reflectance to predict wheat kernel texture. A strong peak around 1,940 nm in the distribution of variance in the second and third factors indicated that moisture content also strongly influenced the effectiveness of the calibration. Distribution of the weights involved in PLS prediction of kernel texture in whole kernels was quite different, indicating that moisture content had more influence than in the prediction of kernel texture in ground wheat.

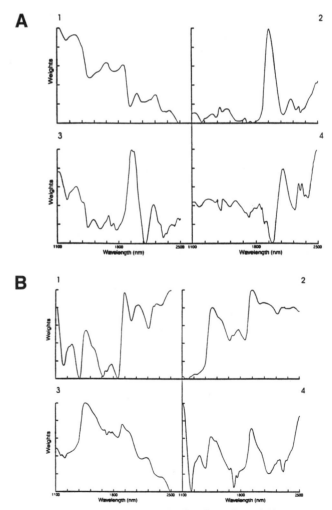

Fig. 7. Distribution of weights for the first four factors in partial-least squares prediction of wheat kernel texture. **A,** ground wheat; **B,** whole-wheat kernels.

In this type of work, the determination of what is contributing to variance in areas which are *not* associated with known constituents is likely to be most fruitful in researching complicated puzzles such as what contributes to, or detracts from, functionality parameters such as wheat "strength." These peaks of variance likely represent the interactions between and among components, which affect physical, rather than chemical, behavior. It is not uncommon for a peak in weights for one factor at a particular wavelength (e.g., corresponding to water) to appear as a valley in the weight distribution associated with a subsequent factor. This is interpreted as meaning that (in this case) water had a strong influence on the computation of the earlier factor, but the opposite influence on the later factor. The reason for these apparent anomalies lies in the fact that, after the first factor, each subsequent factor is associated with residual variance not accounted for by previous factors.

The peaks and valleys observed when displaying the weights often coincide with wavelengths selected by stepwise MLR. This explains why, for many applications, PLS and stepwise MLR regression approaches tend to give similar evaluation statistics in terms of r and SEP (Williams, 1999). Positive influences of a weight at a particular wavelength area are usually associated with positive k constants, and vice versa. The F value corresponding to a k constant indicates the importance of that constant relative to the others and to the equation.

ACKNOWLEDGMENTS

The valuable comments of John Antoniszyn, Philip Downie, and Bill Hruschka are acknowledged with deep gratitude.

LITERATURE CITED

American Association of Cereal Chemists. 1995. Approved Methods of the American Association of Cereal Chemists, 9th ed. Method 55-30. Am. Assoc. Cereal Chem., St. Paul, MN.

American Association of Cereal Chemists. 2000. Approved Methods of the American Association of Cereal Chemists, 10th ed. Am. Assoc. Cereal Chem., St. Paul, MN.

Anonymous. 1996. Ebby's Column. Q/A88. NIR News 7(1):9.

Association of Official Analytical Chemists International. 1995. Official Methods of Analysis of AOAC International, 16th ed. Assoc. Official Anal. Chem., Arlington, VA.

Davies, A. M. C., and Grant, A. 1987. Air-conditioning-generated noise in a near-infrared spectrometer caused by fluctuation in atmospheric water-vapor. Appl. Spectrosc. 41:1248-1250.

Hildebrand, F. C., and Koehn, R. C. 1944. Sources of error in the determination of the protein content of bulk wheat. Cereal Chem. 21:370-374.

Hruschka, W. R. 1987. Data analysis: Wavelength selection methods. Pages 35–55 in: Near-Infrared Technology in the Agriculture and Food Industries. 1st ed. P. Williams and K. Norris, eds. Am. Assoc. Cereal Chem., St. Paul, MN.

Malley, D. F., and Williams, P. C. 1993. Application of near-infrared reflectance spectroscopy in the measurement of carbon, nitrogen and phosphorus in seston from oligotrophic lakes. Can. J. Fish. Aquat. Sci. 50:1779-1785.

Martens, H., and Naes, T. 1987. Multivariate calibration by data compression. Pages 57-87 in: Near-Infrared Technology in the Agriculture and Food Industries. 1st ed. P. Williams and K. Norris, eds. Am. Assoc. Cereal Chem., St. Paul, MN.

Norris, K. H., and Williams, P. C. 1984. Optimization of mathematical treatments of raw near-infrared signal in the measurement of protein in hard red spring wheat. I. Influence of particle size. Cereal Chem. 61:158-165.

Pomeranz, Y., and Williams, P. C. 1990. Wheat hardness: Its genetic, structural, and biochemical background, measurement and significance. Pages 471-548 in: Advances in Cereal Science and Technology. Y. Pomeranz, ed. American Association of Cereal Chemists, St. Paul, MN.

Rosenthal, R. D. 1971. Introducing: The Neotec Grain Quality Analyzer. Address to the Kansas Wheat-Growers' Association, Hutchinson, KS.

Starr, C., Morgan, A. G., and Smith, D. B. 1981. An evaluation of near-infrared reflectance analysis in some plant-breeding programmes. J. Agric. Sci. 97:107-118.

Tong, J., Meurens, M., and Noël, H. 1995. Detection of breast cancer by near-infrared absorption and excitation. Pages 334-336 in: Near-Infrared Spectroscopy: The Future Waves. Proc. Int. Conf. Near-Infrared Spectrosc., 7th. A. M. C. Davies and Phil Williams, eds. NIR Publ., Chichester, England.

Williams, P. C. 1974. Errors in protein testing, and their consequences. Cereal Sci. Today 19:280–282, 286.

Williams, P. C. 1984. Grinders for use in sample preparation for grain analysis. Cereal Foods World 29:770-775.

Williams, P. C. 1987. Variables affecting near-infrared reflectance spectroscopic analysis. Pages 143-167 in: Near-Infrared Technology in the Agriculture and Food Industries. 1st ed. P. Williams and K. Norris, eds. Am. Assoc. Cereal Chem., St. Paul, MN.

Williams, P. C. 1992. Sampling, sample preparation and sample selection. Pages 281-315 in: Handbook of Near-Infrared Spectroscopy. Practical Spectroscopy Series, Vol. 13. D. A. Burns and E. W. Ciurczak, eds. Marcel Dekker Inc., New York.

Williams P. C. 1999. Comparison of calibratrons based on partial least squares and multiple linear regression for near-infrared prediction of composition and functionality in grains. Pages 287-293 in: Proc. Int. Conf. Near-Infrared Spectrosc., 9th. A. M. C. Davies and R. Giangiacomo, eds. NIR Publ., Chichester, England.

Williams, P. C., and Cordeiro, H. M. 1985. Effect of calibration practice on correction of errors induced in near-infrared protein-testing of hard red spring wheat by growing location and season. J. Agric. Sci. (Camb.) 104:113-123.

Williams, P. C., and Sobering, D. C. 1993. Comparison of commercial near-infrared transmittance and reflectance instruments RPD for analysis of whole grains and seeds. J. Near-infrared Spectrosc. 1:25-32.

Williams, P. C., and Sobering, D. C. 1996. Observations on the use, in prediction of functionality in cereals, of weights derived during development of partial least squares regression. J. Near-infrared Spectrosc. 4: 175-187.

Williams, P. C., Norris, K. H., and Zarowsky, W. S. 1982. Influence of temperature on estimation of protein and moisture in wheat by near-infrared reflectance. Cereal Chem. 59:473-477.

Williams, P. C., Norris, K. H., Gehrke, C. W., and Bernstein, K. 1983. Comparison of near-infrared methods for measuring protein and moisture in wheat. Cereal Foods World 28:149-152.

Williams, P. C., Norris, K. H., and Sobering, D. C. 1985. Determination of protein and moisture in wheat and barley by near-infrared transmission. J. Agric. Food Chem. 33:239-244.

Williams, P. C., Cordeiro, H. M., and Harnden, M. F. T. 1991. Analysis of oat bran products by near-infrared reflectance spectroscopy. Cereal Foods World 36:571-574.

Williams, P. C., Sobering, D. C., and Antoniszyn, J. 1998. Protein-testing methods at the Canadian Grain Commission. Pages 37-47 in: Wheat Protein Production and Marketing. Proc. Wheat Protein Symp. Canadian Grain Commission, Winnipeg, Canada (also available on www.cgc.ca).

CHAPTER 9

Variables Affecting Near-Infrared Spectroscopic Analysis

PHILIP C. WILLIAMS
Canadian Grain Commission
Grain Research Laboratory
Winnipeg, Manitoba
Canada

KARL NORRIS
Private Consultant
Beltsville, MD
U.S.A.

I. THE PHILOSOPHY OF ERROR

What is error? Error exists in everything. It represents the difference between reality and perfection. Errors are unavoidable, and the best thing that we can do is to ensure that they are rigorously monitored and minimized.

Errors are the differences between computed or measured near-infrared (NIR) values and the true values. Before discussing the sources of error in NIR testing, we should discuss error and the philosophy of error. There are four important concepts concerning errors.

1. Unless errors are detected, no one knows they are there. The most usual method of detection of errors is by some kind of system for monitoring accuracy of analysis. This can be achieved by using check samples, "blind" duplicates (tests carried out on duplicate samples included in different batches of analysis unknown to the laboratory), or, in the case of NIR analysis, by checking the NIR results against standard laboratory data. Another way of finding out that errors exist is the "hard" way, in which errors lead to revised specifications, improperly formulated products, or some other industry-related route. In this type of error detection, the discrepancies are usually brought to the operator's attention by the quality assurance manager or, in less welcome cases, by the customer.

2. The significance of the error is directly proportional to the true result. For example, an error of 0.4% (absolute) in protein testing is considered high for wheat analysis for protein but not for soybean analysis, in which the respective means for protein content are about 12 and 40%.

3. The significance of the error is also directly proportional to the significance or impact of the true result. An error of 0.4% in protein testing is serious when the accuracy demanded is plus or minus 0.2%. By the same reasoning, an NIR error of 0.3% is acceptable when the accuracy of standard analysis by a reference method (Chapter 3) is no better than 0.3%, and the operation can proceed efficiently with analysis of this degree of accuracy. Errors in screening operations, such as plant breeding, are generally more tolerable, since the main objective is the rapid, inexpensive sorting of good material from bad. For example, an error of plus or minus 3% in digestibility of a forage crop would be more than adequate for screening genotypes in a breeding program. NIR testing is much cheaper and faster than digestibility tests.

Errors in determination of constituents that directly affect price are the most serious. When "cut-off" levels occur, a difference of 0.1% in, for example, protein content of a commodity, may involve thousands of dollars, so that accuracy to within 0.1% is sought.

A basic rule concerning the degree to which errors may be acceptable is that the degree of accuracy needed is that which is sufficient to maintain an efficient operation. For example, there is no need to strive for a standard error of prediction (SEP) of 0.1 when an SEP of 0.25 is satisfactory for a successful operation, such as in screening plant breeders' material. Notwithstanding this type of practical philosophy, it is always preferable to achieve the highest accuracy possible in any test procedure.

Errors can be detected only by comparison with reliable, known values. The best way to detect the possibility of errors is to include check samples of proven composition (and functionality where necessary).

There are two main types of errors: random, or unbiased, errors, which usually reflect precision as well as accuracy, and biased errors, which are all higher or lower than the true values. These biases reflect a persistent change in the system. They may not reflect the precision of testing, since the precision may be excellent, while the results are all 2% high (and, in this case, all precisely wrong!). Biases are the most serious of errors, since an operation can proceed for long periods with results all either higher or lower than the true results. These can have very important consequences where payments are involved. Under these circumstances, the standard error of a single test (SET) data derived from the testing of check samples may indicate that the precision of testing is satisfactory, which can mislead the operator in the absence of accurate knowledge of the true results of check sample analysis.

Sometimes erroneous results are associated with outliers. Several versions of modern NIR software include systems for detection of outliers, and outliers will be discussed later in this chapter.

4. The reasons for the errors have to be sought and corrected. This chapter will identify more than 30 sources of errors in NIR testing, together with precautions to avoid them and measures to correct them.

II. SOURCES OF ERROR IN NIR TESTING

Where do errors come from? At first glance, the NIR technique appears to be relatively free from error sources. It is a simple process. In the early years, an NIR test involved grinding and mixing a sample, loading a cell, and reading the result. For many applications, the grinding step is no longer necessary. No weighing is needed, and there are no reagents to prepare. Nevertheless, over 30 sources of error have been identified. All of them can affect the results, but the most important thing to remember is that *excellent results, with coefficients of variability of 1–1.5%, are consistently achieved. The precision of modern NIR instruments is often superior to that of the reference method with which they are compared.*

Error sources can be divided into factors associated with the instruments, factors associated with the samples, and factors associated with the operator. The most troublesome sources of error are likely to be instrument-to-instrument variability with respect to wavelength, sample selection for calibration, sampling and sample preparation, wavelength selection, and reference laboratory analysis. The sources of error are summarized in Table I. Factors identified by an asterisk also affect precision.

This list is not exhaustive. The factors are not listed in order of importance, since all of them contribute toward the total error. Individual error sources will be described briefly in the remainder of the chapter. More information on individual sources can be gleaned from the references cited. Some are not well documented in refereed literature, mainly because the people that know most about them correct the problems, rather than write about them.

A. Factors Associated with the Instrument

An instrument using NIR radiation to measure the composition or other properties of a sample involves both a photometric and a

TABLE I
Sources of Error in Near-Infrared Technology

A	Instrument sources
1*[a]	Wavelength scale
2	Photometric scale
3*	Instrument temperature control
4	Cell covers
5*	Relative humidity of atmosphere
6*	Instrument-instrument differences
7*	Sample presentation system
B	Sample sources
1	Chemical composition
a	Interactions among constituents
b	Influence of chemical constituents on physical condition of material
c	Moisture status of material
2*	Bulk density
3*	Physical texture of sample
4	External factors (weather, etc.)
5*	Sample temperature
6*	Ambient temperature
7	Conversion factors
8	Whole grain application
a	Kernel (seed) size
b	Pat-length
c	Sample access
d	Color
e	Moisture content
f	Foreign material
g	Temperature
C	Operational sources
1	Calibration practice
a	Number of samples
b*	Sample selection
c	Accuracy of reference analysis
2	Sample preparation
a*	Sampling and subsampling
b*	Grinder type
c*	Grinder condition
d*	Blending after grinding
3	Sample storage
a	Before preparation
b	After preparation
4	Sample cell loading
a	Mixing
b*	Packing
c*	Cleanup between samples
5*	General carelessness

[a] An asterisk indicates that the factor also affects precision.

wavelength scale. These interact, but it is convenient to discuss them independently, because the factors that disturb the instrument's performance act differently on the two scales.

WAVELENGTH SCALE

The NIR instruments used for food and agriculture applications most often use either narrow-band filters or gratings to define the wavelengths used for a given measurement. The important parameters are wavelength, band-pass, purity, efficiency, and stability. The most important parameter is stability over time. High accuracy of the wavelength, narrow band-pass, low stray light, and high efficiency are desirable, but calibration procedures can overcome deficiencies in these parameters as long as they do not change. For multiple instruments to use the same calibration, the instruments must replicate the one used for the calibration, and they must also be stable over time. Errors from small differences among instruments can be minimized by including data from many instruments in the calibration. This procedure sacrifices performance over what could be obtained for individual calibrations, but is widely used to produce a more robust calibration.

The factors that affect the stability of the wavelength parameters are robustness of construction, mechanical wear, mechanical vibration, expansion and contraction from temperature changes, dust on optical components, and deterioration of optical components. Temperature of the environment around the instrument is often the most important factor for stability of the wavelength scale, because all of the components in the instrument respond to temperature changes. Typical filters and grating assemblies will change about 0.5 nm for a temperature change of 1°C. Therefore, temperature control is used within most instruments to improve stability.

The magnitude of the analyte prediction error from a change in wavelength is very dependent on the sample and the calibration algorithm. A change of 0.1 nm causes significant error in some cases, and a change of 2 nm can be tolerated for other measurements.

PHOTOMETRIC SCALE

Accuracy and linearity are more important for the photometric scale than for the wavelength scale, but again, stability of the scales is necessary for good performance. For diffuse transmittance measurements the photometric signal is normally converted to $\log(1/T)$, in which T is the transmittance of the sample at a specified wavelength. $\log(1/T)$ is fairly linear with changes in concentration. The major source of nonlinearity is from stray light, and there are several sources of stray light. The first source is inherent in all filter and grating spectrometers, that is the radiation transmitted by the filters or reflected from the grating that is of a different wavelength from the desired wavelength. This radiation is typically less than 0.1%.

The second source is from radiation leaking around the sample and reaching the detector without going through the sample. This can be a particular problem when attempting to measure the transmittance of a single object such as a soybean seed. To avoid this stray light, a good seal around the seed is required. The third source of stray light is from room light getting to the detector because of inadequate sealing of the sample compartment. The transmittance of the sample may be as low as one part in a million, so extreme care must be used to block any stray radiation.

For diffuse reflectance measurements, the photometric signal is normally converted to $\log(1/R)$ or Kubelka-Munk function, $(1 - R^2)/2R$, in which R is the reflectance at a specified wavelength. Neither of these functions changes in a linear fashion with changes in composition over a wide range. Therefore, for diffuse reflectance spectroscopy, we have two types of nonlinearity. The nonlinearity of the instrument and the nonlinearity of the absorption signal combine to make it more difficult to make accurate measurements of composition. The $\log(1/R)$ value becomes very nonlinear with concentration at high $\log(1/R)$ values (greater than 1.0) because of the surface reflectance from the sample. The diffuse reflectance carries the information about the composition, and the surface reflectance carries little or no information about the composition.

Therefore, when the measured reflectance includes a high proportion of surface reflectance, it is not a correct signal.

Most instruments are designed to reduce the amount of surface reflectance included in the measurement, but no practical method has been developed to eliminate this problem. A typical diffuse reflectance measurement will include a surface reflectance signal of 1%. Therefore, the log($1/R$) value of 1.0, which is 10% reflectance, is really 9% diffuse reflectance plus 1% surface reflectance. On the reflectance scale, this is an error of 10%, and, on the log scale, the error is 4.6%. The effect of the nonlinearity can be readily observed on the reflectance spectrum of canola seeds in which the C-H peaks in the 2,300-nm region are small compared with the peaks in the 1,700-nm region (Fig. 1). The opposite should be the case with the long wavelength peaks several times larger than the shorter wavelength peaks. At the high log($1/R$) values, the surface reflectance is dominating the signal.

Photometric noise is an important parameter in all NIR measurements. The noise may be classified as random and systematic. Random instrument noise limits the measurement of low level constituents and causes random errors in the measurement of any constituent. Systematic noise is a more serious problem, because it causes errors in the accuracy in all constituents. The random noise usually comes from the detectors and electronic amplifiers, with the lamp and its power supply being the second most common source of noise. This is particularly true of a well-designed instrument, because these are noise sources that can be minimized but not eliminated. One of the methods to reduce this noise is the coadding of signals. This reduces the noise but extends the time to make a measurement. Coadding signals does not reduce and can increase the effects of systematic noise. Therefore, it is very important to keep systematic noise to a minimum.

Examples of systematic noise include the noise generated by worn bearings, changes in the humidity in the instrument, electrical interference, poor ground connections, light leaks, inadequate power supplies, changes in temperature, and many others.

Another important parameter in an NIR instrument is the optical geometry of the interface between the instrument and the sample, and this includes the stability of the sample. Since most NIR measurements involve the collection of diffuse radiation, the distance between the sample and the collection optics is a sensitive parameter. It is possible to build an optical geometry to minimize the effect, but this is not common.

Air is normally used as the reference for transmittance measurements, but, for reflectance, a reference standard is required. This can be a source of instability if the reflectance of the standard changes from dust and films collecting on the standard. Differences among standards in different instruments represent a potential problem in transferring a calibration from one instrument to another.

As stated above, useful calibrations can be developed for a single instrument with distorted spectra as long as the instrument has sufficient stability in all the wavelength and photometric scale parameters. Instrument manufacturers have concentrated on developing the needed stability into their instruments, and now they are moving toward improving the match among instruments, so that a single calibration can be used on many instruments.

INSTRUMENT TEMPERATURE CONTROL

Internal fans achieve cooling in most NIR instruments. If these stop working for any reason, results can become erratic due to temperature fluctuation and overheating of some components. This can occur quite quickly and will be revealed if regular precision tests are carried out. On the other hand, results may also be affected if the ambient temperature falls too low. Generally speaking, if the atmosphere is comfortable for the operator, it will also be suitable for the instrument. If variations in ambient temperature are expected under conditions of future operation of the instrument, similar fluctuations should be introduced during the period when the calibration is being developed.

CELL COVERS

The sample cells of most NIR instruments interpose an optical glass or quartz cover between the sample and the detector to en-

sure uniformity of the surface. The thickness and refractive index of the cell cover can affect the accuracy of testing, since the covers are rarely completely planar or of uniform thickness. Often a change of covers, caused by breakage, can cause a linear offset in analytical results. It is particularly important to determine the consistency of results between sample cells when more than one cell is to be used in high throughput operations. It is also important to maintain clean quartz or glass cell covers, especially when analyzing oilseeds or other commodities with more than 8% oil. A case could be made for an open cell with no covers, in that such a sample presentation system would theoretically be less susceptible to specular reflectance. The changes in the micro-orientation and degree of hydration of particles at the immediate surface caused by evaporation are likely to incur errors that would more than counterbalance the benefits of an open reflectance surface.

Glass Versus Quartz. Figure 2 shows the band at 2,198 nm that occurs in glass. This is close enough to the important absorber associated with protein, at 2,178 nm, to affect the prediction of protein content in calibrations in which this band is used. In a recent study, a calibration was developed for the prediction of protein content in a set of ground barley samples. Using multiple linear regression (MLR), the equation used five wavelength points, one of which was 2,172 nm. A quartz window was used in the standard sample cell. Replacement of the quartz window with one of glass increased the SEP from 0.26 to 0.43.

Glass covers vary in thickness both among covers and within individual glasses. This can affect results recorded when rotating the filled sample cell through 180°. This phenomenon is far more noticeable in glass than in quartz windows.

RELATIVE HUMIDITY

The relative humidity (RH) of the atmosphere is not really an instrument factor, but is included in this section, since it may have a strong influence on instrument performance. This phenomenon was first noted by Davies and Grant (1987), who noted that changes in the RH of the atmosphere changed the noise level of a NIR spectrophotometer, particularly in the areas of maximum absorption by water bands. Since it is difficult in practice to achieve operation of an NIR instrument at constant RH, calibrations should ideally be developed under conditions whereby the RH can be varied, for example, by using a room humidifier, and the RH can be monitored during the period of calibration development.

INSTRUMENT-TO-INSTRUMENT VARIABILITY

Two instruments are unlikely to be exactly alike, no matter what make or model of instrument. Inherent differences among discrete filters have been recognized from the outset of NIR technology, and small, but significant, differences occur among diode array instruments and even among monochromators.

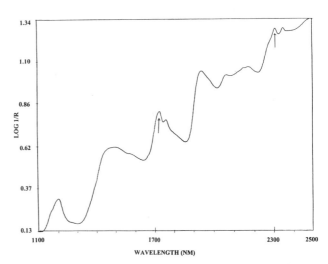

Fig. 1. Spectrum of whole canola seed. Arrows indicate main oil bands (C-H bands).

The net result is that, while a single calibration may allow several instruments of the same type to operate accurately and precisely, it is unlikely that all such instruments will perform with the same efficiency, using the single calibration. Careful monitoring will reveal that small slope/bias corrections are necessary. This is no longer associated with the stigma with which slope/bias correction was formerly associated. Neural network calibrations may eliminate the need for slope/bias corrections in a series of instruments of the same type and when large numbers of spectra have been incorporated into the calibration.

MATHEMATICAL TREATMENT OF THE LOG(1/R) SIGNAL

Several NIR instruments have been used successfully for the analysis of various commodities for protein, moisture, and other constituents using the log(1/R) signal directly after amplification, but without further mathematical treatment. Mathematical treatments of optical signals prior to analysis were suggested by Morrey (1968) as a method for correction for band overlap. They are employed in NIR instruments to resolve band overlap and minimize baseline drift. The most successful mathematical treatments employed to date have been the smoothed log(1/R) itself; the first or second derivative of the log(1/R); the delta log(1/R), as employed in the early Neotec/Pacific Scientific instruments (now Foss/NIRSystems, Inc., Silver Spring, MD); the normalized first or second derivative; and the Kubelka-Munk equation.

All of these have been discussed in earlier chapters. Other treatments, such as curve fitting (Hruschka and Norris, 1982) and the full-spectrum procedures, also described earlier (Chapters 3 and 4), have been used in computerized spectrophotometers with some success. In general, the raw log(1/R) and the first or second derivative of the log(1/R) have proved to be of most practicable use in

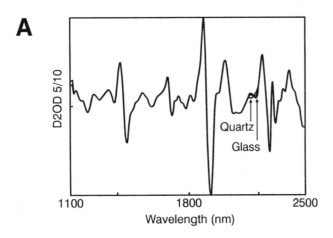

A

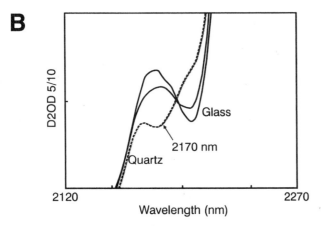

B

Fig. 2. Second-derivative spectra of ground barley scanned with quartz and glass cell windows. **A,** Full-range spectrum and **B,** spectral area around 2,170 nm to illustrate the difference between glass and quartz windows.

bench-type instruments and the most consistent in performance. Bands that appear as peaks or shoulders in the log(1/R) spectrum occur as "cross-overs" or zero points from positive to negative values in a first-derivative spectrum, and as negative, but much sharper "valleys," in a second-derivative spectrum.

Smoothing. Spectral data are often smoothed to reduce the noise. The most common procedures are the Savitzky-Golay and "boxcar" procedures described in Chapter 3. The term "segment" was introduced by Neotec (now Foss/NIRSystems, Inc.) to describe their smoothing procedure, which is the boxcar method. The segment value is the number of data points expressed in nanometers included in the boxcar average. The amount of smoothing to be applied is determined in the calibration process to minimize the error.

Derivatives. Spectral data are often converted to derivatives to enhance the appearance and improve calibrations for constituents. The most common procedures are the use of Savitzky-Golay equations and the "gap" procedure described by Norris and Williams (1984). The size of the gap is an important parameter for optimum calibrations. For narrow-band absorbers, a small gap is best and, for wide-band absorbers, a large gap is best. In addition, the zero-crossings created by derivatives make it possible to choose wavelengths that have little or no sensitivity to an interfering absorber, while maintaining a high sensitivity to the absorber to be measured. Gap variations change the wavelengths for the zero-crossings, making it possible to optimize the sensitivity to the measured absorber by changing the gap size. Smoothing can be applied with derivatives to provide the optimum performance in predicting the constituents in samples.

Table II illustrates the effects of changing segment and derivative size with a computerized spectrophotometer. The test substance was ground hard red spring wheat. Forty-eight samples were used in the calibration and twenty for the analysis. The accuracy of the analysis tended to improve with increased segment size for both first and second derivatives, although there was no significant difference between segment sizes of 8 and 20 nm. The optimum derivative or gap sizes were 30 and 20 nm, respectively. Segment and gap sizes should be optimized for the most reliable calibrations. Both segment and gap sizes are generally larger for whole than for ground grains. This is illustrated in Table IX of Chapter 8 in this monograph.

The engineering section of an instrument manufacturing company has to recognize the needs of both the instrument and the operator. The sample has to be presented to the detector(s) so that the instrument can scan enough material to enable accurate and precise analysis without damaging the sample or affecting the uniformity of the scan. At the same time, the sample size should not be prohibitive. Although there is plenty of grain available at a grain silo or other receiving point, samples are often sent to and from laboratories for calibration development and monitoring. Most samples of grains and seeds are shipped in envelopes that hold about 400 g, and instrument sample presentation systems should be designed with this in mind. Transportation of large samples is

TABLE II
Influence of Segment and Derivative Size on Accuracy of Protein Determination in Wheat

	Segment									
	4 nm		8 nm		12 nm		16 nm		20 nm	
Derivative	SEP[a]	Bias	SEP	Bias	SEP	Bias	SEP	Bias	SEP	Bias
First	0.20	0.11	0.16	0.07	0.16	0.08	0.14	0.07	0.14	0.07
Second	0.29	0.46	0.22	0.13	0.21	0.13	0.20	0.14	0.20	0.16

	Derivative ("Gap")									
	4 nm		20 nm		50 nm		70 nm		90 nm	
Derivative	SEP	Bias	SEP	Bias	SEP	Bias	SEP	Bias	SEP	Bias
First	0.20	0.04	0.16	0.05	0.15	0.08	0.16	0.24	0.16	0.19
Second	0.29	0.13	0.19	0.22	0.24	0.15	0.21	0.56	0.21	0.25

[a] Standard error of prediction.

expensive and troublesome at both points, due to their weight and bulk. The actual sample scanned by the instrument is rarely in excess of 100–150 g, so that it is the dimensions of the sample presentation pathway that requires careful design.

Some instruments have alternative sample presentation options. For use at grain receiving points, a hopper form is used. The hopper is filled with grain (a major advantage of NIR spectroscopy has always been that there is no need to weigh the sample). The grain is accessed to the instrument in increments of a few grams and is stationary during each actual scan. The ingress of each increment is controlled by an impeller or a rotating brush. It is important that the sample stops during scanning. When the rotating brush approach is used, small seeds such as canola or poppy seeds may pass through the brush, and the sample may not be stationary during scanning. This is a source of error. For use with smaller samples, a sample transport option is available. This allows the operator to fill a sample cell that enters and is discharged from the instrument. The sample cell moves either horizontally or vertically during scanning, but the grain itself is stationary inside the cell.

Ground and powdered samples are presented to the instrument in cells of several shapes and sizes. The cells should be deep enough to prevent irradiation from passing right through the sample. A sample depth of about 2 mm will effectively prevent this. Some instruments automatically rotate their sample cell during scanning. This is more important when the sample cell is small and particularly if the instrument is using only a single detector. Ease of cell cleanup between samples is important to the operator. Sample throughput is reduced if the operator has to devote 3–4 min to cleaning the cell between samples, especially when the instrument takes only 30 or 40 sec to scan the sample.

While grains and seeds can be accurately analyzed by NIR spectroscopy without grinding, this is more difficult with materials such as forages. Even here, cells have been developed that allow NIR analysis of whole samples. Predrying and grinding can improve the accuracy of analysis, particularly with regard to high moisture forages and silages, but care must be taken in the drying preparatory to grinding in order to avoid losses of volatile components other than water.

Most NIR measurements on agricultural samples are being made with filter instruments or scanning grating instruments, although diode array instruments have gained a significant market. Fourier transform NIR instruments are not common for agricultural applications, although they are very common for industrial applications. Fourier transform NIR instruments are typically designed to measure a small volume of sample, and the inherent within-sample variability of agricultural samples requires the measurement of a large sample to cancel out this variability. A similar statement can be made for the use of Raman spectroscopy on agricultural samples. Raman spectroscopy may be very valuable to measure the contents of a single cell in a sample, but it is not a choice for measuring the composition of an agricultural product to determine the market value of that product.

B. Factors Associated with the Sample

The most important lesson in NIR spectroscopy is that the sample is what you have to work with. Everything that has happened to the sample up to the time that it is scanned in the instrument will be embodied in the sample and recorded in the spectrum. This includes the influences of genetics, environment, processing, preparation, handling, storage, and any other factors. You can change the optical signal before computing the calibrations, but you can't change the sample.

Since the first edition of this monograph, the trend in NIR analysis of many agricultural materials including grains, pellets, and fresh and dry forages has been toward intact material, thereby eliminating the burden of grinding and its associated errors. This section is included for the benefit of laboratories that still operate reflectance instruments that require a ground sample. Ground samples are still necessary to reference analysis, and the majority of the factors described apply equally to NIR and reference testing.

Most factors associated with the sample concern particle size and shape and bulk density, all of which are a function of the relationship between the texture of the commodity and the grinding procedure. The texture of plants and commodities derived from them is affected by over 20 factors, summarized in Table III.

The analysis of whole kernels of grains and seeds introduced a further set of error sources, but kernel (seed) size remained as an important factor, analogous to the particle size of powdered commodities. Factors associated with industrial commodities and materials are discussed in Chapter 10.

CHEMICAL COMPOSITION

The chemical composition includes the constituents of the commodity and the absorbing groups present in them, any of which may interact with each other. The constituents in agricultural and food commodities most likely to cause interactions are moisture, protein, oil, and fibrous carbohydrates. By Avogadro's Law, there are 6.025×10^{23} molecules in a 1-g molecule, so that, in 1 g of wheat with 12% moisture and protein and 2% oil contents, there will be about 4×10^{21} molecules of water, about 1.6×10^{17} molecules of protein (assuming an arbitrary value of 500,000 for the molecular weight of wheat protein), and about 1.3×10^{19} molecules of oil—plenty of molecules to cause interactions, but proportionately more of those of smaller molecular weight.

The material to be analyzed is a series of molecules all in a constant state of movement. The shapes of these molecules will also be changing, as different groups of atoms carrying different charges approach each other. Certain areas of individual molecules, particularly the larger and more complicated molecules, will come into contact with different areas of other molecules with different reactivity. Factors affecting the functionality, such as the degree of hydration and oxidation status, will all be reflected to a certain extent in changes in the absorbances of the individual molecules. The total absorbance pattern of the material is likely to be affected, and, as a result, so would the loadings and "scores" or "weights," across the whole wavelength range. The velocity of all of these interactions among and within molecules is affected by temperature, so that temperature is likely to have a significant influence on the efficiency of NIR calibrations.

There are three main types of interaction. The first type is the influence of one or more constituents on changing the optimum wavelength for the determination of another. The second is an effect of the chemical constituents on the physical nature of the commodity and on the pattern of reduction during grinding. The third type is the effect of moisture alone on the hydration status of all other constituents.

Interactions Among Chemical Constituents. The most common interactions occurring in food, feed, and forage commodities in-

TABLE III
Factors Affecting Physical Texture of Agricultural and Food Materials

Chemical composition
Moisture content
Plant species
Plant variety
Stage of maturity (forages, grasses, straws, etc.)
Lignification
Growing environment, especially during growing season
Soil fertility
Weather damage during harvest period
Frost during maturation (cereal grains)
Storage conditions
    Length of time
    Temperature
    Atmospheric conditions
    Type of container
    Moisture content at beginning of storage
Fungal infestation
Insect infestation
Processing conditions, e.g., heat, pressure, freezing
Drying conditions
Temperature of material

clude interactions between oil and protein, starch and protein, fiber and protein, oil and fiber, starch and fiber, and water with everything. Some commodities, such as some varieties of corn with high oil content (above 12%), exist when both oil and starch are present in high concentration. The presence of a constituent that is in the highest concentration, is the strongest absorber, or a combination of both of these will tend to "mask" the absorbers of the constituents present in lower concentrations or with weaker absorbers. The instrument may then become less sensitive to variations in concentration of the "lesser" constituent. In instruments with discrete filters, the accuracy of analysis may be affected due to the fact that the fixed wavelengths are not optimum. In a computerized spectrophotometer, this will often result in the selection of alternative wavelength areas for the determination of the "lesser" constituents.

Computerized spectrophotometers may select wavelengths that correspond to higher overtones of the original wavelength. For example, an absorber caused by a second overtone of a -CH- stretching frequency at 2,280 nm may be optimum for a fiber determination in a commodity low in oil. This absorber may be ignored by a computerized spectrophotometer in favor of a higher overtone of the same absorber in another area if the search for "fiber" wavelengths is conducted in a commodity high in oil. When a constituent such as fiber is determined using a primary wavelength that does not correspond to a major absorber, the wavelength may be associated with an area of minimum interference by other constituents rather than with the absorber of the constituent being determined.

The determination of oil, protein, and moisture in many commodities has been well researched and the optimum wavelengths and mathematical treatments established. Table XXVIII of the Chapter 8 gives the predominant wavelengths for the most common constituents of materials of plant and animal origin. Shenk et al (1992) provide a more comprehensive table that includes the functional groups responsible for the absorbances. For expansion of the technique to new commodities and constituents, flexibility in wavelength and mathematical treatments is essential.

Influence of Chemical Constituents on Physical Condition. Oil, water, fiber, protein, and, to a certain degree, mineral matter may all affect the physical state of a commodity. Most commodities must be reduced to a smaller and more uniform particle size before analysis by reference methods (and for analysis by some NIR instruments). The easiest commodities to reduce to a powder are starchy, low to medium (8 to 12%) in protein, and low in moisture and oil.

Oil acts as a pasting agent, and, when a commodity high in oil is pulverized, it tends to form a paste rather than a powder. Oil contents of up to 5% have relatively little influence, but, when the level reaches 7–8%, the effect of the oil becomes noticeable, and the resultant ground grains and seeds are no longer free-flowing. The degree of agglomeration increases directly with oil content until the oil content reaches 20–30%, when few discrete particles exist. At oil contents of 45% and higher, the ground seeds may take on the texture of a paste rather than a powder, depending on the fiber content. The manner of comminution (degree of grind-

ing) of the commodity affects the accuracy and precision of analysis, since the packing of sample cells is affected.

The physical nature of the surface, and hence the diffuse reflectance from it, is strongly influenced by oil content. When ground oil seed samples are stored, even under refrigeration, they tend to agglomerate and the particle characteristics change. Re-analysis may result in large discrepancies. It is recommended that oilseeds, or any type of seed with an oil content above 10%, be analyzed within a day or two of grinding for the most precise results. Whole-seed residues can be stored if further NIR analysis is necessary.

Fibrous constituents may cause significant changes in the pattern of grinding and bulk density of a commodity. The most fibrous substances are mature straws, particularly of crops such as faba beans, cotton, flax, and even wheat, especially the part of the straw nearest the ground. Fibrous commodities have low bulk densities. The amount and pattern of diffuse reflectance from the surfaces are quite different from those of the surfaces of ground grains. The orientation of light, fibrous particles is affected to a greater degree by static electricity at the sample cell window than is that of the denser, more uniform-shaped particles of ground grains.

Interactions among constituents that are highly correlated with each other can result in measurements of a constituent being highly dependent on variations in the composition of another. This is not critical unless the concentration of both needs to be determined, for example, in the case of constituents such as amino acids, when NIR determinations may be reflecting variations in protein rather than the actual amino acids. Wavelengths selected will assist the operator in determining whether the wavelengths conform to the required constituent.

Change in the composition of a single constituent can affect the functionality of the material containing the constituent and the pattern of absorbances from the constituent. For example, the amylose/amylopectin ratio of the starch in wheat flour exerts a strong influence on starch paste viscosity, an important factor in noodle quality, and in areas such as the thickener industry (soups, gravies, etc.). In wheat flour, the lower the amylose content, the higher the starch paste viscosity. These changes in composition cause subtle differences in the spectral characteristics of these materials, which can be utilized in the development of calibrations to predict not only the amylose content but also, and more importantly, the changes in functionality caused by shifts in the composition.

Moisture Status of the Sample. Moisture has been referred to in the Chapter 8. The moisture status of the sample affects the accuracy of analysis before and after grinding. Samples such as meat or fresh fruit, which are very high in moisture, are impossible to grind and are best presented to the instrument in the original form, as a minced sample, or as a slurry. The term "uniform particle size and shape" when applied to most commodities is mythical. The best that can be hoped for in grinding is to achieve consistency. Particles vary in shape to an even greater extent than in size.

Variability in moisture among samples of the same commodity causes significant differences in mean particle size, particle shape, and particle size distribution. Table IV summarizes the influence

TABLE IV
Effects of Moisture Content on Mean Particle Size (MPS) and Particle Size Distribution in Wheat

Wheat[a]	Moisture	500–1,000 μm	500 μm	354 μm	210 μm	149 μm	74 μm	H$_2$O (%)	MPS (μm)
Mean results (N = 12)									
Durum	Normal	1.6	4.4	15.9	15.2	28.0	34.9	12.6	193
	High	1.4	6.4	15.1	16.9	23.5	36.7	16.2	197
HRS	Normal	2.0	4.6	11.6	11.6	22.0	48.8	12.2	178
	High	2.2	6.3	11.6	13.6	21.0	45.2	17.0	185
SWS	Normal	3.4	5.7	8.7	8.5	15.8	57.4	12.7	176
	High	4.8	5.0	9.4	8.4	15.9	56.4	15.4	189
Extreme results									
Durum	Low	1.6	5.0	17.4	19.6	29.4	27.0	7.2	202
	High	5.1	11.7	22.6	18.0	20.6	22.0	15.9	267
HRS	Low	2.2	6.6	15.2	16.0	21.4	38.7	8.7	203
	High	4.4	11.5	18.7	14.9	21.3	29.2	15.4	253
SWS	Low	3.8	5.8	8.3	8.5	15.1	58.4	10.0	181
	High	5.2	6.4	9.4	9.4	15.3	54.2	15.7	200

[a] HRS = hard red spring wheat, and SWS = soft white spring wheat.

of moisture on the mean particle size and particle size distribution of three types of wheat when tempered to different moisture contents and ground on a Cyclone grinder (Udy Corporation, Fort Collins, CO) fitted with a 1.0-mm (0.04-in.) screen.

These figures represent the means of several samples each of hard red spring, soft white spring, and durum wheats. The bottom part of Table IV illustrates the more extreme differences in particle size characteristics that occur.

The moisture content of a sample also affects powdered material in that the hydration status of all components is directly affected by the amount of moisture present, as well as by the relative affinity of each constituent for water. The degree of hydration may influence the optimum area of the spectrum where the absorbers of specific constituents occur. Components of many agricultural and food commodities contain constituents that differ widely in affinity for water (e.g., cellulose has a much higher affinity for water than does starch). When two such constituents occur at low moisture levels, the constituent with the highest affinity for water will tend to reduce the hydration status of other constituents. This may cause changes in the spatial orientation of the molecules with the net effect of different absorbers becoming optimum for the moisture-deprived constituents. This is an area that has not been widely researched, but it has been observed that the testing of commodities such as wheat and barley, which are very low in moisture (7–8%) is usually fraught with errors much larger than can be attributed simply to the mathematical variation in moisture content.

The conception that molecules of the same substance all have identical vibrations may not be true in the case of large molecules of plant origin. For example, molecules of the same protein in different samples with different moisture content may become hydrated to different degrees. It is important to visualize the sample in the same way as the instrument sees it, not as a solid mass, but (as mentioned above) as a conglomeration of atoms assembled into molecular arrays, constantly in motion, but retaining the general overall structure of the molecule. It is possible that even different areas of the same molecule (especially a large, complex molecule such as protein) may become oxidized, reduced, or hydrated to different degrees than are other areas of the same molecules in the same material. When this differential hydration occurs, differences occur in the degree of hydrogen-bonding within the molecule, and the shapes of the molecules may be altered. The molecular vibrations may be changed to the extent that wavelengths optimum for the determination of protein in some samples may not be optimum for others.

BULK DENSITY

The bulk density of Cyclone-ground wheat has been found to vary by over 30% among samples of wheat of different types and over 12% within a type. Variation of this magnitude has a significant effect on surface area and, as a result, on the diffuse reflectance signal. Bulk density influences the amount and orientation of the material that can be packed into a sample cell. Since the "tightness" of the packing is affected, the surface from which reflectance occurs is also affected. Bulk density is affected mainly by chemical composition and, in general, the higher a commodity is in constituents with low density such as cellulose, the lower the bulk density and the more difficult it will be to attain an even and reproducible surface. Differences in bulk densities among commodities result in commensurate differences in the surface characteristics and the optical signal from the surface. Errors in NIR testing of commodities such as straws, dried grasses and other forages, wheat bran, soybean hulls, and similar substances can be improved by grinding with a hammer mill, such as the Christy-Norris hammer mill (Christy Ltd., Scunthorpe, England).

PHYSICAL TEXTURE OF THE SAMPLE

Factors that affect physical texture also affect diffuse reflectance. The main factors have been summarized in Table III. Physical texture means the composition or structure of the commodity. The texture of a substance influences the way it can be reduced to

a fine particle size. The large seed size means that seeds such as faba bean or palm kernels will not freely enter the grinding chambers of grinders such as the Cyclone sample mill. To obtain a uniform sample, this type of seed can be ground with a two-stage reduction process. The best type of grinder for the first stage is the Christy-Norris 8-in. hammer mill, using a special screen with 8- to 10-mm holes. This rugged mill produces a meal with coarse consistency, which can easily be reduced to a fine powder by passage through a Cyclone, Retsch (Brinkman Instruments, Inc., Chicago, IL), Falling Number (Perten Instruments, Springfield, IL), or similar grinder.

ADDITIONAL FACTORS AFFECTING PHYSICAL TEXTURE

Plant species affects the texture of a powdered forage sample by changing the leaf-to-stem ratio and the overall degree of lignification. Stems are more highly lignified than leaves, and a higher proportion of leafier species means that a dried forage will reduce to a fine powder more readily. Stage of maturity also affects texture, since lignification usually increases during the maturation stage.

Plant variety has an important effect on the texture of grains and seeds. Wheat, barley, anc corn varieties differ widely in physical hardness and sufficiently enough for the optical signals in the NIR to be used to determine their texture with a high degree of precision in intact kernels (particularly in the case of wheat).

Wet weather during the harvest period can affect texture of grains by causing bleaching and possibly sprouting, both of which tend to soften grains. Wet weather can affect the texture of forages by inducing the development of molds and other fungi. Frost during the maturation of grains causes the grains to become significantly harder. These changes in texture can be detected in intact kernels by NIR and can be used to predict the degree of change caused by these grading factors.

Poor storage can also affect texture due to the development of molds or bacterial colonies, with associated temperature buildup. Storage effects are related to the temperature, length of time, and atmosphere. Ideal storage conditions combine low temperature with good aeration. Such conditions have no adverse effects on grains, forages, or any other type of commodity.

Samples of animal origin present a further variety of textures and packing factors, many of which are affected by temperature (e.g., butter). These include dairy products (e.g., butter, cheese), liquid and dried milk, ice cream mix, cream, meat and derived products, blood, bone, fish and meat meal, and many others. Many of the products need no sample preparation beyond careful blending, but do require specialized systems for presenting the sample to the instrument. This aspect has been well researched by instrument manufacturing companies, all of which have devised ingenious and practical cells for the analysis of special products for the analysis of powders, slurries, liquids, stiff pastes, and materials such as butter and margarine.

SAMPLE TEMPERATURE

When samples at various temperatures are analyzed in sufficient quantity, they will change the internal temperature of the instrument. Sensitivity to temperature of the detectors is an important reason why temperature affects NIR instrument performance. If samples of different temperature are to be used to stabilize a calibration against the sample temperature effect, all samples of the same temperature should be scanned consecutively as a batch. All samples of the next temperature should then be scanned, and so on, until sets of samples at all desired temperature levels have been added to the database. This is because the samples must be allowed to affect the internal temperature of the instrument. If samples of different temperatures are scanned one at a time, or randomized, a warm sample will likely be scanned immediately after a cold sample (or vice versa). This will not allow the instrument to react to prolonged exposure to samples of a given temperature. NIR analysis of whole seeds is more susceptible to temperature fluctuations than that of ground seeds, since whole seeds are not subject to the grinding step, which causes substantial warming of cold samples.

The influence of sample temperature on the accuracy of NIR analysis of ground wheat has been documented (Williams et al, 1982). The main effect on wheat protein determination was a nearly linear decrease in observed protein content as sample temperature rose from −30 to 45°C. Industrial processing often results in the finished product having a high temperature. For example, flour fresh from the reduction rolls in a flour mill may be at 40–50°C. The flour will subsequently cool to room temperature. These temperature variations must be incorporated into calibrations.

Cooking and food-processing methods represent yet another variable. Cooking temperatures vary from the 40°C traditionally used in drying pasta products, through the 80–100°C used in modern pasta drying, through boiling and steam cooking, to high temperature baking and other methods of processing. There is a strong interaction between cooking temperature and texture of the commodity, and individual operators should carefully research the effects of processing temperature on the texture and sample preparation technique best suited to their products.

AMBIENT TEMPERATURE

The ambient temperature can also affect instrument performance, since detectors and other components are sensitive to temperature. This is particularly applicable to computerized spectrophotometers, which may not be equipped with the same degree of temperature control as commercial bench model instruments.

CONVERSION FACTORS

Conversion factors are used to convert the constituent that is actually measured into the constituent sought. The best known of these is the nitrogen-to-protein factor arbitrarily established at 5.7 for wheat and flour; 6.25 for other grains, pulses, seeds, and forages; and 6.38 for dairy products. Other tests used as reference methods may not provide reliable information on the constituents that contribute to functionality. An example is the crude fiber test, which does not give a true indication of the true indigestible or digestible fibrous components present in a fibrous food or feed. Such factors and methods were established by our predecessors as chemists who, had they had the facilities of modern laboratories to work with, would doubtless be as skeptical of the accuracy of the factors as we are. These factors are a source of error in NIR testing, since an NIR instrument is really a "super-microscope" which "sees" beyond the sample surface, theoretically detecting every molecule and even parts of molecules present.

The nitrogen-to-protein factor is probably the most important of these conversion factors, since protein is the constituent most frequently determined by NIR. Reference methods determine total nitrogen with accuracy. It is unlikely that the amount of nitrogen in the protein fraction of every sample of every commodity is constant. A variation of as little as 0.2% in the actual nitrogen content

of the protein would cause the nitrogen-to-protein conversion factor to change to the extent of introducing an error of about 0.27% protein in wheat at 13.5% protein or about 1.6% protein in soybeans at 40% protein. Tkachuk (1977) noted changes of up to 0.13 in the nitrogen-to-protein factor for wheat and 0.09 in chickpea. These would be sufficient to cause errors of over 0.2% protein in wheat and 0.3% in chickpea at 13 and 20% protein, respectively.

Conversion factors are "hidden" errors. From a philosophical viewpoint, the excellent accuracy that it is possible to achieve with NIR may imply that the nitrogen-to-protein factor is practically constant. In practice, this is not necessarily so, since it is likely that the NIR and reference techniques measure all of the nitrogenous substances present, so that both techniques are equally affected by changes in the nitrogen-to-protein factor, and neither are truly measuring protein. Nonprotein nitrogen exists in plant and animal material mainly in the form of simple peptides and amino acids, and the main absorbers responsible for NIR measurements are likely the same as those in protein. Accordingly, there is no reason why the NIR technique would not encompass this type of nonprotein nitrogen in the same way as the reference test.

WHOLE-GRAIN APPLICATIONS

In 1977, Stermer et al (1977) were the first to demonstrate that whole kernels of wheat could be analyzed for moisture by NIR spectroscopy. Later, Tkachuk (1979) reported on the prediction of protein content in whole-wheat kernels by NIR reflectance, using a Cary computerized spectrophotometer (Nicolet Instrument Corp., Madison, WI). Both of these undertakings were carried out in reflectance mode, and neither were sufficiently successful for practical application in grain handling. Work carried out at the Instrument Research laboratory at Beltsville, MD (*unpublished data*, 1979), demonstrated that accurate predictions of protein content could be made in whole kernels of several different commercial classes of wheat by using an NIR instrument in transmittance mode.

The introduction in 1980 of the first commercially available instrument for analyzing whole grain operated by NIR transmittance. Excellent results were attained by NIR transmittance analysis of whole grains (Williams et al, 1985), and several instruments subsequently became commercially available for this application. Whole-grain NIR analysis can be carried out with success equal to NIR transmittance in reflectance mode (Williams and Sobering, 1993). An advantage of reflectance analysis is that the same instruments can be used for analysis of powders, such as ground grains or flour, by using suitable sample presentation cells. Pathlength has to be very small to enable analysis of powders by NIR transmittance, and difficulties are inevitable when ensuring uniform sample cell loading and cleanup between samples. Another advantage of whole-grain analysis in reflectance mode is that the wavelength range can be extended further into the NIR area. The area between 700–1,800 nm has been found very useful in whole-grain analysis by NIR reflectance (Williams and Sobering, 1993).

Application of NIR to the analysis of whole grains and seeds is now commonplace by both transmittance and reflectance devices, and instrument manufacturers advocate the analysis of whole seeds when possible. Whole-seed analysis eliminates a host of sources of error, including grinding, cell loading, and several others. On the other hand it introduces some sources of error to which ground materials are rather less subject.

Whole grains are much larger than the particles of ground grains, but the particle size effect operates by the same principles. The light from an NIR instrument penetrates the sample of whole kernels much more deeply. More of the light is scattered within the sample, and, as a result, the spectra of whole grains are characterized by higher optical density than are those of ground samples of the same commodity. Figure 3 shows the spectra of whole and ground wheat and the flour milled from the wheat. Note the relatively small difference between ground wheat meal and flour spectra. The differences induced due to particle size are affected by differences in reflectance between the white, very reflective

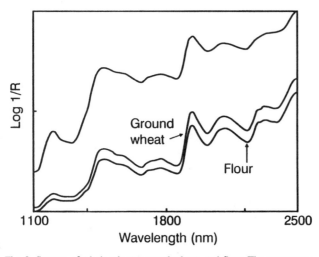

Fig. 3. Spectra of whole wheat, ground wheat, and flour. The upper spectrum is of whole-wheat kernels.

flour and the light brown, less reflective wheat meal. The spectra of flour may even show higher absorbance values than Cyclone-ground whole meal from the same wheat.

Calibrations for whole-grain analysis based on MLR tend to use more wavelength points than do calibrations for ground grains. Similarly, partial least-squares (PLS) calibrations tend to use more factors in whole-grain calibrations than do calibrations based on ground grains of the same commodity.

The main sources of error in whole-grain analysis include seed size, color, pathlength (NIR transmittance only), moisture content, sample temperature, and foreign material.

Kernel Size. This affects NIR whole-grain analysis in the same way as does particle size in ground-grain analysis. Figure 4 illustrates differences in the spectra of whole kernels of wheat of different commercial classes. Samples used in this figure were of approximately the same composition. The particle size index (PSI) is a grinding/sieving test used as a reference for the reporting of wheat kernel texture (American Association of Cereal Chemists, 2000). Using this method, hard wheats give lower PSI values than soft wheats.

Commercial samples of wheat contain kernels with a range of over 100% in size from smallest to largest. Samples of wheat (and other grains) with different average seed size may include seeds, the largest of which may be five or more times the size of the smallest. Seed size affects the way in which light passes through the sample and also how light energy is diffusely reflected from the sample. Figure 5 shows the differences in spectra of whole-wheat kernels of different size. The spectra were derived in reflectance mode from a single sample of hard red spring wheat that had been separated into large, small, and medium-sized kernels by sieving. The spectrum of the original wheat paralleled that of the medium-sized kernels. Large seeds may be sufficiently large to allow clear spaces between seeds in the sample path, so that incident light impinges directly onto the detector in NIR transmittance instruments. This can be overcome by increasing pathlength and is not a factor in NIR whole-grain reflectance analysis.

Small seeds, such as those of weeds, poppy, or canola (oilseed rape) present a denser surface. In NIR transmittance instruments, they may restrict the amount of energy that reaches the detector, when they are present as foreign material in grain samples (Williams et al, 1985).

Pathlength. This may need to be increased or decreased, depending on seed size. Seed size and pathlength are closely associated in instrument design and in the development of calibrations. Pathlength must be optimized during the development of calibrations. For example, a pathlength of 18 mm is optimum for the analysis of wheat and barley, using the Tecator InfraTec (Foss/Tecator AB, Höganäs, Sweden), while 6 mm is required for the analysis of

canola seed, and 30 mm for the analysis of corn (maize) and soybeans. One of the reasons for the changes in pathlength is to prevent gaps between kernels and to ensure that the light passes through a continuous bed of grain. Another reason is that small seeds, such as canola or poppy seed, present a very dense surface, through which it is difficult to pass sufficient light, hence the 6-mm pathlength for canola seed.

Modern NIT instruments automatically change pathlength according to calibrations for different commodities. Whole-grain NIR analysis by reflectance eliminates pathlength as a source of error, since pathlength is essentially constant in reflectance mode.

Sample Access. Some instruments access samples using a small revolving brush, which "steps" grain through the instrument to give several incremental readings. When used for the analysis of small seeds, such as canola or poppy seeds, some seeds may "leak" through the brush at each pause. This has the effect of causing the seeds to move slightly during analysis. This movement induces small changes in appearance of the bed of grain to the light source/detector system, and can cause significant errors in whole-seed NIR transmittance analysis of such commodities.

Very high moisture grains, such as corn, do not flow as freely as drier grain (up to 25% moisture). The grain will not flow through vertical sample presentation systems, and when some sample presentation cells are used, it is difficult to fill the cells consistently, and also to empty them after analysis, without removing the back of the cell. Open transport cells, such as the types available with the Foss/Tecator InfraTec model 1275 or the Perten DA-7000, are convenient designs of cells for this type of application.

Color. Color affects the degree to which light is reflected from, or passed through, the sample. In the case of very bright seeds, some light may be reflected back from the sample and will not carry information to the detector. This phenomenon can be utilized in differentiating among seeds of different color. Large differences in color can cause considerable distortion at low wavelength ranges (e.g., 400–650 nm). This has been discussed in Chapter 8.

Moisture Content. The moisture content of grain and its derived products can vary from close to zero to over 50%. Moisture changes the shapes of spectra, and large changes in moisture content (e.g., from 10 to 30%) may cause shifts in the optimum wavelength for the NIR prediction of moisture and other constituents. NIR analysis of whole grains and seeds, whole fruits and vegetables, meat, fish, and many other materials may call for the analysis of material at an "as-is" moisture content. Unless the moisture range in the grains or seeds is fairly low (e.g., 10–15%), composition in terms of constituents such as protein or oil should be transposed to the original moisture content of the fresh samples to ensure the most accurate results. Table V shows data for the prediction of protein content in wheat on an as-is and constant moisture basis in a series of samples with a large range in moisture content. Data for the prediction of as-is protein were significantly better than the 13.5% data.

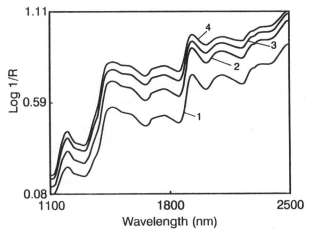

Fig. 4. Effects of wheat kernel texture on spectra of whole-wheat kernels. 1 = Particle size index (PSI) (American Association of Cereal Chemists, 2000) 72 (very soft); 2 = PSI 60 (medium hard/soft); 3 = PSI 53 (hard); and 4 = PSI 40 (extra hard: durum).

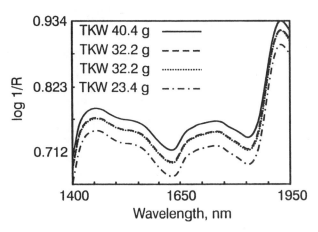

Fig. 5. Effects of wheat kernel size on spectra of whole-wheat kernels. TKW = thousand kernel weight.

Foreign Material. Whole kernels can be analyzed directly from railway cars, farm trucks, or even from the combine. Grain in this condition may contain fairly high amounts of foreign material, such as chaff, pieces of straw, and weed seeds. These can change the effective pathlength. Small seeds affect the transmittance of light through the sample and can cause large errors in NIR transmittance testing (Williams et al, 1985). Foreign material, or at least the bulk of it, should be removed before NIR transmittance testing. NIR reflectance measurements are less susceptible to the errors caused by the presence of small seeds (as foreign material).

Temperature. Temperature affects the accuracy of all NIR testing. Testing of prolonged series of either very cold or very warm samples change the temperature of the instrument and can cause significant biases. Whole grains are more susceptible to the influence of temperature than are ground grains. The heat generated during grinding very cold grains causes the grain to warm up, and the temperature of the ground grains is not appreciably colder than room temperature (Williams et al, 1982). Whole grains are analyzed as they arrive, and, in North American and European winter conditions, the grain can be extremely cold at the time of testing.

It is particularly important in developing calibrations for NIR transmittance or NIR analysis of whole grains that samples be included at extremes of temperature. Sets of at least six samples should be included at temperatures varying in increments of 5°C, and all six samples at each temperature level should be scanned in batches—this will expose the internal aspects of the instrument to the temperature range expected.

C. Operational Factors

CALIBRATION PRACTICE

Number of Samples. During the early years of NIR application to grain analysis, calibrations were established using anywhere between 20–50 samples. The minimum number of samples necessary to establish a stable calibration of the "simple" type (Chapter 8) is now suggested as being 100. Ideally, more samples should be used, depending on the nature of the material to be analyzed, the constituents sought, the sources of variance likely to affect the analysis, and the instrument to be used.

For calibrations involving artificial neural networks (ANN), or "local" calibrations, very large databases are recommended (several thousands, or at least hundreds, of spectra). Spectra for ANN calibrations should have been recorded on several instruments of the same make and model. Ideally, all of these instruments should have been standardized to one "master" instrument.

Sample Selection. The two principles are (i) identification of the likely sources of variance in the commodity to be tested, and (ii) assembly of a set of samples representative of all of the variances, including source, chemical and physical composition, functionality, and spectral identity. The actual number of samples used in a calibration depends upon the anticipated variance. If the variance in the available samples is limited, the efficiency of the calibration in terms of accuracy and precision will not be improved by simply increasing the number of samples. For example, an NIR

instrument may be used to monitor the output of a flour mill, where particle size is relatively uniform, and the range of protein is unlikely to exceed 0.5%. Here, a calibration based on 1,000 samples is not likely to be much more effective than one based on 50.

As a general guideline, the fewer the number of variables likely to be encountered, the smaller the number of samples required for a reliable calibration. For a commodity such as wheat, the variables most likely to affect the wheat are growing season, growing location, planting time, irrigation and fertilizer practice, wheat variety, fungal and insect disease, and weather effects during maturation and harvest periods. For a more complex commodity such as a forage, the variables of plant species and stage of maturity at harvest should be added. For an even more complicated commodity such as a feed mix, the individual components of the mix should be considered. Their proportions in the mix and the interaction effects of all of the above factors on such individual components as grains and dried grasses or forage legumes, which are often used in the mix, all have to be taken into consideration.

For commodities of animal origin such as powdered milk, cheese, butter, meat, and wool, the chief factors affecting NIR analysis include sample composition, texture, temperature, and processing procedure. Fat protein and moisture content are the most important constituent variables and temperature affects the texture and general physical state of most of these commodities. For example, presentation of butter to an NIR instrument using a sample cell that depends on a cut, vertical planar surface would be difficult if the ambient temperature was high enough to melt the butter even slightly, since the surface would not remain flat. Analysis of milk requires constant stirring, since fat globules are constantly rising to the surface, which can bias the results.

Species and stage of maturity (age) affect commodities such as fish, fish meal, meat, and meat meal. Storage is also a factor, since materials of animal origin tend to deteriorate much more rapidly than do those of plant origin. Sample preparation is more complicated with fish and meat products due to their peculiar texture and high moisture content.

All variables should be included at least three times in any calibration, and accumulation of sufficient samples for a stable calibration is often the most critical aspect. If a variable can be identified, it should be easy to accumulate samples with a range of that variable. If an unexpected "new" element appears, such as the effects of a new season or an epidemic of a fungal disease, samples bearing that variable to different degrees can be accumulated and used to update the calibration at the time of its appearance. New elements include a change in processing procedure in a factory or the proposed use of different grain types as the basis for feed mixes, such as switching from barley to wheat due to price differentials. The operator should be aware of the likelihood of such differences in a commodity.

A uniform distribution in composition across the anticipated range reduces errors caused by regression toward the mean from extremes of concentration. This has been discussed in detail in Chapter 8. It is good practice to invest time in accumulating as comprehensive a set of calibration samples as possible before recording any optical data. Samples collected in the early days of the "assembly" period should be scanned at the time of arrival, since this is the way that they will be scanned when the calibration is in full use. They can then be protected from change by careful storage for possible future use in calibration or validation.

Accuracy of Reference Laboratory Analysis. Accuracy in standard (reference) analysis is essential to setting up efficient NIR calibrations and for assembling samples to be used in monitoring an operation. The early days of NIR analysis saw many cases of NIR instruments "blamed" for poor accuracy, when, in most of these instances, it was the reference laboratory analyses used in monitoring that were responsible for high discrepancies and standard errors. The 28 years of NIR application in North America have witnessed significant improvements in the attention paid to standard laboratory analyses. This is reflected in improved NIR statistics, which are due partly to improvements in instrumentation and partly to improvements in reference analyses. Reference

TABLE V
Effects of Large Moisture Range (7.5–18%) on Prediction of Protein Content in Whole Wheat[a]

Parameter	Protein (%)	
	13.5% Moisture	"As Is" Moisture
r^2	0.981	0.988
SEP[b]	0.251	0.227
b[c]	1.018	1.002
a[c]	0.247	−0.087
Bias	0.035	−0.013

[a] All classes combined.
[b] Standard error of prediction.
[c] b = Coefficient of regression, a = intercept of regression.

laboratory accuracy is a critically important factor in the efficiency of NIR analysis, since the SET (precision) of reference analysis is a component of the accuracy of the overall test.

SAMPLE PREPARATION

Sampling and Subsampling Procedure. Sampling is the most important single source of error in any chemical or physicochemical analysis of agricultural commodities and most food products and ingredients. The sample must be truly representative of the total population in every sense, including chemical composition, physical constitution, and the presence of foreign material. Table VI illustrates the number of kernels of wheat in different weights, assuming an average weight per kernel of 35 mg. Similarly, a 70-lb. (32-kg) bale of hay may represent 2,000–10,000 plants, depending on the species and stage of maturity of the plants at harvest.

The larger the population, the more comprehensive the sampling method needs to be. For example, to gain a representative sample of the progeny of a single F2 wheat plant, planted as a single row, all that is necessary is to divide the entire yield in half manually or by using a sample divider. One half can then be used for analysis and the other half replanted, as necessary. At the country elevator or farm level, the practice of diverting by hand into a pail portions of the stream of grain entering or leaving a truck gives a satisfactory sample, provided the entire truckload is sampled in this way every 15–20 sec throughout the loading or unloading.

To sample a vessel, the stream of wheat passing into the vessel should be automatically sampled continuously, ideally using a diverter-type sampler (Bauwin and Ryan, 1974). The sample stream should be divided and a small percentage diverted into a container that will eventually contain the sample that represents the entire shipment. The Kostur sampling system developed and used by the Canadian Grain Commission at terminal elevators is typical of this method (Bryan and Elvidge, 1977). The sample has to be thoroughly blended before analysis and subsampled by passing through a sample divider, such as the Boerner divider (SeedBuro Equipment Co., Chicago, IL), repeatedly until the final half-samples are small enough for analysis.

Representative sampling of forages is even more complicated by the fact that not only is the material usually lengthy and difficult to handle, but care has to be taken to protect the natural ratio of leaves to stems. In a sample such as dried alfalfa, the stem/leaf ratio can easily be altered by careless sampling and sample preparation. The leaves are usually several times higher in protein content and lower in fiber content than the stems. To secure an accurate representation of the composition of fresh forage over an entire field requires a larger number of samples than would be required for grain, and the samples are more difficult to subsample and process. If the forage is field-dried and baled, the task becomes simpler, since a core sample can be taken from every 20th or 100th bale or so, thoroughly blended, and subdivided.

Sampling and sample preparation merits a separate chapter, or even a separate monograph, since different commodities and materials call for a plethora of sampling and sample preparation techniques. Another aspect of sampling concerns access of the sample to the instrument. Instrument designers are beginning to realize the potential improvements in analytical efficiency to be gained by the elimination of the sample cell and its loading error, and a wide variety of sample systems are available to suit individual instruments and materials. The use of fiber optic cables and interactance probes, which effectively bring the instrument to the sample rather than the reverse, has already begun to revolutionize sample access.

Grinder Type. There are two main types of grinding, namely "normal" grinding for sample analysis, and two-stage grinding, which involves prereduction of materials such as forages, feed pellets, or large seeds that cannot enter the grinding chamber, followed by further reduction on a Cyclone or similar grinder. For normal grinding, be it for reference or NIR analysis, it is important to use the grinder optimum for the commodity. Grinders suitable for different commodities are summarized in Table IV in Chap-

ter 8. The Cyclone, Cyclotec (Fisher Scientific, Chicago, IL), Falling Number KT 3100, and Retsch centrifugal grinders are all fitted with 1.0-mm screens and are most suitable for seeds low in oil. The Cyclone grinder with forage head is suitable for grinding up to 20-g samples of forages, or even bigger samples if the receiving jar is replaced with a modified sample receptacle.

For grinding oilseeds, the Retsch centrifugal mill is very effective. Small impeller-type mills, such as the Krups model 75, Braun, Moulinex, or similar mills (all available from department stores that carry kitchen appliances) are also useful. The seeds should be ground in short, 15-sec bursts, with the sample being blended manually between bursts until a uniform meal is obtained. About six to eight bursts are sufficient. For forages, feeds, and very hard seeds such as faba beans, the Christy/Norris 8-in. and the Falling Number KT 3100 hammer mills are most suitable.

Grinder Condition. The mean particle size, particle size distribution, and therefore the diffuse reflectance signal can be markedly affected by the condition of the grinder used in sample preparation. Cyclone sample mills need periodical replacement of the Carborundum ring around the grinding chamber and of the 1.0-mm screen. The surface of the Carborundum becomes progressively polished by the passage of grain samples. At the same time, the holes of the 1.0-mm screen tend to increase in size and the net result is an increase in mean particle size. This can be monitored and the grinder maintained by periodically replacing the Carborundum disk or screen. It is important to ensure that screens are replaced accurately in grinders that employ them. Sometimes a small piece of grain lodges underneath the screen and prevents it from "seating" properly. This can result in differences in particle characteristics and in dramatic errors in testing.

Hulled rice (paddy) is particularly abrasive, and routine grinding will wear impellers, hammers in hammer mills, and screen apertures far more quickly than any other type of grain. Stainless steel impellers can be purchased for some grinders for processing rice (A. B. Blakeney, BRI Australia, personal communication).

Under normal use, the Carborundum disk and screen of a Cyclone grinder should be replaced about every 50,000 samples. The burrs of burr mills become worn with time and need replacing about every 100,000 samples (for grinding rice paddy, the grinder should be inspected after every 1,000 samples or so). Passage of stones or pieces of steel through grinders cause serious damage to screens and burrs and, in the case of the Cyclone grinder, the sparks generated are potentially hazardous. Hammer mills are generally the most rugged of grinders and, barring accidents, need very little attention beside lubrication. The screen should be replaced about every 100,000–250,000 samples, depending on the material ground. Small impeller mills need fairly frequent replacement of the impellers, especially if they are used to grind soybeans or any similar fairly hard commodity.

There are three important points concerning grinding. (i) It is a very boring job. (ii) Staff assigned to grinding samples often feel that they have been identified as being the "lowest on the totem pole" and strive to get it over with as quickly as possible, so as to be assigned less-tedious duties. (iii) It is one of the most important phases of analytical work and should only be assigned to competent and conscientious workers. Poor sample preparation for reference analysis can jeopardize all subsequent analysis, and, as a result, it can have an important effect on NIR calibration development and monitoring.

Blending After Grinding. This is a simple but essential step in improving precision. The ground sample stratifies in the grinder

TABLE VI
Approximate Number of Kernels
of Hard Red Spring Wheat in Different Weights

Weight	Number of Kernels
Kilogram	29,000
Tonne (1,000 kg)	29,000,000
Boxcar (50 T)	$1,428 \times 10^9$
Vessel (10,000 T)	2.857×10^{11}

receptacle and must be thoroughly blended before testing by reference or NIR spectroscopy. The sample jar should be capped and shaken and the contents stirred 15–20 times before loading the sample cell (NIR) or weighing (reference method). This is increasingly important as the sample size increases.

SAMPLE STORAGE

Storage Before Preparation or Use. The most important factors to consider when samples need to be stored for a significant period (more than 2 weeks) before analysis are changes in chemical composition and damage caused by insects, molds and other fungi, and bacteria. In the northern states of North America and in Canada, as well as in dry countries such as Australia, the Indian subcontinent, and the Middle East, samples stored in paper bags or envelopes lose moisture very quickly. They reach an equilibrium moisture level at about 8% in 3 or 4 weeks. In more temperate areas such as northern Europe, parts of the eastern United States and Canada, and the northwest coast of North America, samples stored in envelopes may gain moisture. Fungal and bacterial infestations are more likely to occur in areas where the RH is high, whereas insect infestations are prevalent in hot areas. Storage in paper bags and envelopes for periods longer than a few days is not recommended. Paper is porous and fragile, and many samples have been lost or spoiled due to breakage of paper containers.

Storage in sealed, thick, 4-mil (0.1-mm) plastic bags will protect material from moisture loss, but not from prolonged exposure to insects or from mold or fungal infestation if samples are high in moisture. Generally, forages are less likely to deteriorate in storage because their chemical composition is not so conducive to insect or fungal infestation.

Ideally, samples should be stored refrigerated or frozen, but numbers and bulk may make this impracticable. If samples are tempered to different moisture levels before grinding, they should be allowed to rest in the tempered state for at least 2 days and should be stored in metal or plastic containers, taped to prevent moisture loss (or gain), or in thick plastic bags. If the standing period is likely to be longer than 2 days, the tempered samples should be refrigerated to prevent bacterial or mold development or insect infestation. Samples stored in refrigerated conditions for long periods (1 year or more) may increase significantly in moisture content, to the extent that they may require air drying before use as NIR, calibration, or monitoring samples. To prevent this, the samples should be stored in 4-mil plastic bags, in containers with tight-fitting lids. Plastic containers are preferred, since they are not prone to rust.

Sample Storage After Preparation. The same principles for sample storage before grinding are applicable to preservation of the ground, well-mixed sample after grinding. The main differences are that the samples are usually smaller and that the storage period after grinding is usually shorter, since samples are frequently ground, analyzed, and discarded, and a 24- to 48-hr storage period is usually sufficient. Under conditions in which longer storage is necessary, precautions must be taken to preserve the samples, for example, over weekends. The storage method depends on the size and chemical composition (particularly moisture content) of the sample, the length of storage, the normal atmospheric conditions, and the space and finances available for the storage of samples.

Large samples are more difficult and more expensive to store for long periods than are small samples. Other than samples to be used as check samples or those that may be needed for verification of earlier test results, such as cargo samples, the long-term storage of large samples should not be encouraged. Where it is essential, storage should be in heavy-duty plastic containers with tightly fitting lids, and the containers themselves should be refrigerated to inhibit the development of insect infestation and molds.

Samples of intermediate size (100–1,000 g), for example, flour samples awaiting analysis or physical dough-testing and baking tests, are also best stored in plastic containers, provided they have tight-fitting lids. Round (cylindrical) plastic containers are preferred to square ones, since they are easier to clean. Square containers have corners and are more space efficient. Stacking of plastic containers is not advisable due to possible cracking. The ideal type of storage area is a refrigerated room with air circulation, fitted with movable ("rollaway") shelf units. In this type of storage room, several tonnes of samples can be stored with easy access. Other types of areas include air-conditioned rooms, preferably with no windows, or simply ordinary rooms with air-circulating fans. Aeration, or frequent air change, is very important in sample storage, particularly in the absence of refrigeration.

For samples of 15–25 g, the size normally ground for use in analysis, small, round, flat metal containers of about 20 mm in depth and 60 mm in diameter (two-ounce ointment tins) are the most suitable for short-term sample preservation. They are easy to clean, label, and stack in trays and can be taped to protect against moisture interchange. They are also easy to mix the sample in before weighing. The best alternatives to these are plastic containers with tight-fitting caps. For large-scale analytical operations, turn-around time of sample containers is a significant factor and occupies many staff hours in cleaning and removal of labeling.

Plastic bags and small paper envelopes are useful for conveying ground samples to the laboratory, but can only be used once, and paper envelopes are completely porous to water movement inward or outward. Storage of analytical samples in envelopes or plastic bags makes it difficult to mix the sample before weighing or loading sample cells, and the static electricity accumulation by some plastics may cause further difficulties in transferring samples from the bags.

SAMPLE CELL LOADING

Mixing. Mixing of samples, including flour, before analysis can influence cell loading error. Other factors influencing cell loading include the bulk density of the sample, particle size and shape, the amount of sample loaded into the cell, and static electricity. The mixing of ground samples before loading the cell has been examined (Williams, 1975). No improvement in precision was obtained after 15 mixes of the ground sample in the sample container of a Udy Cyclone grinder (Udy Corporation). A spatula was used, with a rotating movement. Cell loading error has been determined for several commodities including red spring, soft, and durum wheats, barley, oats, and rye. The error due to grinding increased directly to the amount of fiber in the sample, being highest in the case of oats. The sample loading error was significantly greater for the more fibrous materials. In some cases, the sample cell loading error was higher than the grinding error. There was also a sample/cell loading interaction.

Packing. The amount of material packed into the cell can affect the results by influencing the compression of the sample at the surface, which will affect the diffuse reflectance. This variable can be controlled by overfilling the cell and striking the surface level before putting on the cap for all samples. In this way, the average weight of material used is consistent among samples of a single type. The $\log(1/R)$ signals normally increase for dense grains, such as cereals and pulses, as a result of overfilling, and decrease for material with lower packing density, such as straw. The Perten Inframatic 8000 series (Perten Instruments) presents a different type of loading, since it has no sample cell as such. The sample is packed by tamping into a cell formed by a door-type attachment to the front of the instrument. The tamping must be uniform to ensure precision.

Cleanup. Sample-to-sample contamination can occur if sample cells are not efficiently cleaned between samples. For materials low in oil, such as cereals, pulses, and forages, brushing away fine material after emptying the cell is usually sufficient. Quartz windows are subject to buildup of static electricity, and fine particles adhere more closely to quartz than to glass. It is necessary to wipe the window with lens tissues periodically. For oilseeds and some mixed feeds (e.g., poultry feeds that have been sprayed with tallow), the cell window must be lightly brushed free of particles of meal and wiped free of oil using a tissue moistened with alcohol. The window should then be wiped free of alcohol using a tissue soaked in water and finally dried with another tissue.

Cleanup is not limited to the sample cell. The cell carrier and the glass or quartz window that protects the detectors from the atmosphere also accumulate dust and should be cleaned daily or after every 50 samples. Buildup of dust on the interior window can cause the progressive development of bias. An abrupt change in results can occur following cleanup after a lengthy accumulation of dust, and this must be checked after cleaning an interior window.

Carelessness. Carelessness is an annoying, ever-present, and significant source of trouble and includes mistakes in numbering and identifying samples, reading samples out of sequence, recording the wrong data, misplacing decimal points, generally poor quality benchwork, and others. It is usually associated with lack of concentration on the part of the operator. Routine NIR analysis is a boring operation and it is advisable to make it as interesting as possible by keeping the operators au fait with the reasons for, and implications of, the analysis. Good supervision will assist by detecting anomalous results, by maintaining good morale, and by generally running an efficient operation.

D. Outliers

A discussion of outliers belongs in this chapter. In the day-to-day operation of a grain-receiving site, or a processing plant that receives and tests raw materials on arrival, there are no outliers—everything has to be tested. Several software packages include methods for the detection and elimination of outliers. This section is devoted to a discussion of outliers, rather than their removal. An outlier is defined here as a result that differs from the rest of the population by more than three times the SEP. Calibration of an NIR instrument entails developing a stable relationship between the samples and the optics and detector of the instrument.

The danger of outliers is that the results may appear within the range of acceptance, and, unless the outliers are among the samples actually retested by the monitoring procedure, they will not be detected until too late. It is important for instrument software to alert the operator to the fact that a sample that has just been analyzed has appeared as an outlier to the instrument, so that the sample can be rescanned. The main factors causing outliers are summarized in Table VII.

Variance in the NIR signal is affected by particle size, shape, and packing density, by chemical composition, and by the interaction among constituents. Static electricity, which affects the orientation of samples at the cell surface, and also the degree of hydration and hydrogen bonding, sample and ambient temperature, and many other factors that may affect one or more of these can all affect the optical signal. The assembly of samples for use in developing the calibration must encompass all of these variance sources. The only aspects that cannot be included concern variance associated with samples that have not been grown, fabricated, or delivered at the time of calibration.

Outliers are samples that the calibration has not enabled the instrument to recognize, and, as a result, the instrument is forced to "guess" the composition or functionality of that sample.

There are two main types of outlier. These are spectral outliers and analytical outliers. When an outlier appears during the development or validation of a calibration, it is important to verify that it is indeed an outlier. This can be achieved by re-analyzing the sample, including sample preparation, by both reference and NIR methods. Do the NIR test first (it is quicker and cheaper!). If the NIR result changes appreciably, the reason for the outlier can usually be traced to wrong sample identification, faulty sample preparation, or cell loading technique. If the NIR result does not change, the reference analysis must be repeated.

If the sample remains an outlier after retesting by NIR and reference methods, it is a spectral outlier, and the reason for its nonconformity must be investigated more deeply. A true spectral outlier is a sample for which the interaction between the optical signals from the sample does not conform to those of the majority of the samples used in the development of the calibration.

Outliers are individuals and represent the last serious frontier between adequacy and excellence in NIR analysis. They usually

occur irregularly and at up to 2% of the total population. A situation wherein 20 or 30% of the samples display large discrepancies between NIR and reference analyses is not fraught with outliers. This is an NIR situation that calls for investigation of the reasons for the low coefficient of correlation and SEP, of the wavelengths employed, and also of the accuracy of the chemical analyses. Either the calibration has not been properly developed with respect to comprehensive sample assembly, the analytical (or other reference method) is not reliable, or NIR is not applicable in that situation. Outliers occur when all else appears to be working well.

Outliers can occur in calibration sets. This means that the assembly of samples has included a sample or samples carrying a variance that causes the discrepancy in prediction. The outlier will bias the calibration toward itself (Williams and Antoniszyn, 1987). It will improve the capability of the calibration equation to predict what it is supposed to predict in future samples of that ilk, but usually will not improve the efficiency of prediction of the other (and majority of) samples.

Software systems for outlier detection are all based on spectral characteristics and, in most cases, rely on differences in Mahalanobis distances. Most software methods for outlier detection identify the dubious characters from their spectra and indicate that the results of analysis of these samples are likely to be of an unacceptable degree of error. This may not be the case in practice, and the results may not be in error to a greater extent than those of other samples that have not been pinpointed. Conversely, samples for which the error is unacceptable may escape the "eagle-eye" of the outlier detection software.

The "H" statistic introduced by InfraSoft International (Port Matilda, PA) identifies samples with H values of above 3.0 as potential outliers. Experience has shown that the residual errors for such samples are indeed usually higher than are those of most of the population. But residual errors equal to some of those with such high H values have also been observed in the same populations for samples with low and acceptable H values. These samples could also be classified as true outliers, since their residual errors are unacceptably high when their spectral characteristics indicate that they should be otherwise. The reason for such anomalies remains as one of the enigmas of NIR technology.

There is a tendency to remove outliers, and several software combinations include methods for automatic recognition of outliers and removal of them from the calibration. If these options are exercised in association with statistical evaluation of the calibration using cross-validation, there is a danger of exaggeration of the efficiency of the calibration. This can be misleading, since the fact remains that the instrument/math treatment/wavelength selection combination has not accommodated the samples that have been excluded. There are some logical reasons for the removal of a single spectral outlier, or up to 1% of a large population of samples. But if these samples are from a source likely to occur to a significant degree in future operations, the outliers should not be removed and efforts should be made to accommodate them. This can be done either by incorporating more samples of the same type or by accumulating enough samples of that type to generate a separate calibration.

The difference between the optical signals from a set of samples at a given wavelength point and the signal from an individual

TABLE VII
Factors Causing Outliers

Chemical composition
Interaction among constituents
Wavelength selection
Moisture status
Physical texture
Particle size and shape
Bulk density
Sample preparation technique
Sample orientation in cell
Sample temperature
Number of wavelengths use in calibration
Mathematical treatment of log(1/R) data

sample at the same wavelength point that causes it to appear as an outlier may be quite subtle. Optimization of the mathematical treatment can cause the outlier to cease to be an outlier. Again, the addition of an extra wavelength may affect the removal of several would-be outliers. On the other hand, the extra wavelength may change the relationship between other samples and the wavelength combination that has existed up to the removal of the one or two outliers affected by the extra wavelength. Here, the operator has to decide whether to accept one large error or to live with a few samples in the validation set with errors that have increased from what they were before the addition of the extra wavelength point.

Errors in chemical laboratory analyses are frequently cited as the principal source of discrepancy between NIR and standard data. The chemical laboratory is free from blame in the case of *true* outliers, since discrepancies between NIR and chemical data persist in the face of (i) repeated determinations by both methods and, more significantly, (ii) accurate analyses by the NIR calibration of all other samples in the series tested.

When outliers occur in calibration sets, they can improve the integrity of the calibration with regard to future samples that resemble the outlier, but affect the magnitude of the constants. In the case of calibrations of computerized spectrophotometers, they may influence wavelength selection, particularly the second and subsequent wavelength points. "Sleeping" outliers can be detected in what appear to be "normal" calibration sets by computing the residuals using calibration constants developed from a different set of samples. The outliers will emerge as samples with very large deviations.

Several pieces of evidence have emerged that give clues as to the causes of outliers.

1. The precision of analysis of outliers is usually acceptable and re-analysis of them by both NIR and standard procedures does not change the accuracy. This tells us that the enigma is characteristic of the individual outlier samples themselves and not the entire population, the commodity, the constituent, the calibration, or the laboratory.

2. When different mathematical treatments are used to analyze the same sample sets, different individual samples may appear as outliers, and outliers by one mathematical treatment of the optical data may be analyzed accurately (and cease to be outliers) with another. Changing the manipulation of the optical signal can cause a spectrum to behave, or cease to behave, as an outlier, An interesting observation is that the addition of wavelength points in the development of stepwise MLR calibrations often results in emphasizing the degree to which an outlier is an outlier—the residual error increases sometimes by an order of magnitude by the addition of an extra wavelength point. This has been observed with both "raw" and derivatized log($1/R$) signals.

3. The optimum wavelengths for the analysis of a constituent in different sample sets of the same commodity may differ. Table VIII illustrates the three optimum wavelengths for the analysis of five sets of grain sorghum for protein. The sets all originated from different sources. The series E samples represented the combination of all other samples in one large sample set. Series A, B, C, and D were all from different sources. The 1,660-nm and 1,460–1,470-nm wavelengths selected for analysis of the first two sets dominated the combined set.

A slight increase in one constituent such as moisture or fiber can cause a shift in optimum wavelengths for specific samples. The primary or first wavelength in a calibration series is the most important in terms of analysis of a constituent and is rarely affected by one or two individual samples. The second and subsequent wavelengths are more sample sensitive, and a change in one or more of these can affect the accuracy of testing other samples, especially if the regression coefficients (calibration constants) are large.

4. Calibrations based on samples of the same commodity taken from different sources may lead to different individual samples from a single verification series becoming outliers. Since outliers can also occur at any level of composition and there is no consistent pattern for their appearance on the basis of composition, these two pieces of information indicate that the physical characteristics of the individual sample may be the fundamental cause of the outlier.

5. Gross and fairly consistent errors (in terms of unilateral bias) can arise when samples prepared by one grinding procedure are analyzed by NIR using a calibration based on samples that had been prepared by another procedure. These are not true outliers. The outlier may display a large residual error despite no difference in sample preparation (e.g., whole grains) and no apparent difference in particle characteristics. Outliers of the same type tend to be biased in one direction, although again this is not an infallible rule (there are no infallible rules with outliers!). Furthermore, calibrations based on 40–50 outliers of the same type give "normal" regression coefficients and intercepts. This is because they are no longer outliers. All of the "normal" samples will not necessarily be outliers when this type of calibration is used to analyze them, since the outlier calibration will have been derived from samples that represent maximum variance in spectral characteristics.

6. Repacking sample cells does not materially improve the accuracy of outlier analysis, so that packing density and orientation inside the cell are not the primary reasons for discrepancies. Sample bulk density contributes about 10% of the variance in discrepancies between NIR and standard protein results. This is not sufficient to cause errors of the magnitude displayed by outliers.

Outliers are caused by differences in the diffuse reflectance signal from the sample, relative to the optical signals from all other samples in the population. These differences are not necessarily related to chemical composition in terms of the constituent being measured. Interactions between the signal of the constituent to be measured and that of some other constituent that causes interference at the wavelength at which the measurement is being made can also affect the diffuse reflectance signals.

Differences in optical data between an outlier and a "normal" sample may be less apparent than differences in optical data between two "normal" samples. A difference of only 5% in signal received by the detectors of an instrument that has not been calibrated to receive signals of that order of magnitude could lead to deviations of over 0.6% in protein at a mean level of 12.5%. This is an oversimplification of the situation, but illustrates the type of factor that can cause NIR outliers.

Outliers may be caused by the moisture status of the individual samples. Moisture status includes moisture content and degree of hydration. The degree of hydration of complex molecules such as proteins may cause changes in the spatial orientation of the molecules. This could cause samples containing these molecules to interact with the irradiating light differently from other samples. Since errors have been reported from the testing of immediately freshly ground samples as distinct from samples analyzed several hours after grinding, it is possible that temperature and oxidation status may also be factors. This is an area that has to be researched in view of recent advances in testing of whole grain while it is actually being harvested by a combine harvester.

E. Possible Origin of Outliers

The NIR computer or microprocessor does not recognize outliers. It can describe them to a degree, since an outlier will possess most of the attributes of a more familiar member of the population, but there will be features it will not be able to describe. As a result, depending on the degree of uniqueness, the description, in terms of accurate description or analysis, will not be accurate.

It is likely that physical rather than chemical reasons are the most important factors underlying outliers.

TABLE VIII
Optimum Wavelengths (nm) for Protein Determination
in Five Sets of Sorghum Samples

Wavelength No.	A	B	C	D	E[a]
1	1,664	1,660	1,766	2,144	1,662
2	1,476	1,464	1,470	1,636	1,468
3	2,324	1,744	2,330	2,348	1,940

[a] Made up of all four sets combined. Mathematical treatment was the first derivative of the log($1/R$) value.

A sample surface appears to the eye to be more or less dense. In fact, it consists of molecules and, in terms of the size of light photons, the sample consists largely of space.

Molecules are dynamic. The structural forms deployed in organic texts represent the best estimation of the way atoms fit into a configuration that is in accordance with the chemical behavior of the substance. The physical form, in terms of shape, of the molecules is constantly changing and, the larger and more complex the molecule, the more susceptible is the shape of the molecule, or parts of the molecule, to change. The presence of certain molecular combinations or radicals may affect the shape of complex molecules, such as the starch, cellulose, protein, and even oil molecules present in all materials of biological origin. This, in turn, could affect the frequency and intensity of vibrations and, therefore, of absorbances at particular wavelengths. This could happen in individual samples to the extent that the calibration equation would not be able to recognize the sample with the same degree of assurance as other samples of the same general type.

PLS regression should theoretically afford better protection against outliers, since it does not depend so heavily on absorbance at individual wavelengths. In practice, outliers still occur, even with PLS calibrations.

WHAT TO DO ABOUT OUTLIERS

The objective of calibration is to enable the instrument to become reliable for analysis, not merely to generate an attractive set of statistics! Outliers can actually assist in the development of calibration equations by showing areas of deficiency in the sample sets.

So—what to do?

1. It is important to ensure that the outlier is an outlier, and then to define the outlier, find out what causes it to be different, and then find how likely it is to occur again.

2. If further samples of the same outlier material are likely to occur in the future, assemble more of them, add them to the calibration/validation, and then evaluate the degree to which the new calibration has accommodated the alleged outliers. It may be necessary to develop a calibration specifically for the type of samples that appear as outliers in larger sample sets.

3. If it is unlikely that the problem samples will reappear and the sample sets are sufficiently large to represent everything that will appear, determine the SEP and other statistics with the outlier eliminated. If the statistics for the validation or prediction sets are acceptable, the operator can use it—there is no need to eliminate the outlier, since the calibration is doing a good job on the samples that matter.

4. For practical purposes in long-term analysis, outliers should not be removed, unless it has been shown that the recalcitrant sample definitely does not belong (e.g., a sample of ground grain inadvertently included in a calibration set intended for forage analysis). Removal of a number of outliers, as identified by several software assemblies, may not improve the long-term reliability of the calibration. Outliers are caused by differences in spectral variance from that of the overall population. By removing all of the alleged outliers, variance that is beneficial to the on-going stability of the calibration may be removed.

Ideally, calibration equations should be developed with and without the software-identified outliers. Modern NIR and NIR transmittance instruments enable the operator to obtain results by several calibrations simultaneously. Evaluation of both sets of predicted data over a few weeks will enable the operator to decide which is the best calibration to retain. If seasonal effects (e.g., on materials of biological origin) are anticipated, the two calibrations should be retained until sufficient data from a new season have been accumulated to permit a decision to be made. In industrial settings, the same approach can be taken with new deliveries of raw materials, changes in processing, and other factors.

ACKNOWLEDGMENTS

The helpful suggestions of David Hopkins and Philip Downie are acknowledged with deep gratitude.

LITERATURE CITED

American Association of Cereal Chemists. 2000. Approved Methods of the American Association of Cereal Chemists, 10th ed. Method 55-30. American Association of Cereal Chemists, St. Paul, MN.

Bauwin, G. R., and Ryan, H. L. 1974. Sampling, inspection, and grading of grain. Pages 115-134 in: Storage of Cereal Grains and Their Products, 2nd ed. C. M. Christensen, ed. American Association of Cereal Chemists, St. Paul, MN.

Bryan, J. M., and Elvidge, J. 1977. Mortality of adult grain beetles in sample delivery systems used in terminal grain elevators. Can. Entomol. 109:209-213.

Davies, A. M. C., and Grant, A. 1987. Air-conditioning-generated noise in a near-infrared spectrometer caused by fluctuations in atmospheric water-vapor. Appl. Spectrosc. 41:1248-1250.

Hruschka, W., and Norris, K. H. 1982. Least squares curve-fitting of near-infrared spectra predicts protein and moisture in ground wheat. Appl. Spectrosc. 36:261-265.

Morrey, J. R. 1968. On determining spectral peak positions from composite spectra with a digital computer. Anal. Chem. 40:905-914.

Norris, K. H., and Williams, P. C. 1984. Optimization of mathematical treatments of raw near-infrared signal in the measurement of protein in hard red spring wheat. Cereal Chem. 61:158-165.

Shenk, J. S., Workman, J. K., and Westerhaus, M. O. 1992. Application of NIR spectroscopy to agricultural products. Pages 383-431 in: Handbook of Near-infrared Analysis. D. A Burns and E. W. Ciurczak, eds. Marcel Dekker, New York.

Stermer, R. A., Pomeranz, Y., and McGinty, R. J. 1977. Infrared reflectance spectroscopy for estimation of moisture of whole wheat grain. Cereal Chem. 54:345-351.

Tkachuk, R. 1977. Calculation of the nitrogen to protein conversion factor. Pages 78-82 in: Nutritional Standards and Methods of Evaluation for Food Legume Breeders. J. H. Hulse, K. O. Rachie, and L. W. Billingsley, eds. IDRC-Ts7e. The International Development Research Center, Ottawa, Canada.

Tkachuk, R. 1979. Whole kernel analysis with a computerized double beam Cary 17I spectrophotometer. (Abstr.) Cereal Foods World 24:456.

Williams, P. C. 1975. Application of near-infrared reflectance spectroscopy to analysis of cereal grains and oilseeds. Cereal Chem. 52:561-576.

Williams, P. C., and Antoniszyn, J. 1987. The significance of outliers. Pages 249-264 in: Near Infrared Diffuse Reflectance/Transmittance Spectroscopy. Proc. Int. NIR/NIT Conf. J. Hollo, K. J. Kaffka, and J. L. Gonczy, eds. Akademiai Kiado, Budapest, Hungary.

Williams, P. C., and Sobering, D. C. 1993. Comparison of commercial near-infrared transmittance and reflectance instruments for analysis of whole grains and seeds. J. Near-Infrared Spectrosc. 1:25-32.

Williams, P. C., Norris, K. H., and Zarowsky, W. S. 1982. Influence of temperature on estimation of protein and moisture in wheat by near-infrared reflectance. Cereal Chem. 59:473-477.

Williams, P. C., Norris, K. H., and Sobering, D. C. 1985. Determination of protein and moisture in wheat and barley by near-infrared transmission. J. Agric. Food Chem. 33:239-244.

CHAPTER 10

Method Development and Implementation of Near-Infrared Spectroscopy in Industrial Manufacturing Support Laboratories

PAUL J. BRIMMER
Foss NIRSystems, Inc.
Asia-Pacific Operations
Oatley, NSW
Australia

JEFFREY W. HALL
Foss NIRSystems, Inc.
Silver Spring, MD
U.S.A.

I. INTRODUCTION

The first near-infrared (NIR) laboratory analyses were performed on food and agricultural commodities as described in this book. Most of these analyses were performed on solid samples for constituents such as moisture, protein, and fat, which were either difficult or laborious to analyze by other analytical techniques. As a result, NIR spectroscopy has enjoyed widespread use in the food and agriculture industries due to its ability to perform rapid, nondestructive analyses on samples with little or no sample preparation. Analytical, quality control, and research and development laboratories that support industrial manufacturing processes for fine and specialty chemicals, pharmaceutical formulations, petrochemicals, and solid or liquid resins share similar needs with the food and agriculture industries for rapid, nondestructive testing. However, in contrast to the test and measurements requirements in the food and agricultural industries, industrial manufacturing support laboratories employ NIR spectroscopy to determine a vast number of product constituents and physical properties in a widely varying range of sample matrices. Furthermore, the justification for industrial NIR analysis is often very different than for food and agricultural industries, which affects how NIR methods are used in these industries.

There exists significant opportunities for NIR spectroscopy in both the industrial manufacturing support laboratories and directly in the industrial manufacturing environment. In this chapter, we will discuss practical considerations that are both useful and often necessary when developing and implementing NIR spectroscopic methods of analysis in industrial manufacturing support laboratories. The potential for implementing NIR spectroscopy directly in the industrial manufacturing environment also exists. In the following chapter, the practical aspects for developing and implementing process NIR analyzers and methods directly in industrial manufacturing processes are discussed.

A. Laboratory NIR Measurements

In industrial manufacturing support laboratories, NIR spectroscopy has been used to address a wide number of analytical problems on a wide variety of sample types. Generally, laboratory NIR measurements have major utility in quality control/quality assurance (QC/QA) laboratories for quantitative analysis of some important component in a manufactured material. Qualitative NIR methods have also provided significant benefit for incoming material inspection, performed in the laboratory, in an analyzer room, or near a receiving area. For both qualitative and quantitative NIR methods of analysis, there are certain test and measurement requirements that an NIR technique can address that make it highly desirable over traditional methods of analysis. There are also very specific method development requirements that must be addressed to ensure the success of an NIR method.

Even though it may be possible to perform a specific analysis by NIR spectroscopy, it may not be advantageous to do so from an industrial perspective. Thus, the specific advantages NIR analysis offers to a particular industry must first be understood and secondly be realizable before deciding whether NIR spectroscopy should be used for a particular analysis.

Once the industrial manufacturing requirement has been defined, the requirements for successful method development and implementation of NIR analysis in the industrial laboratory environment can be addressed. In order to successfully implement an NIR method, consideration must be given to the presentation of the sample to the NIR spectrometer, the choice of the optimum spectral manipulation/calibration model for the desired analysis, and calibration maintenance issues. The sample handling requirements for NIR measurements are discussed for the various types of samples that are analyzed in the industrial manufacturing support laboratory. Finally, the requirements for successful calibration development, the validation issues for both quantitative and qualitative applications, and model maintenance for NIR methods of analysis in industrial manufacturing support laboratories are discussed.

B. Industrial Manufacturing Requirements

Any method of analysis used in industry must be addressing some perceived need, or it simply serves as a meaningless activity. Often in the industrial environment, this means the analysis must provide some information that is useful in controlling the manufacturing process or in verifying the identity or quality of a product or material within some predetermined specifications. By combining the speed, accuracy, and precision of the analysis with the nondestructive measurement capabilities of NIR spectroscopy, NIR analysis can be used to improve process control, which improves the overall quality of the product. A good example comes from the polymer industry, in which getting the polymer into a form that can be easily analyzed is often extremely difficult and time consuming due to the nature of polymeric materials. Some polymer analyses can take over 4 hr to perform, due mainly to the time it takes to prepare samples for the analytical measurement. NIR spectroscopy can be used to measure polymers in their natural form and can provide analytical results as accurate as the methods currently in use, but in a time frame on the order of minutes rather than hours. This allows control of industrial processes that is not possible with traditional laboratory methods of analysis.

The most significant test and measurement advantage that NIR spectroscopy offers to the industrial laboratory is the requirement of little or no sample handing for the analysis, so that the analysis can be done reproducibly by a number of different analysts in a quicker time frame than traditional wet chemical, separation, or spectroscopic techniques. Not only does this feature of NIR spectroscopy provide faster analysis and more sample throughput for the laboratory, it also yields improved measurement reproducibility over other analytical methods of analysis. The benefit to industrial manufacturers is the ability to tighten the manufacturing specification limits of the product as the standard deviation of the laboratory tests decreases and, therefore, reduce the overall variability (i.e., improved quality and consistency) of their manufactured products.

Little or no sample handling also allows the laboratory to perform analyses faster, which frees up valuable analyst time. Oftentimes, NIR can perform several analyses on the same sample (e.g., percent hydroxyl, percent esterification, and moisture content in cellulose esters), which reduces the number of different analytical tests that must be performed on a particular sample. Also, the level of expertise needed to conduct routine analyses with NIR instruments is low and requires only a minimum amount of training, so analysts do not need to be highly skilled for successful implementation of this technique.

In this "greener" age, in which the environmental impacts of our actions are getting much closer (and much needed) attention, many analyses are being replaced for environmental or health concerns. Many wet chemical tests use solvents that have been demonstrated to be hazardous to our health and the environment. Chlorofluorocarbons have detrimental effects on stratospheric ozone, allowing harmful UV radiation to threaten the earth's biosphere. Many chlorinated solvents are known or suspected carcinogens. Thus, the disposal costs for most solvents are often very high and can, therefore, affect the profitability of many industries. Many routine analytical tests utilize such solvents/reagents, which the chemical industry would like to replace with more environmentally acceptable methods, without compromising analytical performance. Oftentimes, no suitable solvent is available to replace those currently in use. Tests incorporating the use of harmful solvents are now being replaced by NIR methods of analysis, since NIR techniques can perform the analysis directly on the sample, without extraction, dissolution, or chromatographic separation prior to the analysis. Many industrial manufacturing support laboratories are justifying the implementation of NIR methods of analysis by the cost savings to be realized from the reduction in solvent disposal, improvement in worker safety, and compliance with environmental and regulatory guidelines.

In the industrial manufacturing support laboratory, NIR methods of analysis provide many advantages over traditional methods of analysis that allow manufacturing sites to improve their manufacturing processes, product quality, and quality of the conditions for their workforce and to lower the costs incurred in the analysis of the materials they produce. Furthermore, the rapid, nondestructive nature of NIR spectroscopy affords the opportunity for a manufacturing site to divert very capable technical resources to more challenging manufacturing support issues. For example, industrial manufacturing support laboratories can now offer additional analyses that were not possible with the former workloads and/or laboratory space.

C. Industrial NIR Measurement Requirements

As with any inferential method (i.e., requires a model to be developed and maintained), there are specific requirements that must be satisfied to ensure that satisfactory results are obtained from the derived NIR method. NIR calibrations require typically dozens or more samples to adequately characterize the various chemicals and polymers manufactured in industry, which means it will only find utility for materials produced often and in sufficient quantity. Generally, NIR techniques will not find use in characterizing a newly synthesized chemical or to follow reactions in pilot plant operations in which the composition of the reactions to be monitored is changed frequently.

Laboratory NIR measurements are utilized to support both continuous and batch production processes. In a continuous process, the raw materials are added continuously to the manufacturing process, whereas in batch processes, all the raw materials are initially added together and the manufacturing process continued until the desired product is obtained. NIR spectroscopy is utilized for monitoring continuous processes, which require analysis for process control and to ensure product quality. Changes in the constituents of interest can be modeled by an NIR method, if the NIR model adequately characterizes the variability in this component, as well as other interfering components within the matrix. For continuous processes, it is fairly easy to obtain samples for calibration that not only have variability in the component of interest, but also other components in the matrix necessary for the development of robust NIR models. It is also possible to obtain variation in the physical properties of the sample (e.g., particle size, turbidity) and in the raw materials (e.g., different suppliers) that go into the process. Seasonal variations (e.g., temperature, humidity) can sometimes affect NIR models and, therefore, these variations should also be present in the calibration sample set.

NIR spectroscopy can be used to monitor batch production processes for materials that are made on a frequent basis. If only several batches are manufactured in a year, it is difficult to obtain the necessary variability in the material to develop a robust NIR model and can, therefore, take a great deal of time before the samples necessary to develop a suitable NIR model are collected. If only a few batches are represented in the calibration population, the variability in the samples may not adequately characterize batch-to-batch variability. This may require the NIR model to be frequently adjusted. In this case, the analytical performance of the NIR method is normally verified at the beginning of each new batch, by comparing the NIR results with the reference analysis. The number of analyses conducted with the primary analysis to maintain the NIR calibration can approach a significant proportion of analyses that would be run in the absence of an NIR method, negating the benefits of the NIR measurement. For infrequently run batch processes, therefore, the maintenance requirements of the calibration equation far outweigh the benefits the NIR analysis affords. Thus, implementation of a laboratory NIR method of analysis may not be economically justifiable for infrequently manufactured materials.

If the same material is manufactured frequently in a batch production process, suitable samples are readily available for the development of robust NIR models that can be used to monitor and control batch production processes. Usually, the measurement requirement is to detect the "end point" of the batch process and to verify that the product is within acceptable quality limits. Since NIR analyses are performed quickly, with little or no sample preparation, the end point of the batch process can be detected more quickly than with more time-consuming traditional methods of analysis. As a result, batch processes can be finished more quickly and the next batch started much sooner, resulting in more batches completed per unit time, with no capital outlay for additional manufacturing equipment.

Some industrial processes produce materials by what is often called "continuous-batch" processes, which are continuous processes that can be used to manufacture a number of different, but similar, products. The manufacturing campaign for a given product will continue until enough material is made to meet the demand for that material, and perhaps not be manufactured again for several months. The process is then adjusted (transitioned) to produce a different material. This new product is then manufactured until the desired quantity of that material has been produced. For continuous-batch processes, it is important to monitor transitions between products to minimize the production of out-of-specification product. For materials that are made over fairly long continuous campaigns or made frequently, NIR methods of analysis can be successfully implemented. For materials that are produced on an infrequent basis, NIR spectroscopy will find little utility, considering the time necessary for the calibration and maintenance for each material.

II. SAMPLING REQUIREMENTS

Presentation of the sample to the NIR instrument is usually the largest source of error in an NIR measurement and, therefore, an area in which the largest measurement benefit can be gained. Improving the reproducibility of the sample presentation and dealing with other parameters that can affect the measured NIR spectrum is of paramount importance in developing a reliable NIR method. In this section, methods of reproducibly presenting the sample to the NIR instrument and how to reduce the deleterious effects of other physical parameters (e.g., temperature) for a variety of types of samples commonly seen in industrial laboratories are discussed.

A. Liquids

For liquid samples, NIR analyses in the laboratory are usually performed in transmission. Although transmission measurements are generally easier to perform and more repeatable than reflectance measurements, there are several key issues to consider when developing an NIR method of analysis for liquids. In this section, we will address the sampling requirements in terms of sample pathlength, temperature effects, and the turbidity of the liquid sample.

PATHLENGTH

The first thing to consider when collecting an NIR spectrum of a liquid is to select the pathlength that optimizes the absorbance for the analyte of interest. By collecting the NIR spectra of the pure components of a particular product or material, we can identify regions in the spectrum where the absorption features of the component of interest are strong and the spectral characteristics of the other components (interferences) are weak or nonexistent. Rarely in NIR spectroscopy can we find a unique and baseline-separated absorption band for the component of interest, but rather regions where the influence of the component of interest is maximized and the interferences from the other components are at a minimum. There may be several regions that appear to be good candidates, which could result in different optimal pathlengths.

The most appropriate pathlength for a particular analysis can be determined by considering the analyte concentration, sample characteristics, and measurement requirement. First, at longer wavelengths, the absorption bands are generally stronger, which will provide greater sensitivity, and are important if the constituent of interest is present at low (less than 1%) concentration levels. However, the stronger absorptions in this spectral region will limit the measurement to short pathlengths (0.5–2 mm). Shorter wavelengths will have weaker absorption bands, which will allow for longer pathlengths (4–20 mm) and are generally used for the analysis of fairly high analyte concentrations (i.e., percentage levels).

Finally, the analyte selectivity required for a given analysis must be considered. At longer wavelengths, the bands are sharper and more unique than those at the shorter wavelengths, so the analyte selectivity can be assumed to be better at longer wavelengths. Thus, if the NIR analysis requires differentiation between similar materials, it might be preferential to work in regions of the NIR spectrum where the bands are better resolved and more unique (i.e., longer wavelength region). By contrast, if the NIR method will measure a combination of components within the matrix, such as total aromatics (rather than individual aromatic compounds) or a physical property, which is related to more than one chemical moiety (i.e., octane number), the more overlapped bands in the shorter wavelength region often provide a better correlation between the NIR spectral features and the sample property of interest.

For sample handling requirements (i.e., ease of use, sample cell cleaning), it is preferential (especially for viscous materials) to use the longest pathlength possible. Another practical consideration concerns the use of fiber optics. Low hydroxyl silica fibers, which are currently the only economically feasible fiber optics that can be used in the NIR region, begin to attenuate NIR energy above 2,200 nm. Therefore, if NIR measurements are to be made with optical fibers, measurement wavelengths should be limited to those regions below 2,200 nm, unless the length of fiber is very short (<25 cm).

By considering the pathlength issues discussed, a suitable spectral region can be identified. At this point, an NIR spectrum of a "typical" real sample should be scanned to determine the optimal measurement pathlength. The optimal pathlength will provide an absorbance level that is within the limit of the instrument's linear range and sufficient to optimize the sensitivity of the measurement. The linear range of absorbance for a spectrometer is determined by the amount of stray light, as shown in Figure 1. As stray light increases, the usable upper limit of absorbance decreases and the range of linearity is decreased. Generally, for NIR instrumentation currently available, the linear range will be between about 0 to 1.5 absorbance units (AU). Thus, the absorbance of the samples (in the spectral regions used for the calibration) should be kept below the absorbance levels in which nonlinear behavior is expected. Also, from a practical aspect, if samples have the possibility of being turbid, one should ensure that the most turbid samples will have absorbance levels in the spectral region of interest below approximately 1.5 AU. When the absorbance level for the region of interest is optimized, some parts of the spectrum may be nonlinear, and some parts may be totally in saturation. This is not a concern as long as those regions are not being used in the calibration model for this or any other component to be measured.

PHYSICAL EFFECTS

There are some physical parameters of the liquid sample that may affect the performance of an NIR analysis. Two of the most common sample-related effects are variations in temperature and turbidity. Temperature changes will affect not only the density of liquids, but also the hydrogen bonding of polar functional groups,

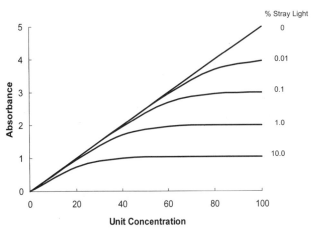

Fig. 1. Effect of stray light upon the linearity of response for a spectrophotometric measurement.

which will cause the position of NIR absorption bands to change. The various means of dealing with these temperature-induced changes that achieve optimum repeatability of the NIR analysis, are described below. Turbidity changes are usually compensated for by use of various spectral manipulation techniques (discussed below) that account for the resultant spectral scattering variations induced by the turbid media.

Temperature Effects. Temperature changes affect the density of a liquid, which has the spectral effect of changing the baseline of the NIR spectrum. In addition, there may be changes in the position of the absorption bands of polar functional groups related to temperature fluctuations. Furthermore, for polar functional groups that participate in hydrogen bonding, the vibrational energy for these polar functional groups will be decreased (relative to the non-hydrogen-bound band position). As temperature increases, the amount of hydrogen bonding decreases, increasing the amount of energy associated with the vibrational bands of this functional group. This increased energy is detected in the absorbance spectrum as a shift in the absorbance maxima for these functional groups to higher energy (lower wavelength, higher wavenumber). Normally, under laboratory conditions, temperature changes are on the order of approximately 10°C if there is no temperature control of the sample, which can induce a shift in polar group absorption bands by only a few nanometers. This shift in absorption band position can still significantly affect the analytical performance of NIR models, especially for industrial applications in which the product attributes have tight manufacturing specification limits.

By preventing this deleterious effect upon NIR absorption bands from occurring (by controlling temperature) the best analytical performance and long-term reproducibility (Hansen et al, 2000) can be achieved. For measurements in which this is not possible, suitable compensatory schemes can be effectively utilized. However, the best analytical option is to not let the offending variation occur, since a correction scheme will only "compensate" for the observed behavior and never eliminate the resultant spectral effect.

Fortunately, for laboratory measurements, control of the sample temperature is usually attainable. Normally, the temperature is held as near to ambient temperature as possible, provided the melting point of the sample is below this temperature. For measurement of temperature-sensitive materials, NIR instrumentation must be able to adequately control temperature, usually to within 0.1°C, and have efficient thermal contact with the sample for quick and even temperature equilibration.

Lower sample temperatures reduce the loss of volatile components in the sample matrix; reduce sample degradation (e.g., oxidation), and reduce temperature equilibration time.

Sometimes, the sample cannot be held within the sample compartment of the NIR spectrometer for the analysis, and the temperature may not be able to be accurately controlled. For example, the sample may have a higher melting point than the instrument components can easily tolerate or the sample could be extremely viscous, such that placing it into a sample cell (or more importantly, getting it out in the cleaning step) makes a remote measurement more preferable. One laboratory configuration that is useful for measuring these types of samples is to interface fiber optics to a remote, temperature-controlled sample holder, which oftentimes allow the use of "disposable" sample cells and can accommodate very high temperatures.

For test and measurement situations in which the sample temperature cannot be controlled, chemometric methods can be utilized to compensate for, but not fully negate, the deleterious effects of temperature variation upon the NIR spectrum. First-order effects of temperature affect the spectral baseline, which can be compensated for in nonscanning instruments (e.g., filter photometers) by subtraction of a wavelength or by the use of spectral derivatives, which is the usual procedure when using full-spectrum NIR instrumentation. The advantages and disadvantages of baseline correction routines will be discussed later, but suffice it to say at this point that this first-order effect must be compensated in order for an NIR method to provide useful analytical information. Another spectral effect due to temperature is the shift in the absorption band position for polar functional groups as temperature is varied. This is usually handled through the use of full-spectrum modeling techniques such as partial least squares (PLS), which can characterize the spectral changes attributable to temperature. By doing so, the temperature effect upon the spectrum can be characterized and allows the modeling of the analytical component of interest to be relatively independent of temperature.

Another method that has been discussed to compensate for temperature-related spectral effects is the addition of the measured temperature of the sample to the calibration model. This addition to "isothermal" NIR calibrations is an attempt to correct for the spectral changes caused by the temperature variation. The disadvantage of this approach is that the correction in a calibration equation is typically performed as a linear correction and will not adequately characterize the temperature effect upon the NIR spectrum, since temperature-induced changes in band position do not alter the absorbance at a given wavelength in a linear fashion. Overall, incorporation of the measured temperature as a correction factor into NIR models has not provided adequate results in the industrial setting, and it is seldom used.

In summary, if temperature can be controlled for an analytical measurement and can be performed in a practical manner for the sample of interest, then using temperature control provides the most accurate analytical results. When this option is not practical, temperature must be incorporated in the calibration model. This can be done by measuring the temperature and applying it as a correction factor to an isothermal calibration equation, but this approach has had limited success. A more effective approach is to include the temperature variation in the calibration sample set and use multivariate calibration modeling algorithms to characterize the spectral effects related to temperature. Once calibrated in this manner, the effects of temperature are "effectively ignored" by the model, and analytical results that are temperature insensitive (within the modeled temperature range) can be obtained.

Turbidity. Some liquid samples can vary in the amount of light scattering, or turbidity, from sample to sample. The suspended material that causes visible light to be scattered (resulting in our eyes detecting the lack of clarity) also scatters NIR energy. As NIR energy is scattered, less energy reaches the detector, and a change in the spectral baseline is observed (first-order effect). As illustrated in Figure 2, as the level of solids within a slurry matrix is increased, the measured absorbance baseline (log[1/T]) increases. In addition to changes in the baseline, scattering variations can also change the average pathlength a photon travels through the sample (second-order effect). This can be seen in Figure 3, which shows the second-derivative spectra for a slurry matrix in which

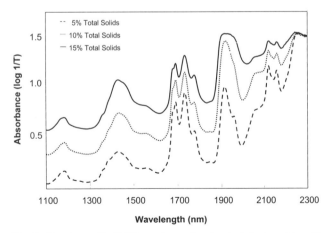

Fig. 2. Absorbance (log[1/T]) spectrum of a chemical slurry matrix with varying levels of total solids. Baseline offset is related to changes in light scattering. The change in the shape of the water band centered at 1,904 nm is due to signal attenuation.

the level of solids varies. As the scattering level increases, the effective pathlength traversed through the sample by the average photon becomes longer. This results in increases in the absorption band intensity for all absorption bands in the spectrum, as seen in Figure 3.

There are several ways of compensating for the effects of light scattering upon the NIR spectrum. First, if there is some way to eliminate the haze from occurring in the samples (e.g., filtering suspended material, elevating the temperature to keep some material from solidifying), the best accuracy will be obtained. Mathematical pretreatment of the spectra, such as multiplicative scatter correction, can also be employed to reduce the spectral effects related to changes in sample turbidity. The spectrum can also be "normalized" by dividing the NIR spectrum by a specific wavelength, usually that of a solvent band (the concentration of which is nearly constant).

A full-spectrum modeling method, such as PLS or principal component regression (PCR), can be used to characterize the effective pathlength changes within the calibration model. However, if simpler mathematical pretreatment of the spectra can adequately handle this type of behavior, it is recommended to preprocess the spectral data with one of the simpler correction schemes, such as a derivative or normalization, prior to performing the calibration. This allows these more powerful chemometric modeling procedures to characterize the spectral variations due to other parameters (sample composition, temperature, etc.) more effectively.

Finally, it should be noted that any mathematical correction is only compensation for, and never totally negates, the effects upon the NIR spectrum related to variations in light scattering. Elimination of the variations in light scattering, if possible, will provide better accuracy and precision over scatter-correcting spectral techniques, but in most cases, this is not possible, and mathematical corrections are a common tool to improve the reproducibility of the NIR measurement.

B. Solids

In contrast to transmission measurements through liquids, reflectance measurements of solids are a more significant challenge in terms of reproducible sample presentation. Rather than simply passing light through a sample, reflectance measurements rely on a combination of reflection, refraction, and diffraction in and around solid samples to produce the NIR spectrum. The challenge with solids is to make the variation in sample presentation low enough to provide an acceptable standard deviation of repeated measurements so that (i) the analysis is accurate enough to ensure the final products are within some defined product specification limits, and (ii) the analysis is precise enough for process operators to control the process within acceptable control limits. Since sample presentation is oftentimes the limiting factor in NIR analyses of solids, nontraditional methods of sample presentation are normally used.

Some of the early uses of diffuse reflectance NIR spectroscopy for analyzing whole grains and forages suggested that perhaps NIR technology could be used to analyze some of the more difficult industrial samples, such as polymer pellets, tablets, granules, and fibrous or woven materials. Furthermore, the experience gained in applying NIR spectroscopy in the agricultural industry clearly demonstrated that it was possible to measure the NIR spectrum of nonpowdered solids, but it required different methods of sample presentation for different types of solids to improve the reproducibility in presenting a sample to the NIR instrument. In order to make NIR technology amenable to the analyses required in industrial laboratories, sampling accessories have been developed to reduce the variation in sample presentation. In this section, the methods used to improve the reproducibility of sample presentation are described for a variety of solids.

POWDERS

Powdered samples are one of the easiest and most reproducible solid matrices to measure by NIR spectroscopy. Powdered samples with a small particle size (at least when viewed from the scale of the size of the NIR beam) and narrow particle size distribution that can be uniformly compacted will yield the most reproducible NIR spectra. There are many powdered samples that are produced in the industrial manufacturing environment that meet these NIR measurement criteria. These samples are introduced to the NIR instrument using circular sample cups ("powder cells") similar to those that have been used for powdered samples in agricultural applications.

Powdered sample cells can be rotated while the NIR spectrum is collected to present a greater area of sample to the spectrometer, presenting more of an "average" of the energy being reflected from the sample. This type of approach becomes more necessary when the sample morphology is more of a granule than a powder. If one tries to measure a solid sample matrix that consists of a blend of a fine powder and granules (e.g., powdered detergents) rather than a finely divided powder like flour, a combination of spinning the sample and using some form of spectral baseline compensation is usually necessary to achieve acceptable NIR results. If the sample is inhomogeneous on a scale smaller than the dimensions of the NIR beam, rotation of the sample can also reduce the observed spectral variability caused by this sample inhomogeneity.

If powdered samples have a large particle size distribution, or if they have an inhomogeneous distribution of their components within the sample matrix on a scale larger than the NIR beam, sample rotation may not adequately compensate for the variations in the NIR spectrum. For example, if the analyte of interest is present only as fairly large particles and is dispersed in a matrix of much smaller particles, there will be significant variations in the number of particles at the surface of the sample (which imparts the greatest effect on the spectrum in reflectance measurements). This causes the NIR signal and the analytical results to vary to a great degree as the same sample is reloaded into a sample cell and rescanned. Even if the sample is rotated, in many cases, this shows no improvement in the reproducibility of the NIR results. Unfortunately, for these solid sample matrices, the most appropriate solution is to grind a large volume of the sample to reduce the particle size distribution and make the sample itself more homogeneous, which improves the reproducibility of the NIR signal. However, these requirements for sample preparation will not only reduce the benefits of NIR spectroscopy for measuring these types of samples, but can also affect the parameters being measured.

PELLETS, FLAKES, AND TABLETS

One of the more difficult types of samples encountered in industrial laboratories are pelletized flakes, or tableted materials. Typical pellet or flake samples are polymers, which are difficult to analyze by traditional wet chemical or spectroscopic techniques

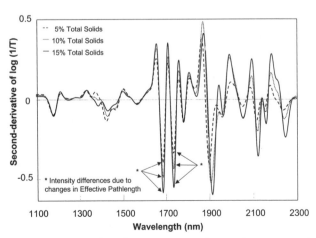

Fig. 3. Second derivative of log(1/T) spectra of a chemical slurry matrix with varying levels of total solids. Changes in band intensity related to changes in effective pathlength related to light-scattering differences (multiplicative scatter effect) are identified (*). The change in the position of the water band centered at 1,904 nm is due to signal attenuation.

due to the difficulty in dissolving, grinding, or melting these materials in preparation for analysis. Typical tableted samples are pharmaceutical solid dosage forms that are generally tested for various parameters such as coating thickness or active ingredient levels (i.e., content uniformity). Most of the traditional analytical methods for these types of samples are very time consuming, and often only a few pellets, flakes, or tablets are measured that are assumed to be representative of the entire batch production. The use of NIR spectroscopy is advantageous for analyzing these types of samples, since it is faster, nondestructive, requires no sample preparation, and measures a larger number of tablets, flakes, or pellets, which is a better representation of the entire batch.

The major difficulty in measuring the NIR reflectance spectrum of flakes, pellets, or tablets is that sample presentation is very nonuniform as compared with homogeneous powdered samples. This sampling nonuniformity produces large variations in the spectral baseline and can cause some effects similar to the multiplicative scattering variations described earlier for highly scattering liquid samples. When introduced into a sample cell, individual flakes, pellets, or tablets will be oriented in many different positions with respect to the NIR beam, and there will be many "gaps" between individual particles, which can let the NIR energy pass through the sample or sample cell entirely and not interrogate the actual solid sample. With this type of variation, it seems, at first glance, almost impossible to obtain useful quantitative information from an NIR spectrum collected on solids of this form. Furthermore, grinding is usually not possible, since it both is too difficult, as in the case of polymers, and will often affect the parameters being measured. Therefore, a method of dealing with these types of samples in their natural state is necessary for the NIR analysis to be of any utility.

There are several approaches that can improve the reproducibility of the collected NIR spectra of powder, flakes, and tablets. For example, a very large area of sample can be presented to the spectrometer and effectively average scans over an area of 60 cm² or more of the sample. This can be achieved in a number of ways, from moving a large area sample cell past the NIR beam as the scans are collected or by sweeping the sample across a window and averaging a number of scans. Sometimes multiple reloads of the same sample are averaged in order to reduce the variations in the packing density of samples in the sample cells. In general, the tighter the samples are packed into a sample cell, the better the overall reproducibility.

Long-term changes in the sample shape or size can also affect the NIR results for these types of samples. For example, in polymer pellets, the pellet size can change as the cutter blades in the extruders become dull or are replaced or as the backpressure in the extruder changes. Sometimes air bubbles can be introduced into the pellets. These types of changes will introduce a multiplicative scattering variation, which will cause the NIR results to vary as these physical parameters vary unless they are taken into account during the modeling process. Spectral compensation approaches discussed later in this chapter can be employed to compensate for these types of physical changes. For pellets and flakes, such scatter correction methods are good insurance so that, when production changes occur that affect the light-scattering properties of the samples, the NIR results are not adversely affected.

For certain products, the analysis of interest is not the average composition of a batch or material, but rather the uniformity of the analyte within the matrix. For this measurement, a number of solid dosage forms or pellets are analyzed individually. Uniformity measurements are normally performed in the pharmaceutical industry, where the distribution of the active ingredient in individual tablets is of great importance for both manufacturing and regulatory compliance requirements. Single tablets are generally much smaller than typical NIR sample cells, and normal presentation schemes are inadequate to provide results that are of sufficient accuracy to be relevant. One approach developed to improve this situation was the design of a single tablet cell (Drennen and Lodder, 1990), which used a concave mirrored surface situated below the tablet that was suspended approximately 1 cm above the mirrored surface. This allowed light passing around the tablet to be refocused on the bottom of the tablet and directed the diffusely reflected light toward the instrument detectors. This provided a better average measure of the entire tablet, since it allowed the NIR beam to interrogate the top and bottom of the tablet and reduced the scattering effects from intagliations and other markings common on pharmaceutical dosage forms.

A significant concern in the analysis of pharmaceutical solid dosage forms is whether the NIR energy is interrogating the entire dosage form, in case there is some variation in concentration from the surface to the core of the tablet. To address this concern, transmission, rather than reflectance, is employed to interrogate the entire bulk of the solid dosage form (Gottfries et al, 1996). A practical comparison of the merits of both measurement modes for analyzing solid dosage forms is provided in Figure 4. Making measurements by transmission through tablets rather than by reflectance allows the tablet to be thoroughly interrogated, rather than just analyzing the surface of the tablet. This alone makes transmission, rather than reflectance, more favored by the government agencies that regulate the pharmaceutical industry. Transmission measurements are more insensitive to intagliations or other markings on the surface of the tablet, so positioning has less effect on the analytical results. With properly designed sample holders and optics, lower detection limits can be routinely achieved by transmission rather than by reflectance. For this reason, reflectance measurements of individual tablets are generally only used to measure dosage levels, where content uniformity (which requires much greater accuracy and repeatability) is performed by transmission.

In order to transmit NIR energy through a solid dosage form, the sample illumination must be maximized, stray light must be minimized, and the signal detection must be optimized since the resultant NIR signal levels generally begin with a baseline above 3 AU. Therefore, only the 700–1,500-nm region of the NIR spectrum is available for use. Since most pharmaceutical active ingredients possess aromatic moieties, which have absorbance bands in the 1,100–1,300-nm region, and most excipients consist of aliphatic moieties, useful quantitative and qualitative measurements can be obtained within the 700–1,500-nm region.

FIBROUS MATERIALS

The major difficulty encountered with sample presentation for fibrous materials is that there can be a large variation in the NIR spectra as the orientation of fibers change. Depending on whether the fibers are oriented in one direction or randomly oriented, the results obtained by the NIR analysis will vary significantly. For nonwoven textile fibers, keeping the fibers oriented in a single direction and scanning in a large area sample cell can reduce the

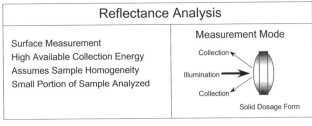

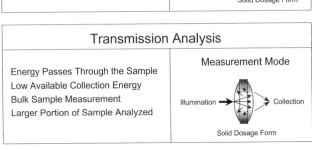

Fig. 4. Comparison of near-infrared transmission and reflectance modes for analyzing pharmaceutical solid dosage forms.

observed spectral variation considerably. The NIR results can also be improved if a large volume of sample is used, so that the sample is tightly packed into a sample cell and scans over a large area are coaveraged to reduce the effects of imperfect orientation of the individual fibers.

For randomized fibers (e.g., wood pulp), samples are usually placed into a circular sample cell and packed as tight and uniform as possible. It is usually a good idea to rotate the sample to average any sample inhomogeneities, since these samples can be somewhat nonuniform. It is preferable that samples such as wood pulp be dry, since, when wet, they tend to clump, which causes them to be extremely inhomogeneous and difficult to place in a cell.

SHEETS

Samples such paper, woven fabric, and metal foils are examples of this type of sample. Paper and fabric samples tend to have fewer problems with sample orientation, and samples can be cut for infrared analysis in either the roll direction or machine direction. These samples are typically placed into a large area sample cell in order to average scans over a fairly large area of sample, since the surface features may be nonuniform. Since these samples are thin, they generally will not fill a cell's entire thickness, so oftentimes a backing is placed behind the sample to press it flat against the window of the cell, making the surface interrogated by the NIR beam uniformly flat. The backing is usually made of a fairly good diffuse reflector (such as a piece of aluminum, Teflon, or white ceramic), since some of the NIR beam may penetrate through these thin samples. The second pass of the NIR beam through the sample keeps the variation in the NIR signal fairly constant from sample to sample and improves the sensitivity of the NIR analysis.

Metal sheets can be analyzed for the coatings that are applied to them. These types of samples are generally placed in sample cells in a similar manner to other types of sheets (described above). However, one difference with metals is that they do tend to have marks on them (machine or roll marks) that scatter light differently depending on their orientation. Therefore, the orientation of samples should be kept constant for these types of samples. An additional caution when measuring metal sheets is that as these marks change with time or, for different rolling machines, they can drastically affect the NIR spectra, especially if the coated material being investigated has a thickness on the order of the wavelength of energy used to investigate the sample (1.1–2.5 microns). For extremely thin coatings, interference fringes and the effects of anomalous dispersion upon the NIR spectrum can drastically affect the NIR results. In such cases, the spectral variations due to variations in machine marks can obscure the features for the coating and make it impossible to obtain an NIR measurement.

C. Slurries

A final type of sample that can be analyzed by NIR spectroscopy is either a mixture of solids suspended in a liquid or an emulsion of two immiscible liquids. Examples of "slurried" samples are opaque salves, fermentation broths, and latex rubber. These types of samples are generally strongly light scattering, and they are generally measured in reflectance. (Samples, which are only slightly light scattering, can usually be treated as liquids.) Slurries can vary in viscosity and may pour like water or be thicker than molasses. Provided the materials can be easily poured and there aren't significant losses of volatile compounds from the sample upon exposure to air, it is generally easy to handle these samples in standard cuvettes.

Strongly scattering slurries are generally very uniform (although some media may experience sedimentation), so averaging over a large area of sample is seldom necessary. The cuvettes are simply placed as close to the reflectance detectors as possible for the reflectance measurement (as the distance from the detectors increases, light collection decreases and, therefore, the sensitivity decreases). Generally, very large optical pathlength cuvettes (20–30 mm) are used to improve the ease of sample loading and unloading and to ensure that the sample is of "infinite" thickness for the reflectance

measurement. Infinite thickness is defined as the thickness at which any increase in thickness causes no change in the NIR spectrum (Kortum, 1969).

The amount of energy scattered from samples such as opaque salves is related to the amount of light-scattering material (e.g., TiO_2) present in the matrix. Fortunately, the level of this material added to the matrix is usually tightly controlled and, therefore, compensation for scattering differences within these matrices is normally not required. For fermentation broths, however, the amount of light scattering is affected by the amount of biomass, which can change as the fermentation proceeds. For the measurement of bioprocess samples, either multiplicative scatter correction or normalization in the calibration model needs to be performed to offset the significant changes in "effective pathlength" observed for these types of samples (Brimmer and Hall, 1993).

III. QUANTITATIVE ANALYSIS

Having determined a method of reproducibly presenting the sample to the spectrometer, a robust quantitative NIR method can be developed. Development of an NIR method will consist of collecting samples that possess suitable matrix variability that can be used as calibration standards, obtaining accurate reference analytical data and NIR spectra for each of the samples, and deriving an appropriate quantitative algorithm. To ensure that the derived model is representative of the analyte under investigation, the model will be validated using another set of representative samples not used in the calibration process. To ensure that the performance of the routinely used method is maintained, model maintenance, which involves periodic verification of the instrument performance as well as verification that the NIR model is still valid for the material currently being manufactured, must be performed. This section will cover the practical aspects necessary for robust quantitative NIR method development.

A. Calibration Development

To develop a reliable NIR calibration model, the calibration standards must represent the variations to be expected in the samples that will be analyzed in routine operation. This includes having a representative distribution of the analytical parameter of interest, over the analyte range to be measured (exceeding to some extent the upper and lower control limits for the product), as well as a variation for all other chemical and physical parameters that could vary in the manufacture and handling of these materials. This calibration sample distribution is often described as a "boxcar" calibration sample distribution (Fig. 5). In general, a boxcar calibration sample distribution will contain a relatively even distribution of samples over the expected analyte range and a consistent number of samples at each analyte concentration level. Since the

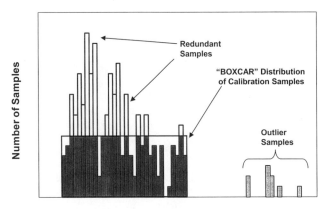

Fig. 5. "Boxcar" distribution for a calibration sample population. Additional samples with similar constituent concentrations are considered redundant, and those separated from the "normal" samples are considered outliers.

variation in matrix properties observed for industrial manufacturing samples tends to be much less than that for natural products analyzed in the agricultural and food industries, the number of calibration samples required to adequately characterize an industrial process is typically on the order of dozens, rather than hundreds. Furthermore, since calibration samples express the variation observed for actual process samples, the samples contained within the calibration population must be "real" production samples.

In the industrial manufacturing environment, collecting suitable calibration samples can take a great deal of time, especially since samples on the high and low end of the analyte range are "out-of-specification" materials that are produced rarely and production operators are usually reluctant to create these samples simply to aid in calibrating an instrument. Thus, to collect samples that contain variation in all components, including that of the constituent of interest, other process and environmental variables will require that the samples be collected from the manufacturing process over an extended period of time.

For this reason, retained samples are sometimes used to develop a calibration model, since it is quicker to sort through retained samples to collect a representative calibration set rather than waiting for the optimum calibration samples to be produced. Even though retained samples can contain the range of concentrations of the constituent of interest necessary to derive a calibration model, the impact of several negative attributes of retained samples must be considered when using retained samples to derive an NIR model. First, the concentration of the constituent of interest may have changed over time, necessitating a new laboratory analysis. This concern is easily addressed by performing a reference analysis on the samples. More difficult to accommodate are changes in the properties and composition of the samples (oxidation, changes in sample morphology, moisture content), which may differ from newly manufactured samples. These matrix changes can sometimes be addressed by combining both retained and new samples in a single calibration set. However, it can be difficult to incorporate both old and new samples easily into a single calibration equation, especially if there are fundamental differences (new process, different starting materials) between retained and new samples.

Another concern when using retained samples is that these samples may have equilibrated to the current environment (e.g., the samples will all have the same moisture levels), so that no variability in these components will be expressed in these samples. Sometimes these parameters can be artificially varied (e.g., moisture levels, sample temperature) in an effort to incorporate these variables into the calibration equation. By this approach, retained samples can sometimes be used to represent sufficient variation in the sample variables to obtain adequate calibration models (especially for simpler analyses). Sometimes, due to the nature of the sample and the desired analysis, retained samples simply cannot be used, and one must exercise patience and remain confident that over some period of time an acceptable range of sample variation will occur. To assist in these situations, it is often useful is to develop a "starter" calibration using retained samples that express as much variation as is initially possible, which can be continually improved over time as the sample variability and range of analyte concentration is expressed in the samples collected from the process.

Another method of obtaining a broader variation in matrix properties for a particular calibration set is to include several individual "products" into a common calibration set. These "products" will generally have the same molecular structure and differ only in their specification (analyte) range. A practical example is polyols, which vary in their hydroxyl value (a measure of molecular weight). By adding several different polyols (with the same molecular backbone) into one calibration set, wider sample variation can be obtained more quickly than by developing a calibration for each "product" separately, usually without serious degradation in the analytical performance of the NIR method.

Once the calibration standards have been collected, accurate analytical data for the desired parameter must be generated, as well as accurate NIR spectra. The NIR analysis and the reference analysis must correspond to the same sample, which can sometimes be a bit difficult if conditions of the sample change rapidly (e.g., volatile solvents, moisture) or if the samples are inhomogeneous. Ideally, the reference analysis should be performed and the NIR spectrum collected as close in time as possible to ensure that the sample being assayed by the reference analysis is representative of the material that is scanned by the NIR instrument. Thus, the sample scanned by the NIR instrument should be analyzed by the reference method, rather than measuring different aliquots of the same sample.

If there is some concern about the homogeneity of the sample being scanned on the NIR instrument, several aliquots of the sample should be measured by the reference analysis and the average used for calibration development. This is sometimes the case with analyzing polymer pellets, in which typically only two to three pellets are used for the reference analysis, but the NIR beam interrogates hundreds. The sampling error of the NIR measurement of only a few pellets is too large to be practical. If the pellets are inhomogeneous in terms of composition, the best option is to average several reference analyses and use this average value for comparison with the NIR spectral data.

Sometimes, a less labor-intensive solution for dealing with the problem of inhomogeneous samples is to alter the material scanned by the NIR instrument so that it better represents the sample analyzed by the reference method. For example, if the NIR analysis involves averaging scans over a large area of a sheet of material, and the reference analysis requires only a sample covering a small fraction of the area scanned by the NIR instrument, it is preferable to measure the NIR spectra of only a small area of material for calibration model development and perform the reference analysis on this same sample. Once the sampling differences are resolved and the calibration is developed, the NIR instrument can be used to scan over larger sample areas and to produce a more representative measurement.

B. Spectral Manipulation

Not only is it important to have accurate reference analytical data for the calibration standards, but the collected NIR spectra must also be relatively "noise free." For many types of samples, it is the presentation of the samples that is nonreproducible and contributes to the most significant variation in the collected NIR spectrum and must, therefore, be corrected. Improvements in sample presentation (discussed earlier) improve the overall sample reproducibility, but oftentimes the residual amount of spectral variability must be compensated for. Variations in how the light interacts with a sample will impart changes in the spectral baseline (first-order effects) and can also produce changes in the intensity of the absorption bands (second-order or multiplicative effects). In addition to the optimum sample presentation methods discussed earlier, spectral baseline variations and changes in the effective pathlength/penetration depth of the NIR energy through/into the sample must be compensated for to obtain accurate and reliable NIR results.

Variations in light scattering for reflectance and transmission measurements can be compensated for by calculating the second derivative of the NIR absorbance data that is collected by full-scanning instruments, or by baseline subtraction if discrete wavelength instrumentation is used. Derivative transformations provide a superior compensation for the variation in the baseline over baseline subtraction procedures, because the NIR spectra usually do not have areas that are affected only by scattering. It has been shown that a second-derivative spectral pretreatment can provide adequate baseline correction, which can easily be performed as a preprocessing step before calibration development or routine analysis (O'Haver and Begley, 1981).

As the scattering of the NIR energy through the sample increases, the "effective" pathlength that an average photon travels through the sample (for transmission) or the penetration depth (for reflectance) can vary, with a concomitant variation in the intensity of all absorption bands in the spectrum (assuming all absorption bands are in the linear range of the spectrometer), as shown in Figures 2 and 3. This observed change in NIR absorption band intensity that changes as the light-scattering properties of the sample changes (a second-order effect) is often referred to as the multiplicative scatter effect (Martens et al, 1983; Geladi et al, 1985). This effect is essentially a change in the amount of sample interrogated by the NIR energy and must be compensated for in order to obtain accurate NIR results.

Many multiplicative scatter correction routines are available to compensate for both first- and second-order scattering effects (Martens et al, 1983; Geladi et al, 1985; Barnes et al, 1989). All techniques require some wavelength range selection to perform a suitable scatter correction. Once selected, these same wavelength ranges are used on subsequent routine samples as a pretreatment of the spectral data. It is beyond the scope of this chapter to discuss the merits of each method of scatter correction, which are discussed in more detail in Chapter 4 of this book.

Another approach that is used to compensate for "effective" pathlength differences is to divide the entire NIR spectrum by a wavelength in which the spectral variation is related only to changes in effective pathlength. This approach has been described for reflectance measurements of solids (Norris, 1983; Norris and Williams, 1984) and successfully applied to reflectance measurements performed on highly light-scattering liquids (Brimmer and Hall, 1993). This approach attempts to normalize the changes in effective pathlength, or the penetration depth, by dividing the NIR spectrum by a wavelength that is relatively independent of concentration. For liquids, this may be a solvent absorption band. For simple mixtures, this can also be an isosbestic point where the individual components contribute equally to the NIR signal at this wavelength.

Both normalization and multiplicative scatter correction procedures require selection of either the best single wavelength or the optimum spectral region, respectively. The simpler normalization routines tend to be more stable when encountering variations not represented in the calibration data set, but are not as powerful as the multiplicative scatter correction routines that use all or large portions of the spectrum for correction. While multiplicative scatter corrections routines are more able to perform a suitable correction than are normalization procedures on complex systems, they can be more susceptible to matrix variability that is not present in the calibration population. The decision of which procedure to use in practice will depend upon which correction routine provides the necessary accuracy and repeatability for a given application and requires the least maintenance.

A combination of optimizing the sample presentation and spectral scatter correction routines is essential in obtaining NIR results that provide the accuracy and precision required for industrial laboratory analyses. The increasing development of improved methods of sample presentation combined with the development of more effective spectral scatter correction routines will allow NIR methods of analysis to be successfully developed and implemented for increasingly more difficult samples and analyses.

C. Calibration Models

Once we are assured of the accuracy of the reference analytical data and the quality of the NIR spectral data, calibration models can be developed. There are several different chemometric techniques that can be used, ranging from relatively straightforward methods such as multilinear regression (MLR) to more complex full-spectrum methods such as PLS and PCR. More exotic modeling techniques that utilize Fourier transform techniques (Giesbrecht et al, 1981; McClure et al, 1984) or artificial neural networks (Long et al, 1990; Gemperline et al, 1991) have been described in the literature, but these techniques have yet to be implemented for routine use in industrial manufacturing support laboratories.

A common question in the industrial application of NIR spectroscopy is "With all the different modeling techniques, which is best for a particular analysis?" The criteria for selecting an appropriate modeling technique will be based upon the ability of each

technique to derive a reliable method of analysis that provides adequate accuracy and long-term reproducibility and requires minimal maintenance. An evaluation of these attributes for using multiple-wavelength and/or full-spectrum calibration modeling methods for industrial manufacturing applications is discussed below.

The most common modeling algorithms used to derive NIR models for industrial samples are multiple-wavelength calibration methods, such as MLR, or full-spectrum modeling techniques, such as PLS or PCR. A comparison of the attributes of these modeling techniques is provided in Table I. Generally, multiple-wavelength modeling techniques will require less matrix variability to be present in the calibration population (i.e., the number of samples and degree of sample variability required) than will more complex, full-spectrum modeling techniques. For chemically complex matrices, multiple-wavelength models may be less accurate, but tend to be more forgiving (less sensitive to variation), when predicting the composition of samples that were not present in the calibration samples. Full-spectrum models will have more demanding requirements for calibration sample sets and usually require more calibration maintenance than simpler models. The advantage of complex modeling methods rests with their ability to characterize more complex sample matrix variables. For example, physical measurements (such as intrinsic viscosity of a polymer or the boiling point of crude oil) that are influenced by a number of components in the sample can be more fully characterized by these complex full-spectrum modeling methods. Similarly, other physical parameters that can cause a spectral interference (e.g., temperature variations) can also be characterized more effectively with full-spectrum algorithms so they do not interfere with the NIR analysis.

A good rule of thumb in developing NIR analytical methods is to use the simplest calibration modeling method that provides adequate accuracy and precision for a particular analysis. In all modeling procedures, the fewest number of wavelengths or spectral factors that provide suitable analytical performance should be selected to reduce the possibility of "overfitting." Simpler calibration modeling procedures like MLR are less susceptible to overfitting than the more complex modeling methods, and they are usually affected to a lesser degree by variations that are not exhibited in the calibration set (raw material vendors, product improvements). However, these simpler models may not provide the same degree of accuracy as the complex models, but they may provide sufficient accuracy to perform their intended function. Since a simpler method is less affected by changes in the sample matrix (which is common in many industrial processes), the calibrations will have to be updated less frequently. A method that provides reliable data routinely, and requires less calibration maintenance, will be considered more reliable than a method that requires frequent recalibration (even if the accuracy is better!).

To achieve suitable analytical results by NIR spectroscopy in the industrial manufacturing support laboratory, full-spectrum mod-

TABLE I
Comparison of Multiple-Wavelength
and Full-Spectrum Modeling Methods Used
to Derive Quantitative Near-Infrared Models

Characteristics of Methods	
Multiple-Wavelength Methods (MLR)[a]	**Full-Spectrum Methods (PLS, PCR)[a]**
Used to characterize simple chemical systems	Used to characterize complex chemical systems
Easier to interpret spectral information	More difficult to interpret spectral information
Requires a smaller calibration data set	Requires a large calibration data set
Methods more straightforward to transfer	Methods more difficult to transfer
Less sensitive to unmodeled matrix variation	More sensitive to unmodeled matrix variation

[a] MLR = multilinear regression; PLS = partial least squares; and PCR = principal component regression.

eling procedures are often necessary. This may be due to the complexity of the sample matrix or to variations in the physical properties of the sample (e.g., temperature, pH) that affect the position of absorption bands. In such cases, more complex modeling procedures like PLS and PCR have proven beneficial and extended the applicability of NIR spectroscopy to a widely varying number of industrial matrices and analytes. However, it is very important to ensure that, when using these powerful routines, the requirements for developing stable models are observed. Generally, complex modeling methods will require larger calibration sample sets, with all the possible variation expected within the samples to be expressed (independently) so that the modeling techniques can characterize all variations within the samples. The number of calibration "factors" must also be kept to a minimum to avoid overfitting. It is also important to use only those spectral regions that are in a linear range, since these mathematical routines are linear operations, and to ignore spectral regions that may have variations not present in the calibration set (e.g., water), or instrument variations (e.g., grating artifacts or detector crossover).

One final consideration when using multivariate modeling procedures such as PLS or PCR is that more reliable models arise from using small windows of the spectrum rather than the entire spectrum. A multivariate model that encompasses a large portion of the NIR spectrum will be more susceptible to changes in the product (such as a new raw material supplier) than a model that uses only a small portion of the spectrum, since a change in the sample matrix may not affect the particular region of the NIR spectrum that is being used to perform the analysis. If, by using only narrow segments of the NIR spectrum, a calibration model can be derived that has approximately the equivalent standard error of calibration to a model that uses the full NIR spectrum, the calibration models derived from the narrow segments of the NIR spectrum will tend to be more robust and therefore be preferred.

In conclusion, simpler multiple-wavelength models are generally preferable if they can provide adequate accuracy for a given analysis, since they are easier to develop and maintain in an industrial laboratory environment. Generally, these simpler modeling techniques are limited to performing analyses on simpler chemical matrices. The applicability of NIR can be broadened through the use of more complex, full-spectrum modeling techniques, which are better able to model more complex parameters. It must be remembered, however, that these more complex modeling techniques require more effort in development and more frequent maintenance. These types of calibration models usually require the continued support of the method development chemist/spectroscopist to maintain and adjust calibrations to a greater extent than the simpler calibration modeling procedures.

D. Validation

Once a calibration is developed, a new set of samples from the process is measured to ensure that the derived NIR model is representative of the parameter of interest and provides acceptable results for samples for which the calibration was intended. The NIR spectra should be collected exactly as they will be in routine analyses (i.e., on fresh material collected from the manufacturing process). Since materials can change with time as they age, pick up moisture, lose volatile components, or undergo other changes, problems in validating a calibration that was derived using retained samples can be encountered. For example, when deriving a calibration model for a component in polymer pellets, older pellets may have picked up moisture or changed their degree of crystallinity or a slip agent may have bloomed to the surface, which could affect the NIR spectrum and give erroneous results for the component of interest. Instead of developing a calibration model to incorporate all of these variations, it is best to remember how the NIR calibration is to be used (usually to predict the composition or properties of freshly manufactured samples). Unless it is necessary to utilize the NIR method to analyze both new and retained samples (e.g., if the NIR method will be used to analyze customer complaint samples as well as to analyze newly produced material), only freshly manufactured material should be used to derive the calibration model and validate the model. In instances in which the model will be applied to old and new samples, the data sets for both the calibration model and the validation model will be collected and measured by both the NIR instrument and the reference analytical technique over a period of time, instead of all at once, which helps incorporate instrument and environmental variability into the data sets.

E. Calibration Maintenance

Once a calibration equation is developed and validated, the method can be incorporated into routine operations. This doesn't mean that no further work on the NIR model will be required. One usually tries to make the calibration as "robust" as possible, but with process improvements, changes in raw materials, and other "unknown" causes, calibrations may not provide accurate results over an instrument's lifetime. In addition, after an instrument is serviced (e.g., lamp changes), changes in the performance of the calibration may be observed. In many cases, especially for the simpler models, instrument servicing only imparts a bias to the calibration equation, which is easily remedied by altering the y-intercept of the calibration equation. For more complicated models, the effects can be more drastic and may require additional samples to be introduced to the calibration data set to maintain the analytical performance of the NIR method.

In either case, the performance of the NIR method must be checked periodically to ensure that the results are within established control limits. NIR results can change based on either changes in the instrument or changes in the sample. It is usually advisable to control chart both the instrument performance and the difference between reference analysis and the predicted NIR results for newly manufactured samples to ensure the NIR results remain valid (Fig. 6). This routine check should be performed on a daily or weekly basis as part of routine operations for the NIR method.

For the instrument, there are several parameters that could change over time or after servicing. If one knows the instrument is changing outside the normal variation (by use of a control chart), the proper corrective action can be effected (Fig. 6). The parameters that are typically measured are the photometric response (noise or signal-to-noise ratio), wavelength accuracy and precision (short- and long-term changes in the wavelength axis), and bandwidth (resolution). For discrete wavelength instruments, changes in the wavelength axis are not expected, and only the response from each filter is normally checked using absorbance standards.

Even though the instrument is operating within established control limits, the calibration could still be inaccurate due to changes in the sample. Control samples are therefore measured on a fre-

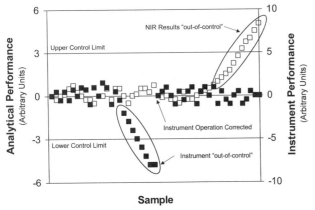

Fig. 6. Control chart of analytical performance and instrument performance. Encircled regions indicate when the instrument and/or method are operating out of control and corrective action is necessary.

quent basis (daily to weekly) to ensure no significant process changes have occurred that have affected the performance of the NIR method. If these control samples are not routinely measured, the NIR method will only be checked when the results are outside of the specification limits for the product (which leads the process operators to lose confidence in the NIR technique) or when customer complaints about samples come in (i.e., when the NIR method missed the fact that the product was outside of the product specification limits). Neither analytical situation is desirable and, therefore, the analytical performance of the NIR method is routinely compared with the reference method to ensure the NIR method is operating "in control."

Evaluating the performance of an NIR method is usually performed by control charting the difference between the reference analysis and the predicted NIR result for a control sample, as shown in Figure 6. As long as the results are within the upper and lower control limits and the points appear randomly distributed above and below the zero line in this chart, the NIR results are in control. If the results are nonrandom or appear to trend as in the right hand side of the control chart, the NIR results are not in control, and sources of the problems, whether it is in the instrument, laboratory, or new variations in the samples, must be found.

One caution to be exercised is to ensure that freshly manufactured samples, handled in the same manner as routine samples, are used as check samples. It is generally not a good idea to use a so-called "Gold Standard," which is usually a large quantity of well-characterized sample whose analysis obtained some time in the past is taken as "absolute." There are several erroneous situations that can arise when using this approach. First, the sample may change over time, either in the constituent of interest or some other component, which may alter the NIR spectrum and the predicted result for this standard but not for actual process samples. Hence, if the predicted results have changed significantly and the NIR calibration is adjusted, the NIR method will then be providing incorrect analytical data for actual process samples. In addition, the "Gold Standard" may not contain any information about the material currently being manufactured and could be significantly different from the original calibration samples. Control charting the difference between the reference analysis of a newly manufactured sample and the predicted NIR result is the best way to ensure the NIR method is operating in control. Furthermore, the reference analysis should be done as close in time as possible to the spectral collection to ensure no significant changes in the sample occur between the NIR method and the reference analysis.

By maintaining control charts for the instrument and analytical results in this manner, changes can be detected quickly and a plausible course of action can be effected rapidly to address the cause of the deviation to minimize the time the NIR method is out of control and "off-line."

IV. QUALITATIVE ANALYSIS

Qualitative analysis by NIR spectroscopy is beginning to find wider utility, particularly in the pharmaceutical industry for raw material inspection. First, let it be stated that the qualitative use of NIR spectroscopy is not generally used to answer the question "What is it?" but rather "Is this material similar to some material that was measured previously?" This is an altogether different question, and the method of answering this question will be somewhat different than how a typical analytical chemist would try to determine the identity of an unknown material.

Qualitative NIR analysis can be used to verify the identity of a material and perform a qualitative assessment. To identify a material, the NIR spectrum of the material in question must have been collected previously. By comparing the NIR spectrum of the "unknown" material to a library of spectra collected for many materials, the material can be determined to closely resemble one of the materials in a library, or unlike any of them. To perform a qualitative assessment, the NIR spectrum of the material in question is compared with the composite spectrum and spectral deviation that is observed for several different lots of a material of "acceptable" quality. Thus, a qualitative NIR determination will address whether the NIR spectrum of the sample is similar to a material present in a library and also whether it fits within the spectral variation observed for past lots of the material.

Qualitative NIR analysis began in earnest in the pharmaceutical industry, which requires a quick raw material identification/verification. The benefits of using NIR spectroscopy for material identification over the traditional physical, wet chemical, and chromatographic methods of analysis are the reduction in the costs and the time required to perform these analyses. More rapid raw material inspection can also eliminate production delays related to the length of time necessary to perform traditional methods of analysis. A qualitative NIR analysis is usually performed in either the receiving area (or loading dock) or in the laboratory using grab samples. Furthermore, being a nondestructive sample analysis technique, material identification by NIR spectroscopy will usually not require highly trained analysts to obtain accurate qualitative information.

This same rationale for material identification is beginning to be used in the chemical industry, especially for chemical companies that supply raw materials to the pharmaceutical companies. Reducing the time for analysis in the chemical industry not only reduces the costs of the laboratory tests, but can also reduce some other incurred costs of incoming raw materials, such as demurrage charges on railcars. These charges can be in the hundreds of dollars per hour per railcar, which can become very significant while the railcars are held until the laboratory results verify the product identity and quality so the railcars can be off-loaded to storage tanks. Also, due to the incompatibility of many chemicals used in the chemical industry, an NIR analysis can provide extra protection against an inadvertent mixing of incompatible chemicals, which could have potentially catastrophic results. Similar benefits are also being realized for identification of materials in pilot plant operations. The ease in presenting a sample to the spectrometer allows this analysis to be performed by less technically skilled personnel at remote analyzer houses or near railcar off-loading stations far from a laboratory.

In practice, qualitative NIR analyses tend to be less demanding and more straightforward to develop and maintain than quantitative methods, and they can provide information that is very useful in the qualitative assessment of incoming raw materials and the quality checks for in-process and final products. In the industrial manufacturing environment, these types of simple quality checks help ensure that products are kept within specification, which reduces the amount of off-specification material produced.

A. Library Development

There are several different qualitative algorithms that are currently being used for material identification/verification, as well to measure the "quality" or acceptability of incoming materials. It is not in the scope of this chapter to compare these methods, other than to say that the full-spectrum approaches are generally favored, especially in the pharmaceutical industry, where missing a possible contaminant in an incoming raw material is unacceptable. For these full-spectrum methods, one can make some judicious decisions on spectral regions to use in order to tailor the method for the task at hand.

First, the spectral regions used for the analysis should all be within the linear range of the NIR spectrometer. When the absorbance ($\log[1/T]$ or $\log[1/R]$) is outside the linear range of the instrument, absorption bands can become distorted and nonreproducible. Not only will the information be compromised, but also as instrument changes occur (e.g., lamp changes), these nonlinear regions can significantly change, which can result in a large number of "false negatives" being reported from the NIR method.

If a simple identification is required, without a measure of quality, it is often advantageous to remove certain regions of the spectrum that might cause a large number of misidentifications. For example, in pilot plant operations in the chemical industry, the operators simply want to verify that materials contained in barrels are labeled correctly. In this case, the presence of small amounts of impurities are of no interest. In developing a library for this type of application, it would be beneficial if the spectral regions near the water absorptions were not included, since changes in the moisture content from barrel to barrel (or season to season) will cause a large number of false negative results. When operators experience false negative results with a test, the confidence they have with a method will be diminished. Alternatively, a library could be created over time to introduce enough variation in the moisture levels of all the materials used in a pilot plant operation, but no greater advantage in terms of identification will be gained and the economic justification for the measurement will be compromised. It is also useful to guard against misidentifications that may arise from instrument changes, such as lamp replacement, which could cause small changes in the position of grating artifacts (e.g., Wood's anomaly). Usually, elimination of these small spectral regions is beneficial in the long run.

For qualitative analyses that require a measure of quality and in which changes in minor components are important to the acceptability of materials, use of smaller spectral regions is not appropriate. If one wishes to be sensitive to the presence of contaminants (which is especially important for pharmaceutical ingredients), as much of the spectrum should be utilized as possible. In these types of applications, more samples are required to represent the acceptable variability in these materials.

B. Validation

In qualitative NIR applications, as in quantitative analysis, the method should be validated and the method needs to be maintained. Validation involves scanning materials not in the original library and verifying that they are correctly identified. If a measure of quality is also used, it is important to verify that all acceptable quality material passes the validation test. In addition, it is often useful to use some "off-class" material to ensure this material is flagged as outside the acceptable range of variability for this material.

C. Maintenance

Generally, the maintenance of qualitative methods is performed in a similar approach to quantitative methods. As always, prior to performing NIR analyses, one should check the instrument diagnostics such as signal-to-noise ratio and bandwidth to ensure that false negative results, attributable to analyzer performance, are not obtained. If the instrument is working within specification and there haven't been any hardware changes in the instrument, then any false negative results are probably due to changes in the sample. This may be due to changes in the moisture content or other minor components or, for solids, to changes in physical parameters, such as particle size, that may cause the material to be flagged as outside the tolerance limits. If these samples are acceptable under normal testing procedures, then this variation probably wasn't present in the original library, and these samples should be added to the library. Before adding these samples to the library, care should be exercised to ensure that this new variation is indeed representative of the actual materials.

V. CONCLUSIONS

In this chapter, we have discussed the opportunities for using NIR spectroscopy in industrial manufacturing support laboratories. In order for NIR techniques to be beneficial, the NIR analysis must address some analytical requirement better than existing methods of analysis to offset the time and effort required for developing NIR methods. The justification of NIR spectroscopy for laboratory use was discussed to illustrate the perceived benefits of NIR methods of analysis over existing analytical methods. We have outlined the practical considerations for NIR method development, including sample presentation, spectral manipulation, calibration development, validation, and maintenance to demonstrate what is involved in developing and implementing NIR methods of analysis in the industrial manufacturing support laboratory. Hopefully, these practical aspects discussed will help those developing NIR methods to avoid some common pitfalls, which others have previously encountered and which have occasional resulted in NIR spectroscopy being considered as a "nonviable" technique for use in the industrial manufacturing support laboratory.

In the industrial manufacturing support laboratory, NIR spectroscopy has proven very useful for performing many different analyses on a widely varying number of sample matrices. As a result, NIR methods of analysis have been able to provide more timely information for control of manufacturing processes, improved incoming material inspection, in-process testing, and final product quality testing. By providing more timely and cost-effective process and product information, the chemical, polymer, petrochemical, pulp and paper, textile, and pharmaceutical industries have been able to realize significant cost savings through improvements in manufacturing efficiencies, product quality, and consistency. The use of NIR will certainly continue to expand in utility within industrial manufacturing support laboratories.

LITERATURE CITED

Barnes, R. J., Dhanoa, M. S., and Lister, S. J. 1989. Standard normal variate transformation and de-trending of near-infrared diffuse relectance spectra. Appl. Spectrosc. 43(5):772.

Brimmer, P. J., and Hall, J. W. 1993. Determination of nutrient levels in a bioprocess using near-infrared spectroscopy. Can. J. Appl. Spectrosc. 38(6):155-162.

Drennen, J. K., and Lodder, R. A. 1990. Non-destructive NIR analysis of intact tablets for determination of degradation products. J. Pharm. Sci. 79:622-627.

Geladi, P., MacDougall, D., and Martens, H. 1985. Linearisation and scatter correction for NIR reflectance spectra of meat. Appl. Spectrosc. 39(3):491.

Gemperline, P. J., Long, J. R., and Gregoriou, V. G. 1991. Nonlinear multivariate calibration using principal components regression and artificial neural networks. Anal. Chem. 63:2313.

Giesbrecht, F. G., McClure, W. F., and Hamid, A. 1981. The use of trigonometric polymonials to approximate visible and NIR spectra of agricultural products. Appl. Spectrosc. 35:210.

Gottfries, J., Depui, H., Fransson, M., Jongeneelen, M., Josefson, M., Langkilde, F., and Witte, D. 1996. Vibrational spectrometry for the assessment of active substance in metoprolol tablets: A comparison between transmission and diffuse reflectance near-infrared spectrometry. J. Pharm. Biomed. Anal. 14:1495-1503.

Hansen, W. G., Wiedermann, S. C. C., Snieder, M., and Wortel, V. A. L. 2000. Tolerance of near infrared calibrations to temperature variations; A practical evaluation. J. Near Infrared Spectrosc. 8(2):125-132.

Kortum, G. 1969. Reflectance Spectroscopy. Springer, New York.

Long, J. R., Gregoriou, V. G., and Gemperline, P. J. 1990. Spectroscopic calibration and quantitation using artificial neural networks. Anal. Chem. 62:1791-1797.

Martens, H., Jensen, S. A., and Geladi, P. 1983. Multivariate linearity transformations for NIR reflectance spectroscopy. Pages 205-234 in: Proc. Nordic Symp. Appl. Statist. Stokkand Forlag, Stavanger, Norway.

McClure, W. F., Hamid, A., Giesbrecht, F. G., and Weeks, W. W. 1984. Fourier analysis enhances NIR diffuse reflectance spectroscopy. Appl. Spectrosc. 38:322-329.

Norris, K. H. 1983. Extracting information from spectrophotometric curves. Predicting chemical composition from visible and NIR spectra. Pages 95-113 in: Food Research and Data Analysis. H. Martens and H. Russwurm, Jr., eds. Applied Science, London.

Norris, K. H., and Williams, P. C. 1984. Optimization of mathematical treatments of raw near-infrared signal in the measurement of protein in hard red spring wheat. I. Influence of particle size. Cereal Chem. 61: 158-165.

O'Haver, T. C., and Begley, T. 1981. Signal-to-noise ratio in higher order derivative spectrometry. Anal. Chem. 53:1876-1878.

CHAPTER 11

Method Development and Implementation of Near-Infrared Spectroscopy in Industrial Manufacturing Processes

PAUL J. BRIMMER
Foss NIRSystems
Asia-Pacific Operations
Oatley, NSW
Australia

FRANK A. DeTHOMAS
ABB Automation
Analytical Division
Woodstock, MD
U.S.A.

JEFFREY W. HALL
Foss NIRSystems, Inc.
Silver Spring, MD
U.S.A.

I. INTRODUCTION

In today's manufacturing environment, there is an ever-increasing need for real-time process analytical chemistry. The driving force behind this transition from traditional laboratory analysis to process analysis is the need for more rapid process control information, as well as for economical, safety, and environmental considerations.

For some of the very same reasons that near-infrared (NIR) spectroscopy was considered as an alternative laboratory technology in the 1980s and 1990s, NIR spectroscopy is now receiving significant attention as an alternative process analyzer technology. Specifically, with NIR technology, the opportunity exists to obtain rapid (analysis time on the order of seconds to minutes), nondestructive, multiconstituent analysis of process samples, with minimal or no sample preparation or pretreatment. Furthermore, for process analysis, the rugged design of the current generation of process NIR instrumentation, combined with the availability of high optical quality fiber optics, affords an intrinsically safe, highly flexible interface between the process analyzer and the process sample. As a result, NIR spectroscopy is now routinely considered a viable process analyzer technology. However, only if a process measurement provides data that is relevant in the control of a process, or offers substantial advantages over existing methods of analysis, will it be required or even considered.

In the previous chapter, we discussed the practical considerations that are both useful and often necessary when developing and implementing NIR spectroscopic methods in industrial manufacturing support laboratories. In this chapter, the practical aspects for developing and implementing NIR analyzers and methods of analysis for monitoring and controlling industrial manufacturing processes are discussed.

II. PROCESS MEASUREMENT REQUIREMENTS

To determine the efficacy of using any analytical technology (including NIR spectroscopy) in a process, several critical factors must be considered. For example, is continuous real-time chemical information necessary for process control? The type of process (e.g., continuous, batch), as well as the process characteristics (e.g., process time constant), will dictate the type of process control information required. Process measurements may also be desirable if they eliminate difficulties encountered for laboratory analysis, such as safety considerations concerning sample collection and analysis, sample stability problems, or if the laboratory analyses are cost prohibitive.

A. Process Type

The first factor to consider when deciding whether compositional information is needed is the type of information required to control the process. Each type of manufacturing process will have different process monitoring and control requirements. Once the process control requirements are established, the appropriate method of analysis can be determined. For each type of manufacturing process, real-time NIR analysis affords unique opportunities for process monitoring and control.

The first type of manufacturing process to consider is a continuous process, which can be defined as a manufacturing process in which the material is handled as a continuous stream moving through the manufacturing process from beginning to end. For example, individual ingredients are constantly fed to a polymerization reactor, where components react with one another and then polymerize as they move forward in the polymerization system. The control of a continuous process initially involves establishing a proper "set point" as quickly as possible and then maintaining operation at this particular set point for the duration (i.e., process life cycle) of the manufacturing run (which can be days to weeks to months in duration).

To improve control for continuous processes, compositional information must be supplied at a rate that is faster than the rate at which changes can occur in the process (i.e., the process time constant). However, if changes induced in a continuous process will affect product composition only hours after the changes were implemented, then direct process analyses offer little advantage over laboratory analyses of "grab" samples in controlling the process. Fortunately, for continuous processes, the process time constant is on the order of minutes, and, therefore, direct process measurements are especially useful in controlling continuous processes. For example, if a process suddenly begins to drift out of specification due to a problem related to reactant flow, temperature, pressure, or some other process variable, a real-time compositional measurement will rapidly detect the problem and enable quick rectification.

The second type of manufacturing process to consider is called a batch process. In batch production processes, a complete set of materials are initially added and processed in the same vessel until the process is finished. For example, a set recipe of materials is added to a drier and dried until a certain moisture level is reached. For batch production processes, the process control requirements are quite different than those for continuous processes. Batch processes can last several hours to over a day, and the need for process analysis is to detect the process end point. If off-line analysis is used to determine the end point, samples may be taken too infrequently and the reaction end point may be overestimated. This can lead to increased manufacturing costs and inefficient use of manufacturing resources (i.e., power requirements and equipment). Therefore, process measurements can be used as an alternative to laboratory testing, when laboratory analyses are available too infrequently.

There is also a hybrid of these two manufacturing process types, often called a continuous-batch process. A continuous-batch process is basically a continuous process that is used to manufacture many different types of materials, using the same equipment. The length of time used to manufacture one product (a "batch") can be several hours to several days. The process is then "transitioned" to the next product to be manufactured until the end of the production run for that product. This transition can be as simple as changing a set point to a new value, or it may require the introduction of different feed stocks to the reactor. Common production examples are changing the grade and additive packages in a polymer process and the production of various cellulose esters, all using the same continuous process equipment, but with different feed stocks (depending on ester type) and different process operating conditions. For continuous-batch processes, NIR spectroscopy is used predominantly to adjust the process to the proper set point as quickly as possible for these relatively short-duration manufacturing processes. Continuous-batch manufacturing processes will often require different calibrations for each different product, which adds additional complexity to NIR method development and implementation strategies for these processes.

B. Sample Collection and Analysis

Process measurements can be preferable to laboratory analyses if there are problems in obtaining a representative sample for analysis. For example, there can be sample collection problems associated with samples containing volatile components in process streams having elevated temperatures and pressures. The sample might also change in some way when removed from the process (e.g., oxidation, moisture variations), which makes it difficult to obtain a representative process sample. Therefore, while a laboratory analysis could provide suitable analytical data, the inability to collect a representative sample impedes its utility for process control.

For many industrial processes, there are safety-related issues for sample collection, which can make laboratory measurements undesirable. Some examples are the possibility of sample ignition, accidental exposure to toxic materials, or potential hazards from handling materials at elevated temperatures and pressures. In these instances, direct process measurements are highly desirable.

The costs associated with routine laboratory testing may also justify process measurements. For example, the requirement for 24-hr/day laboratory testing necessitates that personnel be available 24-hr/day to collect samples and perform the analyses. The analysis may require the use of solvents (and their subsequent disposal) or other sample preparatory materials, which may be costly. By eliminating routine, laborious testing, reagent and materials costs can be reduced and personnel can be utilized more effectively to help improve, support or optimize manufacturing processes.

Having determined that a real-time NIR process measurement is justifiable and is necessary for process monitoring and control, the process analyzer and the process sample interface can be considered.

III. PROCESS SAMPLE INTERFACE

The interface of an NIR instrument to a process can vary widely, depending primarily upon the type of sample being measured. The process sample interface for liquids will be vastly different than that for solids, and the types of implementation issues will also vary with the different sample characteristics. In this section, we discuss considerations that are necessary for interfacing a process NIR analyzer to a process stream.

A. Liquids

Process NIR measurements of liquids are generally made in transmission, and spectroscopic measurements are usually straightforward to perform in a liquid manufacturing process stream. That is not to say that process NIR measurements of liquids do not have their own difficulties in producing reliable results, however.

The major sampling considerations for process liquid samples are associated with pathlength, sample interface (i.e., flow cell or probe), and whether a sample bypass loop is needed. The pathlength will be determined by the absorbance level in the spectral region where the measurement is made, as well as by the physical properties of the sample (e.g., viscosity). The nature of the process sample and the analyzer installation/maintenance philosophy of the process engineering personnel will determine whether the process sample interface should consist of a flow cell, a direct fiber-optic probe interface, or a bypass stream that allows for sample cleanup and temperature control. A sampling port near the process sample interface for calibration/validation and model maintenance is highly recommended.

PATHLENGTH
Theoretically, the optimum absorbance in the spectral region used for an analysis should lie between approximately 0.1 and 1.0 absorbance units (AU). As with laboratory measurements, the sample pathlength that provides an absorbance level within this range ensures that the signal is in a linear range and maintains the sensitivity of the desired analysis. However, the physical characteristics of the sample (e.g., viscosity) must be considered as well, since the material must be able to flow through the selected pathlength for the NIR measurement to be practical. Thus, the ideal pathlength will be a compromise between the theoretical optimum and the practicality of making the measurement, given the sample characteristics. In practice, the longest possible optical pathlength is selected to ensure that the process dynamics are not impeded and to ensure that the sample in the process sample interface is easily replenished with fresh material.

Knowledge of the spectral regions that contain useful spectral information about the analyte of interest must be obtained in the initial stages of developing a process NIR analytical method. This is usually achieved by comparing the NIR spectra of the major components within the process sample. Spectral regions can then be isolated where the analyte has strong absorption bands and where other matrix components possess minimal overlapping absorption bands. Next, a spectrum is collected for a typical process sample in a short pathlength cell (1–2 mm), which provides a spectrum that is on scale throughout the entire NIR region. The maximum pathlength that can be tolerated in each of the spectral regions can now be calculated (keeping the absorbance level below 1.0 AU). Generally, one finds several promising spectral regions. These will include both short wavelength regions (700–1,500 nm), which can tolerate longer pathlengths (5–30 mm) but yield poorer spectral resolution, and longer wavelength regions (1,500–2,500 nm), which require short optical pathlengths (0.5–5 mm) but provide better spectral discrimination (Fig. 1).

The optimum spectral region will depend upon the physical nature of the process sample and the required sensitivity and selectivity for the analysis. For example, viscous samples will not flow easily through a short pathlength cell or probe, so longer pathlengths and use of the shorter wavelengths will be necessary for process analysis. Furthermore, longer pathlengths reduce the possibility of sludge and other contaminants in the process stream from clogging the cell or probe. On the other hand, the measurement of a component whose bands are not well resolved from bands of other components in the matrix, or is present at low concentrations, is more easily accomplished in the longer wavelength region, which requires use of shorter pathlengths. The NIR absorption bands occurring at longer wavelengths will be stronger, sharper, and better resolved than their corresponding higher overtone bands, which occur at shorter wavelengths. By balancing the physical properties of the sample, the process conditions, and the required analytical performance, the optimal spectral region and, hence, the optimum pathlength for a process NIR measurement can be determined (Fig. 1).

Finally, it is important to remember that when using fiber optics, affordable fibers currently available begin to attenuate the NIR energy beyond 2,200 nm, which limits the practical spectral region for process NIR analysis to wavelengths below 2,200 nm (Fig. 2). The distance between the instrument and the process stream (i.e., the fiber-optic length) will also impact upon the available spectral region (Fig. 2).

TEMPERATURE

As discussed in the previous chapter, temperature can affect NIR measurements by changing the spectral baseline and the position of absorption bands associated with polar functional groups. Since the temperature of manufacturing processes will normally vary, the changes in process temperature and the resultant impact upon a process NIR measurement is an important consideration. Since many industrial processes are operated at high temperatures

(>200°C), an additional concern is the temperature limitations imposed by the materials used for the process interface.

Temperature changes induce changes in the density of a liquid, which alters the spectral baseline. This effect can be corrected for by subtracting a baseline wavelength (when measuring with discrete wavelength instrumentation) or by calculating a first- or second-derivative of the original absorbance spectrum (when using full-spectrum instrumentation). Changes also occur in the position of absorption bands associated with polar functional groups. Temperature changes affect the degree to which polar functional groups participate in hydrogen bonding. This affects the strength of the covalent bonds in polar functional groups, which alters the frequency at which these groups vibrate, which changes the position of the absorption bands. Since process temperatures can vary substantially, the position of polar group absorption bands can vary significantly. Thus, an effective corrective modeling technique will be necessary to ensure the suitable analytical performance is achieved from the NIR method.

The most accurate results will be obtained under isothermal conditions (Hansen et al, 2000), since any spectral correction will not eliminate the effect of temperature upon the NIR spectrum, but only compensate for the observed effect. In the process environment, providing isothermal conditions is often not possible, especially for samples at temperatures above 100°C or for very viscous liquids, which would be difficult to pump through a temperature-controlled cell. There are processes, however, in which the material in a process is in the 30–80°C range and whose temperature can be relatively easily kept to within ±0.5°C through the use of a temperature-controlled water bath. If the sample is not viscous, it can simply be pumped through coiled tubing placed in a water bath, which can be temperature controlled either electrically or using plant steam.

For process streams in which temperature cannot be held constant, the deleterious effects of temperature can be minimized either by selecting spectral regions where temperature-induced changes are minimal or by incorporating temperature into the spectroscopic model. Temperature-induced shifts in band position can be reduced by averaging the response over a short spectral range. For example, when using filter photometers, selection of filters with a broader bandpass (which measure over a wider wavelength region) helps to minimize the effects of changing band positions. However, this approach is usually limited to applications in which the analyte(s) of interest possesses well-resolved bands, since expansion of the spectral window will also increase the possibility of interference by other components present in the process sample.

Most often when performing process NIR measurements, temperature variation is incorporated into the calibration model to compensate for its effects upon the NIR spectrum. This can be

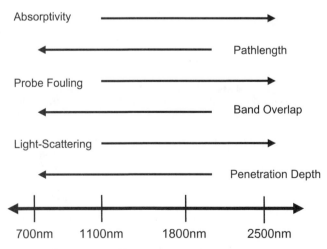

Fig. 1. Characteristics of the near-infrared (NIR) spectral region that vary with wavelength. These characteristics should be considered when determining the most appropriate region of the NIR spectrum to perform an NIR measurement.

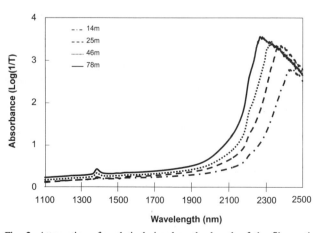

Fig. 2. Attenuation of analytical signal as the length of the fiber-optic bundle increases. As length of the bundle increases, the usable near-infrared spectral region is truncated toward shorter wavelengths. Signal attenuation beyond 2,200 nm is due to fiber-optics absorption characteristics.

accomplished in two ways. The temperature can be measured using a resistive temperature device and added as a linear correction to the "isothermal" calibration model. This approach compensates well for small changes in band position (i.e., a narrow range of temperature change), which can be adequately modeled using a linear correction. Some processes that operate at elevated temperatures (but have narrow temperature range fluctuations) have successfully used this approach to minimize the deleterious effects of temperature changes on the NIR measurement. However, if the process temperature range changes extensively (especially from day to night, and from season to season), a linear correction approach will have limited utility.

An alternative method of incorporating temperature into the calibration model is to allow the temperature of the process samples to vary normally as the NIR spectra of the calibration samples (and validation samples) are collected and to incorporate this spectral variation into the calibration model. If temperature is allowed to vary in the calibration samples, the calibration modeling algorithms can be used to characterize the analyte of interest and the temperature effects (and other interferences). By this approach, the calibration model attempts to characterize the effects of temperature, which effectively negates the effects of temperature (within the range of temperature variation modeled) upon the NIR spectrum when predicting the concentration of the desired analyte(s). Multilinear regression (MLR) techniques are not particularly successful at modeling temperature-induced NIR spectral changes, since adding an additional wavelength is a simple linear correction, much like adding a linear temperature correction term, as discussed earlier. Typically, full-spectrum multivariate regression techniques, such as partial least squares (PLS) or principal component regression (PCR), are employed to characterize the resultant shifts in band position.

PRESSURE

Pressure is another variable that can change in an industrial process. Usually, pressure differences have minimal influence upon spectral baselines or features and, therefore, cause little problem from a calibration standpoint. However, bubbles can form due to a sudden drop in pressure in a pressurized process stream. These problems should be considered in the design of the sampling stream (e.g., maintaining back pressure on the liquid to inhibit bubble formation or the use of debubblers to remove bubbles). Alternatively, suitable spectral compensation can be effected by using spectral normalization algorithms or scatter-correction techniques if the signal-to-noise ratio (S/N) of the NIR measurement is adequate. Also, some industrial processes operate at very high pressures (>5,000 psi), so concerns about the sample interface surviving these conditions is also an issue.

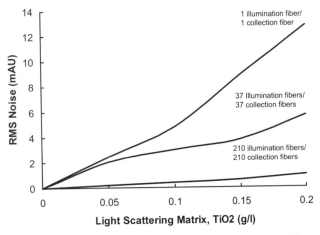

Fig. 3. Influence of light scattering (TiO_2 [g/liter] suspended in tetrachloroethylene) on root mean squared noise of the near-infrared spectra collected with single fibers, small bundles, and large bundles of fiber optics. As light scattering increases, fiber bundle size is increased to maintain analytical performance.

SUSPENDED SOLIDS/BUBBLES

In industrial manufacturing processes, an ideal "clean" process stream does not exist. There are almost always some suspended solids or bubbles in process liquids that scatter the NIR energy used to investigate the process stream. A number of different types of filters are available to remove these contaminants, but none are 100% effective and all require preventative maintenance. While filters may not provide 100% rejection of solids, they most certainly minimize the amount of solids, which prevents the sampling lines from clogging and keeps sample probes and flow cells from fouling. Liquid streams can also have gas bubbles, which also scatter the NIR energy. Debubblers can be installed just before the sample interface to improve the situation (for nonviscous materials), but in flowing streams, new bubbles can form at any irregular surface.

The effect of light scattering by particulates or bubbles in liquid process streams will not affect all process sample interfaces equally. Thus, the process sample interface must be designed to ensure optimal performance under these deleterious conditions. As illustrated in Figure 3, as the light scattering increases, the S/N decreases (i.e., root mean squared [RMS] noise increases), which affects the analytical performance of the NIR analysis. For fiber-optic-based process sample interfaces, as the number of fibers contained within the fiber-optic bundle increases, the S/N of the NIR spectrum will be less affected by particulates and bubbles present in the process stream (Fig. 3). This can be understood by considering the following simplified example. If particulates have a particle size approximately equivalent to that of the diameter of the optical fiber, then one particle passing in front of a single fiber probe will totally block all energy; a fiber bundle with 10 fibers would have 10% of the total energy blocked; and a fiber bundle with 100 fibers would have 1% of the total energy blocked. While most particles are much smaller than the typical fibers used in NIR technology, this analogy is useful. Thus, as the light-scattering properties of the sample stream increases to maintain the analytical performance of the NIR measurement, the number of fibers in the process sample interface should be increased (Fig. 3). However, an unavoidable practical consequence of increasing the number of fibers within the fiber-optic bundle is that the length of the fiber bundle must be reduced to provide greater access to the available NIR spectral region (Fig. 2).

In addition to affecting the S/N of NIR measurements, particulates and bubbles in process streams will affect the spectral baseline (first-order effect) as well as the "effective pathlength" (second-order or multiplicative scatter effect) (Martens et al, 1983; Geladi et al, 1985). Provided the S/N is high, spectral correction schemes can be used to compensate for the first- and second-order effects of light scattering. As described in the previous chapter, correction schemes can be as simple as compensating for first-order effects by subtracting a baseline or performing a derivative on the spectral data. Second-order light-scattering effects can be compensated for by dividing the NIR spectrum by a (specifically chosen) wavelength or by using more powerful mathematical multiplicative scatter correction algorithms.

TRANSMISSION MEASUREMENTS

There are four possible configurations by which NIR measurements of liquids can be obtained in a process stream or reactor (Figs. 4–7). One approach is to use a flow cell (Fig. 4), with the energy being delivered through this cell by conventional optics or fiber optics. Similarly, direct insertion of a pair of transmission probes aligned at 180 degrees to each other provides a true transmission measurement (Fig. 5). A third way to collect spectra of liquids is by the use of an interactance/immersion probe (Fig. 6). In this configuration, the energy delivered by the illuminating fiber(s) passes through the sample where it contacts a mirrored surface and is reflected back through the sample to the collection fiber(s). A final design (Fig. 7) provides true transmission measurements in a single probe body. This transmission probe design (i.e., single-sided trnsmission) encompasses optics within the probe body to redirect energy from a single illumination fiber 180 degrees, collimate the beam passing through the sample, and refocus the energy to a second single-collection fiber.

Each of these transmission measurement designs requires practical considerations for implementation. The interactance/immersion and single-fiber transmission probes have the benefit of being easier to install in a process line, since they require only a single entry point. Furthermore, a single entry point is the only practical interface that can be installed directly into reactors, mixers, and other types of large containers where a transmission probe pair or transmission flow cell are not possible or too cumbersome.

The pathlength is fixed by the distance between the mirrored tip and the main probe assembly for the interactance/immersion probe or by the distance between the windows for the transmission probe pair. The gap or pathlength should be fairly wide (several millimeters), especially for viscous liquids, to ensure that (i) the material will flow through the gap; (ii) the process flow dynamics are not impeded; and (iii) the sample present between the probe(s) or gap is continuously rejuvenated.

Interactance/immersion probes (Fig. 6) tend to be more susceptible to being clogged by debris in the process stream than in a true transmission measurement, since the sample gap is approximately half the transmission pathlength (for the same absorbance level). Surface coating on the probe (i.e., probe fouling) will reduce the energy reaching the sample and will influence the actual measurement, which in extreme cases results in a constant assay of the coated material, not the actual process sample. In addition, coatings on the mirrored tip will reduce the amount of light reflected back to the collection fibers, decreasing the S/N of the measurement and reducing the precision of the analysis. Single-sided transmission probes are not as susceptible to clogging or probe fouling. However, since current designs utilize single fibers, they have limited utility in strongly light-scattering process streams.

Transmission measurements are less susceptible to probe fouling because there is a doubling of the gap between the incident and collected light compared with immersion probes with a comparable optical pathlength. Transmission probe interface designs (Fig. 5) can incorporate single fibers or fiber-optic bundles. These types of probes collect the energy that has passed through the sample, and, therefore, light coatings on these probes (i.e., probe fouling), even though they block some of the NIR radiation, do not prevent interrogation of the bulk of the sample. Transmission probes tend to be more efficient at measuring highly scattering

liquids, such as emulsion polymerizations. As long as there is a sufficient S/N in the collected signal, the effects of the light scattering from the matrix or from a coating on the probes can be minimized by math treatments (e.g., spectral derivatives, normalization routines, and multiplicative scatter correction). The disadvantage of transmission probe pairs is that they are not readily installed in reactors or mixers, unless a bypass stream is installed to circulate material through a pipe where the probes can be installed.

True transmission measurements can also be accomplished using a liquid flow cell (Fig. 4), with light illumination/collection performed either with conventional optics or fiber optics. This approach has all the advantages of transmission probes discussed above, but suffers from problems associated with sample cell cleaning. Transmission measurements using flow cells find use in clear, nonviscous liquid streams and, to facilitate sample cell cleaning, are usually connected to a second stream to allow the cell to be flushed with fresh solvent. Since material must be pumped through these types of flow cells, this design does not find use for direct in-line measurements, but relies on a bypass stream to enable the NIR analysis of the process sample.

Practical limitations exist for the placement of transmission probe pair interfaces, since they cannot normally be installed directly into reactors or mixers. The exception is the single-sided transmission probe design (Fig. 7), which provides a true transmission measurement but requires only a single point of entry to the process stream. Transmission probe pairs and transmission flow cells are normally installed after the process reactor or mixer in a transfer line. This is practical for continuous production processes, but for batch production processes, the material must be pumped through a circulating sample loop, since these types of interfaces cannot be directly inserted into a batch reactor. This configuration can be difficult to implement and may not be possible due to the physical characteristics of the process sample.

FLOW RATE

The flow rate of a process stream can affect the NIR spectrum if the flow rate is sufficient to produce turbulent flow conditions. Turbulent flow can cause a "ripple" to appear in the spectral baseline, which is irreproducible and difficult to compensate for by using baseline correction routines or by characterization in the calibration equation. It is preferable to keep the flow laminar. This allows the NIR baseline to remain stable, which allows the precision of the NIR measurement to be kept within acceptable limits. For direct in-line measurements, the flow rate is dictated by the process and cannot be adjusted. For processes with very high flow

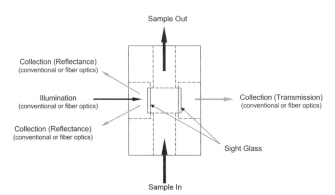

Fig. 4. Schematic of side-stream flow cell utilizing conventional or fiber optics.

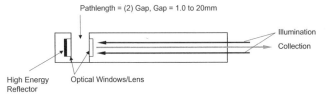

Fig. 6. Interactance immersion probe for analyzing clear to slightly light-scattering liquid process streams. A concentric ring of fiber optics illuminates the sample (outer core) and collects (inner core) the resultant light.

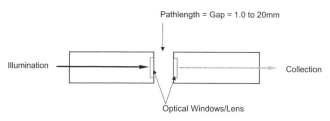

Fig. 5. Transmission probe pair for analyzing clear to strongly light-scattering liquid process streams.

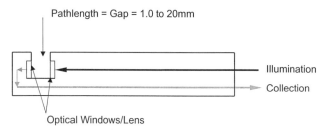

Fig. 7. Single-sided (single fiber optic) transmission probe for analyzing clear liquid process streams.

rates, the best viable option is to use a side-stream, in which the flow rate can be kept low enough to ensure laminar flow.

IN-STREAM VERSUS SIDE-STREAM INTERFACES

Liquid systems can be measured either directly in a process stream or in a side-stream. A direct interface to the process requires minimal hardware, since there is only a single entrance point for an immersion probe or two entrance points for a fiber-optic transmission pair. However, with a direct interface configuration, the probe(s) cannot normally be accessed until the production process is shut down. Furthermore, sample conditioning cannot be effected with a direct interface. In addition, a sampling port may not be available to collect a sample near this measurement point, which can make calibration model development and maintenance difficult. Also, as noted in the previous section, in-stream flow rates may cause the flow to be turbulent, which adversely affects the measured NIR spectrum.

To avoid the practical limitations of direct interfaces, a high-speed side-stream sample loop can be utilized. In this design, filters and debubblers can be readily added to cleanup the sample prior to the NIR measurement. These devices not only improve the precision of the NIR measurement, but can also reduce the frequency of probe fouling from contaminants, which can coat windows on sample probes and cells. Temperature control is also easier to perform in bypass sample streams. Sampling ports can be added directly to bypass streams to ensure representative samples are collected for calibration development and maintenance. Furthermore, bypass streams can be isolated at any time to allow access to the process sample interface (e.g., for cleaning). A key additional advantage offered by bypass streams is the ability to pump calibration standards through the sample interface or to pump cleaning solvents through the system if the probe(s) or flow cell becomes fouled.

While additional costs are normally incurred for installing a bypass stream due to the extra piping, valves, filters, and any heat tracing or insulation required, the initial costs are quickly recovered by the overall convenience of operating and maintaining the NIR process analyzer. There is a concern, however, with this design that filters require regular preventative maintenance or that they will eventually clog or otherwise fail. While bypass streams provide the means for optimum sample conditioning, for certain process samples, a reliable side-stream configuration may not be possible due to the nature of the process sample (e.g., hot molten polymer). In these cases, a direct in-line measurement provides the only possible method for obtaining reliable, real-time process NIR measurements.

B. Solids

Performing process spectroscopic measurements on solids presents a much greater challenge than liquids, since these measurements are generally made in reflectance mode, which has lower energy-collection efficiency than transmission measurements. Generally, reflectance measurements are made in the 1,100–2,500-nm spectral region, since light-scattering efficiency is higher in this region than at shorter wavelengths, which helps to offset the less efficient light collection. In reflectance measurements, the sample scatters the NIR energy that is used to produce the measured NIR spectrum. Therefore, the spectral characteristics can change dramatically as physical properties of the sample (e.g., particle size, shape, orientation, surface roughness, packing density, sample thickness, etc.) vary. As a result, performing reflectance measurements on solids in the process environment sometimes requires a sample interface that can be adapted to the changing physical characteristics of the process sample. This ensures the best reproducibility of the NIR spectrum, which yields better overall analytical performance for the analysis, lower detection limits, and better long-term stability.

Since process solids can exist in a variety of forms, different types of process sample interfaces are necessary to optimize the sensitivity of the analysis. Depending upon the type of solid, the measurement challenge rests with obtaining a representative sample (due to inhomogeneity of the material), ensuring consistent presentation of the sample (e.g., changing sample height, sample voids), and accommodating for a limited sample thickness. The approaches taken to ensure a representative measurement of the different types of solid samples is discussed in this section.

POWDERS AND GRANULAR MATERIALS

Powders are manufactured using batch and continuous production processes. This affects the method of sample presentation. In batch production processes, the NIR measurement of powders will typically be used to identify the end of the batch process, such as uniformity of a blend or the dryness of a powder. The process sample interface can be made either by a fiber-optic bundle and probe inserted directly into the dryer or mixer (Fig. 8) or by a sampling "window" that is installed on the side of the process vessel through which the process status is measured. In each of these process sample interface designs, the NIR spectrum is collected of the material that is in contact with the probe or sight-glass window. For probe interfaces, the position of the probe is critical because the movement of the powder within the reactor or drier must rejuvenate the sample being measured. Knowledge of the dynamics of the movement of solid samples within process vessels is, therefore, very important in determination of the optimum placement of the probe interface. For observations through a window, care must be taken to ensure that the observation window is located in a region where the powder is in continual movement, which allows continual sample rejuvenation and therefore ensures a representative process measurement is obtained. For process installations in which rejuvenation of the sample is difficult to ensure, the probe interfaces and/or sight-glass windows can be cleaned by blowing nitrogen or instrument air over the probe window periodically (White, 1990).

Screw conveyers, belt conveyers, air, or vacuum can be used to transport powders. Quite often, identifying a location where a packed sample exists that can reproducibly scatter sufficient NIR energy can be difficult. Generally, optimal sample measurement locations will be found near transition points, such as the end of a free-fall in a pipe, in a hopper, or on a conveyer belt where the material is deposited.

In transport pipes, the NIR measurement can be made through a sample window or by a fiber-optic probe (Fig. 8). As in the batch processes, the main concern is rejuvenation of the sample, which at transition points is not a problem so long as the fiber-optic probe or sample window is not recessed such that the sample can accumulate in the resulting crevice. Alternatively, measurements of heterogeneous powders or more coarse granular materials in transport lines and hoppers can be performed using a noncontact analyzer configuration (Fig. 9), in which the detector head is mounted flush to the inside surface of a hopper or transport line and the energy is transmitted to the sample and collected from the sample through a sight glass. A final challenge for this process NIR analysis is to ensure that the density of the powder flowing through a pipe or hopper is sufficient and uniform enough to reproducibly scatter the NIR energy back to the process NIR interface.

Powders, as well as more coarse granular materials (i.e., pellets), can also be measured as they are transported along on conveyer belts. Measurements on conveyer belts are typically made

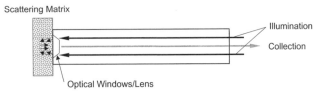

Fig. 8. Interactance reflectance probe for analyzing solids or strongly light-scattering fluid (or homogeneous solid) process streams. A concentric ring of fiber optics illuminates the sample (outer core) and collects (inner core) the resultant light.

using a noncontact analyzer configuration (Fig. 10), in which a sensor head is mounted at some predefined distance above the sample stream. The height that the sensor head is mounted above the conveyor belt is a function of the reflectivity of the sample, the collection efficiency of the instrument optics, and the desired area of sample to be illuminated. For noncontact measurement configurations, the change in the distance between the sensor head and the sample or sample voids can adversely affect measurement performance. The sample height can be coarsely adjusted using a leveling device (e.g., a flat piece of metal) placed just before the measurement. The residual minor height variations can be compensated for by averaging NIR measurements over a large amount of sample or by introducing a baseline correction or spectral normalization variable in the calibration model. Small sample voids can introduce a minor degradation in the analytical performance of the process NIR measurement, but for many processes, this is of minimal concern. In contrast, large sample voids or empty conveyer belts can significantly affect the analytical performance of the process NIR measurements and must be negated. The most practical means to compensate for large sample voids is to use pattern recognition routines (when using full-spectrum instruments) or to use null-sample (i.e., empty belt) models to detect when an NIR spectrum is being collected of a sample void and not report a result.

Since measurements on conveyer belts are made in an open environment, ambient light can affect the measurement performance. For this reason, instruments designed for these applications are normally postdispersive rather than predispersive to minimize the effects of ambient light. If a predispersive instrument is used, the ambient light must be blocked with a suitable light shield.

WEBS AND SHEETS

Products produced as webs and sheets provide a fairly uniform sample for an NIR measurement. An example of this type of sample is found in paper manufacturing, in which large sheets of paper (over a meter in width) are moved through the manufacturing process over rollers. These sheets provide a fairly flat sample surface, which visibly looks very uniform, and simplifies the NIR measurement. These samples are typically measured using noncontact reflectance configurations, as illustrated in Figures 9 and 10. The challenge for making reproducible measurements on sheets and webs is to eliminate sample flutter. Sample flutter can be largely eliminated by making the NIR measurement at the rollers, rather than between them, since the flutter is minimal near these transition points.

There are practical limitations to the type of process analytical information that can be provided due to the width of the sample, as well as the speed at which the material is moving. These types of processes have sheets more than a meter in width and can move at speeds on the order of meters per second, which allows an average measure of some sample parameter but impedes defect analysis. For this reason, process NIR measurements for webs and sheets are usually used to monitor the average composition of the material or the average thickness of a substance deposited on its surface. The uniformity of the material across the width of the material is an important process monitoring or control variable and, to address this concern, some process NIR instruments can be moved across the web or can be configured with several sensing heads across the width of the web.

FILMS AND FIBERS

For reflectance measurements to be independent of sample thickness, they need to be thicker than the "infinite penetration depth" of the NIR energy used to interrogate the sample. Infinite penetration depth is the depth in the sample at which any increase in the sample thickness causes no change in the measured spectrum. In the NIR region, infinite penetration depth is typically several millimeters for most powders and webs. Samples such as thin translucent films and fibrous strands do not conform to this requirement and, therefore, their NIR spectra will be affected by changes in the thickness or bulk density of material being measured. In addition, since NIR absorptions are characteristically weak, for films, the limited amount of material will produce weak NIR signals, which limits the accuracy, precision, and practical detection limits for this measurement. Thus, these types of samples provide a big challenge in developing a process sample interface that yields suitable analytical performance.

One approach that can be used to increase the signal from these types of samples is to provide a "double pass" through the sample by reflecting the light that has passed through the sample back through the sample a second time before being collected (Ghosh and Brodmann, 1993). This can be done using a diffusely reflective material to increase the effective pathlength through the sample. This also tends to reduce the effect of other variations, such as nonuniformity in fiber strands and height variations. In many instances, it is necessary to compensate for considerable baseline variations and/or multiplicative scatter variations (due to bulk density variations) by employing baseline correction and normalization techniques.

C. Suspensions and Emulsions

The spectroscopic analysis of highly light-scattering suspensions and emulsions presents unique challenges for NIR spectroscopy that are different from either liquid or solid samples. These types of process samples are generally composed either of a solid suspended in a liquid or of an emulsion made up of two immiscible liquid phases. Depending upon the optical characteristics of the sample and the type of analysis required, these measurements can be performed in reflectance by utilizing the light-scattering nature of the samples or in transmittance mode, if the degree of light scattering is low enough. Alternatively, for strongly light-scattering matrices, transmission measurements can be performed in shorter wavelength regions (700–1,100 nm) where light scattering is less efficient.

If the analyte of interest is in the liquid phase of a suspension, NIR spectral analysis is more appropriately performed in reflectance mode. The reason for this is that the NIR energy will mainly interrogate the liquid (continuous) phase due to back-scatter by the "particles" and, therefore, not penetrate much of the suspended solids or dispersed phase. Conversely, if the analyte of interest exists in the suspended or dispersed phase, it is prudent to collect transmission spectra, since a greater proportion of the energy will interrogate the suspended or dispersed material than in the case of a reflectance measurement.

Reflectance measurements of emulsions and suspensions can be performed by using either conventional optics or fiber optics.

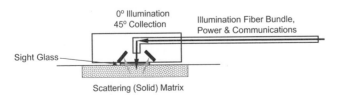

Fig. 9. Noncontact process sample interface for analyzing solid materials in hoppers or transport lines by reflectance. Detector head is normally in intimate contact (using a sight glass) with the process stream.

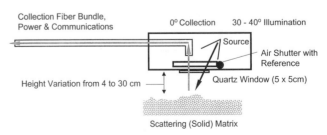

Fig. 10. Noncontact process sample interface for analyzing solid materials transported on conveyors by reflectance.

With conventional optics, the measurement is performed through a sight-glass, which can be problematic, since the sample at the surface of the glass may not be readily rejuvenated or the glass may become coated by the suspended material, blocking interrogation of the bulk of the material and providing a nonrepresentative measurement of the sample properties. Direct insertion of a fiber-optic probe can minimize concerns about sample rejuvenation, since the probe is placed into a reactor or process stream where the sample flow is greater than near the walls of a vessel. However, if the suspended material will adhere to the probe surfaces, the surface coating could again decrease the efficacy of the measurement. For reflectance probes, coatings inhibit the ability to measure the bulk sample. If heavily coated, these probes simply measure the coating on the probe, which is not desirable.

Similarly, transmission measurements of emulsions and suspensions can be performed by using either conventional optics or fiber optics. However, transmission measurements may require a side-stream sample loop, since both an illumination and collection probe or window will be needed. Transmission measurements tend to be less problematic than reflectance concerning material coated on probes or windows, since the NIR energy passes through the bulk of the material and thus contains information about its total composition. That is not to say that coatings cause no effect upon the measured spectrum, since they will reduce the amount of light collected and could affect the precision of the NIR measurement.

In some processes (e.g., fermentation broths), the sample is relatively transparent in the early stages of the process, but can become quite opaque as the reaction proceeds, requiring the use of both reflectance and transmission measurements during the course of a fermentation. Instead of using two probes (either multiplexed to one NIR instrument or connected to two separate NIR instruments), a single probe that utilizes fiber optics can be configured to perform both reflectance and transmission measurements (Fig. 11). By alternately measuring the transmission and reflectance spectrum throughout the batch, the optimum spectrum can be chosen and the appropriate calibration equation (based on reflectance or transmission data) can be applied.

IV. PROCESS INSTRUMENTATION

To facilitate process control, process NIR instrumentation (including the instrument, process sample interface, and process data communication) must provide timely, reliable compositional information. The overall system must have a minimum downtime and require infrequent maintenance (e.g., cleaning the sample interface, instrument component replacement) for the process operators to rely on the instrument and trust the information provided by the instrument. A process analyzer must also be able to survive in the process environment, be maintainable by process instrument analysts, and be able to communicate results in a manner that is understandable by process control systems.

To comply with these process requirements, the proper hardware, including rugged instrumentation, the most suitable process sample instrument interface, and the appropriate data communication options, must be available in the process sensor. A practical description of the NIR technologies that are currently available and

the typical process analyzer configurations employed to monitor or control industrial manufacturing processes are provided below.

A. Process Analyzer Configurations

REMOTE SENSOR/REMOTE SENSING

Early process measurements (e.g., gas chromatography) were performed with instruments that were located in environmentally controlled analyzer houses on samples that were pumped from the actual process stream to the inlet port of the instrument. While this sampling approach could measure changes occurring in the process stream, the sample lines were easily clogged and there was often a time delay associated with sampling. As a result, this approach was limited to the samples that could be easily pumped to the analyzer (e.g., nonviscous stable liquids).

With developments made in the 1990s in NIR instrumentation and fiber-optic technology, the opportunity now exists to move NIR instrumentation directly into the process rather than taking the process sample to the laboratory instrument. As shown in Figures 12 and 13, process NIR analyzers can be incorporated into a process manufacturing environment in either a remote sensor or remote sensing configuration. Both remote sensor and remote sensing configurations are effectively utilized in industrial process applications. The sensor configuration is ultimately dictated by both the process requirements and the necessary process sample interface, as discussed earlier.

In a remote sensing configuration (Fig. 12), the instrument is typically located in a relatively well-controlled environment, such as an electrical room or an analyzer house designed for analytical equipment. These environments are fairly similar to a typical laboratory environment, with environmental control to keep the temperature and humidity from changing to any great degree. This configuration may also enable the analyzer to be located outside of an electrically classified, or otherwise safety classified area.

A remote sensing configuration (Fig. 12) utilizes long lengths of fiber optics (20–200 m) to allow the instrument to be separated from the process sample interface. The use of remote sensing configurations will be limited to applications that can be measured by using single or small bundles (10–50 individual fibers) of fiber optics. Thus, remote sensing applications are usually limited to clear liquid streams that under normal process conditions contain minimal or no bubbles or suspended particles.

In a remote sensor configuration (Fig. 13), the analyzer is located very close to the measurement point and is, therefore, situated directly in the harsh manufacturing environment. Therefore, the instrument must be designed to withstand the process environment and have suitable enclosures for this environment. NIR process instruments used as remote sensors must be able to withstand the vibration, temperature, and humidity of the process environment where it is located.

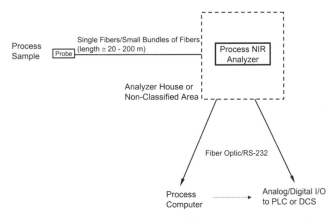

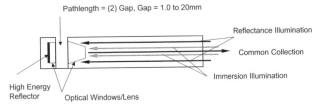

Fig. 11. Combined interactance reflectance and immersion probe (U.S. patent no. 6137108) for analyzing fluid process samples that vary from clear to strongly light scattering. A tricentric ring of fiber optics illuminates the sample for reflectance (outer core) or immersion (inner core) and collects (center core) resultant light.

Fig. 12. Remote sensing near-infrared (NIR) analyzer configuration. Process analyzer is located in a well-controlled environment, and long lengths of fiber optics are used to interface the instrument to the process stream. Information from the analyzer is conveyed to the process control system using RS-232 or digital communication protocols.

A remote sensor configuration (Fig. 13) utilizes short lengths of fiber optics (1–20 m) to enable adequate signal collection, which is necessary for the relevant process measurements. Remote sensor applications are generally associated with highly scattering matrices, which require large bundles (50-400 individual fibers) of fiber optics. Thus, remote sensor applications will be associated with solids, emulsions, and suspensions or fluid streams that under normal process conditions contain bubbles or suspended solids.

Furthermore, process NIR analyzers configured as remote sensors are typically enclosed in a suitable (secondary) environmental enclosure to provide additional protection to the instrument itself and to also isolate the instrument electronics and ignition sources from the environment, which is often an "intrinsically safe" area. The interior of the instrument enclosures are usually kept slightly pressurized with instrument air or nitrogen to preclude potentially explosive vapors from getting inside the enclosure. Since many process NIR analyzers are installed outdoors, secondary enclosures can also provide additional protection from the elements (i.e., heat, rain, snow, wind). Some temperature control is also possible with environmental enclosures, which is useful to limit the temperature swings seen from day to night and season to season. Environmental enclosures add additional complexity and cost to the process instrumentation, but provide needed protection for the analyzer and compliance with regulations regarding placement of electronic equipment in intrinsically safe environments.

MULTIPLEXING

For particular process monitoring and control situations, it may be both desirable and practical to monitor several process streams with a single process analyzer. Multiplexing becomes advantageous when both the cost per measurement point and the total implementation cost are reduced. However, while the cost per measurement point is drastically reduced with multiplexing, the liability per measurement point is significantly increased (i.e., if the analyzer performance is compromised) and, therefore, careful consideration should be given as to the whether multiplexing provides a real advantage for the desired process installation.

Multiplexing can be performed using either remote sensing or remote sensor configurations, but how it is accomplished will differ for these two types of process analyzer configurations. For remote sensing configurations, multiplexing is usually performed using separate channels of single-fiber or small bundles of fiber optics and probes that are interfaced to a single, centralized NIR instrument. This is the most common type of multiplexed configuration and is used primarily for nonscattering to slightly scattering liquid streams. Remote sensors can be multiplexed in a similar manner and can also provide multiple-point measurements of different process streams (from transmission measurements for clear liquids to interactance reflectance measurements of solids) by combining different fiber optic bundle sizes and different process sample interfaces. However, each sample stream must be in close proximity (10 to 20 m) to the process NIR analyzer.

Alternatively, a single-point process NIR analyzer can be used to measure multiple process streams by moving the sample from several individual process streams through a single flow cell or manifold that is interfaced to a single process sample interface. To be functional, the flow cell or manifold must have the previous sample swept out of the sample interface, and a new sample introduced to the sample interface, before a subsequent measurement can be taken. Another consideration when using a single-sample interface for each product is that the type of probe (and pathlength) is fixed for each sample, which may compromise the measurement performance for some process samples.

One concern about multiplexed systems is that, when there is an instrument problem, process operators lose the ability to monitor not only one, but many different, process streams. Sometimes during these periods, laboratory analyses on samples collected from all the process streams can provide the needed analysis, but this places a significant additional burden on the support laboratory, and the analyses will not come as frequently or as quickly as the on-line analysis. For this reason, the instrument problem must be quickly rectified, and the NIR instrument needs to have minimum downtime for a multiplexed system to be useful for controlling the process.

When considering multiplexing several streams to a single instrument, the trade-offs between reduced cost per measurement point must be compared with the risks of running several process streams without the benefit of continuous monitoring and perhaps having several streams go out of control simultaneously.

B. NIR Instrumentation

A variety of technologies exists that can be used to separate the polychromatic NIR spectra into monochromatic frequencies that can be used for quantitative and/or qualitative purposes. Broadband, discrete filter photometers, or instruments using light-emitting diodes (LEDs) provide NIR spectral information at specific wavelengths. To more fully capture NIR spectral information, technology such as diffraction gratings, interferometers, diode arrays, or acousto-optic tunable filters (AOTFs) are used.

Following is a brief description of the practical aspects of each type of technology applicable for use in process NIR analysis in terms of the selectivity, sensitivity, and reliability that can be realized in the industrial manufacturing environment. There is no particular technology that is superior in all cases. Generally, the type of instrumentation chosen will be the one that provides adequate analyses for a component that is important for process control, and can do so at the lowest cost. The fact that another instrument, which utilizes a different technology, can provide more information is usually meaningless (and therefore not economically justifiable) for process analyzers, unless that additional information is both useful and required for controlling a process. The different technologies are, therefore, compared with reference to their probability of success for implementation in industrial manufacturing processes and their ability to provide information that is relevant to process operators.

DISCRETE WAVELENGTH PHOTOMETERS

Filter Photometers. Some of the first NIR instruments used for process analysis were filter photometers, due to the ruggedness of this type of instrument. Filter photometers use bandpass filters mounted on a filter wheel that rotates each filter into the NIR beam to measure the response at several specific spectral wavelengths of interest (Fig. 14). The number of filters in a filter wheel will vary from two to three to a dozen or more. Data collection is usually taken by coaveraging the signals from multiple rotations of the filter wheel, which can produce high S/N signals in less than 1 min. Since the only moving part in these types of systems is the motor, which rotates the filter wheel, these instruments are very stable and hold up well in process environments. Filter photometers are usually installed as remote sensors since they can operate effec-

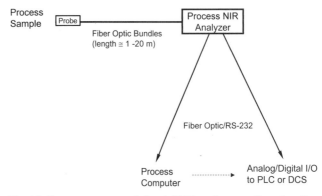

Fig. 13. Remote sensor near-infrared (NIR) analyzer configuration. Process analyzer is located near the process stream in typical process environment. Short lengths of fiber optics are used to interface the instrument to the process stream. Information from the analyzer is conveyed to the process control system using RS-232 or digital communication protocols.

tively in the process environment, although some have been interfaced to single-fiber optic probes, and therefore act as remote sensing devices. Filter photometers have a longer history for performing process measurements than any other type of NIR technology, and they have a reputation of providing reliable results for a variety of different types of samples.

LED-Based Photometers. A relatively new approach to generate discrete NIR wavelengths for process measurements utilizes LEDs. Instead of using a broadband source and separation into discrete wavelengths with bandpass filters, these systems use LEDs, which only emit a small band of spectral wavelengths (Fig. 15). By using a dozen or so LEDs emitting at different wavelengths, a profile of a specific NIR spectral region can be constructed. In operation, the diodes are cycled on and off sequentially, and the light returning from the sample is sent to a common detector. One significant advantage to this type of system is that there are no moving parts. Currently, these types of instruments are limited to the 700–1,100-nm region, due to the wavelengths that can be produced by LEDs. The penetration depth of light through samples in this region is quite deep, allowing solids and light-scattering samples to be measured with this type of instrumentation. LED-based instruments to date have been outfitted with fiber optics, either single fibers or bundles, and can, therefore, be used either as a remote sensor or for remote sensing. There are concerns, however, with this type of instrumentation regarding the wavelength repeatability of the diodes and how closely the wavelengths of different diodes match, which may affect the performance when diodes are replaced.

Practical Considerations for Photometers. Filter photometers and instruments utilizing LEDs as light sources provide measurements at discrete wavelengths, rather than a continuous scan. For applications in which analyte bands are well resolved or for analytes of significant concentrations, these types of instruments can provide suitable analytical results. However, since NIR spectra are composed of broad, highly overlapped absorption bands, it is often difficult to isolate unique spectral regions using these types of instruments.

The applications of discrete wavelength instruments are generally used when the spectrum of the analyte has very distinct and unique absorptions, in which the selectivity of the instrumentation is not as important. For example, filter photometers are often used to measure percentage levels of moisture, since water has very strong absorptions that are usually well resolved from the absorption bands related to other components in the matrix.

Since NIR spectra are also frequently affected by variations in light scattering from the sample (either from particulates or bubbles in liquid streams, or particle size or inhomogeneities in the surface of solid samples), suitable compensation will be necessary. Discrete wavelength instruments usually have additional wavelengths besides the "analytical" wavelength for the analyte being measured that are used to compensate for the changes in the "baseline" of the NIR signal. This usually requires a region of the spectrum in which there are no significant NIR absorption bands. Some of the more sophisticated chemometric techniques, such as PLS or PCR, are not as useful with discrete wavelength measurements as they are with full-spectrum measurement techniques, in which subtle spectral differences can be used to develop quantitative and qualitative models. Thus, the overall analytical performance can be impeded due to the inability to use more complex algorithms. Therefore, discrete wavelength instruments will most frequently be used to measure materials that are present at percentage levels, and when the precision and accuracy levels required, are typically greater than several tenths of a percent.

FULL-SPECTRUM SPECTROPHOTOMETERS

The remaining technologies that are utilized in process NIR instruments fall into the category of full-spectrum spectrophotometers, in that they provide a continuous scan, rather than a few discrete data points. These full-spectrum instruments provide more spectral data and have broader utility in monitoring and controlling industrial manufacturing processes than discrete wavelength instruments. The different types of full-spectrum NIR instruments are described below.

Grating-Based Spectrophotometers. Diffraction gratings have long been used in laboratory spectrophotometers in the UV, visible, and infrared spectral regions. Diffraction gratings separate the radiation from a continuous source into its component wavelengths, and, by sequential movement of this grating, wavelengths can be sequentially passed through an exit slit, producing a scan over a certain spectral range (Fig. 16). Until recently, gratings were scanned at fairly low speeds (a full scan produced on the order of

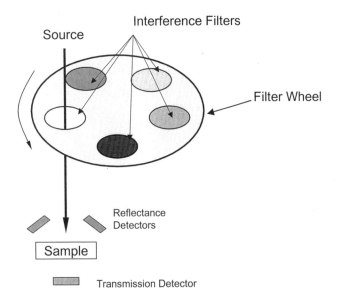

Fig. 14. Schematic of the optics of a filter photometer-based near-infrared analyzer.

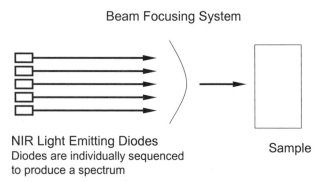

Fig. 15. Schematic of the optics of a light-emitting diode-based near-infrared (NIR) analyzer.

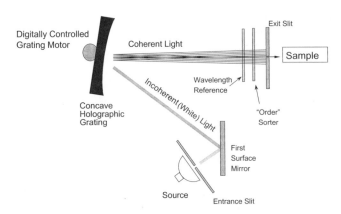

Fig. 16. Schematic of the optics of a digitally-controlled dispersive (holographic) grating-based near-infrared analyzer (Foss NIRSystems, Inc., Silver Spring, MD, U.S.A.).

a minute or more) and were fairly sensitive to vibration. Thus, grating-based spectrometers were only found in the laboratory environment and were considered inadequate for implementation into a process environment.

Today's grating-based NIR instruments are either fast-scanning (i.e., several complete spectra collected per second) or fixed grating spectrometers and are designed to ensure that their performance is not affected by vibration. Furthermore, grating drive mechanisms are now typically controlled using optical, rather than mechanical encoders to measure the position of the grating and are, therefore, less sensitive to vibration (Fig. 16). This development has allowed grating-based NIR instruments to be used in nonlaboratory environments. With these and other improvements in design (shock mounting the optics in the instrument, continuous wavelength correction, and automatic detector gain settings), grating-based process NIR instruments are finding wide use in the industrial process environment for measuring all types of sample matrices.

Grating-based NIR spectrometers can also be coupled with an array of individual detectors into what is known as a diode array spectrometer, as shown in Figure 17. Diode array spectrometers are designed with the grating held in a fixed position that disperses discrete wavelengths onto a photodiode array detector. Since none of the critical optics components move (the only moving part is a shutter to alternately block a reference and sample channel), mechanical wear on the optical components and sensitivity to vibration is virtually eliminated. These instruments also provide very fast spectral acquisition rates (microseconds) so these types of instruments can be invaluable for monitoring processes that require very fast measurements. To date, most diode array-based analyzers scan in the short-wave NIR region, but as newer, more cost-effective diode array detector materials become commercially available and economically viable, increased measurement opportunities for this technology platform for monitoring industrial processes will arise.

Spectrometers that use diffraction gratings to disperse NIR energy are usually interfaced to the process using fiber optics. Grating-based spectrometers can be coupled with both single fibers and fiber-optic bundles, making them useful in both remote sensor and remote sensing configurations. Data collection generally re-

quires 5–30 sec (except diode array-based analyzers) to obtain a spectrum with a S/N that is high enough to yield suitable qualitative or quantitative NIR analysis.

Interferometers. Interferometers are commonly used in mid-IR instruments, where their improved performance over grating-based spectrometers made dispersive mid-IR spectrometers obsolete in the 1980s (Griffiths and DeHaseth, 1986). Interferometers are also finding use in NIR spectrometers because of their ability to supply high wavelength precision data in a relatively short spectral acquisition time period.

Interferometers do not physically separate white light into its component wavelengths. Rather, spectra are generated mathematically from an interference pattern, created by a device called an interferometer. Three different types of interferometers used in NIR spectroscopy are shown in Figures 18–20. The interferometer splits a collimated beam of energy into two equivalent beams (using a beamsplitter) that are sent to two equidistant mirrors and redirected back to the beamsplitter, where they are recombined. At this point, the two beams undergo constructive interference, since the two beams passed equal distances, such that all wavelengths are in phase. By making the distances traveled by the two beams unequal (e.g., by moving one of the mirrors or crystal), the result is that some wavelengths will be in phase, and others out of phase. By combining the two beams, both constructive and destructive interference of the energy occurs. The amount of constructive and destructive interference varies as the moving mirror position is changed, resulting in an interference pattern called an interferogram. This interferogram can be converted to a spectrum through the use of a mathematical process known as a Fourier transform.

Interferometers replaced grating-based instruments in the mid-IR region because of the higher S/N obtainable from interferometric measurements made using detector-noise limited instrumenta-

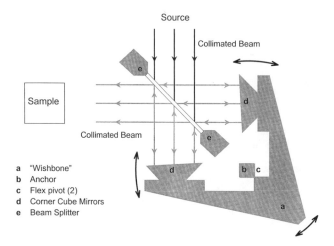

a "Wishbone"
b Anchor
c Flex pivot (2)
d Corner Cube Mirrors
e Beam Splitter

Fig. 19. Schematic of the optics of a Wishbone interferometer-based Fourier transform near-infrared analyzer (ABB Bomem, Quebec City, Quebec, Canada).

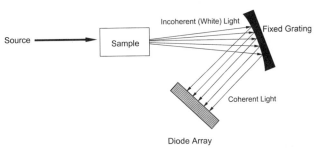

Fig. 17. Schematic of the optics of a fixed grating-diode array analyzer.

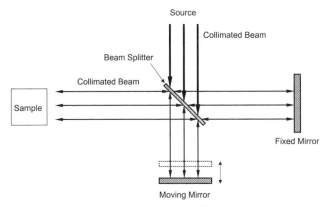

Fig. 18. Schematic of the optics of a Michelson interferometer-based Fourier transform near-infrared analyzer.

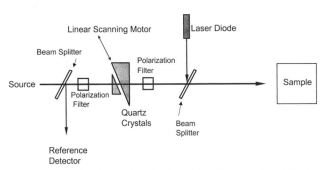

Fig. 20. Schematic of the optics of a Crystal interferometer-based Fourier transform near-infrared analyzer (adapted from Bran+Luebbe, Buffalo Grove, IL, U.S.A.).

tion. NIR instrumentation is not detector-noise-limited, and the advantages in S/N are not observed in NIR interferometers (Coates, 1994). Noise levels for commercial NIR interferometers are typically 10 times higher than that of a well-designed grating spectrometer. An additional concern is in the error in mirror alignment in an interferometer, which is more critical for the shorter wavelength NIR region than in the mid-IR region. The major advantage of interferometric measurements is the high wavelength precision, which is a practical benefit for applications in which the absorption bands are very narrow.

Interferometers can experience stability problems in the process environment due to vibrations, which affects the positioning of the moving mirror in the interferometer. In general, interferometer-based instruments are interfaced to the process via single fibers, which allows the instrument to be placed in a more stable, vibration-free environment. For these reasons, interferometer-based NIR instruments are generally used for remote sensing. This limits these types of instruments to those applications that can be performed with single-fiber sample interfaces (transmission through nonscattering media) or materials that can be pumped to the analyzer house through a flow cell (nonviscous liquids).

AOTFs. A relatively new technology for collecting a full NIR spectrum comes about through the use of a birefringent filter, whose transmission characteristics can be altered by varying the frequency at which the crystal is made to oscillate. In AOTF technology a piezoelectric crystal, which is attached to the birefringent filter, is excited by a radio frequency signal, producing a vibration of that frequency within the filter. A schematic of an AOTF system is shown in Figure 21. If a beam of NIR energy is directed through the filter while it is vibrating, a limited band of optical frequencies will pass through the crystal (due to constructive interference), and the remainder will undergo destructive interference and not be transmitted. A spectral scan can be obtained by ramping the frequency of the vibration, which can be done in on the order of milliseconds for the collection of one complete spectral scan.

These very fast scan speeds allow spectra with an adequate S/N to be collected in less than a 100 msec. This technology is amenable to monitoring very fast reactions or those that need to updated very frequently, which may not be observable with slower scanning instrumentation. The fast spectral acquisition rate will allow the spectra to be coadded for longer periods of time, which improves the S/N, thus improving the precision of the analysis. Furthermore, since data points in the spectrum are collected almost simultaneously, spurious bubbles and particulates will affect all wavelengths at about the same time, resulting in an improvement in the ability of spectral scatter correction techniques to compensate for the effects of scatter upon spectral data collected on rapid scanning instruments over spectral data collected on slower scanning instruments.

There is some concern whether or not the effects of the electric fields created by large pumps and motors in the process environment would adversely affect the AOTF crystal and whether vibra-

tions in the process environment would affect the vibration of the AOTF crystal. For these reasons, AOTF-based analyzers are usually placed in an environment remote from the sample measurement point. AOTFs are typically interfaced to the process via single fiber optics and are, therefore, used for remote sensing. The applicability of AOTF-based NIR analyzers are therefore limited to those applications in which single-fiber-optic interfaces are suitable.

Practical Considerations for Spectrophotometers. A full-spectrum analyzer provides more spectral data to be collected from a process measurement than does discrete wavelength instruments. More spectral information can enable better characterization of the analyte(s) of interest, as well as the effects of interfering components in the matrix when developing quantitative and/or qualitative models. Since NIR spectra consist of broad, highly overlapped spectral features, a continuous scanning instrument can enable an NIR analysis to be performed where the utility of a discrete wavelength instrument is limited. By applying multivariate chemometric analysis techniques on a whole spectrum or large spectral regions, a greater degree of selectivity is available with full-spectrum instruments. An additional benefit of having more spectral data is that the possibility of performing multiple constituent analyses on a single NIR spectrum exists.

With a more complete NIR spectrum, mathematical pretreatments of the NIR spectrum such as spectral derivatives, which compensate for baseline variations, can be performed more efficiently than with discrete wavelength spectral data. Furthermore, most of the full-spectrum NIR instruments collect the spectrum very quickly (several to dozens of scans per second), which permits spectral corrections like normalization or multiplicative scatter correction techniques to be applied. (Scanning speed is important since scattering changes in the process stream due to bubbles or particulates need to be affecting the whole spectrum simultaneously if spectral correction routines are to be able to correct for these effects. Slow scanning instruments would record an "event" such as a transient bubble as a "spike" in the spectrum.)

Since there is enhanced selectivity and superior ability to correct for deleterious NIR spectral effects, full-spectrum instruments can analyze samples with greater precision and routinely measure analytes at levels below 1%. There are some additional benefits to having a full-spectrum scan. Full-spectrum measurements provide better outlier detection and can also ensure that the correct calibration is being used for the product being measured using pattern recognition routines. In addition, full-spectrum instruments can be used for qualitative testing (e.g., identification of materials), which is not as feasible with discrete wavelength measurements.

These advantages and additional capabilities of full-spectrum instruments come at a cost, since full-scanning process NIR instruments are more expensive than discrete wavelength instruments. There are, however, many process applications in which the benefits of using full-scanning instruments easily offset the additional cost of the analyzer.

C. NIR/Process Operator Interface

Integration of process NIR technology into an industrial manufacturing process only begins with the installation of the analyzer. To complete the implementation, the compositional information must be made available for integration into the process manufacturing control strategy (Figs. 12 and 13). Therefore, once the NIR analysis is performed on a process sample, that information must be transmitted back to the process operators in a form that is recognizable to them. The protocols most commonly utilized are those used with existing temperature, pressure, and flow control devices. For these devices, information is sent to a distributed control system (DCS) by use of 4–20-milliamp current loops. In such current loops, the amount of current sent to the DCS is proportional to a signal within a range preset for high (20 milliamp) to low (4 milliamp) values for the measured parameter. Thus, quantitative NIR information can be sent to the DCS with the results for each analyte transmitted on a separate current loop. Autonomous process control is more desirable than having a per-

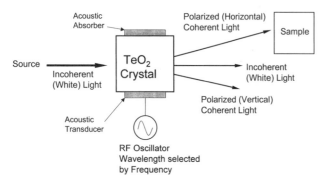

Fig. 21. Schematic of the optics of an acoustic optical tunable filter-based near-infrared analyzer (adapted from Brimrose Corporation, Baltimore, MD, U.S.A.). RF = radio frequency.

sonal computer located in the control room that provides analyses from each process analyzer, since process operators concentrate primarily on the DCS monitors and their built-in alarms.

There is an ever-increasing demand for more information to be transferred from the process analyzer, such as instrument error messages and instrument diagnostic information, which indicates the performance status of the analyzer. Furthermore, information about the relevance of the process sample (outlier detection and product quality) is often desirable. This type of information is most easily transferred to a process logic controller (PLC) via digital rather than analog communications. To meet these process communication requirements, process NIR instrument manufacturers supply digital communication, which allows integration of their process analyzers and sampling systems into control strategies that require diagnostic information about the process NIR instrument. With the introduction of more advanced digital communication protocols such as ModBUS, FieldBUS, or OPC (OLE for process control) (Kamen, 1999), even more process NIR information, such as complete spectra, can now be accessed for process analysis or seamlessly transferred through a network to data management systems for data archival and retrieval.

V. QUANTITATIVE ANALYSIS

Quantitative information that is obtained from a process NIR measurement can be used to control different types of manufacturing processes. However, the approach taken for method development, including collection of calibration standards, the type of calibration model, and method validation and maintenance issues, differs greatly depending upon the type of manufacturing process and the desired analysis. Thus, different types of manufacturing processes will, as a general rule, require different strategies for method development and implementation of process NIR analyzers.

For continuous processes, it is necessary to control the start-up of the process so that the optimum process set point is reached as quickly as possible. It is also necessary to detect when the process is going out of control so corrective action can be taken. Therefore, to reduce the start-up time and to get the process "lined-out" (i.e., the time in a process in which the product(s) concentration reach and are maintained at the desired set point), a continuous analysis that is very accurate is important. In contrast, to detect when the process is going out of control, the analyses must be frequent and precise so that changes in the process trend can be detected. Therefore, to follow the operating trend of a continuous process, the precision of the process NIR measurement is more important than the accuracy of the process NIR measurement.

For batch production processes, end point detection (i.e., when chemical changes in the process are no longer occurring) is the process control parameter. Therefore, information at the beginning of the batch is generally of little importance, and calibrations are normally developed that focus upon the latter part of the batch process to ensure accurate detection of the process end point. For end point detection of batch production processes, both the accuracy and the precision of the process NIR measurement are important for identifying when the process is truly finished.

Regardless of the process type, it is important to remember that when developing a quantitative method for process analysis, the simplest method that adequately performs the desired analysis (including type of instrument, calibration modeling method, etc.) should be the most robust. As the complexity of the method increases, the method can become more susceptible to changes in the instrument, raw materials, process conditions, or other environmental factors. If any of these factors compromise the analytical performance of the process NIR analyzer, the confidence in the results being produced by the NIR analyzer will be reduced, sometimes to an extent that the process operators will no longer use this information. Many process analyzers have suffered this fate and now hang lifeless on countless walls in manufacturing areas. Furthermore, the derived NIR methods must be able to be maintained by personnel who are responsible for process instru-

mentation and are usually not trained in spectroscopy or chemometrics. Thus, if the process NIR analyzer and/or the calibration model are either too complex or difficult to be easily maintained by plant personnel, the method (and the analyzer) will eventually fall into disuse.

A. Sample Selection

When developing an NIR method, representative process samples must first be collected to develop a calibration equation. For process NIR analysis, this will require that the samples be collected directly from the manufacturing process. Calibration samples must not only cover the range of analyte levels expected in the process, but must also include all of the other variations (other components in the matrix as well as physical variables, such as temperature) that might be expected. Very seldom can all of these types of variations be obtained from laboratory-generated samples, unless the process matrix is a very simple mixture.

When a continuous production process is operating under control, the sample composition does not change significantly. Therefore, it will be difficult to obtain samples that contain the variation necessary to derive robust NIR calibration equations. A similar situation will exist for batch production processes when most of the required calibration samples are collected toward the end of the batch process.

Therefore, to collect calibration samples that are representative of the variations to be encountered in an actual manufacturing process will take time. The length of time necessary to collect these samples will depend upon the type of manufacturing process. To allow the benefits of NIR spectroscopy to be realized more quickly, a method development strategy commonly utilized is to get as much of the variation as possible, perhaps 60–80% of the changes in the concentrations of the matrix components, and use these data for a "starter" calibration. For batch processes, this can be accomplished by sampling over the latter part of several different batches. For continuous processes, if the process varies over the course of several days, most of the needed sample variation can be observed within a couple of weeks. If the process doesn't change much over the course of a week or more, it may take several months to get samples that possess the required degree of variation. (For continuous processes, one way that often is used to obtain the required sample variation is to collect samples during process start-up or shut-down, since the process will vary to a greater extent during these time periods.) Collection of the NIR spectra (and samples for reference analysis) over at least two cycles of a continuous production process (both above and below the process set point and control limits), or five to seven different batches for batch production processes, is usually adequate for developing starter calibration models.

Once a starter calibration is derived, the unrepresented process sample variation can be obtained over time by looking for not only spectral variation, but also analyte (and interferent) concentration variations. Spectral outliers can be identified using pattern recognition algorithms, while concentration outliers can be found by control charting the reference analysis against the predicted NIR results obtained by using the starter calibration. By collecting the NIR spectra of the "flagged" outliers, these outlier samples can later be added to the calibration set to construct the final calibration model. Once all the process sample variation that is to be expected for the manufacturing process is represented by the calibration sample population, the final calibration can be developed.

B. Calibration Modeling Methods

There are a number of calibration modeling methods available to describe the relationship between the observed variation in the NIR spectrum and the concentration of the analyte of interest. Each of the methods possesses their own strengths and weaknesses, and none are universally better than the others. The calibration methods used commonly for process NIR analysis are either multiple-wavelength methods or full-spectrum techniques.

Multiple-wavelength methods are those that use the absorbance measurements at only a few wavelengths to predict the analytical result. The most commonly used method of this type is MLR. Full-spectrum modeling techniques are those that use the entire spectrum, or truncated spectral regions, as input into the model. The most commonly used techniques of this type are PLS and PCR analysis. A detailed comparison of the merits of the individual methods can be found in Chapter 4 of this book. Here, we will discuss the major advantages, limitations, and applicability of the different modeling methods for process NIR analysis.

To ensure that the derived process NIR models will be robust, there are practical compromises to be considered between the simplicity (or complexity) of a chemometric technique, the ease of use and implementation of the models, and the resultant applicability of the method. As a general rule, the simplest model that provides adequate information that is suitable for process monitoring and/or control should be used. Using a more sophisticated model may improve the calibration model, but may compromise the prediction results or make long-term reliability and method maintenance problematic. Finally, since the selection of the "optimum" number of wavelengths or spectral factors is not always intuitively obvious during calibration development, it is often useful to develop several calibration models and compare the merits of each model based upon the predictive ability of each of the derived models during the method validation stage.

MULTIPLE-WAVELENGTH REGRESSION TECHNIQUES

Multiple-wavelength regression techniques are generally used for discrete wavelength measurement systems and are commonly used for full-spectrum measurement systems. The analytical wavelengths are selected to characterize spectral features related to the constituent of interest and to correct for spectral interferences due to scattering or overlapping absorption bands.

For discrete wavelength measurement systems, the primary wavelength selected is usually related to the analyte of interest. Subsequent wavelengths are chosen to correct for baseline variations, related to variations in the light-scattering properties of the sample, or to correct for overlapping interfering absorption bands due to other matrix components. For discrete wavelength measurement systems, since discrete data points represent the spectral data rather than a continuous scan, some of the simpler baseline correction schemes, such as calculating a derivative, may not be possible and, therefore, a baseline correction for discrete wavelength measurement systems is normally performed within the calibration model. (It should be noted that the addition of "correction" wavelengths is more aptly described as "compensation" wavelengths, since they cannot totally correct for baseline changes, but only attempt to reduce the impact of these spectral variations upon the derived NIR models.) Care should be exercised when developing models for discrete wavelength analyzers (or any analyzer platform), as it is particularly easy to incorporate multiple discrete points (wavelengths) into the model to overcompensate for the lack of wavelength specificity and therefore, overfit the model, which produces both nonrepresentative NIR results and instability in the method.

When using multiple-wavelength regression techniques with full-spectrum NIR instruments, spectral math pretreatments are typically performed prior to developing the calibration model to compensate for baseline offsets and for resolution enhancement (or reduction). Some of the common pretreatments are spectral derivatives and multiplicative scatter correction. By using these spectral pretreatments, the wavelengths selected can be optimized for the analyte of interest and to minimize the contribution of interfering absorption bands related to other components within the matrix.

Multiple-wavelength modeling methods combined with discrete wavelength instruments tend to work well for simpler applications, require fewer calibration samples (10–20), and in general, are less sensitive to variations in the spectra that were not represented in the original calibration sample population. When multiple-wavelength modeling methods are used with full-spectrum measure-

ment systems, the utility of these methods can be extended to more difficult applications, but will generally require a larger calibration population (30–50 samples). These models also tend to be fairly insensitive to matrix variations that were not represented in the calibration sample population, since only a few wavelengths are utilized. However, multiple-wavelength regression methods tend to have limited utility for analyzing very complex matrices and highly overlapped NIR spectral features or for characterizing physical parameters, which are due to a number of matrix components (e.g., crystallinity and intrinsic viscosity of polymers).

FULL-SPECTRUM REGRESSION TECHNIQUES

More complex, full-spectrum regression techniques, which utilize all or large portions of an NIR spectrum, can more fully characterize subtle spectral differences. Thus, these techniques are commonly employed to model analytes whose spectral features are highly overlapped by features due to other components within the matrix or can be affected by changing process conditions (e.g., temperature). Full-spectrum techniques are also useful for characterizing "physical" properties that are related to multiple components within the process sample matrix. Depending upon the parameter to be measured and the process under investigation, full-spectrum regression methods can require a larger calibration population (50–100 samples) compared with simpler modeling techniques. Full-spectrum regression techniques will often tend to be more susceptible to spectra effects that arise from unmodeled variations in the process sample (e.g., process improvements, changes in raw materials).

C. Validation

Once a calibration model has been developed, is must be tested to ensure that the model is representative of the analyte under investigation and will produce accurate predictions on samples not used in the development of this model.

The individual modeling approach used to derive the process NIR model, combined with the process manufacturing type, will dictate the validation population sample size. The validation samples should be collected over a period of time corresponding to several manufacturing cycles in a continuous process or to several batches for a batch process. For continuous processes, this can be done in conjunction with the routine sampling for laboratory analyses. For example, NIR spectra corresponding to the time when routine samples are collected can be used (along with the laboratory results) as a validation set. In this way, the sample variability that is needed can be obtained without adding additional sampling requirements upon manufacturing personnel.

For validation sets that possess a similar sample population distribution to the calibration set, the standard error of prediction (SEP) should closely match the standard error of calibration (SEC). Having an SEP that is much larger than the SEC may indicate that the derived calibration model has overfit the calibration set and is not truly representative of the actual analyte (and other) variation in the matrix under consideration. This may be due to the inclusion of too many wavelengths in multiple-wavelength regression models or too many spectral factors in full-spectrum regression models. An unusually high SEP (compared with the SEC) may also indicate that the model is not truly representative of the analyte or property being modeled in the calibration samples (or validation samples).

Finally, all outliers should be investigated to identify the reason for disagreement between the reference assay and the predicted NIR result. The observance of spectral outliers can indicate the presence of unmodeled variation that needs to be represented in the calibration data set before the model is universally applicable for this manufacturing process matrix. Alternatively, samples that possess a difference between the reference analysis and the predicted NIR result that is greater than 3σ can also indicate the presence of unmodeled variation in the calibration population or a problem with sample collection or the reference analysis.

D. Maintenance

Once the method has been validated, the method is implemented, and data are provided continuously to the operators who control the manufacturing process, usually via the DCS located in a remote process control room. However, to instill confidence in the process NIR measurement and to verify that the process NIR analyzer is producing accurate results, routine analyzer and NIR method maintenance are critical components in the implementation of a process NIR method.

There are two basic changes that can occur in the measured NIR spectrum that can affect the predicted NIR results. Either the instrument can drift (or change after servicing an instrument component) or the samples may reflect some variation that wasn't present in the original calibration sample population. To ensure that the instrument is functioning reliably, standard instrument diagnostics such as wavelength accuracy, noise, bandwidth, and the response to reference standards can be plotted on a control chart. To verify the analytical performance of the derived NIR method, the difference between the reference analysis and the predicted NIR result obtained for routinely collected control samples can be plotted on a control chart. Control samples are typically collected and measured every 4–8 hr, so this method can, in many cases, quickly verify if the instrument and calibration are "in control." Sometimes there are difficulties in performing this essential comparison. For example, the samples may not be easy to obtain at the sampling point or there may be significant delays in getting the reference analysis for a particular material. (Until the NIR method proves to be reliable, the process operators will most likely wait for the reference analysis to confirm the changes seen in the predicted NIR result before adjusting the process.)

Control charting the instrument diagnostics, as well as the difference between the reference analysis and the predicted NIR results, is an effective means of monitoring the stability and performance of both the analyzer and the derived NIR method.

With full-scanning process NIR analyzers, the spectrum itself can be used to confirm that the newly collected spectrum is the same as the spectra collected for the calibration and validation samples. By using pattern recognition techniques, or outlier detection, variations in the instrument or spectral variations due to a change in the sample can be quickly detected. However, this diagnostic tool only detects that there are differences in the spectrum and does not indicate whether the change is attributable to the instrument or the process sample. This additional diagnostic test that is available with full-scanning process NIR analyzers can be very helpful in assisting with process analyzer troubleshooting.

The ability to detect outliers (in routine operation) and flag them so the process operators would know to ignore a particular reported analysis, to store the NIR spectra, and to collect a sample for reference analysis enables samples to be added to the calibration data set to improve the existing NIR model. However, these types of "events" are not desirable from the point of view of a process operator, since it means the process analyzer is not operating in control. For this reason, it is useful to use a starter calibration to flag outliers for a period of time before the method is validated and fully implemented, so the calibration model can be made as immune as possible to these variations before the process operators begin to observe this data routinely. Once the method is finally implemented, outlier detection routines are helpful in alerting the process operators that something is amiss in the process.

For process measurements, it is preferable to perform analyzer and method maintenance procedures automatically, rather than require a specialist to diagnose problems. Process operators are not spectroscopists or chemometricians, and they should not be expected to perform this function. Furthermore, since processes may run continuously, 24 hr/day, 7 days/week, the "experts" may not always be available to troubleshoot when problems occur. Hence, the more outlier detection and control charting capabilities for instrument diagnostics and analytical performance indication that can be implemented automatically, the more attractive an NIR analysis will be to the process operators.

VI. QUALITATIVE ANALYSIS

A. Process Requirements

Qualitative analysis by NIR can be helpful in maintaining quantitative methods by detecting samples that are unlike those used for quantitative calibration development. There are other needs in the process for which a qualitative assessment can be readily performed by NIR spectroscopy. For example, by using pattern recognition algorithms, process NIR qualitative methods of analysis can provide a qualitative assessment of a process sample, insuring that the NIR spectrum is within an acceptable variability of what is considered a good "quality" product. Examples in which qualitative NIR methods of analysis can be utilized in industrial manufacturing processes are discussed below.

PRODUCT IDENTIFICATION
In continuous-batch manufacturing processes, different materials may be produced at different times, using the same manufacturing equipment. In these instances, different NIR models are developed for each product, especially if the products have very different molecular structures. To ensure that the proper NIR model is applied to a particular process matrix, an NIR identification of the product can be performed to identify the particular matrix, and then the appropriate quantitative methods can be applied.

To perform this analysis, the same NIR spectrum that is used for the product identification is used to derive the quantitative analysis for that product. This approach is very useful for process analysis, since the "solution" to the situation of applying the incorrect calibration equation may not be solvable by process operators. Performing an identification analysis prior to performing a qualitative analysis is also very advantageous in situations in which autonomous selection of quantitative NIR methods can improve the confidence in the process NIR measurement. Additionally, a qualitative analysis can ensure that the material is not an outlier within a given product type and, hence, avoid reporting a potentially erroneous result. Since process operators can readily lose confidence in an analytical device that provides an incorrect analysis, even if the problem is as simple as using the wrong calibration equation, implementing a product identification test into a process NIR method of analysis can ensure that the full economic benefit of a process NIR analyzer is realized.

PRODUCT COMPOSITION
Many products that are manufactured in an industrial process are produced using a number of individual steps or processes. These can be a number of batch processes that are operated in a series or a combination of batch and continuous processes. In this manufacturing situation, a quick test to ensure the product in one step in the manufacturing process is correctly formulated before moving it on to the next stage can reduce the overall time it takes to manufacture the final product (i.e., cycle-time reduction).

For example, many processes require several materials to be mixed, which can be done using an automatic weighing system, mass flow devices, or manually. Adding the same material twice, not adding one of the components, or adding the wrong material will introduce formulation errors. These types of errors happen on an infrequent basis, but when they do, the material must be reworked, made into "off-class" material, or simply discarded. Furthermore, if this formulation error is not caught immediately, the material may be processed through several more manufacturing stages before the formulation error is detected. The costs associated with these types of errors are very high, and if these types of events occur several times a year, it can result in significant losses in productivity and an overall increase in manufacturing costs.

To confirm product composition, an NIR spectrum can be collected in a manner appropriate for the particular type of sample. By using a pattern recognition algorithm, the NIR spectrum can be used to verify that the material produced is within the observed spectral variation for previously manufactured material (i.e., the new batch includes all the components or the correct reaction products have been formed). An in-process NIR test reduces the time delays experienced with traditional methods of analysis and can, as a result, increase manufacturing capacity. This process NIR measurement is particularly effective for monitoring the end point product composition for batch production processes to ensure that the product is formulated correctly before moving on to the next stage in the manufacturing process.

In industrial manufacturing processes, there are also many "undefined" process "upsets" that can produce either an off-class product or render the formulation potentially unusable once it reaches the final manufacturing step. By performing continuous, rapid screening of materials at different points in the manufacturing process, changes in the qualitative nature of the NIR spectra can assist in quickly identifying process upsets. A real-time process NIR measurement will help to avoid making off-class product for several hours between "scheduled" sample collection times. By enabling process operators to isolate the location and time of a process upset, a more rapid diagnosis and resolution of the cause of the process upset can be found, leading to fewer similar occurrences in the future.

PRODUCT UNIFORMITY

A final qualitative NIR application in the industrial manufacturing environment is to ensure the manufacturing uniformity of a product. For example, in a pharmaceutical blending operation, the content uniformity or blend homogeneity of the components within a solid dosage form formulation must be confirmed (Sekulic et al, 1996). Blending is complete when the solids are evenly dispersed throughout the matrix (and may become unevenly mixed if overblended). A qualitative process NIR analysis can ensure that not only is the material similar to previously manufactured material but that, from spectrum to spectrum, the composition of the mixture isn't changing (which would indicate that parts of the overall mixture are not blended). This type of determination is not possible by collecting samples for laboratory analysis, since only a small portion of the overall mixture is tested for homogeneity. A qualitative process NIR analysis provides a simple and rapid product uniformity measurement that can be difficult to perform quantitatively for all of the components in the mixture.

VII. CONCLUSIONS

In this chapter, we discussed the many facets that must be considered to ensure success in implementing a process NIR analyzer for monitoring or controlling industrial manufacturing processes. Advances in fiber optics, instrumentation, and chemometric routines for data analysis have made the analysis of industrial process streams by NIR spectroscopy a reality.

The configuration of the sample interface is highly dependent upon the physical characteristics of the sample. Consequently, the type of sample interface will determine whether the optimum system will be configured as a remote sensor or a remote sensing configuration. The accuracy and precision necessary to control the process, as well as the complexity of the sample matrix, will determine the type of instrumentation needed (photometer or spectrophotometer) and the complexity of the chemometric routine used (multiple-wavelength or full-spectrum methods). Finally, a successful process NIR method will supply the NIR spectroscopic information in a format that is usable by process operators and process control systems.

The capabilities of process NIR measurements have progressed significantly within the 1990s in providing timely information for many different types of processes to enable improved monitoring and control of industrial manufacturing processes. The process information available through the use of process NIR analyzers yields not only quantitative information, but qualitative information as well. The chemical, polymer, petrochemical, pulp and paper, textile, and pharmaceutical industries, as well as the food industries, have been able to realize significant economic benefits through improvements in manufacturing efficiencies, product quality, and consistency. The potential for using NIR spectroscopy for monitoring or controlling industrial manufacturing processes will undoubtedly expand even further in capability and breadth of applicability in the future.

LITERATURE CITED

Coates, J. 1994. Designing the ideal process analyzer—Or at least making the attempt. NIR News 5(2):7-9.

Geladi, P., MacDougall, D., and Martens, H. 1985. Linearisation and scatter correction for NIR reflectance spectra of meat. Appl. Spectrosc. 39(3):491-500.

Ghosh, S., and Brodmann, G. L. 1993. On-line measurement of durable press resin on fabrics using NIRS. Textile Chem. Colorist 25(4):11-14.

Griffiths, P. R., and DeHaseth, J. R., eds. 1986. Fourier Transform Infrared Spectrometry. John Wiley & Sons, New York.

Hansen, W. G., Wiedemann, S. C. C., Snieder, M., and Wortel, V. A. L. 2000. Tolerance of near infrared calibrations to temperature variations; A practical evaluation. J. Near-Infrared Spectrosc. 8:125-132.

Kamen, E. W. 1999. Introduction to Industrial Controls and Manufacturing. Academic Press Series in Engineering. Academic Press, New York.

Martens, H., Jensen, S. A., and Geladi, P. 1983. Multivariate linearity transformations for NIR reflectance spectroscopy. Pages 205-234 in: Proc. Nordic Symp. Appl. Statist. Stokkand Forlag, Stavanger, Norway.

Sekulic, S. S., Ward, H. W., Brannegan, D. R., Stanley, E. D., Evans, C. L., Sciavolino, S. T., Hailey, P. A., and Aldridge, P. K. 1996. On-line monitoring of powder blend homogeneity by NIRS. Anal. Chem. 68(3): 509-513.

White, J. G. 1990. Near Infrared Spectroscopy for In-Process Monitoring of Pharmaceutical Dry Products. Eastern Analytical Symposium, Somerset, NJ, U.S.A.

CHAPTER 12

Analytical Application to Fibrous Foods and Commodities

F. E. BARTON, II and S. E. KAYS
U.S. Department of Agriculture
Agricultural Research Service
Richard B. Russell Research Center
Athens, GA
U.S.A.

I. INTRODUCTION

The analysis of highly fibrous feeds with near-infrared spectroscopy (NIRS) by diffuse reflectance is different from that of grains in several respects. The components of the plant matrix are more complex and involve numerous discrete interactions. The work of Hruschka and Norris (1982) showed that, for ground wheat, the summation of spectra of the chemical components (protein, starch, cellulose, moisture, and simple sugars) did not adequately reflect the total spectral composition when curve-fitting techniques were applied to the spectra. When the complexities of a forage sample are considered, it becomes obvious that the interaction of protein with lignin and carbohydrate along with minor constituents would make analyses by "pure components" impossible. Alternatively, it is possible to consider analyses based solely on functionality present in the spectrum if their relationship to some measure of quality is known. This requires a much better understanding of both the spectra of forages and what constitutes quality than we currently have. Therefore, the analyses must be made on the basis of the spectra correlated to empirical results. Again, the amount of fiber is much greater, i.e., the fiber is the matrix as opposed to being a component, as in most foods and feeds.

There has been very little work completed on fiber content and fiber structure in foods and grains. Lund and Smoot (1982) studied the dietary fiber content of tropical fruits and vegetables. They found that not only was the amount of fiber small and variable, but the fibrous fraction differed between species and from that of forage plants. Baker (1977, 1978) and Baker and Holden (1981) examined cereals and grains for fiber content and evaluated several methods of fiber analysis to see which one would be most suitable for cereals. Baker (1977) found that a buffered acid detergent fiber (ADF) determination apparently improved the recovery of cellulose. The general viewpoint of these authors was that no one method seemed to be best, as none gave a number that corresponded to an identical fraction in all cereals and grains. The new enzymatic procedures of Asp et al (1983) may provide both an assay and a means of characterizing fiber for monogastrics and ruminants. This chapter will be devoted to the following subjects: the structure and composition of highly fibrous materials with forage grasses as the examples; methods of analyses; and near-infrared (NIR) analysis specifically.

II. STRUCTURE AND COMPOSITION OF FORAGES

The microanatomical differences between warm- and cool-season (C-3 versus C-4) grasses were examined by Akin and Burdick (1973). One of the major differences noted was that warm-season grasses had a more highly developed parenchyma bundle sheath than the temperate grasses. In this study, the sections were stained for lignin with acid phloroglucinol to identify the sites of lignification. In a subsequent study (Akin and Burdick, 1975), the percentage of each type of tissue was reported for temperate and tropical grass species. The warm-season (tropical) grasses contained far more of the less easily degradable tissues than the cool-season (temperate) grasses. Figure 1 shows the cross section of a typical warm-season grass, coastal bermudagrass (CBG) (*Cynodon dactylon* (L.) Pers.). Note the large, thick-walled parenchyma bundle sheath on the outside of the large vascular bundles of CBG. This tissue is degraded slowly by rumen microorganisms in CBG, and, in some cases, can show a positive reaction for lignin with histochemical reagents, such as chlorine-sulfite. The mesophyll of the warm-season grass is composed of small, densely packed cells, whereas a typical cool-season grass, Kentucky-31 tall fescue (KY-31) (*Festuca arundinacea* Schreb.) has more loosely arranged cells. As shown by Akin and Burdick (1975), the cross-sectional area of fescue that comprises the mesophyll is 60%, while that of bermudagrass is only 27%.

Warm-season grasses are recognized to have higher fiber contents than cool-season grasses. The values in Table I taken from Barton et al (1976) reflect average increases of 10 percentage units in neutral detergent fiber (NDF) and 4 percentage units in ADF for the warm-season grasses. These differences persist when only the 4-week warm seasons are compared (NDF average 65.0, ADF average 34.6). The lignin data are of particular interest in that the higher-fiber content, less-digestible warm-season grasses also have a higher lignin content (4.9 versus 4.3%). If one considers only the 4-week regrowth samples, the higher-fiber, less-digestible warm-season grasses are identical in lignin (4.3%) and virtually identical in digestibility (61.6 to 61.5%) to the cool-season grasses, while maintaining an average of 9 percentage units NDF and 4 percentage units ADF higher fiber content. Clearly, compositional differences do not answer the question of quality and animal performance differences for warm-season versus cool-season forages.

One way to try and resolve some of the obvious differences between digestibility data and compositional analysis is to study both by the same method. Akin et al (1973) showed the pattern of forage digestion by rumen microorganisms (Fig. 2). This and subsequent studies (Akin and Burdick, 1975, 1981; Akin et al, 1975, 1977, 1983a, 1983b, 1984a, 1984b; Akin and Barton, 1983) have shown the indigestible residue to be made up of lignified vascular bundles, sclerenchyma, and cuticle for both warm- and cool-season species and that the parenchyma bundle sheath of the warm-season grass is very slowly degraded and not fully digested with a 48 hr in vitro incubation. Microscopic (light and electron) examination of the digested leaf sections allows direct observation of the plant tissues that were attacked and digested by rumen microorganisms. If samples are viewed after specific periods of incubation, the relative rate and mode of microbial attack can be ascertained. This qualitatively describes the fiber digestion of a given species grown under the specific environmental and management conditions for a given sample, but is not necessarily related to composition. To relate microscopic evaluation to a particular fiber analysis, the material represented by the residue after digestion by rumen microorganisms must be determined.

III. THE ANALYSIS OF FORAGES

Fibrous materials traditionally have been analyzed by the Weend Proximate Analysis Procedures as a means of estimating total digestible nutrients. In the Proximate Analysis procedures,

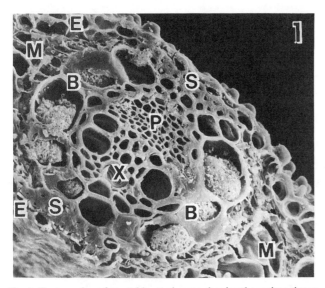

Fig. 1. Cross section of coastal bermudagrass showing the various tissues within the plant: sclerenchyma (S), epidermis (E), parenchyma bundle sheath (B), mesophyll (M), xylem (X), and phloem (P) (×500).

percentage of dry matter is determined by oven drying; percentage of crude protein is expressed as 6.25 × percentage of nitrogen from the Kjeldahl analysis; percentage of fat is determined by ether extraction; percentage of crude fiber, by alternate base and acid treatments; and percentage of ash, by incineration. These procedures continue to be the standard methods in use by many state testing laboratories. The forage analyses are empirical, based on the assumption that the reagents or experimental conditions affect each sample in an identical manner. They are all gravimetric procedures, and the calculated results are relative percentages. Moore and Mott (1973) and Martens and Russwurm (1983) have published excellent reviews that detail the status of gravimetric forage analyses. Since the molecular weight of a forage sample or any constituent therein cannot be determined, these percentages are the only way to express compositional properties quantitatively. The analyses are very dependent on sampling techniques, technician experience, and the environment in which the sample is analyzed, and all of the procedures are destructive.

Peter J. Van Soest published procedures on detergent fiber analyses (Van Soest, 1963a, 1963b; Van Soest and Wine, 1967, 1968) and a means of classifying forage fractions (Van Soest, 1967, 1973). The detergent analysis system defines the sample by separating it into two fibrous fractions, an NDF obtained by extraction in a boiling buffered 2% sodium lauryl sulfate solution and an ADF fraction obtained by extraction in a boiling 2% solution of hexadecyltrimethyl ammonium bromide in 1.0 N sulfuric acid. The NDF test (Van Soest and Wine, 1967) yields data on total fiber or cell walls, while the ADF test (Van Soest, 1963b) gives the lignified portion of the plant cell walls. The ADF procedure is primarily used as a preparatory step to lignin determinations, which are carried out either with 72% sulfuric acid (Van Soest, 1963b) or permanganate (Van Soest and Wine, 1968). The detergent analysis procedures, ADF and NDF in particular, have been used with varying degrees of success to predict the extent of digestion and relative intake, respectively (Rohweder et al, 1978). In the 1970s, these procedures began to replace the crude fiber procedure as the preferred methods of analyses for fiber in forages. The ADF procedure is now a method approved by the Association of Official Analytical Chemists International (AOAC). The literature values for many forages in the above references are comparable to those in Table I, which were obtained by the

TABLE I
Percent Compositional Analysis of Grasses[a]

Grass	IVDMD[b]	Protein	Ash	NDF[c]	ADF[d]	Hemicellulose	Holocellulose	PML[e]
Coastal: 4 week	66.1	19.2	7.8	61.0	29.1	31.8	61.2	4.1
Coastal: 8 week	50.4	11.0	5.4	71.2	40.0	31.2	66.8	6.0
Coastcross-1: 4 week	66.1	18.7	7.2	60.0	31.9	28.1	53.6	3.5
Coastcross-1: 8 week	54.9	13.9	7.2	62.9	39.0	23.9	55.6	5.5
Bahia: 4 week	59.6	15.7	6.2	71.0	35.7	35.3	60.8	3.4
Bahia: 8 week	53.2	9.2	5.9	67.5	35.0	32.5	76.0	5.3
Pangola: 4 week	54.5	7.0	4.7	69.4	41.7	22.7	57.3	6.3
Pangola: 8 week	48.6	5.9	5.1	67.0	29.5	37.6	47.4	4.8
Average tropical	57.8	12.6	6.2	66.2	35.2	30.4	59.8	4.9
Kenhy: 4 week	65.6	13.2	8.3	58.2	33.6	24.7	41.6	3.2
Ken-Blue: 4 week	58.1	15.5	7.1	54.0	30.6	23.5	43.5	4.3
Brome: 4 week	64.2	14.3	8.8	56.2	34.3	21.9	51.1	4.8
Orchard: 4 week	62.8	14.8	8.2	57.9	33.3	24.6	43.9	4.1
Kentucky-31: 4 week	62.7	14.2	8.8	58.4	31.4	25.5	45.2	3.4
Timothy: 4 week	66.8	13.4	8.8	55.6	34.7	20.9	42.0	4.1
Kentucky-31: 4 week (fall)	55.0	12.6	8.4	59.8	31.6	28.2	46.0	5.6
Kenhy: 4 week (fall)	60.1	12.6	8.0	57.3	30.7	26.6	43.8	4.2
Orchard: 4 week (fall)	58.8	17.8	9.7	54.0	29.4	24.6	45.5	4.8
Ken-Blue: 4 week (fall)	61.0	17.3	6.3	57.6	27.5	30.1	40.3	4.1
Average temperate	61.5	14.6	8.2	56.9	31.7	25.1	44.3	4.3
Average standard deviation	2.62	0.13	0.13	1.08	1.00	1.04	1.00	0.30

[a] Source: Barton et al (1976).
[b] In vitro dry matter digestibility.
[c] Neutral detergent fiber.
[d] Acid detergent fiber.
[e] Permanganate lignin.

detergent analyses procedures. However, the assumption is made that the reagents are affecting the same treatment on all samples, regardless of species, environment of growth, and agricultural management practices.

One way to examine this question is with microscopy, i.e., evaluation of leaf sections before and after microbial digestion. Akin et al (1975) examined by scanning electron microscopy the tissues that comprise the residues of NDF and ADF of leaf sections from CBG, a warm-season grass, and KY-31, a cool-season grass. In these experiments, 5-mm sections of the leaf blades were treated with the boiling reagents, prepared for microscopy, and viewed with a scanning electron microscope (Fig. 3). The mild treatment with neutral detergent reagent left the cell walls virtually intact in CBG and slightly distorted the mesophyll in some KY-31 samples. The amount of tissue removed from the leaf section was determined gravimetrically (Table II). It was found that much less tissue was removed from the sections than from ground leaf blades. The NDF conditions were such that the fragile cell wall membranes were not ruptured and cell contents were not removed unless the cell was opened by the knife when the sections were cut (Table II).

Treatment with acid detergent reagents revealed differences both between species and for all species when compared with digestion. For the warm-season CBG, the residue contained portions of the parenchyma bundle sheath. This tissue (Fig. 3b), which resisted the acidic treatment, was slowly degraded by rumen microorganisms. The opposite was true for KY-31: the only tissues remaining after a 1-hr treatment were cuticle, sclerenchyma patches, and pieces of vascular tissue far exceeding the digestion of KY-31 by rumen microorganisms. Thus, as a measure of extent of digestion (Rohweder et al, 1978), ADF would overestimate the digestion of KY-31 and underestimate the digestion or nutritive value of CBG. Direct comparisons of quality estimated from ADF values between temperate and tropical (cool- and warm-season) grasses must be made with caution. The differential response of the plant cell wall to these analytical reagents reflects differences in their availability to rumen microorganisms, and a linear response suitable for all species should not be expected.

Barton et al (1976) determined correlations between rumen bacterial digestibility and chemical data of warm- and cool-season grasses. It was found that protein and lignin content were the most significant factors. There was no correlation between digestibility and NDF if both the temperate and tropical grasses were considered as one set. Clearly, the empirical analyses do not truly measure quality when different and diverse species are being compared.

The digestibility of ADF by rumen microorganisms was examined by electron microscopy (Barton et al, 1981). The patterns of degradation or removal of tissue by ADF reagents and of digestion by rumen microorganisms paralleled those observed by Akin et al (1975). In these experiments, the isolated acid detergent treatment, followed by bacterial digestion, removed more fiber than either alone (Fig. 4a). The reverse dual treatment (Fig. 4b), i.e., digestion followed by acid detergent treatment, consistently removed even more tissue. The gravimetric results given in Table III showed about a 5–12-percentage-unit difference for the two treatments when the order was reversed. The qualitative data from the micrographs greatly enhanced the interpretation of the numerical gravimetric results. Moreover, it showed where overinterpretation of apparent fiber digestion values can be made if no consideration is given to the effect of the dual treatment on the plant cell wall.

Traditionally, the major inhibitors to forage quality and therefore animal performance are considered to be the amount of cell wall fiber and the extent of lignification (Moore and Mott, 1973; Wheeler and Mochrie, 1981). Barton and Akin (1977) examined the effect of lignin removal from the cell wall on the remaining tissues and the digestion of those tissues by rumen microorganisms. Potassium permanganate, used to oxidize lignin from ADF, was used to delignify the cell walls of temperate and tropical grasses (Barton and Akin, 1977). The results of those experiments showed that the lignified tissues were disturbed (Fig. 5). The parenchyma bundle sheath was separated from the highly lignified inner bundle sheath and the sclerenchyma was, in some cases, separated into

individual cells. In the temperate grasses, some of the mesophyll was removed. When the permanganate treatment was applied to NDF residue of ground forage, about 17% of the material was removed. Only about a fourth of this amount was lignin, and the rest of the material removed was carbohydrate. The rate and extent of digestion increased when the delignified leaf sections were incubated with rumen microorganisms. Tissues that were usually degraded between 24 and 48 hr were extensively degraded after only 1 hr (Fig. 6). Some tissues (sclerenchyma) that usually are not degraded also were digested. Thus, it is not just the amount of lignin, but the extent to which it is tied to the plant cell wall, that determines the rate and extent of digestion and forage quality. These subjective criteria are not obvious to the analyst in the usual empirical analyses. It is quite possible to overinterpret the results of those analyses. Other methods will give analytical results that are more definitive, but the cost in time and resources would be markedly high and the number of analyses that could be done is quite limited.

Two techniques that give more definitive results are high-pressure liquid chromatography (HPLC) and carbon 13 nuclear magnetic resonance spectroscopy (^{13}C-NMR). These techniques have been used to characterize hemicellulose, lignin, and lignin–carbohydrate complexes. Each has some distinct advantages over the other, but they also complement each other.

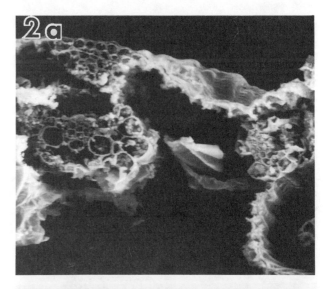

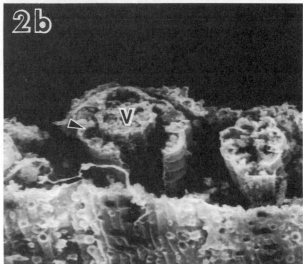

Fig. 2. **a,** Cross section of Kentucky-31 tall fescue incubated with rumen microorganisms for 48 hr; lignified tissues, cuticle, and portions of bundle sheaths remain (×150). **b,** Cross section of coastal bermudagrass incubated with rumen microorganisms for 48 hr; lignified vascular tissue (V) are partially degraded (×150).

Albersheim et al (1967) published the alditol acetate procedure for the analysis of plant cell wall hydrolyzates. The standard gas-liquid chromatograph (GLC) method had one disadvantage in that sample preparation was laborious and time-consuming. The HPLC method of Barton et al (1982) and the modifications by Windham et al (1983) have decreased the number of steps and time required as well as provided an improved recovery and good materials balance.

The procedure is essentially a 2.0 N trifluoroacetic acid hydrolysis of NDF followed by (HPLC) analysis of the hydrolyzate. The method is precise, i.e., a 1–2% error in each component sugar (Table IV)—and small differences in component sugars are readily apparent ($P < 0.05$) (Table V). Perhaps the most significant factor of these analyses is the materials balance. When simple sugars or commercial polysaccharides were hydrolyzed and analyzed, the recovery through HPLC was approximately 96.7–98.5%. For a sample of NDF, about 50% of the fiber was hydrolyzed and present in the hydrolyzate and 50% was present in the fibrous residue (Windham et al, 1983). Of the fraction hydrolyzed, only 40–70% (20–35% of the NDF) was recovered. That fraction not recovered as sugar either precipitated in the round bottom flask during washing or was retained in the Waters C-18 Sep-Pak1 (Waters Corp., Milford, MA) that was used as a final cleanup step. The acetone soluble material, which was yellowish brown, resembled a lignin carbohydrate complex (LCC). The yellowish brown material had been analyzed by ^{13}C-NMR. The work of Barton et al (1981) on ADF digestion and of Akin et al (1975) on the NDF and ADF of

TABLE II
Percent Residue of Neutral Detergent Fiber and Acid Detergent Fiber from Whole, Wiley-Milled Forage and Intact Leaf Samples of Coastal Bermudagrass and Kentucky-31 Tall Fescue[a]

	Neutral Detergent Fiber		Acid Detergent Fiber	
Grass	Whole[b]	Leaf[c]	Whole	Leaf
Coastal bermudagrass	59.4 ± 0.3	78.3 ± 1.6	29.1 ± 0.8	25.3 ± 0.7
Kentucky-31 tall fescue	50.7 ± 0.6	79.1 ± 2.3	28.6 ± 0.2	27.8 ± 1.3

[a] Source: Akin et al (1975).
[b] Average of 12 determinations plus standard deviation for whole, ground samples.
[c] Average of three determinations plus standard deviation for leaf samples.

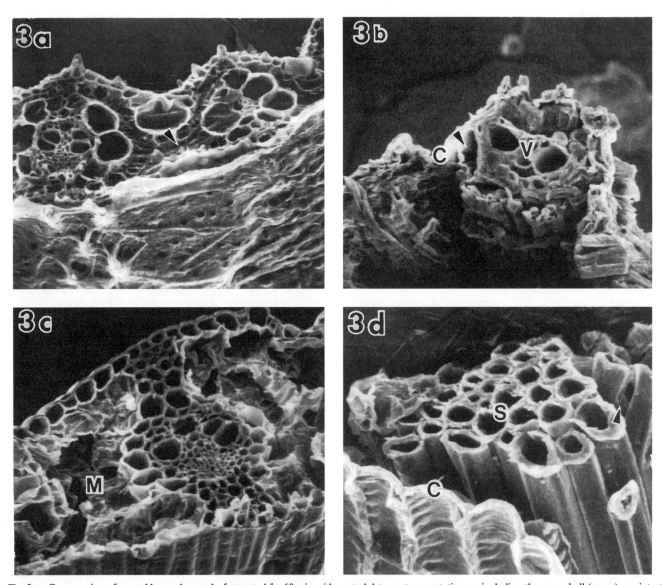

Fig. 3. **a,** Cross section of coastal bermudagrass leaf extracted for 60 min with neutral detergent reagent; tissues, including the mesophyll (arrow), are intact as in the control sections (×240). **b,** Cross section of coastal bermudagrass extracted for 60 min with acid detergent reagent; lignified vascular tissue (V) and cutinized epidermis (C) remain, and portions of outer bundle sheath (arrow) are infrequently observed to remain (×256). **c,** Cross section of Kentucky-31 tall fescue leaf extracted for 60 min with neutral detergent reagent; tissues are similar to those in control samples, and all are intact except for the mesophyll (M) (×224). **d,** Portions of tissues remaining in Kentucky-31 tall fescue after 60 min of extraction with acid detergent reagent; cuticular layers of epidermis (C) and sclerenchyma (S) separated into individual cells (arrow) are seen (×650).

forage grasses and earlier work (Akin et al, 1973) on the grasses themselves suggest that the readily hydrolyzable fraction could be the slowly degraded portion of the plant cell wall, and the extent to which it is tied to lignin or plant phenolics could determine its rate of digestion. The highly lignified portion of the plant cell wall that is essentially indigestible establishes by difference the maxi-

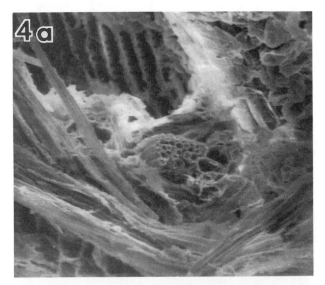

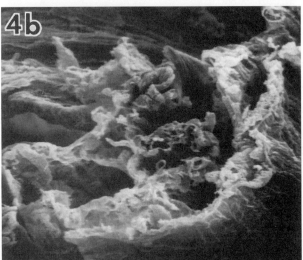

Fig. 4. **a,** Cross section of Kentucky-31 tall fescue extracted for 60 min with acid detergent reagent and then incubated with rumen microorganisms for 48 hr; only cuticle, portions of vascular bundles, and sclerenchyma remain (compare with Figs. 2a and 3d). **b,** Cross section of Kentucky-31 tall fescue incubated with rumen microorganisms and then extracted with acid detergent reagent for 60 min; small pieces of vascular bundles, cuticle, and some sclerenchyma remain (compare with Figs. 3a and 4a).

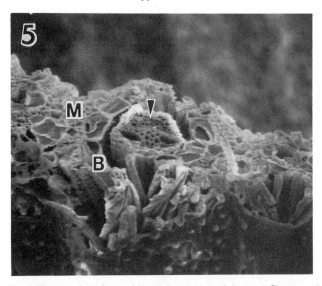

Fig. 5. Cross section of coastal bermudagrass neutral detergent fiber treated for 30 min with buffered $KMnO_4$ reagent and demineralized for 10 min. Mesophyll (M) and phloem (arrow) are intact, but the parenchyma bundle sheath (B) is almost completely separated from the rest of the vascular bundle. Sclerenchyma has been divided into individual cells and separated from the vascular bundle (×384). (Reprinted, with permission, from Barton and Akin, 1977)

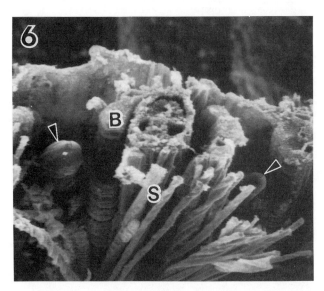

Fig. 6. Delignified coastal bermudagrass neutral detergent fiber digested. Mesophyll is completely removed, and phloem and parenchyma bundle sheath (B) are beginning to be removed. Sclerenchyma (S) is broken apart into fiber cells, and rumen protozoa (arrows) are present between vascular bundles (×384). (Reprinted, with permission, from Barton and Akin, 1977)

TABLE III
Percentage of Digestibility of the Five Grasses, Acid Detergent Fibers, and Combined Dry Matter Disappearance (CDMD)[a]

Grass	% IVDMD	% IADFD	% ADER	CDMD-1[b]	CDMD-2
Coastal bermudagrass	57.41	5.66	46.56	77.24	67.08
Coastcross-1 bermudagrass	61.17	5.91	41.17	77.16	68.53
Kentucky-31 tall fescue	64.54	5.76	44.73	80.40	76.93
Kenhy tall fescue	66.87	7.18	47.15	82.49	75.50
Orchardgrass	64.53	6.98	53.86	83.63	71.35

[a] Source: Barton et al (1981). Isolated acid detergent fiber (ADF) disappearance (IADFD) is a dry matter removed from isolated ADF by rumen microorganisms; acid detergent extracted residue (ADER) is the percentage of dry matter removed by extraction with acid detergent reagent of in vitro residue.

[b] CDMD-1 = (NDF × ADER/100) + IVDMD; CDMD-2 = (% ADF × IADFD/100) + (100 − ADF), in which CDMD is combined dry matter disappearance, NDF is neutral detergent fiber, and IVDMD is in vitro dry matter digestibility.

mum energy a ruminant can derive from a given forage sample. The largest compositional difference between tropical and temperate grasses is the amount of "hemicellulose." The tropical grasses contained 30–35% more hemicellulose as estimated by the percentage of NDF minus the percentage of ADF of the respective forage (Goering and Van Soest, 1970; Barton et al, 1976). Van Soest (1967) also stated that hemicellulose appears to have a special relation to lignin. The above results would support that statement.

Since the 1950s, NMR spectroscopy has been one of the most definitive structural elucidation tools and has provided chemists with a

TABLE IV
Precision of Method[a]

Injection No.	Sugar Components (mg) After Repeated Injections[b]			
	Xylose	Arabinose	Glucose	Galactose
1	37.13	11.57	13.88	TR[c]
2	37.28	11.74	13.73	TR
3	37.82	11.53	14.45	TR
4	38.14	11.76	14.16	TR
5	37.98	11.46	14.09	TR
6	37.19	11.08	13.91	TR
7	37.35	11.76	13.75	TR
8	37.89	11.55	14.47	TR
9	37.13	11.32	13.51	TR
10	37.64	11.17	13.43	TR
Mean	37.55	11.49	13.94	TR
Std. dev.	0.38	0.24	0.35	
Coefficient of variance	1.02	2.09	2.55	

[a] Source: Barton et al (1982).
[b] Milligrams recovered per 100 mg of coastal bermudagrass neutral detergent fiber hydrolyzed.
[c] TR = trace; amount present was inadequate for quantitative purposes.

TABLE V
Precision of Sample Preparation
for Coastal Bermudagrass (CBG) Neutral Detergent Fiber[a]

Sample	Xylose	Arabinose	Glucose	Galactose
CBGa	36.67 ± 0.44[b]	11.58 ± 0.37	14.06 ± 0.26	TR[c]
CBGb	37.47 ± 0.36	11.46 ± 0.34	14.04 ± 0.36	TR
CBGc	37.35 ± 0.26	11.64 ± 0.39	13.92 ± 0.34	TR

[a] Source: Barton et al (1982). Data presented as milligrams recovered per 100 mg of fiber hydrolyzed.
[b] Each value is the mean of five observations (i.e., five injections of three sample preparations) ± the standard deviation.
[c] TR = trace; amount present was inadequate for quantitative purposes.

wealth of structural and analytical information. Since 1975, the utility of this spectroscopic technique has been extended to more complicated and biological samples by ^{13}C-NMR. The technique was used by Barton et al (1982) and Windham et al (1983) to assist in the analysis of the hydrolyzable fraction of plant cell walls previously discussed. In this case, the hemicellulose of forages was obtained by alkali extraction (Routley and Sullivan, 1958). The NMR spectra (Fig. 7) shows structures that have been interpreted as glucoarabinoxylans. The spectrum contained no resonance (signal) for the C-6 carbon of glucose, the only carbon different from the first five carbons of the sugar xylose. It is known from HPLC analysis that the hemicellulose fraction contains 5–10% glucose. One may reasonably assume that the C-6 of glucose is bound and shifted under another resonance. This assumption was verified by adding the totals for xylose and glucose obtained from the NMR (Table VI) and comparing the compositional values with the HPLC results (Table V). The agreement was within ±5% when each was compared as percentages of whole grass. From these data, a structure for grass hemicellulose that has an arabinoxylan attached to a 1:4 glucan (Fig. 8) was suggested. The work of Albersheim et al (1967) describes the hemicellulose of sycamore as a xyloglucan and an arabinoxylan. The evidence described above from the ^{13}C-NMR indicates a single structure for grass hemicellulose, since no evidence for multiple C-6 carbons was observed.

The NMR method has been used to elucidate the structure of forage lignin (Himmelsbach and Barton, 1980). Lignin is a poly-

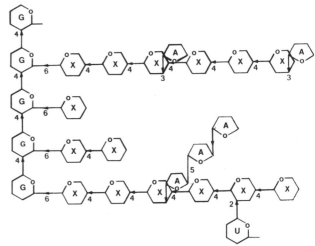

Fig. 8. Possible structure for polysaccharide portion of grass hemicellulose.

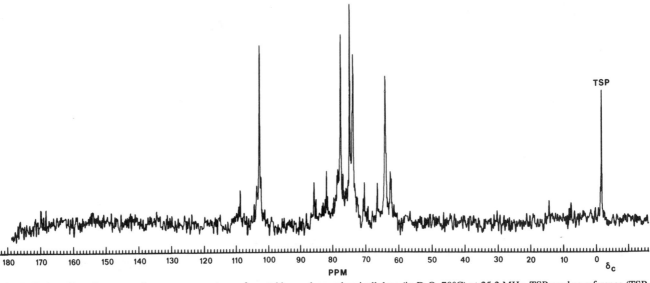

Fig. 7. Carbon-13 nuclear magnetic resonance spectrum of coastal bermudagrass hemicellulose (in D_2O, 70°C) at 25.2 MHz. TSP used as reference (TSP = sodium 3 trimethylsilylpropionate, 2,2,3,3, d4).

mer of phenylpropanoid (C-9) units and is quite complex. It differs only in the 3- and 5-position substitution of the phenyl ring by methoxyl groups, i.e., *p*-coumaryl (none), coniferyl (monomethoxyl), and syringyl (dimethoxyl). The work of Ludemann and Nimz (1973, 1974a, 1974b), Nimz (1974), Nimz and Ludemann (1974, 1976), and Nimz et al (1974a, 1974b, 1975) has characterized the lignins of hardwoods and softwoods. Acetal lignins extracted and described by Bolker and Teraschima (1966) contained some carbohydrate, as did the lignin extracted from bermudagrass by the milled wood lignin procedures (Fig. 9). Grass lignins cannot be extracted free of carbohydrates. There were differences noted by Himmelsbach and Barton (1980) in the spectra of the lignins from different species of grass. The cool-season (temperate) fescue contained about equal amounts of the three lignin moieties, while the warm-season (tropical) bermudagrass contained predominantly the unsubstituted *p*-coumaryl moiety. This unsubstituted lignin would be the one considered most likely to be attached to carbohydrates (Sarkanen and Hergert, 1971) and would have the greatest chance to tie up more carbohydrates. This tie-up of more of the warm-season grass carbohydrates through the *p*-coumaryl unit would render them less available for digestion by rumen microorganisms than the cool-season grass carbohydrates. The nature of this binding of carbohydrate and lignin had not been discretely determined. Harkin (1973) has

TABLE VI
**Relative Percent Composition of Coastal Bermudagrass (CBG)
and Orchardgrass (OG) Forage Hemicelluloses
by C-13 Nuclear Magnetic Resonance**

Carbohydrates	CBG	OG
Xylose + glucose	84.1	77.5
Arabinose	15.9	22.5

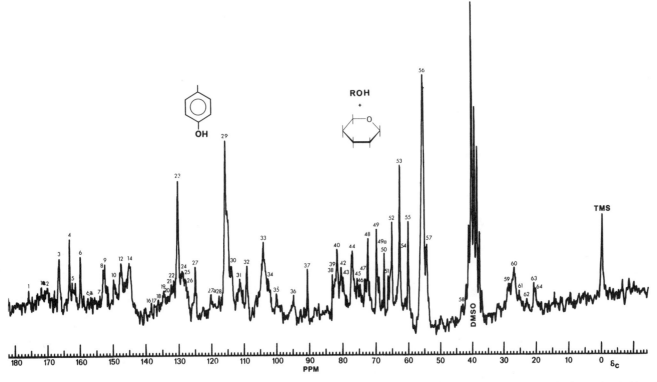

Fig. 9. Carbon-13 nuclear magnetic resonance spectrum of coastal bermudagrass lignin at 25.2 MHz. Unsubstituted aromatics predominate (signals 23 and 29). Presence of carbohydrate is shown in the 60–90-ppm region. DMSO = dimethylsulfoxide, and TMS = trimethylsilane.

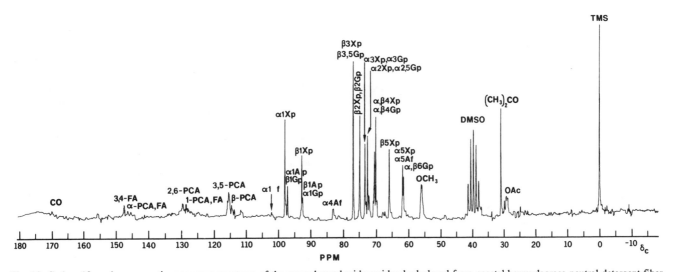

Fig. 10. Carbon-13 nuclear magnetic resonance spectrum of the nonpolysaccharide residue hydrolyzed from coastal bermudagrass neutral detergent fiber. Appears to be a lignin-carbohydrate complex. DMSO = dimethylsulfoxide, and TMS = trimethylsilane.

suggested a covalent linkage that would be consistent with the work of Himmelsbach and Barton (1980). Evidence can be found for the existence of an ether or a carbonyl ester resonance in the lignin spectra and in those of the LCC shown by Himmelsbach and Barton (1980). The large resonances at 162 and 172 ppm (Fig. 1 in Himmelsbach and Barton, 1980) indicated the presence of carbonyls in esters.

Examination of the spectra of the LCC isolated from the HPLC analysis of the NDF showed an arabinoxylan associated with an unsubstituted lignin moiety and the presence of an ester linkage (Fig. 10). Subsequent HPLC analysis of the carbohydrate portion gave a xylose-to-arabinose ratio of 1:1 (F. E. Barton II, W. R. Windham, and D. S. Himmelsbach, *unpublished data*). This suggested that the lignin was preferentially associated with the arabinose moiety, because normal xylose-to-arabinose ratio is in the range of 3–4:1.

The use of ^{13}C-NMR and the interpretation of the spectral data along with HPLC data gave a fairly clear picture of the structure of the molecular entities that comprise the more fibrous and lignified portions of the plant. It can be said with some certainty what comprises the plant cell wall and how some of the pieces fit together, but the methods lack several important criteria for good routine analytical methods.

By definition, a good analytical method is one that is accurate, precise, rapid, low cost, easily interpretable, and simple in sample preparation, and that examines the structure in situ with no chance

for sample preparation artifacts to alter the results. Certainly, the NMR method is accurate, precise, and interpretable, but it is neither rapid nor low cost. The HPLC method is an improvement in speed over the gas chromatography alditol acetate procedures and reasonable in cost, but still not fast enough to handle large numbers of samples. The standard gravimetric methods are inexpensive, and large sample numbers are accommodated by tailoring laboratory methods. Many state testing laboratories run 2,000–3,000 samples a year. What is needed is a faster, cheaper method. Three recent texts have given updated and expanded treatments of the standard methodology (Ulbricht and Theander, 1980; Wheeler and Mochrie, 1981; Martens and Russwurm, 1983). An article by Asp et al (1983) described an enzymatic method of fiber analysis and a method of describing soluble and insoluble dietary fiber. During the latter part of the 1980s, a lot of effort was expended worldwide to develop standardized enzymatic procedures for dietary fiber. These culminated in the adoption as an "Official Method" that is known as the "Prosky Method" by the AOAC International (Prosky et al, 1992). These procedures may be refined and automated, but they still retain the high labor costs involved with gravimetric analyses. Li and Cardozo (1992) continued research on dietary fiber by chemical means to reduce the time and effort required by the enzymatic procedure.

IV. NIR AS AN ANALYTICAL METHOD

NIRS is a rapid, inexpensive method to analyze forages for compositional, bioassay, or quality estimates. The method was developed and used for the rapid determination of oil, protein, and moisture in grains and oilseeds (Norris and Hart, 1965; Ben-Gera and Norris, 1968; Hymowitz et al, 1974; Rinne et al, 1975). It has been used to predict the nutritive value of feedstuffs (Norris and Barnes, 1976; Shenk et al, 1977). More recently, the composition of forages was investigated (Norris et al, 1976; Shenk and Hoover, 1977; Shenk et al, 1977, 1978; Barton and Burdick, 1979, 1980, 1983; Barton and Coleman, 1981; Burdick et al, 1981).

The NIR method requires an NIR spectrometer equipped for diffuse reflectance, a minicomputer for data handling and calculation of calibration equations, and suitable samples to be used for calibration. A set of samples (n) is selected for calibration. There are three basic ways to select samples for the calibration: (i) a percentage or some number of samples from the set taken randomly, (ii) every ith sample, or (iii) a subset that covers the analytical range and is qualitatively similar to the entire sample population to be analyzed. In the case of the particular system that is part of the National Near-Infrared Research Project (U.S. Department of Agriculture), spectra are taken and 700 data points for each sample are stored in a computer file. The data for several compositional factors (X) are added to the spectral information. The file then contains n samples of 700 data points and X constituents. Multiple linear regression techniques are used to generate an analytical equation as shown by

$$\%X = C0 + C1(I1) \pm C2(I2) \pm Cj(Ij) \qquad (1)$$

in which $C0$ is a constant representing the intercept and Cj are the constants multiplied by the intensity function at j. The regression techniques can be stepwise or multiple stepwise (Shenk et al, 1977); they can involve derivatives of the logarithm of reciprocal reflectance (Norris et al, 1976; Shenk et al, 1978) and combinations of techniques such as ratios of derivatives (Norris, 1983a, 1983b). It can involve data treatments such as the use of Gaus-Jordan linear algebra as a means of obtaining calibration equations (Honigs et al, 1983). An equation can be derived for each compositional factor. The spectra of unknown samples are then taken and the analytical result calculated. It is possible to obtain up to 10 analyses per sample in a few minutes. Although the instrument is expensive, the per-sample cost is quite low. Mobile systems are available to take the analysis to the samples (J. S. Shenk, personal communication). Once properly calibrated, the instruments can be used to analyze several hundred samples a day.

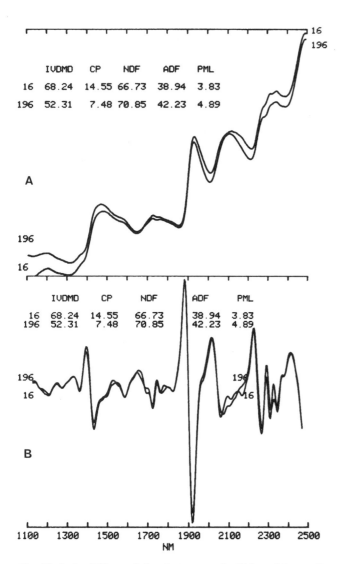

Fig. 11. **A,** Log(1/R) near-infrared spectrum of a high- and low-quality Old World Bluestem grass. **B,** Second derivative of log(1/R) near-infrared spectrum of the same samples. IVDMD = in vitro dry matter digestibility, CP = cross-polarization, NDF = neutral detergent fiber, ADF = acid detergent fiber, and PML = permanganate lignin.

Calibration of NIR instruments has been, and will continue to be, the source of much research (Barton and Burdick, 1979; Barton and Coleman, 1981; Hruschka and Norris, 1982; Honigs et al, 1983). Calibration depends on three factors: the appropriateness of the calibration set, the quality of the calibration data, and the mathematical treatment of the spectral data. The appropriateness of the samples chosen for calibration has been the subject of considerable statistical debate. As mentioned above, some researchers believe that the selection should be random or at least arbitrary from the total population (Shenk and Hoover, 1977); others would choose the samples to reflect a wide range of compositional data (Barton and Coleman, 1981). However a set is chosen, once its spectra are recorded, the choice of the mathematical data treatments must be considered next.

Norris et al (1976) reported that the logarithm of reciprocal reflectance and its second derivative were the two most promising mathematical treatments for forages. The logarithm of reciprocal reflectance gives a spectral result analogous to absorbance. The second derivative (D2OD) can be obtained by successive derivatization over typically 20–30 nm with some smoothing or by using three equal spectra lengths (A, B, and C) of 20–30 nm and the equation

$$D2OD = \log(1/R)A - 2\log(1/R)B + \log(1/R)C \qquad (2)$$

Maddams (1980) reviewed curve fitting and derivative techniques to show both their utility and their limitations. Most of the forage work has been done using second derivative. This mathematical treatment generally gives better-resolved spectra with minima occurring for each maximum that was resolved. The data treatment can also remove differences from a nonuniform particle size from the spectra. If the appropriate derivative segment is selected, a multicomponent absorption signal (Fig. 11) can be resolved into its individual absorption with a minimum at the wavelength where the maximum in $\log(1/R)$ (absorbance) spectra would have been had the band resolved.

The spectra in the NIR are broad, featureless, often overlapping absorbances. The signals arise from carbon-hydrogen stretching vibrations, their harmonics, overtone, and combination bands. Early work by Kaye (1954, 1955) characterized the spectra for simple molecules and described basic instrumentation necessary for acquiring spectra.

The signals in this region, 800–2,500 nm, are weak (100–1,000 times less intense than the mid-infrared [MIR]), but the radiation sources are more intense and the detectors more efficient. Since the amount of absorbed light is small, the distortion is also small. The NIR is an optically thin region, i.e., dispersion is weak and refractive indices are relatively constant. The weak signals, which would normally be considered a disadvantage, are in reality more of an advantage, since there is less variation in signal intensity compared with either the MIR or ultraviolet.

The third calibration consideration is the quality of the analytical data of the samples themselves. The data in Table VII were obtained from 100 Old World Bluestem grass samples that were used to calibrate an NIRS instrument (Coleman and Barton, 1982). These data were obtained from three separate calibration attempts. First, the laboratory gravimetric values from the samples that were used in the initial (I) calibration were obtained 2–3 years before the spectra were taken. The results were not good, as indicated by low R^2 values and high standard errors of calibration (SEC) (Table VII, rows corresponding to run I). The R^2 and SEC improved when all the samples were reanalyzed and the regression made to recalibrate the instrument (Table VII, rows corresponding to run R). There were still some samples that were identified as having suspect data, i.e., they failed the t test, or had spectra that differed from those of other samples with similar data. These samples were reanalyzed and their new data, if different, substituted in the calibration file to give the third set of R^2 and SEC values (Table VII, rows corresponding to run C), which were more acceptable. Good standard analytical data are absolutely essential to calibration. The subsequent NIR analyses can only be as good as the total of all laboratory errors.

The calibration and analyses of forages have been accomplished in our laboratory on three instrument systems. The first was a tilting-filter reflectance spectrometer, the second was on an early commercial scanning monochromator, and the third was on the spectrometer common to the U.S. Department of Agriculture Near-Infrared Network (Barton and Burdick, 1979; Barton and Coleman, 1981; Burdick et al, 1981; Barton and Burdick, 1983). Although these instruments differ in design and levels of sophistication of the spectrometer, they all yield spectral data with 300–2,000 data points per spectrum. The tilting-filter and the network

TABLE VII
Effect of Laboratory Data on Calibration[a]

Analysis	Run[b]	Mean + Std. Dev.	λ[c]	R^{2}[d]	SEC[e]	Repeat[f]
Dry matter	I	93.3 ± 1.82	3	0.63	1.11	0.03
	R	95.3 ± 1.16		0.73	0.61	0.06
	C	94.7 ± 1.06		0.84	0.43	0.02
Protein	I	12.2 ± 1.96	3	0.84	0.80	0.03
	R	12.4 ± 2.00		0.87	0.71	0.04
	C	12.3 ± 2.01		0.94	0.49	0.05
Neutral detergent fiber	I	67.7 ± 3.12	5	0.65	1.86	0.16
	R	68.2 ± 2.91		0.73	1.52	0.08
	C	67.6 ± 2.71		0.82	1.15	0.27
Acid detergent fiber	I	38.8 ± 2.55	3	0.63	1.55	0.10
	R	38.6 ± 2.95		0.81	1.27	0.25
	C	38.8 ± 2.90		0.87	1.04	0.09
Permanganate lignin	I	4.9 ± 1.14	3	0.38	0.90	0.05
	R	3.7 ± 1.23		0.66	0.71	0.01
	C	3.7 ± 0.87		0.61	0.54	0.01
IVDMD[g]	I	60.4 ± 4.10	3	0.68	2.33	0.21
	R	60.2 ± 5.78		0.65	3.40	0.33
	C	62.5 ± 3.34		0.83	1.36	0.11

[a] Source: Coleman and Barton (1982).

[b] I = file data were several years old; R = samples were all reanalyzed routinely; and C = samples from R that were statistically outliers were reanalyzed and new data incorporated into file.

[c] Number of wavelengths used in equation.

[d] Coefficient of determination.

[e] Standard error of calibration.

[f] Repeatability error.

[g] In vitro dry matter digestibility.

spectrometer use 700 data points (the network spectrometer produces 1,873, but only 700 data points are centered and averaged data points are saved). The older monochromator initially took only 300 data points, but currently takes 2,000. This capability is important when multiple analyses and mathematical transformations (derivatives, ratios, etc.) of the spectral data are required.

Fixed-filter instruments give one data point per filter. As such, the scanning systems are capable of producing two orders of magnitude more spectral data. With this increase in available data comes the requirement for computational hardware and software to utilize it. The inherent computational capabilities of the tilting-filter instruments are quite limited. Barton and Burdick (1979) and Burdick et al (1981) showed that while one "product" (file) would hold six equations based on six wavelengths with a possibility of using one or more of five mathematical treatments, it was necessary to use a different set of wavelengths for each constituent to be analyzed (Table III in Burdick et al, 1981). To change derivative segment width, a new EPROM (erasable programmable read only memory) chip (Neotec, Silver Spring, MD) had to be installed. This constituted a major revision (Table V in Barton and Burdick, 1979). The current network instrumentation, which incorporates a holographically ruled grating monochromator, is interfaced to, in our case, a Digital Equipment Corporation PDP 11/34A minicomputer (Digital Equipment Corp., Maynard, MA). This computer and its attendant software (which was written at Pennsylvania State University, University Park) will run either the monochromator or the tilting-filter spectrometer. The computer stores all the spectral data points and performs smoothing and mathe- matical transformation. In addition, all of the laboratory data are stored so that the spectral transformation parameters can be changed and the regressions rerun without having to re-enter data. The data files can also be edited, which has led to improvement in SEC and standard error of performance (SEP) as shown in Table IX (Barton and Coleman, 1981). Norris (1983a) prepared an excellent review of the process of extracting compositional data from spectral data. The U.S. Department of Agriculture, Agricultural Research Service, National Near-Infrared Reflectance Research Project published the results of its collective studies in Handbook 643 (Barton, 1985).

There have been three major improvements in the NIRS method as described above. First, NIRS is now an accepted official method of the AOAC International for protein, ADF, and moisture (Barton and Windham, 1988; Windham and Barton, 1991).

Second, the statistical data treatments have become more sophisticated and robust. Principal component analysis (PCA) and partial least squares (PLS) and the regression models developed for predictive analyses are used (Cowe and McNichol, 1985; Haaland et al, 1985; Geladi and Kowalski, 1986a, 1986b; Martens and Martens, 1986; Robert et al, 1986; Bertrand et al, 1987; Downey et al, 1987; Devaux et al, 1986, 1987, 1988a, 1988b). The statistical technique of "clustering analysis" was introduced as a means of separating populations and detecting outliers (Tormod, 1987; Martens and Naes, 1989). The most salient advantage to these data treatments is a reduction in the magnitude of the SEP and models that are more broadly applicable. This is not without some cost. The interpretation of models based on PCA and PLS are more complex, since they use all wavelength information in each component or PLS term, yet, when examined, they are rich in information (Bertrand et al, 1985, 1987, 1991; Robert et al, 1987, 1989a, 1989b). The use of these statistical techniques has been advanced by the increase in power and capacity of personal computers.

Third, the current generation of instruments can accommodate many sample presentation devices other than a spinning cup and reflectance detectors. There are bulk sample cells, liquid cells, fiber optics, and flow through cells that use reflectance, transmission, and transflectance detectors. These devices have been highlighted at recent Pittsburgh Conference shows and their acceptance verified

TABLE VIII
Variation Resulting from Technique

Constituent	Conventional Mean	Conventional SE[a]	Fibertec Mean	Fibertec SE
Neutral detergent fiber	68.5	0.30	64.9	0.24
Acid detergent fiber	39.1	0.39	36.4	0.34
Permanganate lignin	6.54	0.45	5.43	0.74

[a] Standard error between replicate mean squares.

TABLE IX
Effect of Laboratory Data on Calibration

Analysis	Type[a]	Range	$\overline{SD}$[b]	λ[c]	R^2[d]	SEC[e]	Repeat[f]
Protein	A	7–21	0.62	1	0.94	1.07	0.1676
	B	8–23	0.18	1[g]	0.95	1.05	0.1434
	R	7–22	0.15	1	0.94	1.10	0.1789
Neutral detergent fiber	A	43–76	1.99	3	0.97	1.45	0.2750
	B	43–75	0.45	3	0.96	1.76	0.2693
	F	44–75	0.42	3	0.98	1.24	0.2458
Acid detergent fiber	A	30–44	2.03	3	0.93	1.18	0.2627
	B	29–45	0.25[h]	3	0.85	1.84	0.2994
	F	28–43	0.43	3	0.87	1.49	0.2684
Permanganate lignin	A	3–12	0.51	3	0.92	0.71	0.1662
	B	2–12	0.21	2	0.71	1.50	0.1029
	F	…[i]	…	…	…	…	…
Acid detergent lignin	A	3–10	0.56	3	0.87	0.70	0.1552
	B	3–10	0.11	3	0.90	0.67	0.1201
	F	1.5–10	0.30	3	0.94	0.57	0.1386
In vitro dry matter digestibility	A	42–76	2.57	4	0.95	2.00	0.6426
	B	47–69	1.04	4	0.95	1.38	0.5251
	F	…	…	…	…	…	…

[a] A = average of the four best laboratories (conventional method); B = conventional method data from Athens laboratory; R = rerun percentage of protein in Athens laboratory; and F = Fibertec analysis data.
[b] $\overline{SD}$ = standard deviation.
[c] Number of wavelengths used in equation.
[d] Coefficient of determination.
[e] Standard error of calibration.
[f] Repeatability error.
[g] Using three degrees $R^2 = 0.98$, SEC = 0.64.
[h] Six replications per sample.
[i] … = No Fibertec equivalent.

by the attendance at technical sessions at Pittsburgh Conferences, Eastern Analytical Symposiums, and other scientific forums.

Two areas of the NIR method need further study. First, the precision, and with it the accuracy of the fiber and other laboratory measurements, must be improved. Second, the physical significance of the analysis must be determined. It is essential that methods be developed that measure components in the plant that influence its digestion and subsequent utilization by ruminant animals.

The Tecator Fibertec System (Foss North America, Silver Spring, MD) is a semiautomated extraction apparatus that enables improvements in accuracy and precision of gravimetric fiber determinations. The word "extraction" is used here in the context of solubilization of materials from the sample whether they are hydrolyzed, digested, solubilized by reflux, or complexed. The accurately weighed sample is introduced into a sintered glass crucible and placed into the Fibertec hot-extraction unit (Foss North America) (up to six crucibles at a time). The reagents are pumped into the crucibles (preheated) and boiled in the crucible extraction column. After the allotted time, the samples are filtered and washed with no sample transfer. The Fibertec has an excellent system for washing the sample free of all detergent, which eliminates one large source of error. The vacuum filtration and washing are very uniform for all samples, further helping to reduce errors. The precision of analysis is improved overall as shown in Table VIII. The standard deviation for triplicate determinations is roughly three-fourths that of the conventional boil-and-stir procedure.

Improvement of reference analysis data can be demonstrated by using the data to calibrate an NIR spectrometer. Table IX shows the results of three different wet-chemical calibration data sets on the same sample (Barton and Burdick, 1983). The first (A) is the average of triplicate analyses by four different research laboratories (12 replicates); the second (B) is a single-laboratory triplicate

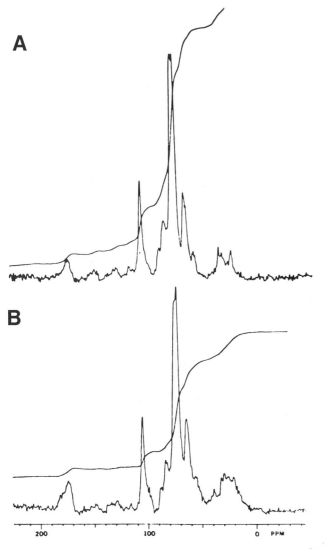

Fig. 12. Solid-state carbon-13 nuclear magnetic resonance spectrum of **A**, coastal bermudagrass and **B**, Kentucky-31 tall fescue at 50 MHz.

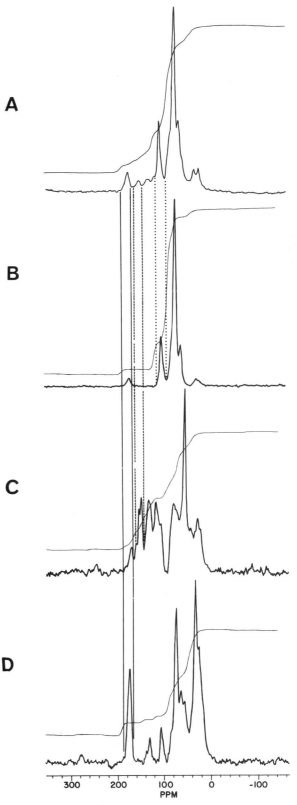

Fig. 13. Solid-state carbon-13 nuclear magnetic resonance spectrum of **A**, coastal bermudagrass (CBG); **B**, CBG holocellulose; **C**, CBG lignin; and **D**, CBG protein at 50 MHz. These spectra are used to define the regions used for quantitative determinations of carbohydrate, lignin, and protein.

determination from the authors' laboratory. The third is a triplicate determination using the Fibertec in our laboratory. The precision (as determined by lower standard deviation of triplicate) of the Fibertec (F) results are generally better than the average of the four laboratories and always better than the single-laboratory determination. This is reflected in the calibration of an NIR instrument by the R^2 and SEC values.

Note in Table IX the large, between-laboratory errors as indicated in the standard deviation column for the average of the combined results from four laboratories. The Fibertec results are obviously more precise, but since the results appear as good or better than the averages for the data calibration from the four laboratories, the results may also be more accurate and yield a better calibration. Data in Table IX provides an internal check for within-laboratory consistency. Since crude protein was not run on the Fibertec, those values labeled "R" are a repeat of the single-laboratory analysis at a different time used for an internal check.

The second major limiting factor to the NIR method is the relationship of the compositional data to the spectra. Work by Himmelsbach et al (1983) has shown that solid-state ^{13}C-NMR can be used to compare the relative ratio of plant constituents. This method may be able to provide accurate data on constituents as they are in situ. Historically, the one great limitation to the use of ^{13}C-NMR for structural and quantitative results is that the sample must be dissolved in a suitable solvent before the spectrum can be taken. This limitation means that solid samples that are insoluble (i.e., whole grass) are not compatible with the NMR technique. Again, the extraction of hemicellulose, lignin, or LCC modifies the in situ structure. Lignin, once isolated, is very susceptible to condensation reactions upon exposure to moisture and air. Interpretations must be made with the knowledge that artifacts of isolation are likely to exist. The NMR technique of cross-polarization/magic angle spinning (CP/MAS) (Schaefer and Stejskal, 1979) has made it possible to record the spectra of rigid materials, including fibrous residues and even whole-grass samples.

Basically, there are three important constituents of plant fiber: carbohydrate, lignin, and protein. Solid-state ^{13}C-NMR spectra of these constituents have been obtained separately or in combination by several groups of researchers (Atalla et al, 1980; Earl and Vanderhart, 1980, 1981; Rutar et al, 1980; Dixon et al, 1981; O'Donnel et al, 1981; Schaefer et al, 1981) with the CP/MAS technique (Schaefer and Stejskal, 1979). Several researchers have also reported that the ^{13}C-NMR signals caused by protein carbonyls were proportional to values obtained for protein by standard chemical methods.

Using this technique, Himmelsbach et al (1983) reported the application of CP/MAS ^{13}C-NMR for the determination of the ratio of carbohydrate, lignin, and protein between grass species from single spectra of whole-plant material. Figure 12 gives the spectra of two common forage grasses, KY-31 tall fescue and CBG. Both spectra contain relatively narrow lines that correspond to the carbohydrate, lignin, and protein in the plant. The spectra in Figure 13 show how the individual constituents in the whole grass can be divided into their individual components for quantitative analysis.

Since no molecular weights can be determined for the grasses, and the packing densities and a number of physical parameters cannot be easily obtained or held uniform, direct measurement of a constituent is not possible by this technique. Ratios were determined to compare the quantitative results for these two grasses. The data in Table X are results obtained from several instruments and at two field strengths. The 50-MHz carbon data appeared to be the best and agreed closely to NIR and quite well with the wet chemical analyses data. A probable explanation for this is that both spectral methods, the NIR and NMR, are measuring responses that result in signals from the constituents in the entire plant material, whereas the wet chemical methods measured entities that may not have been representative of the entire plant material. In the chemical methods, as in all empirical methods, it has been assumed that all the plant materials respond to the chemical reagents identically, which is unlikely to be the case. It is possible that more low-molecular-weight carbohydrates may be washed away as cell solubles from a cool-season grass species such as KY-31 than from a warm-season grass such as CBG. In the case of NDF, this would result in a higher value for the CBG/KY-31 carbohydrate ratio. It could be caused by structural differences between the plants, and not true differences in carbohydrate contents of the intact plants, or by differences in response of soluble carbohydrate to the solid-state technique.

Structural differences in lignin could also cause "lignin" in one species to be more susceptible to permanganate treatment. This would be especially true for the type of "lignin" in a warm-season plant like CBG in which p-coumaryl units have been suggested to occur as side chains on polymeric lignin or even on polysaccharides as α- and β-unsaturated esters (Himmelsbach and Barton, 1980). These C-9 phenolic units would be more susceptible to hydrolysis by the acid detergent reagent and would be removed to a greater extent with the hydrolyzable carbohydrate than would the true polymeric lignin. As a result, there would be a slightly lower value for the CBG/KY-31 ratio for "lignin" by the chemical method than was actually present in the plant.

For protein, the instrumental techniques might produce a different ratio of protein value because the standard Kjeldahl chemical method actually determines total reduced nitrogen. The Kjeldahl method could be influenced by nonprotein nitrogen in

TABLE X
Ratio of Fiber Constituents in Coastal Bermudagrass vs. Kentucky-31 as Determined by CP/MAS ^{13}C-NMR[a], Near-Infrared Reflectance, and Chemical Analysis[b]

Technique	Constituent		
	Carbohydrate	Lignin	Protein
Nuclear magnetic resonance			
37.7 MHz	1.27	1.57	0.68
50.0 MHz	1.27	1.66	0.59
Near-infrared reflectance[c]	1.27	1.70	0.58
Chemical analysis[c]	1.33	1.60	0.65

[a] Cross-polarization/magic-angle spinning carbon-13 nuclear magnetic resonance.

[b] Source: Himmelsbach et al (1983).

[c] Each ratio is the result of at least triplicate replications with standard deviations of ±1 in percentage of composition.

TABLE XI
Calibration of Near-Infrared Reflectance Instrument for Broad Spectrum of Samples[a]

Analysis	N[b]	SEC[c]	R^2[d]	Repeatability Error	SDP[e]	SDA[f]
In vitro dry matter digestibility	3	1.74	0.92	0.16	0.37	1.72
Crude protein	2	0.72	0.98	0.02	0.22	0.09
Acid detergent fiber	3	1.6	0.88	0.08	0.24	0.28
Acid detergent lignin	3	0.78	0.87	0.06	0.12	0.29

[a] Source: Barton and Coleman (1981). Set of 30 samples included 13 bermudagrass, 5 alfalfa, 4 orchardgrass, and 1 timothy grass. Twenty were hays and ten were fresh-frozen and freeze-dried.

[b] Number of wavelengths.

[c] Standard error of calibration.

[d] Coefficient of determination.

[e] Average standard deviation of predicted values of El Reno samples that were duplicated, i.e., five different samples.

[f] Average standard deviation of actual laboratory values for 10 El Reno samples that were duplicated.

the plant, such as reduced nitrogen in free amino acids or as ammonium ion in ammonium nitrate that had been absorbed into the plant from fertilization. The NIR and CP/MAS ^{13}C-NMR methods would tend to discriminate against free amino acids versus protein. NIR discriminates by picking different wavelength sets for protein than for amino acids (Williams et al, 1984, 1985). CP/MAS ^{13}C-NMR can be made to discriminate against smaller and generally more mobile species (Stejskal and Schaefer, 1975) and would not respond to residual ammonium nitrate.

The acquisition of solid-phase ^{13}C-NMR spectra is faster than solution phase for these high-molecular-weight entities (3–4 versus 48 hr per sample). However, the cost of these high-quality research instruments is high ($275,000 to over $500,000). It is currently the only structurally definitive (easily interpretable) method of obtaining quantitative data on plant cell walls.

V. ADVANTAGES OF THE CHEMOMETRIC METHOD

Chemometrics as a discipline within chemistry can be defined as the development and application of mathematical and statistical methods to extract useful chemical information from chemical measurements (Kowalski, 1977). The extraction of compositional information from spectral curves can be considered as a chemometric method (Norris, 1983a, 1983b). The basic impetus for the development of predictive analyses is the increasing cost of performing laboratory analyses and the time required to obtain the results. Chemometric methods have been used for decades. Whenever a standard curve is constructed from a series of standard solutions assuming linearity from Beer's Law and is used to read the concentration of unknowns directly from the scale knowing only its absorbance, a chemometric method has been used. Standard curves used in analyses accomplished with ion-selective electrodes are another example. The great decrease in the cost of computers has made possible the NIR technology to which this text is directed. Because of the statistical routines, calibration equations can be developed and used to provide an analytical result without having performed the analysis in question, simply from taking the spectrum. NIR is essentially a chemometric method.

Currently, commercial laboratories charge $25–50 for a protein and fiber analysis. With current NIR technology, the completion of 1 or 10 analyses on a sample entails the same analytical input, which realizes far more information at the same cost. The cost of NIR equipment ranges from under $10,000 to over $100,000. The cost of the associated computer would range from less than $3,000 to systems over $25,000 for a workstation-based system. Computer capacity and software flexibility are inherently important. The computer must be capable of storing all spectral data, the calibration (chemical) data, and regression equations. The software must be capable of performing derivative calculations and varying the derivative size and smoothing of log(1/R) data. The system requires a sophisticated multiple regression analyses program to ensure that the optimum wavelengths are selected for the calibration equations. It should also contain internal statistical tests to examine the spectra and data of the samples for identification of erroneous data or of samples that may be considered outliers. As such, the systems can also be used to manage laboratory databases and provide internal as well as external checks on accuracy. With the savings in labor resulting from the ability to run several hundred samples in 1 day, NIR becomes very cost effective even for the smaller laboratories (Williams, 1977).

The advantages of the NIR method for managing laboratory data can best be illustrated by examples (Barton and Burdick,

1983). Table XI gives the results of the NIR analysis of a series of samples, five of which were blind duplicates. The column labeled SDP (standard deviation between like samples by NIR) shows that the NIR instrument recognized the identical samples more closely than did the laboratory analyses as shown by the column labeled SDA (standard deviation of like samples by laboratory methods) for everything but crude protein. The numerical difference between SDP and SDA for protein is probably very close to the instrumental error for the method. Overall, it is apparent that precision as well as speed can be improved. These analyses required 5 weeks in the laboratory by gravimetric methods, but only 1 hr by NIR.

The data in Table XII result from in vitro digestibility determination on a set of Old World Bluestem grasses. The samples in the middle with the large negative biases (i.e., predicted values are larger than measured values) were run as triplicates, with acceptable laboratory standard deviations. When these samples were rerun, the new values agreed with the NIR predicted values. This error could have been caused by a restriction in the automatic pipette or in the straining of the rumen inoculum, which lowered the number of rumen microorganisms, or by the lowering of the activity of the inoculum because of residual detergent in the three racks of tubes. It is difficult to check standard laboratory results for an indeterminate error such as this. The results would normally have been accepted and the error of 10–30% would not have been detected. This example illustrates that accuracy can be improved through the use of a chemometric procedure. Table XIII contains data obtained from the determination of dry matter in which both positive and negative biases were found. In the laboratory, the tray used to carry "dry" crucibles from the oven to the desiccator to the balance held 20 crucibles. The data reflects differential moisture absorption among trays of crucibles in a desiccator over time. The chemometric procedure is very sensitive to laboratory practices that involve consistent errors and makes these practices quite easy to check, where before the errors went undetected.

Perhaps one of the most successful chemometric assays has come out of an area that was previously considered too difficult, i.e., the analysis of dietary fiber (Baker, 1983; Horvath et al, 1984; Williams et al, 1991; Kays et al, 1996, 1997, 1998). These studies have shown the possibility of developing models for dietary fiber by NIRS and the possibility of including the unique spectral characteristics of high-fat and high-crystalline sugar cereal products into an NIR calibration for the accurate and rapid determination of dietary fiber. An initial model for total dietary fiber was expanded to include cereal products with high sugar content, including products with large amounts of crystalline sugar (Kays et al, 1997). The standard error of cross-validation for the model was slightly higher than that for the original model (1.88% compared with 1.64%); however, the model predicted samples with a much wider range of sugar content. Second, the model was expanded to include cereal products with high fat content (Kays et al, 1998). The fat-expanded model had a similar level of accuracy to the original model (standard error of cross-validation 1.75%, $R^2 = 0.98$). Third, the model was expanded to include both high-fat and high-sugar products. As with the previous models, the fat- and sugar-expanded model encompassed a broad range of cereal products such as breakfast cereals, crackers brans, flours, pastas, cookies, and a broad range of grains, such as wheat, oat, barley, rye, corn, millet, amaranth, and products containing multiple grains. In addition, the fat- and sugar-expanded model encompassed products with a wide range in fat content, sugar content, and crystalline sugar content. The fat- and sugar-expanded model had a similar standard error of cross-validation to the original model and was found to accurately predict total dietary fiber in an independent group of

TABLE XII
Actual vs. Predicted IVDMD[a]

Sample No.	Average Residual Size	Bias
1–26	2.5	Positive and negative
27–36	10.0	All negative
37–40	4.4	All negative
41–60	1.7	Positive and negative

[a] In vitro dry matter digestibility.

TABLE XIII
Actual vs. Predicted Dry Matter

Sample No.	Average Residual Size	Bias
1–20	1.74	All negative
21–40	0.42	0.0
41–60	1.78	All positive

samples also with a broad range of product types, grains, fat content, sugar content, and total dietary fiber content. The standard error of cross-validation, standard error of performance, bias, slope, and coefficients of determination observed indicated a high degree of precision and reliability in determining dietary fiber using the fat- and sugar-expanded calibration. As such, the dietary fiber database was granted a U.S. Patent (no. 6,114,699) and was the first database for chemometric models to be patented (Barton et al, 2000).

If one is going to implement a chemometric procedure in the laboratory, several factors need to be considered. Equations to produce the analytical result can be generated that give excellent calibration statistics but do not provide acceptable analysis. It is possible to overfit the data, and one must be aware of the sources of laboratory error and the true errors associated with the procedure that produced the data to be used for calibration. This is discussed in detail in an earlier chapter. Since the spectra from solid-state ^{13}C-NMR can be readily interpreted and quantitative data are present in the spectra, it seems logical to use the spectroscopic data as input for the chemometric measurement. NIR instrumentation gives analyses in about 2 min, whereas the solid-phase ^{13}C-NMR requires 2–4 hr as the principal analytical method, which is much too slow. It is also possible to use NMR data to calibrate the NIR instrument. The same "chemical" information that is obtained from NMR is present in NIR spectra and would provide the accuracy and precision needed to measure quality in forages. Chemometrics can be used for basic research into forage, plant, and fiber structure as it relates to microbial degradation. Although NIR instrumentation is currently the front end of the method, other portions of the electromagnetic spectrum and other techniques may be coupled to the computer for chemometric analyses in the future. The work of Renard et al (1987) is an application of multivariate techniques with the MIR spectral region.

A major problem for NIR spectroscopists over the years has been the criticism by the classical spectroscopist of the viability of the NIR measurement without a real interpretation of spectral response in terms of the particular molecular vibration that is being used for the quantitative measurement. The report of two-dimensional correlation techniques has provided the first chance to unequivocally assign the qualitative interpretation to NIR spectral assignments that have been used so successfully for quantitative analysis (Barton and Himmelsbach, 1991; Barton et al, 1992). The initial report concentrated on the assignment of C-H aliphatic stretch and an overall explanation of the technique (Barton and Himmelsbach, 1991). The contour map (Fig. 14) is a two-dimensional projection of correlation of the NIR and MIR spectra. The dark narrow regions are those of the highest correlation activity between the spectral regions for the sample set. In general, these are areas where correlation has rapidly changed from high to low and back over a small wavelength/frequency distance. The regions of most correlation activity in the NIR were 1,385, 1,730, and 2,200–2,450 nm. The places in the MIR that correlated with the above regions were 2,700–2,900 cm^{-1}, and, in the "fingerprint" region, were 1,500–1,700 cm^{-1} and around 800 cm^{-2}. In general, these wavelengths correspond to C-H aliphatic stretch, C=O, C=C, and N vibration C-H deformation in organic compounds. The particular view shown in Figure 14 is at the NIR coordinate of 2,303 nm (upper left window) and the MIR coordinate of 2,850 cm^{-1} (upper right window). There are a number of different C-H stretches, but the ones that exhibit the highest correlations are principally those of waxes and oils and some C-H stretch in carbohydrates, lignin, and protein. The vertical scale on the extreme right is the color-coded (black and white here) coefficient of determination (R^2) scale for the contours. The coordinates for the contour slices can be found on the map. In this case, they are above 0.9. Techniques such as this will help to expand the utility of NIR and chemometric methods for future applications.

The two-dimensional vibration spectroscopy technique allows us to see the correlations between the spectral regions. This technique works well when the basic rules are followed. The spectra of a set of samples (taken as identically as possible on each in-

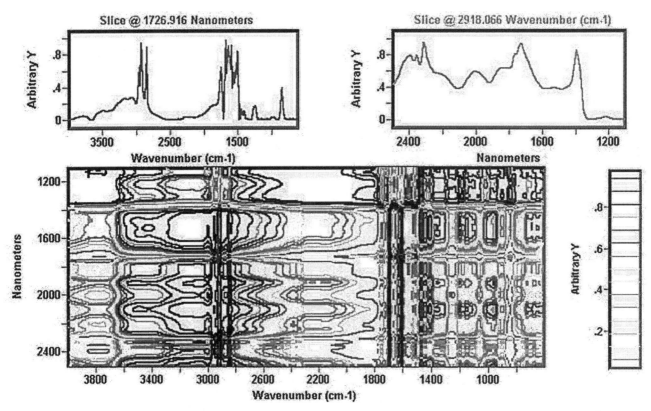

Fig. 14. Two-dimensional contour map of the near-infrared (z-axis from 1,100–2,500 nm) versus mid-infrared (x-axis from 4,000–600 cm^{-1}) with the coefficient of determination (R^2, y-axis) delineating the contours. Individual slices of the map in the small upper windows show the correlation plots for coordinates 1,711 nm (upper left, MIR) and 2,920 cm^{-1} (upper right, NIR) (arrow on contour map). Arrow in the vertical window on the right shows the value of R^2 for the coordinates shown; the aliphatic C-H stretch. (Redrawn from Barton et al, 1992)

strument) is acquired in each region and arranged in a file so that the sample is paired for the two regions being correlated. A point-for-point, simple least squares regression is performed and the resulting contour map plotted. Since it is simple least squares regression, only three samples are required for a valid result; however, a minimum of five are always used. There are four other criteria for the method as outlined by Barton et al (1992). The second is that a high signal-to-noise ratio is needed in both spectral regions. The third is that there must be a wide diversity in the data file. That is, the differences in the data set should be large to preclude spurious correlations, as described by Birth (1985). The fourth and fifth criteria do not apply today, since Pentium computers are more than adequate in speed and have large disk storage to accommodate the 7–30 megabyte maps. In the study by Barton et al (1996), the third criteria was violated. Obviously, there are not large differences between hard red winter (HRW) and hard red spring (HRS) wheats. The correlation maps were made on sets of 10, 25, and 35 samples of HRW and HRS wheats. These maps reveal some expected, as well as some unexpected, results. The magnitude of the correlations was very low for the chemical information. The sets of samples whose spectra were taken on two different instruments have particularly low correlations, but the real information is not the greatest source of variance in the set. A set of 10 samples (five HRW, five HRS) were scanned on the Nicolet 850 FTIR (Nicolet Instruments, Madison, WI) at an 8-cm^{-1} resolution. They were scanned in two blocks, 11,000–2,000 and 7,400–400 cm^{-1}, with different sources and beamsplitters. The best case would have been to use only one instrument setup, but the MIR spectra are very noisy from 7,400–4,000 cm^{-1} with the potassium bromide beamsplitter. These spectra were preprocessed using the standard normal variant and multiplicative scatter correction algorithms in GRAMS32 (Galactic Industries, Salem, NH). The 10 spectra were obtained from the combining of the NIR and MIR from the Nicolet FT-IR overlaid (Nicolet Instruments). The spectra were taken in two blocks and combined using MatLab (MATLAB, Natick, MA). The 10 single spectra represent the closest approximation to taking all of the spectra in one environment. The resulting contour map shows a dramatic improvement in the R^2 values around 0.62 for the argon C-H stretch at 5,900 cm^{-1} in the NIR to the amide bands at 1,480 cm^{-1} and 1,560 cm^{-1} in the MIR.

One of the most recent advances is the use of NIR imaging techniques to measure compositional and quality parameters (Bertrand et al, 1991; Robert et al, 1992). This promises to be most useful for quality and process analysis.

LITERATURE CITED

Akin, D. E., and Barton, F. E., II. 1983. Rumen microbial attachment and degradation of plant cell walls. Fed. Proc. 42:114-121.

Akin, D. E., and Burdick, D. 1973. Microanatomical differences of warm-season grasses revealed by light and electron microscopy. Agron. J. 65: 533-537.

Akin, D. E., and Burdick, D. 1975. Percentage of tissue types in tropical and temperate grass leaf blades and degradation of tissues by rumen microorganisms. Crop Sci. 15:661-668.

Akin, D. E., and Burdick, D. 1981. Relationships of different histochemical types of lignified cell walls to forage digestibility. Crop Sci. 21:577-581.

Akin, D. E., Amos, H. E., Barton, F. E., II, and Burdick, D. 1973. Microanatomical differences of warm-season grasses revealed by light and electron microscopy. Agron. J. 65:825-828.

Akin, D. E., Barton, F. E., II, and Burdick, D. 1975. Scanning electron microscopy of coastal bermuda and Kentucky-31 tall fescue extracted with neutral and acid detergents. J. Agric. Food Chem. 23:924-927.

Akin, D. E., Robinson, E. L., Barton, F. E., II, and Himmelsbach, D. 1977. Changes with maturity in anatomy, histochemistry, chemistry, and tissue digestibility of bermudagrass plant parts. J. Agric. Food Chem. 25:179-186.

Akin, D. E., Barton, F. E., II, and Coleman, S. W. 1983a. Structural factors affecting leaf degradation of Old World Bluestem and Weeping Love Grass. J. Anim. Sci. 56:1434-1446.

Akin, D. E., Wilson, J. R., and Windham, W. R. 1983b. Site and rate of tissue digestion in leaves of C_3, C_4, and C_3/C_4 intermediate *Panicum* species. Crop Sci. 23:147-155.

Akin, D. E., Brown, R. H., and Rigsby, L. L. 1984a. Digestion of stem tissues in *Panicum* species. Crop Sci. 24:769-773.

Akin, D. E., Rigsby, L. L., and Brown, R. H. 1984b. Ultrastructure of cell wall degradation in *Panicum* species differing in digestibility. Crop Sci. 24:156-163.

Albersheim, P., Nevins, D. J., English, P. D., and Karr, A. 1967. A method for the analysis of sugars in plant cell-wall polysaccharides by gas-liquid chromatography. Carbohydr. Res. 5:340-345.

Asp, N. G., Johansen, C. G., Hallmer, H., and Siljestem, M. 1983. Rapid enzymatic assay of insoluble and soluble dietary fiber. J. Agric. Food Chem. 31:476-482.

Atalla, R. H., Gast, J. C., Sindorf, D. W., Bartuska, V. J., and Maciel, G. E. 1980. Carbon-13 NMR spectra of cellulose polymorphs. J. Am. Chem. Soc. 102:3249-3251.

Baker, D. 1977. Determining fiber in cereals. Cereal Chem. 54:360-365.

Baker, D. 1978. Fiber in wheat foods. Cereal Foods World 23:557-558.

Baker, D. 1983. The determination of fiber in processed cereal foods by near-infrared reflectance spectroscopy. Cereal Chem. 60:217-219.

Baker, D., and Holden, J. M. 1981. Fiber in breakfast cereals. J. Food Sci. 46:396-398.

Barton, F. E., II. 1985. Near Infrared Reflectance Spectroscopy (NIRS): Analysis of Forage Quality. G. C. J. Marten, J. S. Shenk, and F. E. Barton, II, eds. U.S. Dep. Agric. Handb. No. 643.

Barton, F. E., II, and Akin, D. E. 1977. Digestibility of delignified forage cell walls. J. Agric. Food Chem. 25:1299-1303.

Barton, F. E., II, and Burdick, D. 1979. Preliminary study on the analysis of forages with filter-type near infrared reflectance spectrometer. J. Agric. Food Chem. 17:1248-1252.

Barton, F. E., II, and Burdick, D. 1980. Prediction of crude protein in dehydrated coastal bermudagrass by NIR reflectance. Pages 103-106 in: Proc. Res. Ind. Conf. Coastal Bermudagrass Processors Assoc., 10th. U.S. Dep. Agric. Agric. Res. Serv., Athens, GA.

Barton, F. E., II, and Burdick, D. 1983. Prediction of forage quality with NIR reflectance spectroscopy. Pages 532-534 in: Proc. Int. Grassl. Congr., 14th. Westview Press, Boulder, CO.

Barton, F. E., II, and Coleman, S. W. 1981. Potential of near infrared reflectance spectroscopy for measuring forage quality. Okla. Agric. Exp. Stn. Misc. Publ. 108.

Barton, F. E., II, and Himmelsbach, D. S. 1991. Near infrared reflectance spectroscopy and other spectral analyses. Pages 240-247 in: Analytical Applications of Spectroscopy II. A. M. C. Davies and C. S. Creaser, eds. The Royal Society of Chemistry, London.

Barton, F. E., II, and Windham, W. R. 1988. Determination of acid detergent fiber and crude protein in forages by near infrared spectroscopy: Collaborative study. J. Assoc. Off. Anal. Chem. Int. 71:1162-1167.

Barton, F. E., II, Amos, H. E., Burdick, D., and Wilson, R. L. 1976. Relation of chemical analysis to in vitro digestibility for selected tropical and temperate grasses. J. Anim. Sci. 43:504-512.

Barton, F. E., II, Akin, D. E., and Windham, W. R. 1981. Scanning electron microscopy of acid detergent fiber digestion by rumen microorganisms. J. Agric. Food Chem. 29:899-903.

Barton, F. E., II, Windham, W. R., and Himmelsbach, D. S. 1982. Analysis of neutral sugar hydrolysates of forage cell walls by high-pressure liquid chromatography. J. Agric. Food Chem. 30:1119-1123.

Barton, F. E., II, Himmelsbach, D. S., Duckworth, J. H., and Smith, M. J. 1992. Two dimensional vibration spectroscopy: Correlation of mid- and near-infrared regions. Appl. Spectrosc. 46:420.

Barton, F. E., II, Himmelsbach, D. S., and Archibald, D. D. 1996. Two dimensional vibration spectroscopy. V: Correlation of mid and near infrared of hard red winter and spring wheats. J. Near Infrared Spectrosc. 4:139-152.

Barton, F. E., II, Kays, S. E., and Windham, W. R. 2000. Prediction of total dietary fiber by near infrared reflectance spectroscopy. U.S. Patent No. 6,114,699.

Ben-Gera, I., and Norris, K. H. 1968. Determination of moisture content in soybeans by direct spectrophotometry. Isr. J. Agric. Res. 18:125-132.

Bertrand, D., Robert, P., and Loisel, W. 1985. Identification of some wheat varieties by near infrared reflectance spectroscopy. J. Sci. Food Agric. 36:1120-1124.

Bertrand, D., Lila, M., Furtoss, V., and Robert, P. 1987. Application of principal component analysis to the prediction of lucerne forage protein content and in vitro dry matter digestibility by NIR spectroscopy. J. Sci. Food Agric. 41:299-307.

Bertrand, D., Robert, P., Melcion, J. P., and Sire, A. 1991. Characterization of powders by video image analysis. Powder Technol. 66: 171-176.

Birth, G. S. 1985. Evaluation of correlation coefficients with a stepwise regression analysis. Appl. Spectrosc. 39:729.

Bolker, H. I., and Teraschima, N. 1966. Infrared spectroscopy of lignins. Pages 110-124 in: Lignin Structure and Reactions. J. Marton, ed. American Chemical Society, Washington, DC.

Burdick, D., Barton, F. E., II, and Nelson, B. D. 1981. Prediction of bermudagrass composition and digestibility with a near infrared multiple filter spectrophotometer. Agron. J. 73:399-403.

Coleman, S. W., and Barton, F. E., II. 1982. Calibration of a near infrared reflectance spectrometer for prediction of forage quality. Okla. Agric. Exp. Stn. Misc. Publ. 109.

Cowe, I. A., and McNichol, J. W. 1985. The use of principal components in the analysis of near-infrared spectra. Appl. Spectrosc. 39: 257-266.

Devaux, M. F., Bertrand, D., and Martin, G. 1986. Discrimination of bread-making quality of wheats according to their variety by near-infrared reflectance spectroscopy. Cereal Chem. 63:151-154.

Devaux, M. F., Bertrand, D., Robert, P., and Morat, P. 1987. Extraction of near infrared-spectral information by fast Fourier transform and principal component analysis. Application to the discrimination of baking quality of wheat flours. J. Chemometrics 1:103-110.

Devaux, M. F., Bertrand, D., Robert, P., and Qannari, M. 1988a. Application of multidimensional analyses to the extraction of discriminate spectral patterns from NIR spectra. Appl. Spectrosc. 42:1015-1019.

Devaux, M. F., Bertrand, D., Robert, P., and Qannari, M. 1988b. Application of principal component analysis on NIR spectral collection after elimination of interference by a least-squares procedure. Appl. Spectrosc. 42:1020-1023.

Dixon, W. T., Schaefer, J., Sefcik, M. D., Stejskal, E. O., and McKay, R. A. 1981. Quantitative chemical composition of materials such as humic soils, lignins, and coals by high resolution carbon-13 NMR. J. Magn. Reson. 45:173-176.

Downey, G., Robert, P., Bertrand, D., and Devaux, M. F. 1987. Near infrared analysis of grass silage by principal component analysis of transformed reflectance data. J. Sci. Food Agric. 41:219-229.

Earl, W. L., and Vanderhart, D. L. 1980. High resolution magic angle sample spinning carbon-13 NMR of solid cellulose. J. Am. Chem. Soc. 102:3251-3252.

Earl, W. L., and Vanderhart, D. L. 1981. Observation by high-resolution carbon-13 nuclear magnetic resonance of cellulose. 1. Related to morphology and crystal structure. Macromolecules 14:570-574.

Geladi, P., and Kowalski, B. R. 1986a. Partial least-squares regression: A tutorial. Anal. Chim. Acta 185:1-17.

Geladi, P., and Kowalski, B. R. 1986b. An example of 2-block predictive partial least-squares regression with simulated data. Anal. Chim. Acta 185:19-32.

Goering, H. K., and Van Soest, P. J. 1970. Forage fiber analysis (apparatus, reagents, procedures, and some applications). U.S. Dep. Agric. Handb. No. 379.

Haaland, D. M., Easterling, R. G., and Vopicka, D. A. 1985. Multivariate least-squares methods applied to the quantitative spectral analysis of multicomponent samples. Appl. Spectrosc. 39:73-84.

Harkin, J. M. 1973. Lignin. Pages 323-373 in: Chemistry and Biochemistry of Herbage. G. W. Butler and R. W. Bailey, eds. Academic Press, New York.

Himmelsbach, D. S., and Barton, F. E., II. 1980. [13]C nuclear magnetic resonance of grass lignins. J. Agric. Food Chem. 28:1203-1208.

Himmelsbach, D. S., Barton, F. E., II, and Windham, W. R. 1983. Comparisons of carbohydrate, lignin, and protein ratios between grass species by cross polarization-magic angle spinning carbon-13 nuclear magnetic resonance. J. Agric. Food Chem. 31:401-404.

Honigs, D. E., Freelin, J. M., Hietje, G. M., and Hirschfeld, T. M. 1983. Near-infrared reflectance analysis by Gauss-Jordan linear algebra. Appl. Spectrosc. 37:491-497.

Horvath, L., Norris, K. H., Horvath-Mosonyi, M., Rigo, J., and Hegedus-Volgyesi, E. 1984. Study into determining dietary fiber of wheat bran by NIR-technique. Acta Alimentaria 13:355-382.

Hruschka, W. R., and Norris, K. H. 1982. Least-squares curve-fitting of near infrared spectra predicts protein and moisture content of ground wheat. Appl. Spectrosc. 36:261-265.

Hymowitz, T., Dudley, J. W., Collins, F. E., and Brown, C. M. 1974. Estimation of protein and oil concentration in corn, soybean and oat seed by near-infrared light reflectance. Crop Sci. 14:713-715.

Kaye, W. 1954. Near-infrared spectroscopy. I. Spectral identification and analytical applications. Spectrochim. Acta 6:257-287.

Kaye, W. 1955. Near-infrared spectroscopy. II. Instrumentation and technique. Spectrochim. Acta 7:181-204.

Kays, S. E., Windham, W. R., and Barton, F. E., II. 1996. Prediction of total dietary fiber in cereal products using near-infrared reflectance spectroscopy. J. Agric. Food Chem. 44:2266-2271.

Kays, S. E., Barton, F. E., II, Windham, W. R., and Himmelsbach, D. S. 1997. The prediction of total dietary fiber by near-infrared reflectance spectroscopy in cereal products containing high sugar and crystalline sugar. J. Agric. Food Chem. 45:3944-3951.

Kays, S. E., Windham, W. R., and Barton, F. E., II. 1998. Prediction of total dietary fiber by near-infrared reflectance spectroscopy in high-fat- and high-sugar-containing cereal products. J. Agric. Food Chem. 46: 854-861.

Kowalski, B. R. 1977. Chemometrics: Theory and Applications. Am. Chem. Soc. Symp. Ser. 52. Washington, DC.

Li, B. W., and Cardozo, M. S. 1992. Nonenzymatic-gravimetric determination of total dietary fiber in fruits and vegetables. J. Assoc. Off. Anal. Chem. Int. 75:372-374.

Ludemann, H. D., and Nimz, H. 1973. Carbon-13 nuclear magnetic resonance spectra of lignins. Biochem. Biphys. Res. Commun. 52:1162-1169.

Ludemann, H. D., and Nimz, H. 1974a. ^{13}C-Kernresonanzspektren von Ligninen, 1: Chemishe Verschiebungen bei monomeren und dimeren Modellsubstanzen. Makromol. Chem. 175:2393-2407.

Ludemann, H. D., and Nimz, H. 1974b. ^{13}C-Kernresonanzspektren von Ligninen, 2: Buchen-und Fichten-Bjorkman-Lignin. Makromol. Chem. 175:2409-2422.

Lund, E. D., and Smoot, J. M. 1982. Dietary fiber content of some tropical fruits and vegetables. J. Agric. Food Chem. 30:1123-1127.

Maddams, W. F. 1980. The scope and limitations of curve fitting (to vibrational band systems). J. Appl. Spectrosc. 34:245-267.

Martens, H., and Martens, H. 1986. Near-infrared reflectance determination of sensory quality of peas. Appl. Spectrosc. 40:303-310.

Martens, H., and Naes, T. 1989. Multivariate Calibration, Chapter 4. Page 254 in: Multivariate Calibration. John Wiley and Sons, New York.

Martens, H., and Russwurm, H., Jr., eds. 1983. Food Research and Data Analysis. Applied Science Publishers, New York.

Moore, J. E., and Mott, G. O. 1973. Structural inhibitors of quality in tropical grasses in anti-quality components of forages. Pages 53-98 in: Anti-Quality Components of Forage Crops. A. G. Matches, ed. Crop Science Society of America, Madison, WI.

Nimz, H. 1974. Beech lignin-proposal of a constitutional scheme. Angew Chem., Int. Ed. Engl. 13:313-321.

Nimz, H., and Ludemann, H. D. 1974. ^{13}C-Kernresonanzspektren von Ligninen, 5: Oligomere Lignin modellsubstanzen. Makromol. Chem. 175:2577-2583.

Nimz, H., and Ludemann, H. G. 1976. Kohlenstoff-13-NMR-Spektren Von Ligninen, 6: Lignin-und DHP acetate. Holzforschung (J. Anim. Res.) 30:33-40.

Nimz, H., Ludemann, H. D., and Becker, H. Z. 1974a. Kohlenstoff-13-Kernresonanzspektren von Ligninen, 4: Die Lignine der europaischen mistel (Viscum album L.). Pflanzenphysiol. 73:226-233.

Nimz, H., Morgharab, I., and Ludemann, H. D. 1974b. ^{13}C-Kernresonanzspektren von Ligninen, 3: Vergleich von Fichten lignin mit kunstlichem lignin nach Freudenberg. Makromol. Chem. 175:2563-2575.

Nimz, H., Ebel, J., and Griesbach, H. Z. 1975. Structure of lignin from soybean cell suspension cultures. Z. Naturforsch. Teil C 30:442-444.

Norris, K. H. 1983a. Extracting information from spectrophotometric curves. Predicting chemical composition from visible and near-infrared spectra. Pages 95–114 in: Food Research and Data Analysis. H. Martens and H. Russwurm, Jr., eds. Applied Science Publishers, New York.

Norris, K. H. 1983b. Multivariate analysis of raw materials. Pages 527-535 in: Chemistry and World Food Supplies: The New Frontiers, CHEMRAWN II (Chemical Research Applied to World Needs). L. W. Shemilt, ed. Pergamon Press, New York.

Norris, K. H., and Barnes, R. F. 1976. Infrared reflectance analysis of nutritive value of feedstuffs. Pages 237-241 in: Proc. Int. Symp. Feed Composition, Anim. Nutr. Requir. Comput. Diets, 1st. P. V. Fonnesbeck, L. E. Harris, and L. C. Kearl, eds. Utah Agric. Exp. Stn., Logan.

Norris, K. H., and Hart, J. R. 1965. Direct spectrophotometric determination of moisture content of grains and seeds. Pages 19-25 in: Principals and Methods of Measuring Moisture in Liquids and Solids. Vol. 4. A. Wexler, ed. Reinhold, New York.

Norris, K. H., Barnes, R. F., Moore, J. E., and Shenk, J. S. 1976. Predicting forage quality by infrared reflectance spectroscopy. J. Anim. Sci. 43:889-897.

O'Donnel, D. J., Ackerman, J. J. H., and Maciel, G. E. 1981. Comparative study of whole seed protein and starch content via cross polarization-magic angle spinning carbon-13 nuclear magnetic resonance spectroscopy. J. Agric. Food Chem. 29:514-518.

Prosky, L., Asp, N. G., Schweizer, T. F., De Vries, J. W., and Furda, I. 1992. Determination of insoluble and soluble dietary fiber in foods and food products: Collaborative study. J. Assoc. Off. Anal. Chem. Int. 75: 360-367.

Renard, C., Robert, P., Bertrand, D., Devaux, M. F., and Abecassis, J. 1987. Qualitative characterization of the purity of milled durum wheat products by multidimensional statistical analysis of their mid-infrared diffuse reflectance spectra. Cereal Chem. 64:177-181.

Rinne, R. W., Gibson, S., Bradley, J., Seif, R., and Brim, C. A. 1975. Soybean protein and oil percentages determined by infrared analysis. U.S. Dep. Agric. Res. Publ. ARS-NC-26.

Robert, P., Bertrand, D., and Demarquilly, C. 1986. Prediction of forage digestibility by principal component analysis of near infrared reflectance spectra. Anim. Feed Sci. Technol. 16:215-224.

Robert, P., Bertrand, D., and Devaux, M. F. 1987. Multivariate analysis applied to near-infrared spectra of milk. Anal. Chem. 59:2187-2191.

Robert, P., Bertin, C., and Bertrand, D. 1989a. Rumen microbial degradation of beet root pulps. Application of infrared spectroscopy to the study of protein and pectin. J. Agric. Food Chem. 37:624-627.

Robert, P., Bertrand, D., Crochon, M., and Sabino, J. 1989b. New mathematical procedure for NIR analysis: The lattice technique. Application to the prediction of sugar content of apples. Appl. Spectrosc. 43:1576.

Robert, P., Bertrand, D., Devaux, M. F., and Sire, A. 1992. Identification of chemical constituents by multivariate near-infrared spectral imaging. Anal. Chem. 64:664-667.

Rohweder, D. A., Barnes, R. F., and Jorgensen, N. 1978. Proposed hay grading standards based on laboratory analyses for evaluating quality. J. Anim. Sci. 47:747-759.

Routley, D. G., and Sullivan, J. T. 1958. The isolation and analysis of hemicellulosis of bromegrass. J. Agric. Food Chem. 6:687-692.

Rutar, V., Blinc, R., and Ehrenberg, L. 1980. Protein content determination in solid organic materials by proton-enhanced magic angle sample spinning carbon-13 NMR. J. Magn. Reson. 40:225-227.

Sarkanen, K. V., and Hergert, H. L. 1971. Classification and distribution. Pages 874-876 in: Lignins, Occurrence, Formation, Structure, and Reactions. K. V. Sarkanen and C. H. Ludwig, eds. Wiley-Interscience, New York.

Schaefer, J., and Stejskal, E. O. 1979. High resolution ^{13}C-NMR of solid polymers. Pages 282-324 in: Topics in Carbon-13 NMR Spectroscopy. Vol. 3. G. C. Levy, ed. Wiley-Interscience, New York.

Schaefer, J., Sefcik, M. D., Stejskal, E. O., McKay, R. A., and Hall, P. L. 1981. Characterization of the catabolic transformation of lignin in culture using magic-angle carbon-13 nuclear magnetic resonance. Macromolecules 14:557-559.

Shenk, J. S., and Hoover, M. R. 1977. Infrared reflectance spectrocomputer design and application. Pages 122-125 in: Advances in Automated Analysis. Technicon International Congress (1976), Vol. 2. Industrial Symposia, Mediad. Technicon, Inc., White Plains, NY.

Shenk, J. S., Norris, K. H., Barnes, R. F., and Fisset, G. W. 1977. Forage and feedstuff analysis with infrared reflectance spectro-computer system. Pages 454-463 in: Proc. Int. Grassl. Congr., 13th. Westview Press, Boulder, CO.

Shenk, J. S., Westerhaus, M. O., and Hoover, M. R. 1978. Infrared reflectance analysis of forages. Pages 242-244 in: Proc. Int. Grain Forage Harvesting Congr. ASAE (Am. Soc. Agric. Eng.), St. Joseph, MI.

Stejskal, E. O., and Schaefer, J. 1975. Removal of artifacts from cross-polarization NMR experiments. J. Magn. Reson. 18:560-563.

Tormod, N. 1987. The design of calibration in near infra-red reflectance analysis by clustering. J. Chemometrics 1:121-134.

Ulbricht, T. L. V., and Theander, O. 1980. OECD Workshop at Uppsala, Sweden, 17-19 Sept. 1980. Agric. Environ. 6:87-348.

Van Soest, P. J. 1963a. Use of detergents in the analysis of fibrous feeds. I. Preparation of fiber residues of low nitrogen content. J. Assoc. Off. Agric. Chem. Int. 46:825.

Van Soest, P. J. 1963b. Use of detergents in the analysis of fibrous feeds. II. A rapid method for the determination of fiber and lignin. J. Assoc. Off. Agric. Chem. Int. 46:829-835.

Van Soest, P. J. 1967. Development of a comprehensive system of feed analyses and its application to forages. J. Anim. Sci. 26:119.

Van Soest, P. J. 1973. The uniformity and nutritive availability of cellulose. Fed. Proc., Fed. Am. Soc. Exp. Biol. 32:1804-1808.

Van Soest, P. J., and Wine, R. H. 1967. Use of detergents in the analysis of fibrous feed. IV. Determination of plant cell-wall constituents. J. Assoc. Off. Anal. Chem. Int. 50:50-55.

Van Soest, P. J., and Wine, R. H. 1968. Determination of lignin and cellulose in acid detergent fiber with permanganate. J. Assoc. Off. Anal. Chem. Int. 51:780-785.

Wheeler, J. L., and Mochrie, R. D., eds. 1981. Ed. Forage Evaluation: Concepts and Techniques. Am. Forage Grassl. Council, Lexington, KY, USA and CSIRA, East Melbourne, Victoria, Australia.

Williams, P. C. 1977. The economics of near infrared reflectance spectroscopy in testing grain for protein and moisture. Pages 131-136 in: Advances in Automated Analysis. Technicon International Congress (1976), Vol. 2. Industrial Symposia, Mediad. Technicon, Inc., White Plains, NY.

Williams, P. C., Preston, K. R., Norris, K. H., and Starkey, P. M. 1984. Determination of amino acids in wheat and barley by near-infrared reflectance spectroscopy. J. Food Sci. 49:17-20.

Williams, P. C., MacKenzie, S. L., and Starkey, P. M. 1985. Determination of methionine in peas by near-infrared reflectance spectroscopy (NIRS). J. Agric. Food Chem. 33:811-815.

Williams, P. C., Cordeiro, H. M., and Harnden, M. F. T. 1991. Analysis of oat bran products by near-infrared reflectance spectrosopy. Cer. Foods World 36:571-574.

Windham, W. R., and Barton, F. E., II. 1991. Moisture analysis in forage by near-infrared reflectance spectroscopy: Collaborative study of calibration methodology. J. Assoc. Off. Anal. Chem. Int. 74:324-331.

Windham, W. R., Barton, F. E., II, and Himmelsbach, D. S. 1983. High-pressure liquid chromatographic analysis of component sugars in neutral detergent fiber for representative warm and cool-season grasses. J. Agric. Food Chem. 31:471-475.

CHAPTER 13

Qualitative Near-Infrared Analysis

HOWARD MARK
Mark Electronics
Suffern, NY
U.S.A.

I. INTRODUCTION

When the first edition of this book was published in the early 1980s, methods for qualitative analysis based on near-infrared (NIR) data were relatively undeveloped. For the most part, the method of interpreting NIR spectra for qualitative purposes was visual inspection of a printout of the spectra, similar to the practice employed in mid-infrared (mid-IR) and other spectral regions. Consequently, the approach taken in the first edition of this book was to list the broad categories of applications of NIR, as then used, for any purpose other than quantitative analysis (mainly, determination of composition at that time).

Since that time, NIR-orientated mathematicians and statistical/ chemometricians have developed several useful methods for the analysis and interpretation of NIR spectra for the purpose of identifying and classifying the materials represented by the samples. NIR users have begun to employ these techniques, variously called "qualitative analysis," "discriminant analysis," "product identification," pattern recognition," or one of a number of lesser-used terms. All of these describe the same general concept, which is to determine from the NIR spectrum the nature of the sample (as opposed to determination of its composition). These advances are normally represented by computerized algorithms that allow the computer to be "trained" to "recognize" the materials used to develop the training, or calibration, set using a concept similar to that used in quantitative NIR analysis.

In a separate, but parallel, effort, manufacturers of mid-IR (in particular, Fourier transform infrared [FTIR]) spectrometers have implemented computerized "library-searching" routines. These attempt to mimic the pattern-recognition capabilities of the human eye/brain combination in order to compare the spectrum of a pure material with spectra stored in master libraries of spectra. In these terms, the "training" methods employed by NIR users can also be viewed as "library-search" methods, but ones in which the users develop their own libraries. Each approach has its own advantages and disadvantages. These are summarized in Table I.

Since the early 1990s, manufacturers of FTIR spectrometers have been introducing instrument versions designed and/or optimized for NIR (as well as mid-IR) operation. Along with this trend, we can expect to see more "library-searching" types of algorithms being used in NIR analysis.

This chapter is organized by describing the algorithms used for the identification of materials, rather than by application areas, and an attempt is made to present a description of "current practice." The 1998 Fundamental Reviews issue of *Analytical Chemistry* includes a review of chemometrics (Lavine, 1998). This review includes methods of classification, pattern recognition, etc. The methods include some exotic and very specialized algorithms, many of which were developed for individual specific studies, and will likely never be used again, even by the developer. Descrip-

tions of such specialized algorithms are not included in this chapter. This chapter embraces only methods that are broadly used for spectroscopic (especially NIR) identification, either because they are widely discussed and disseminated in the literature or because they are methods provided by instrument manufacturers and used by their customers.

II. DATA PRETREATMENTS

As with quantitative NIR analysis, raw spectral data (which in NIR spectroscopy is usually recorded and stored as $\log[1/R]$ or $\log[1/T]$ values) may be transformed into other types of variables, using similar transformations. First- and second-derivative spectra, spectra that have been "smoothed" by any of several specialized functions, spectra that have been "normalized" in various ways (including multiplicative scatter correction), and baseline-corrected spectra have all been used for NIR identification. The reasons for these data treatments are the same as for quantitative NIR analysis—to reduce the amount of extraneous variance in the spectra, so that only "true" spectral features will be used in the identification.

Data transforms do not have the same relationship to qualitative as to quantitative analysis. While some extraneous variations (e.g., particle size and scattering variations) are not true spectral variations and the data transformation will remove them to a greater or lesser degree, in some cases, these variations may be an important characteristic of the materials to be identified. In such cases, removing them from the spectra reduces the identification capacity of the data, rather than enhancing it. Data transforms must be applied with more caution to qualitative than to quantitative analysis.

III. MAHALANOBIS DISTANCES

The first reported application of an algorithm to identify materials based on their NIR spectra is attributable to Rose (Rose, 1982; Rose et al, 1982), who used the linear discriminant function capability of the "canned" routines in the SAS software ensemble (SAS Institute, Inc., Cary, NC) to distinguish among several pharmaceutical materials. This was followed by the initial development and use of an algorithm designed for this purpose, based on Mahalanobis distances (Mark and Tunnell, 1985). Some variations on the theme were reported (Mark, 1986) and shortly thereafter Whitfield et al (1987) published a table of limiting values to use for different conditions.

Mahalanobis distance is defined by a distance measure based on a set of multivariate data (the training data) that are used to describe it and whose Euclidean length varies according to the direction in space in which it is being measured. The equivalent Euclidean length is large in directions (i.e., dimensions) where the data are spread out and small in directions where the data are compact. Since the Mahalanobis distance depends on the data, it changes as the defining data change. A simple example of this is given in Figure 1.

Three different materials, with different spectra, are plotted in a multidimensional space (here, two dimensions). The "B" ellipse is shown with the demonstration data comprising the cluster corresponding to that substance. The differences in the spectra cause the clusters of data points corresponding to the various substances (of which only one is shown) to be offset from one another. It is this offset that forms the basis for distinguishing among the various substances. A new material purported to be one of the materials represented in the training set will lie within one of the clusters of data points.

To form the basis for a computerized algorithm, this must be expressed mathematically. The basis for this expression is to measure the distance from the centroid of each data cluster. Presumably, data that lie within a cluster will be "close to" (i.e., have a small distance from) the centroid of that cluster and "far from" all other clusters.

Simple Euclidean distances are unsatisfactory for this purpose, since their use does not take into account the manner in which NIR data are commonly spread out through multidimensional space. When the characteristics of the data are taken into account, the distance measure is determined by the ellipses shown, and the distance from the centroid of a data cluster to the surrounding ellipse defines the unit distance measure. This distance is computed from the matrix equation

$$D^2 = (\mathbf{X} - \overline{\mathbf{X}})\mathbf{M}^{-1}(\mathbf{X} - \overline{\mathbf{X}})' \qquad (1)$$

In Figure 1, the two lines marked "M" in the "A" ellipse are both one unit long, notwithstanding that they have different Euclidean lengths. When a new sample is measured, the distance is determined by the number of lengths of the vector starting at the centroid of the data cluster and pointing in the direction of the new data point.

IV. THE POLAR QUALIFICATION SYSTEM

The polar qualification system (PQS) presents an interesting extension of the use of specifying points in multidimensional space. This concept was introduced by Kaffka and Gyarmati (1990) and is clearly seen to be an extension of previous work on quality spaces (Kaffka and Gyarmati, 1989), an approach that takes the idea of multivariate measurements in a tangential direction to the Mahalanobis distance approach, although the capability of using Mahalanobis distances is retained when the data are presented and treated in the proposed manner (Kaffka and Gyarmati, 1994).

The concept is to change the axes of an ordinary spectral display from a set of Cartesian (x,y) axes to a polar coordinate (θ,r) system. Instead of simply replotting the spectral data on the new

TABLE I
Comparison of the Characteristics of the Two Categories of Qualitative Analysis Algorithms

Comparison	"Training" Methods	Library Searching
Advantages	"Library" is private: maintains secrecy of proprietary materials	Libraries containing many compounds are available
	Can include spectral variations due to external variables	Standard libraries contain high-quality spectra; attempt to have "perfect" spectra that can be reproduced by other scientists
	Can accommodate special and unusual compounds	Libraries available from several sources
Disadvantages	Time and effort needed to accumulate and measure training samples and create the identification model	Does not include variations of the samples or effect of external conditions
		Standard spectra may not be representative of material to be identified due to differences in physical form, impurities, etc.

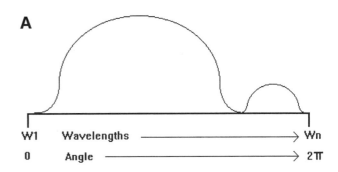

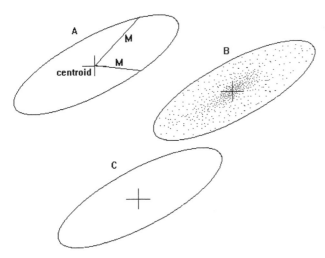

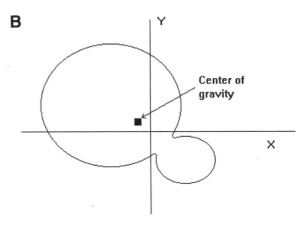

Fig. 1. Mahalanobis distances are determined by an ellipse (or ellipsoid) fitted to the data; the distance from the center of the data to the surrounding ellipse (or ellipsoid) is one Mahalanobis distance, regardless of the equivalent Euclidean distance. Thus, both lines marked "M" in the "A" ellipse are the same length: one Mahalanobis distance. The three ellipses represent the portions of the space occupied by three different materials; the differences in their spectra cause the ellipses to be offset from each other.

Fig. 2. Showing how the polar qualification system converts an ordinary spectral display into a polar display. **A,** The ordinary x,y plot of a (stylized) spectrum, with the x axis having both the Cartesian and polar units. **B,** The angular measure converted to a true polar coordinate system.

coordinate system, the previous x coordinate (wavelength) is equated to the polar angle (θ), and the previous y coordinate (absorbance) is equated to the radius vector. Figure 2 illustrates the transformation.

The stylized spectrum in Figure 2A contains two spectral peaks, one large and one small. The x axis of this spectrum shows both sets of coordinate values; even though Figure 2A is actually plotted in a Cartesian coordinate system, the values of the corresponding polar angles are shown with their equivalency to their wavelength values in the Cartesian system.

Figure 2B shows the effect of converting this plot to a true polar coordinate system: the large spectral peak extending through approximately $^3/_4$ of the spectrum shown in Figure 2A (and, therefore, covering angles equivalent to the range of 0 to $^{3\pi}/_2$) becomes the large lobe of the plot that, in fact, extends, in the polar coordinate system, over the first three quadrants: from 0 to $^{3\pi}/_2$ in Figure 2B. Similarly, the small peak that occupies the last quarter of the spectrum, from $^{3\pi}/_2$ to 2π, is transformed into the small lobe in the fourth quadrant of Figure 2B (i.e., occupying the polar angles from $^{3\pi}/_2$ to 2π). The length of the radius vector at each angle in Figure 2B is equal to the absorbance at the wavelength corresponding to that angle in Figure 2A.

So, how is this used to do identification? To answer this question, one notes the (somewhat exaggerated, to be sure) point in Figure 2B marked "center of gravity." This point is the "center" of the figure corresponding to the spectrum shown in Figure 2A. The value of the center of gravity is measured according to the new x and y axes of the polar plot of Figure 2B, and the values along these two axes are computed as follows.

$$x = \frac{1}{k} \sum_{i=1}^{k} A_i \cos\left(\frac{2\pi \times i}{k}\right) \qquad (2A)$$

$$y = \frac{1}{k} \sum_{i=1}^{k} A_i \sin\left(\frac{2\pi \times i}{k}\right) \qquad (2B)$$

in which A_i = the absorbance at the ith wavelength in the spectrum, and k = the number of wavelengths in the spectrum.

Other samples of that same material will have similar spectra, and, therefore, they will also have centers of gravity near that of the sample shown. Thus, all the samples of this material will give rise to a cluster of points in the neighborhood of the one illustrated, just as the data at individual wavelengths gave rise to a cluster of points.

A different material, having a different spectrum, will generate a different pattern when the spectral plot is converted to polar coordinates in this same way. Thus, the center of gravity for a different material will form a cluster of data points distinct from the one generated by the material shown. This is a similar situation to the one encountered when discussing Mahalanobis distances: each different material gives rise to a cluster of data points, and the spectrum from a new material gives rise to a data point falling within any given cluster and identifies that material as the one corresponding to that cluster. The major difference between the PQS approach and the Mahalanobis distance method is that, with the PQS algorithm, the centroid, used to determine and be compared with the various cluster, is itself defined by the entire spectrum of the materials in the training algorithm, rather than only a small subset as the Mahalanobis distance measure uses.

Having defined the clusters, it is still necessary to provide a means of deciding whether a new data point belongs to any of the given clusters or not. As Kaffka and Gyarmati (1994) points out, this can itself be done using either Euclidean or Mahalanobis measures of distance between the new data point and the given clusters.

V. PRINCIPAL COMPONENTS

The calculation of principal components has found widespread use for NIR quantitative analysis and, for that reason, has been widely described in the NIR literature (e.g., Robert et al, 1987; Mark, 1987, 1992; Williams and Norris, 1987; Naes and Isaksson,

1992; the multiple applications described in the proceedings of every International Conference on Near Infrared Spectroscopy; and this book), as well as in the chemometric and spectroscopic literature in general. With the growth of interest in qualitative analysis through the use of NIR, interest also grew in the use of the same methods of using the spectral data as were applied to quantitative analysis. Principal components were seen as a premier technique for this purpose, since it was expected to show the same benefits for qualitative analysis as it does for quantitative analysis.

In fact, principal components are admirably suited for this purpose. By definition, they maximize the variance accounted for by the calculated factor or factors (e.g., page 272 in Anderson, 1958). Since spectra of different samples of a given material will tend to be the same, the differences creating the variance that will be preferentially maximized will be those due to the spectral differences between different material, and, therefore, the use of principal components enhances whatever spectral differences exist between different materials.

Thus, the principal component approach to identification has become popular since the early 1980s. This can be seen just by the plethora of papers using that approach, selected arbitrarily from the proceedings of two International Near-Infrared Conferences (Hammond, 1994; Legrand et al, 1994; Osborne and Mertens, 1994; Schilling and Ritzmann, 1994; Schilling et al, 1994; Varadi et al, 1994; Bertrand et al, 1995; Jouan-Rimbaud and Massart, 1995) and the variety of applications for which it was used. An illustration of this is shown in Figure 3.

Following Martens and Russwurm (1983), Figure 3A uses a banana-shaped figure to represent a spatial shape with different variances in different directions. Figure 3A shows how the data lie in this space, relative to the axes. This banana-shaped figure lies at the origin of the coordinate system. Real data do not normally

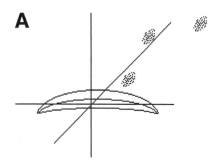

A

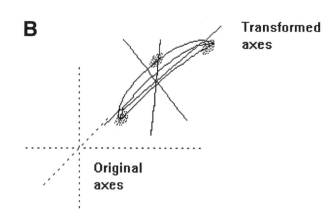

B

Transformed axes

Original axes

Fig. 3. Principal components provide an orthogonal set of axes that have the maximum variance along the first axis, the second-largest variance along the second axis, etc. **A,** The banana-shaped figure illustrates how different variances can lie in different directions. **B,** The mathematics of the principal component calculation assures that the axis corresponding to the maximum variance actually corresponds to the maximum variance of the data by translating and rotating the axes suitably.

lie around the origin, but rather more nearly like the hypothetical data shown in Figure 3A, arbitrarily disposed in the space. The mathematics of the principal component transform translates and rotates the axes so that the principal axes, i.e., those defined by the principal component transform, correspond to the actual data. The axis of maximum variance then corresponds to the direction of maximum variance of the data, the axis of second-largest variance corresponds to the direction of the second-largest variance of the data, and so forth (Fig. 3B).

The proof of the nature of the functions providing these properties is not simple. Anderson (1958) provides rigorous derivations showing that it is an eigenvalue problem based on the variance-covariance matrix of the data spectra. As an eigenvalue problem, it can, therefore, be expressed in the simpler form of generalized eigenvalue problems.

$$\lambda \Psi = E \Psi \qquad (3)$$

in which E = the eigenvector, which comprises the principal component, and λ = the eigenvalue, which represents the amount of variance of the original data that is accounted for by the corresponding eigenvector.

VI. SOFT INDEPENDENT MODELING OF CLASS ANALOGIES

Soft independent modeling of class analogies (SIMCA) has also been extensively described in the chemometric literature (Wold et al, 1983; Sharaf et al, 1986; Beebe et al, 1998). This approach also uses the calculation of principal components, but in a different way than described in the previous section. While the straightforward application of principal components, as described in the previous section, calculates the eigenvectors and eigenvalues using the training samples from all the materials for which that model is being generated, the SIMCA approach calculates multiple sets of principal components. In fact, a different set of principal components is calculated for each different material for which the model is being generated.

The result is that the data corresponding to each material are much more tightly and carefully defined by the principal components, since each material has its own set defined for it. Thus, groups of data points that are close together (i.e., have more nearly similar spectra) are better distinguished from each other; this makes the SIMCA approach better than the simple use of principal components when materials with nearly similar spectra are to be distinguished. Figure 4 illustrates how each group can be specified with its own set of principal axes. On the other hand, if all the spectra have clearly distinguishable spectra, then use of the simpler principal component algorithm will provide an adequate model, while requiring fewer total samples and less work and resources to obtain the model.

VII. K-NEAREST NEIGHBORS

The previous algorithms all required a training step, and an explicit model was created to do the identification. These methods all have a common characteristic and impose a common requirement on the training data in order for the calibration statistics to adhere to the theoretical behavior: they are all parametric methods and the data must follow the multivariate normal distribution. The K-nearest neighbors algorithm, by contrast, is a nonparametric method and does not necessarily require "training," because a model as such need not be created and the data need not be multivariate Normally distributed. Indeed, the K-nearest neighbor algorithm is particularly suited to identifying data that is not multivariate Normally distributed. On the other hand, when one speaks of "nearest neighbors," the use of some sort of distance measure is implied, and if that distance measure should be, for example, Mahalanobis distances, then having the data be multivariate Normally distributed is preferred for that reason.

Figure 5 shows a situation in which the K-nearest neighbors would be a preferred method. The small circles and squares represent data from samples of two different types and whose data are, in fact, separated from each other. For the data shown in Figure 5, any classification criterion based on determining the distribution around the multivariate mean of each data group would surely include samples belonging to the other group, because of the unusual distribution of the sample readings.

An unknown sample such as shown by the "X" in Figure 5 can be identified correctly using the K-nearest neighbor approach. Basically, all that is needed is to calculate the distance from the unknown sample to the samples whose identities are known, sort the samples according to their distance from the unknown, and note the identities of the K-closest known samples ("closest," in this case, usually being measured by Euclidean distance). All that is then needed is essentially a voting procedure: the number of samples of each type within those K samples are counted and the unknown is than identified as being the same as whichever type predominates in its neighborhood, by virtue of having the highest count.

VIII. CORRELATION COEFFICIENT

"Correlation" is defined as the tendency for two sets of numbers to vary together with each other. The correlation coefficient is a numerical measure of that tendency and is defined (or rather, its square is defined, to keep the expression simple) by the equation

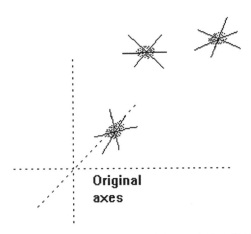

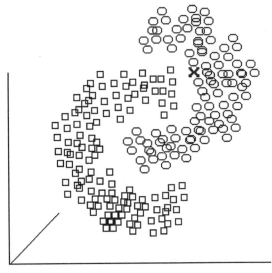

Fig. 4. In the soft independent modeling of class analogies (SIMCA) approach, each group of data points, representing a different set of samples, has its own set of principal component axes calculated for it.

Fig. 5. K-nearest neighbors is suited to data that does not follow a multivariate Normal distribution, such as data distributed as shown here. The circles and the squares represent different materials. The "X" represents an unknown sample.

$$R^2 = \frac{\sum_i (X - \bar{X})(Y - \bar{Y})}{\sqrt{\sum_i (X - \bar{X})^2 \sum_i (Y - \bar{Y})^2}} \quad (4)$$

The correlation coefficient is 1 (unity) if the two sets of numbers are exactly in step, and it can become as low as -1 if the two sets of number vary in exact opposition. The two sets of numbers do not have to be exactly the same; a constant offset, or even a scaling factor, between them will not affect the value of the correlation coefficient. Thus, a higher correlation coefficient indicates that a given spectrum is "more like" the spectrum of one material than another.

Because of these properties of the correlation coefficient, it is well suited to comparing the spectrum of an unknown material with those of several different known materials and to identifying the unknown material as the one with which it matches best, i.e., with which it has the highest correlation coefficient. As with Mahalanobis distances, it is possible to assign a threshold value, such that if the unknown does not match any of the known materials by at least the amount specified by the threshold, one can conclude that the unknown cannot be identified, i.e., it is not any of the known materials.

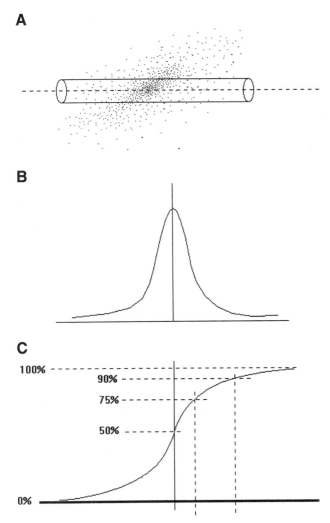

Fig. 6. Development of the quantiles. **A,** Illustrates the density of points corresponding at various locations along the axis of a cylinder running through a set of multivariate data. **B,** Presents the probability density function for the data of **A**; for a multivariate Normal distribution, the probability density function is also Normal. **C,** Illustrates the cumulative distribution function, along with some selected quantiles.

IX. BOOTSTRAP ERROR-ADJUSTED SINGLE SAMPLE TECHNIQUE

This algorithm defines another nonparametric method. It was developed at the University of Indiana (Lodder and Hieftje, 1988a, 1988b) quite a while ago, but has not found widespread use. Nevertheless, it is reported here since it was developed for use in conjunction with NIR spectroscopy and has been used in some specialized NIR applications.

This algorithm is based on the application of quantile (a more formal name for "percentile") analysis to NIR data. A quantile is simply the percentage of the data that falls below a given value. A familiar example can be seen in the Normal (Gaussian) distribution. Figure 6 illustrates a multivariate Normal distribution of data (Fig. 6A) and that the density of data along any selected cylindrical section through it will follow a Normal distribution (Fig. 6B). If one integrates this distribution, a curve is obtained (Fig. 6C) that now represents the cumulative distribution function. The value of the cumulative distribution function corresponding to any value along the x axis represents the fraction of the data falling below that x value. Since the fraction of the data falling below any given x value must be between 0 and 100%, the cumulative distribution function approaches these two values asymptotically (Fig. 6C).

A key point here, however, is that data need not follow the Normal, or even any known, distribution in order to allow the construction of a cumulative distribution function. The cumulative distribution function can be constructed, for any set of data, from the data itself. This is particularly useful when the data at hand do not follow the multivariate Normal distribution function, which, as Lodder and Hieftje point out (Lodder and Hieftje, 1988a), is not an uncommon situation when dealing with NIR data.

While a full exposition is beyond the scope of this chapter, the algorithm entails combining several concepts into a single calculation.

1. The use of the empirical distribution function of the data.
2. Developing this empirical distribution function through repetitive (Monte-Carlo) resampling of the actual measured spectra and, thus, determining the characteristics of the population from the measured samples.

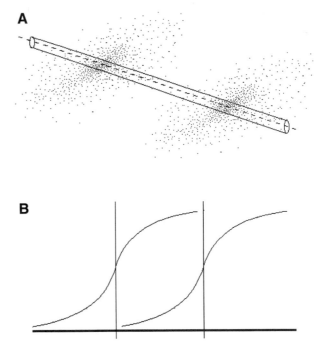

Fig. 7. **A,** For identification purposes, the cumulative distribution function of both materials needs to be developed from a cylindrical sampling space running between them. **B,** Illustrates the two functions; the probability of belonging to either data cluster can be determined for a data point falling anywhere along the cylindrical sampling space.

3. Determining the cumulative distribution functions for pairs of materials to be identified by (conceptually) creating a cylindrical space running through the data clusters corresponding to both materials. This is illustrated in Figure 7, which shows the clusters of data points corresponding to two different materials (Fig. 7A) and the cumulative distribution functions for the two data clusters (Fig. 7B).

4. Assigning probabilities for a given data point belonging to either data cluster.

ACKNOWLEDGEMENTS

I thank Foss/NIRSystems, Inc., and CAMO, Inc., for providing information helpful in preparing this chapter.

LITERATURE CITED

Anderson, T. W. 1958. An Introduction to Multivariate Statistical Analysis. 1st ed. John Wiley & Sons, New York.

Beebe, K. R., Pell, R. J., and Seascholtz, M. B. 1998. Chemometrics: A Practical Guide. John Wiley & Sons, New York.

Bertrand, D., Novales, B., Devaux, M. F., and Robert, P. 1995. Discrimination of durum wheat products for quality control. Pages 430-435 in: Near Infrared Spectroscopy: The Future Waves. Proc. Int. Conf. Near-Infrared Spectrosc., 7th. NIR Publications, West Sussex, United Kingdom.

Hammond, S. V. 1994. The cost avoidance role of NIR in pharmaceutical production. Page 396 in: Leaping Ahead with Near Infrared Spectroscopy. Proc. Int. Conf. Near-Infrared Spectrosc., 6th. NIR Spectroscopy Group, Royal Australian Chemical Institute, North Melbourne, Victoria, Australia.

Jouan-Rimbaud, D., and Massart, D. L. 1995. Wavelength selection for the multivariate calibration of near infrared spectroscopic data. Pages 194-197 in: Near Infrared Spectroscopy: The Future Waves. Proc. Int. Conf. Near-Infrared Spectrosc., 7th. NIR Publications, West Sussex, United Kingdom.

Kaffka, K., and Gyarmati, L. S. 1989. NIR expert system for sample identification. Pages 222-231 in: Proc. Int. Conf. Near-Infrared Spectrosc., 2nd. Korin Publishing Co., Ltd., Tokyo, Japan.

Kaffka, K. J., and Gyarmati, L. S. 1990. Qualitative (comparative) analysis by near infrared spectroscopy. Pages 135-144 in: Proc. Int. Conf. Near-Infrared Spectrosc., 3rd. Agricultural Research Centre Publishing, Gembloux, Belgium.

Kaffka, K. J., and Gyarmati, L. S. 1994. Quality determination in the food industry. Pages 261-269 in: Leaping Ahead with Near Infrared Spectroscopy. Proc. Int. Conf. Near-Infrared Spectrosc., 6th. NIR Spectroscopy Group, Royal Australian Chemical Institute, North Melbourne, Victoria, Australia.

Lavine, B. A. 1998. Chemometrics. Anal. Chem. 70(12):209R-228R.

Legrand, A., Scotter, C. N. G., and Voyiagis, M. 1994. NIR for juice authenticity screening. Pages 307-311 in: Leaping Ahead with Near Infrared Spectroscopy. Proc. Int. Conf. Near-Infrared Spectrosc., 6th. NIR Spectroscopy Group, Royal Australian Chemical Institute, North Melbourne, Victoria, Australia.

Lodder, R. A., and Hieftje, G. M. 1988a. Quantile BEAST attacks the false-sample problem in near-infrared reflectance analysis. Appl. Spectrosc. 42(8):1351-1365.

Lodder, R. A., and Hieftje, G. M. 1988b. Detection of subpopulations in near-infrared reflectance analysis. Appl. Spectrosc. 42(8):1500-1512.

Mark, H. 1986. Normalized distances for qualitative near infrared reflectance analysis. Anal. Chem. 58(2):379-384.

Mark, H. 1987. Studies of principal components as a calibration method for near infrared reflectance analysis. Chimicaoggi Edizione: Teknoscienze Srl. Sept:57-65.

Mark, H. 1992. Data analysis: Multilinear regression and principal components analysis. Pages 107-158 in: Handbook of Near-Infrared Analysis. D. Burns and E. Ciurczak, eds. Marcel Dekker, New York.

Mark, H. L., and Tunnell, D. 1985. Qualitative near infrared reflectance analysis using Mahalanobis distances. Anal. Chem. 57(7):1449-1456.

Martens, H., and Russwurm, H. 1983. Food Research and Data Analysis. Elsevier Science Publishers, New York.

Naes, T., and Isaksson, T. 1992. Computer-assisted methods in near-infrared spectroscopy. Pages 69-93 in: Computer-Enhanced Analytical Spectroscopy. Vol. 3. P. C. Jurs, ed. Plenum Press, New York.

Osborne, B. G., and Mertens, B. 1994. Authentication of Basmati rice by near infrared transmittance spectra of individual grains. Pages 161-167 in: Leaping Ahead with Near Infrared Spectroscopy. Proc. Int. Conf. Near-Infrared Spectrosc., 6th. NIR Spectroscopy Group, Royal Australian Chemical Institute, North Melbourne, Victoria, Australia.

Robert, P., Bertrand, D., Devaux, M. F., and Grappin, R. 1987. Multivariate analysis applied to near-infrared spectra of milk. Anal. Chem. 59:2187-2191.

Rose, J. 1982. Quantitative and Qualitative Analysis with NIRA. In: Annu. Symp. Near Infrared Reflectance Anal., 2nd. Technicon Instrument Corporation, Tarrytown, NY.

Rose, J., Prusik, T., and Mardekian, J. 1982. J. Parenteral Sci. Technol. 36:71-78.

Schilling, D., and Ritzmann, H. 1994. Rapid on-line identification of plastics using a novel ultrafast near infrared sensor. Page 369 in: Leaping Ahead with Near Infrared Spectroscopy. Proc. Int. Conf. Near-Infrared Spectrosc., 6th. NIR Spectroscopy Group, Royal Australian Chemical Institute, North Melbourne, Victoria, Australia.

Schilling, D., Bottlang, L., and Dammertz, W. 1994. Near infrared transmission analysis of intact pharmaceutical tablets. Pages 441-449 in: Leaping Ahead with Near Infrared Spectroscopy. Proc. Int. Conf. Near-Infrared Spectrosc., 6th. NIR Spectroscopy Group, Royal Australian Chemical Institute, North Melbourne, Victoria, Australia.

Sharaf, M., Illman, D., and Kowalski, B. 1986. Chemometrics. John Wiley & Sons, New York.

Varadi, M., Toth, A., and Hajos, G. 1994. Investigation into enzymatic modification of food proteins by NIR. Pages 479-481 in: Leaping Ahead with Near Infrared Spectroscopy. Proc. Int. Conf. Near-Infrared Spectrosc., 6th. NIR Spectroscopy Group, Royal Australian Chemical Institute, North Melbourne, Victoria, Australia.

Whitfield, R. G., Gerger, M. E., and Sharp, R. L. 1987. Near-infrared spectrum qualification via Mahalanobis distance determination. Appl. Spectrosc. 41(7):1204-1213.

Williams, P., and Norris, K. 1987. Near-Infrared Technology in the Agricultural and Food Industries. American Association of Cereal Chemists, St. Paul, MN.

Wold, S., Albano, C., Dunn, W. J., Esbensen, K., Hellberg, S., Johansson, E., and Sjostrom, M. 1983. Pattern recognition: Finding and Using regularities in multivariate data. Pages 164-173 in: Food Research and Data Analysis: Proc. IUFoST Symp. Applied Science Publishers, London. Distributed by Elsevier Science Publishing Co., Inc., New York.

NEAR-INFRARED SPECTRA

The figures presented on the following pages illustrate the near-infrared reflectance spectra of a large number of substances of pure chemicals and agricultural and food products. The second-derivative spectra are displayed above the log(1/R) spectra to accentuate the absorbance peaks. For the second-derivative traces, segment size was one wavelength point (about 1.6 nm) and derivative (gap) was six points (about 10 nm). A key to the spectra is on pages 281–282.

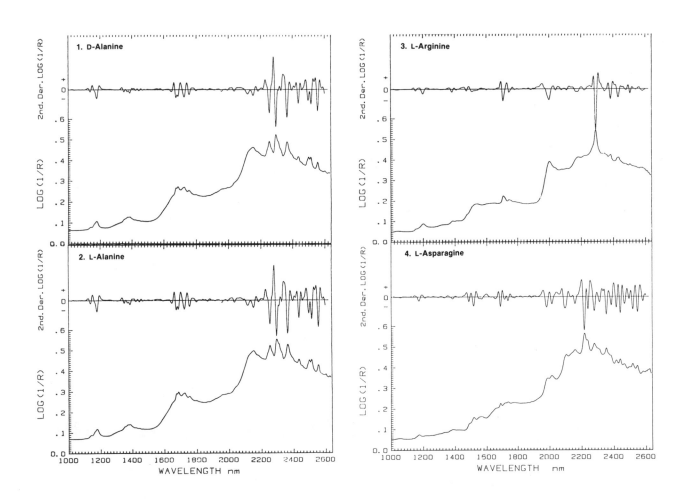

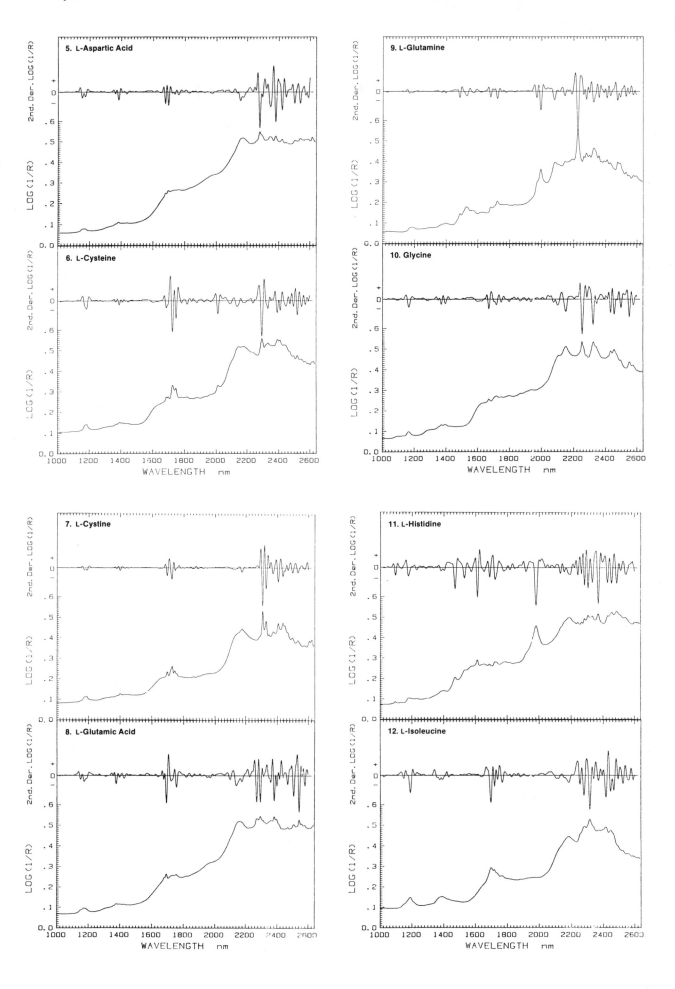

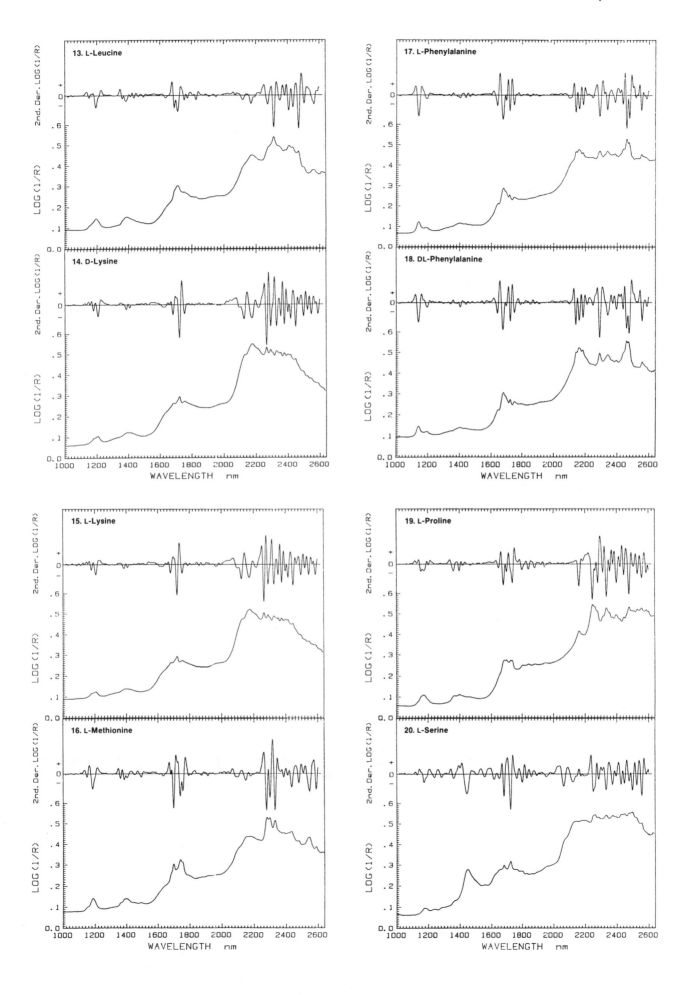

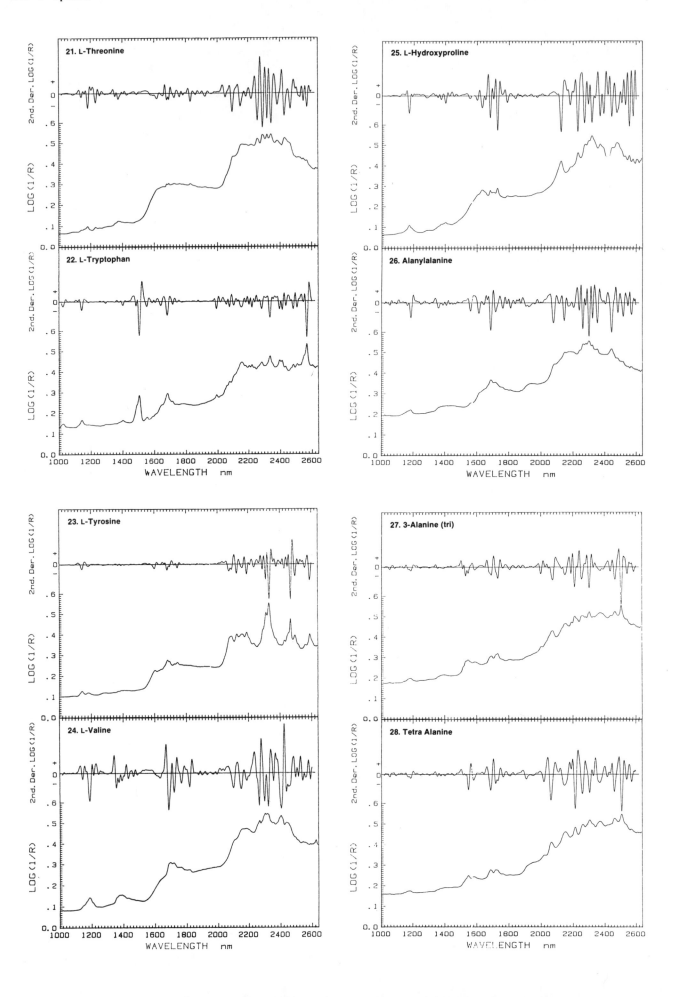

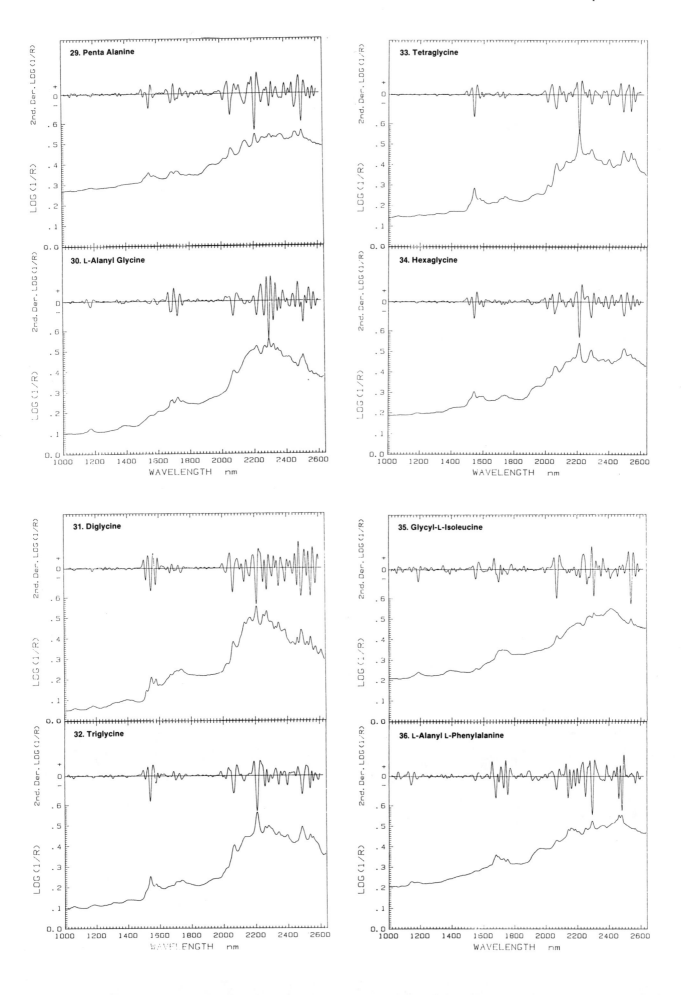

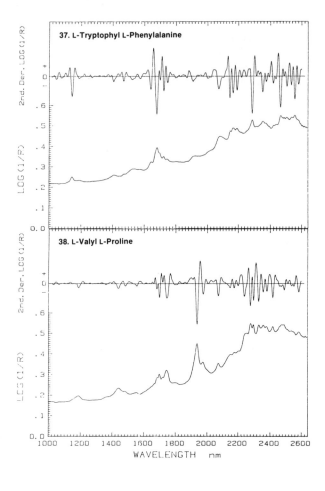

37. L-Tryptophyl L-Phenylalanine

38. L-Valyl L-Proline

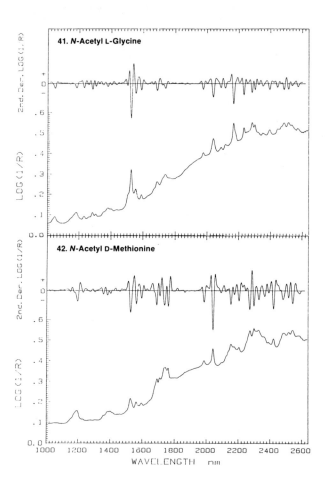

41. *N*-Acetyl L-Glycine

42. *N*-Acetyl D-Methionine

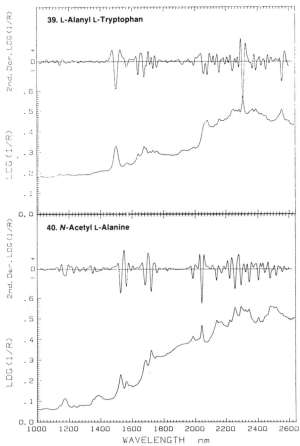

39. L-Alanyl L-Tryptophan

40. *N*-Acetyl L-Alanine

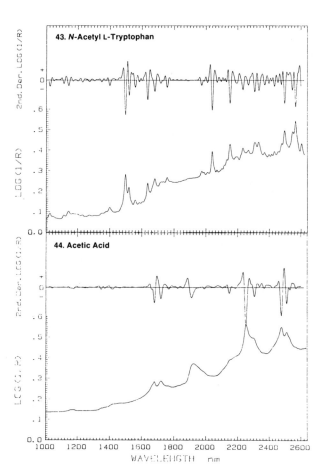

43. *N*-Acetyl L-Tryptophan

44. Acetic Acid

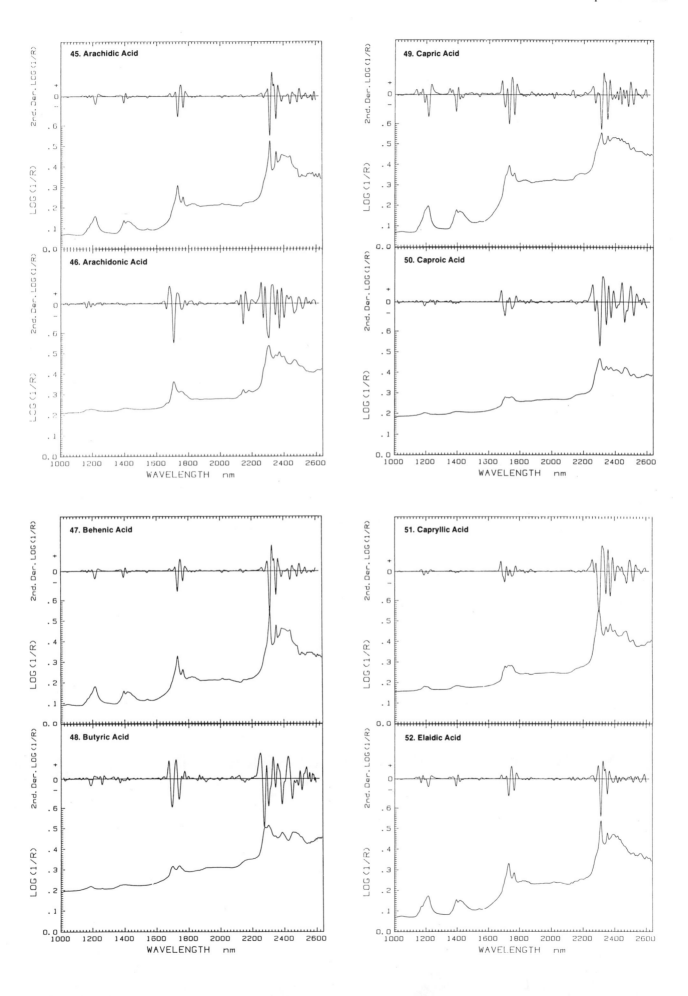

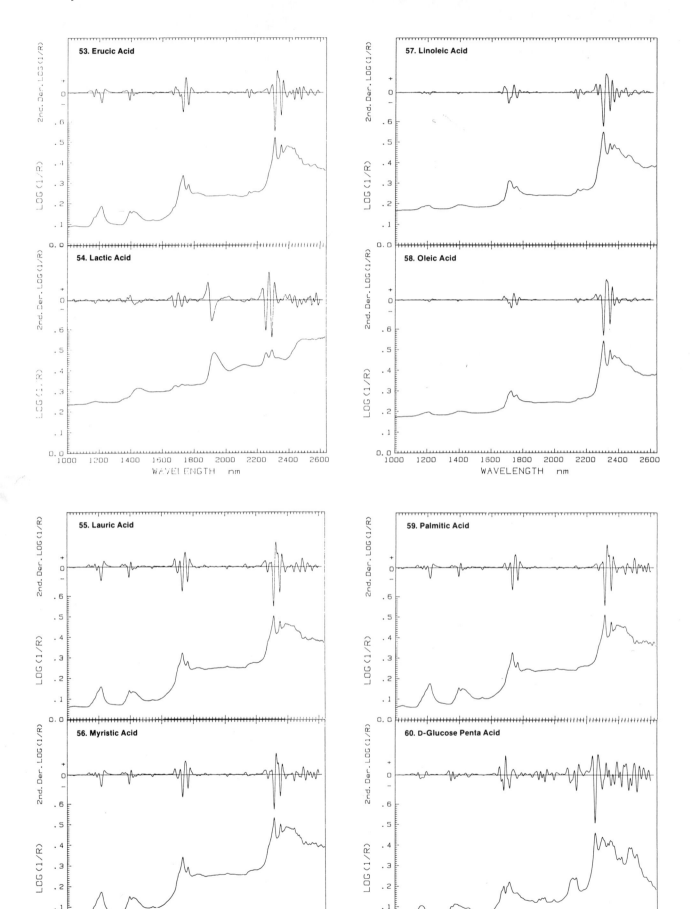

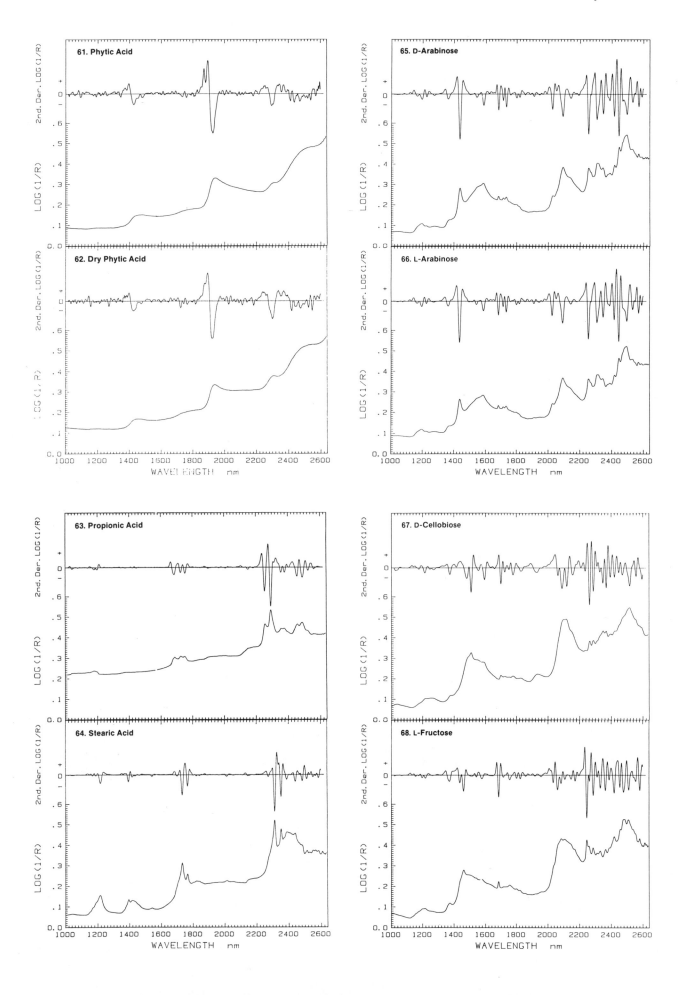

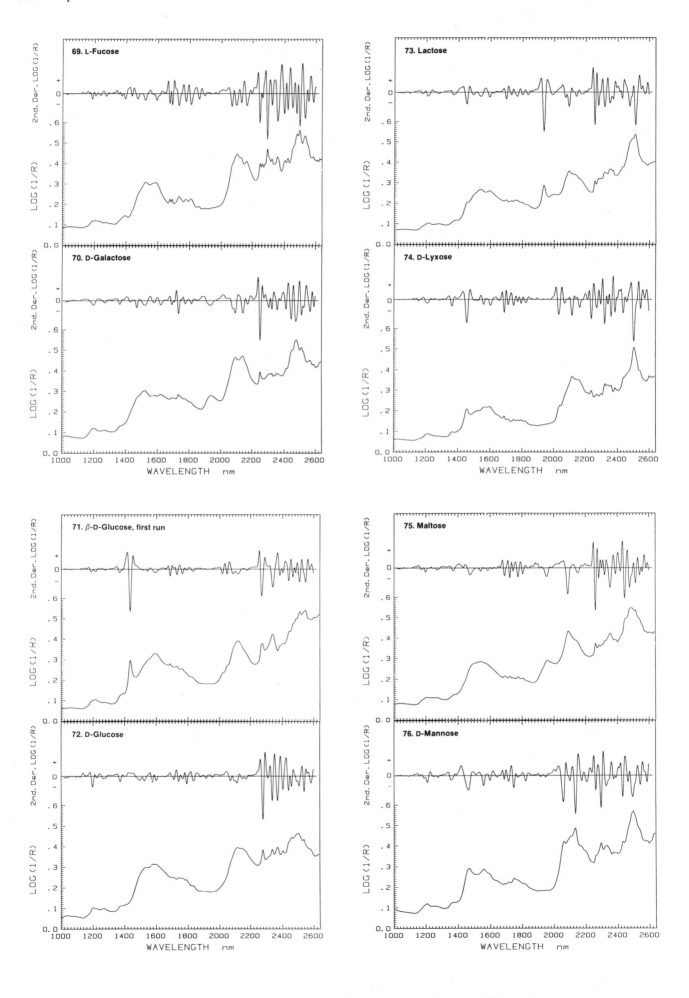

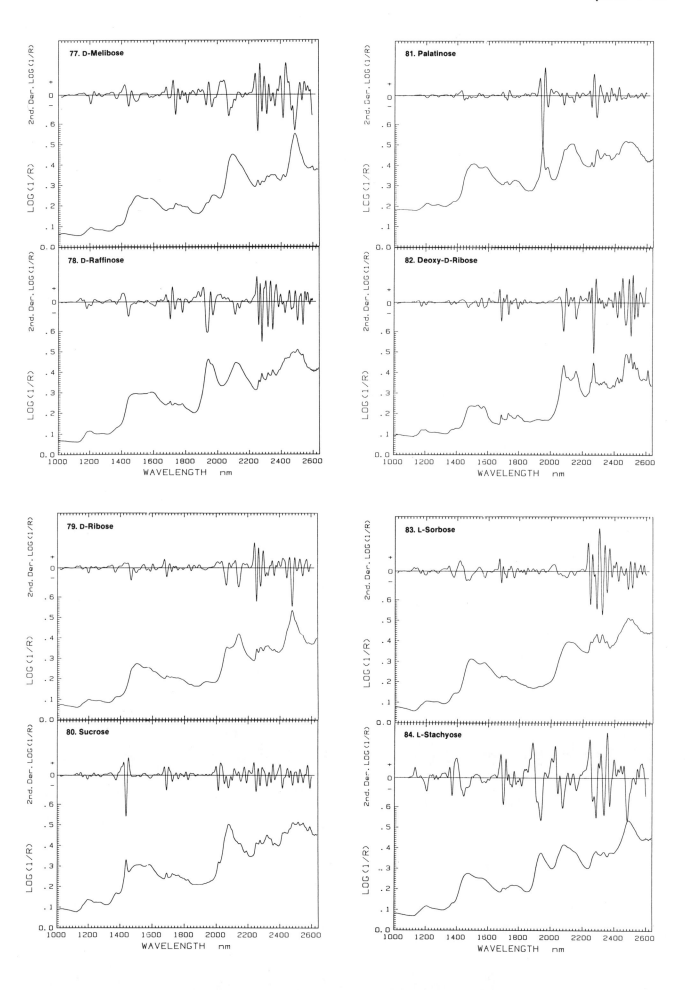

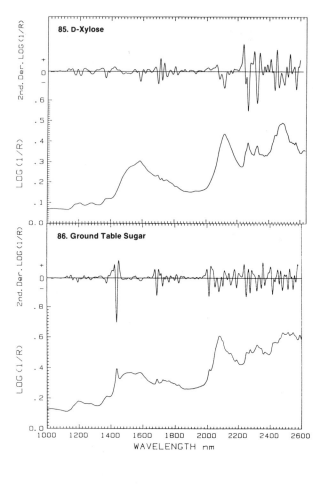

85. D-Xylose

86. Ground Table Sugar

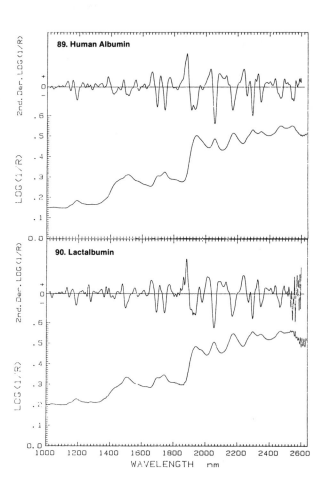

89. Human Albumin

90. Lactalbumin

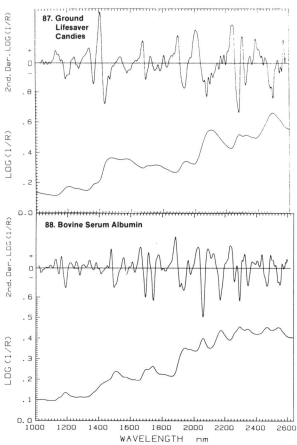

87. Ground Lifesaver Candies

88. Bovine Serum Albumin

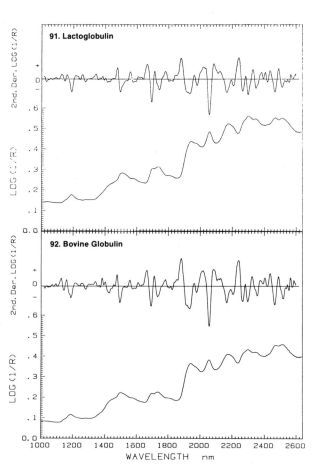

91. Lactoglobulin

92. Bovine Globulin

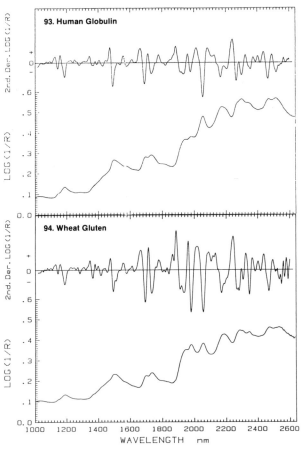

93. Human Globulin

94. Wheat Gluten

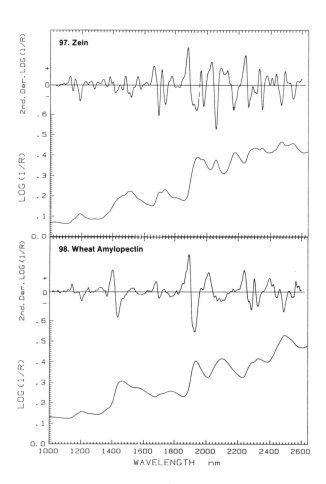

97. Zein

98. Wheat Amylopectin

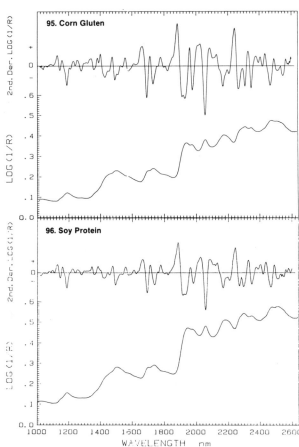

95. Corn Gluten

96. Soy Protein

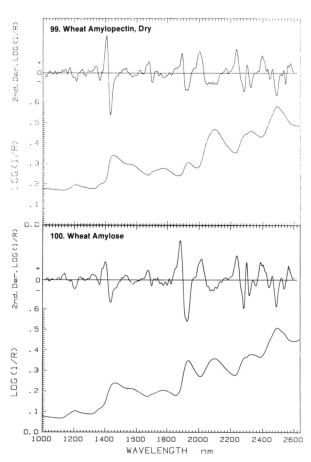

99. Wheat Amylopectin, Dry

100. Wheat Amylose

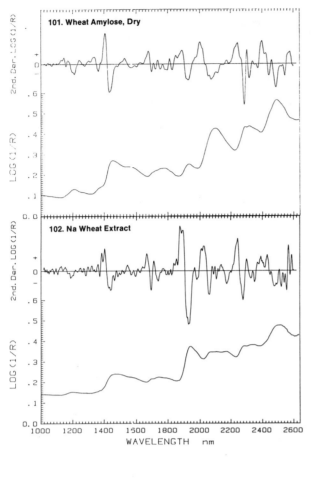

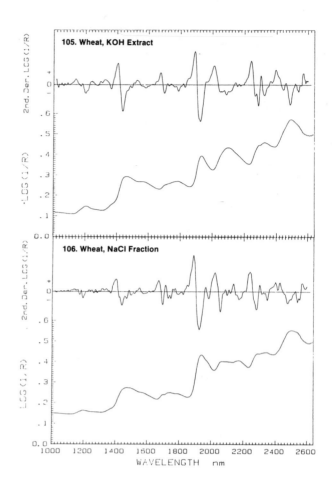

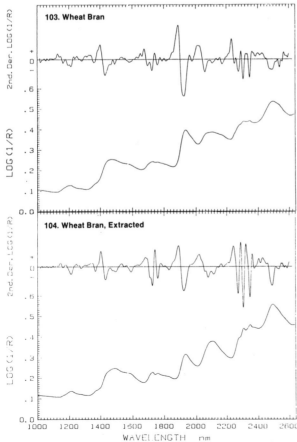

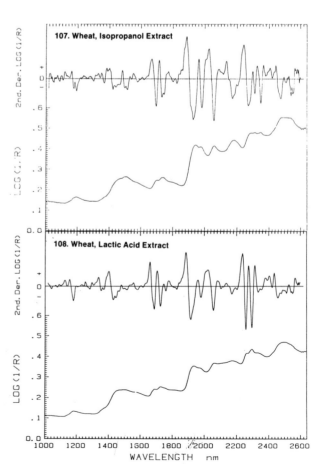

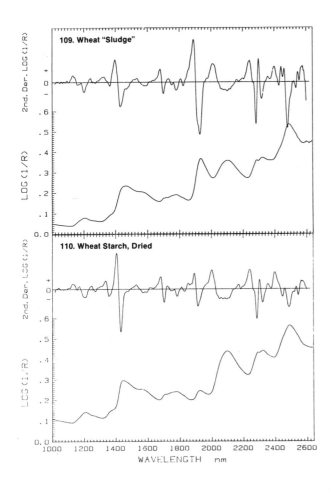

109. Wheat "Sludge"

110. Wheat Starch, Dried

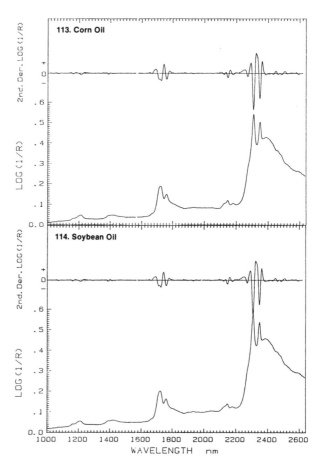

113. Corn Oil

114. Soybean Oil

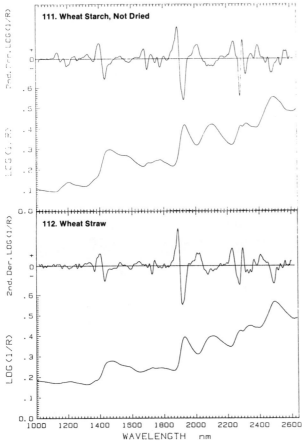

111. Wheat Starch, Not Dried

112. Wheat Straw

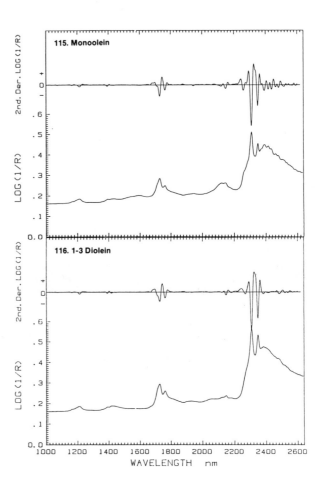

115. Monoolein

116. 1-3 Diolein

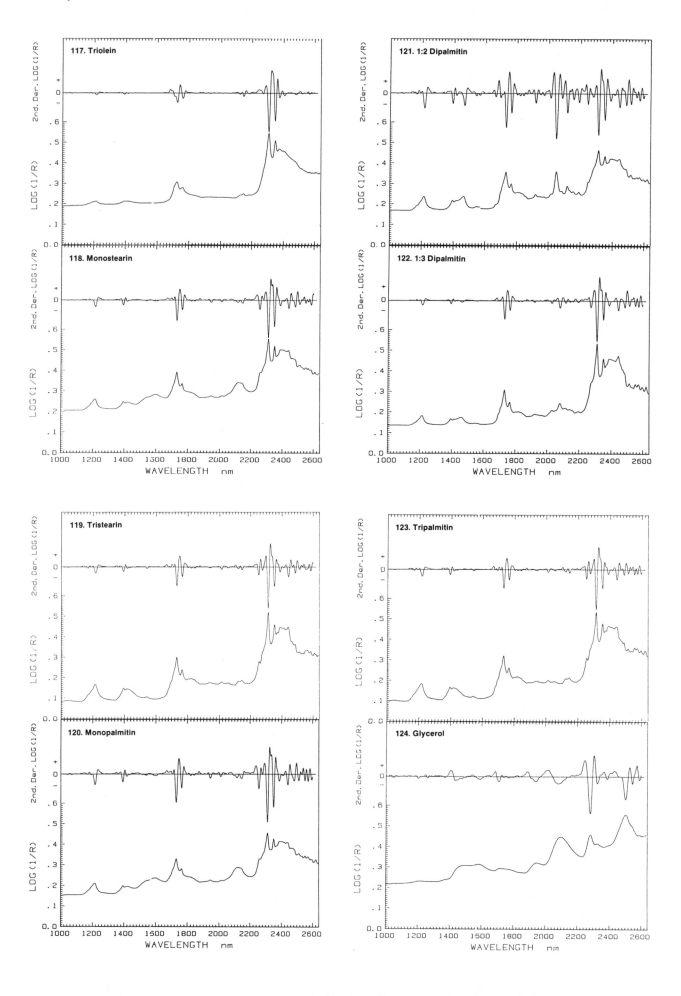

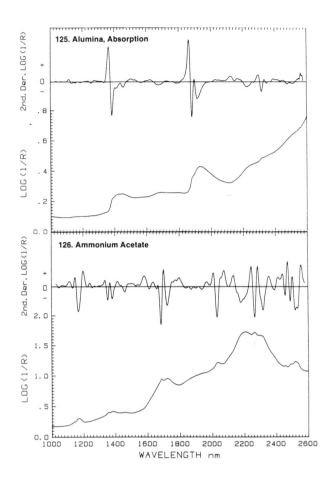

125. Alumina, Absorption

126. Ammonium Acetate

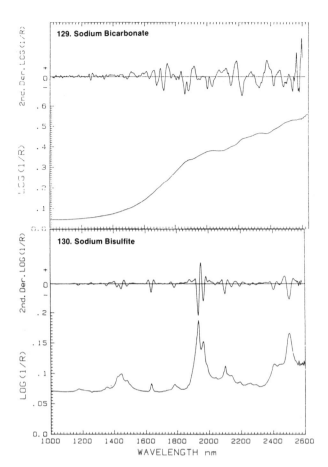

129. Sodium Bicarbonate

130. Sodium Bisulfite

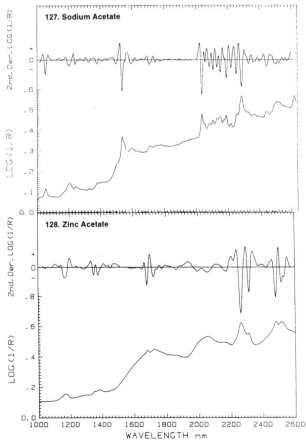

127. Sodium Acetate

128. Zinc Acetate

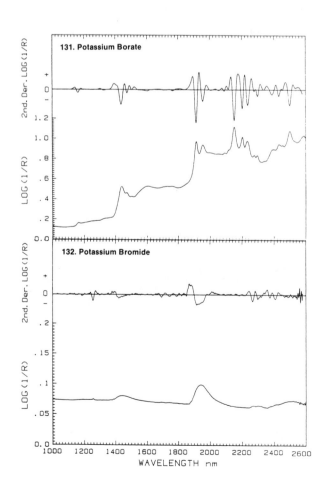

131. Potassium Borate

132. Potassium Bromide

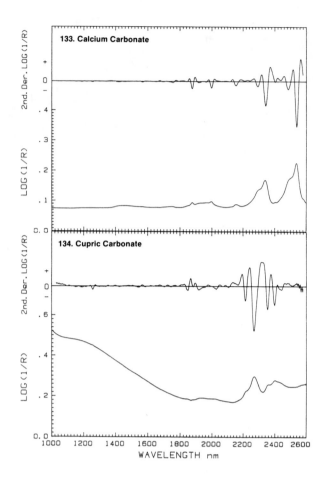

133. Calcium Carbonate

134. Cupric Carbonate

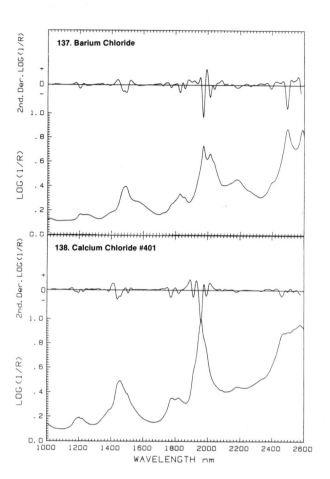

137. Barium Chloride

138. Calcium Chloride #401

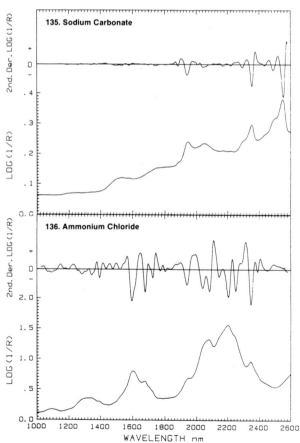

135. Sodium Carbonate

136. Ammonium Chloride

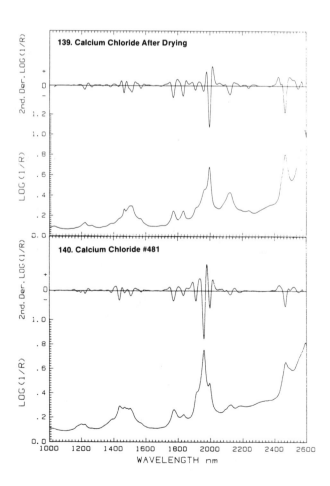

139. Calcium Chloride After Drying

140. Calcium Chloride #481

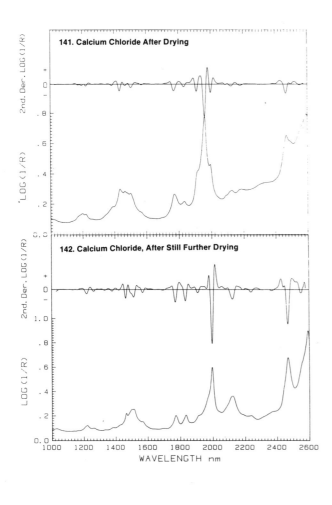

141. Calcium Chloride After Drying

142. Calcium Chloride, After Still Further Drying

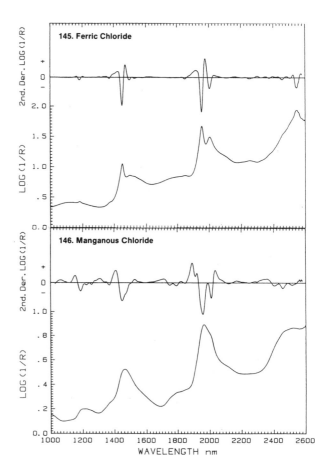

145. Ferric Chloride

146. Manganous Chloride

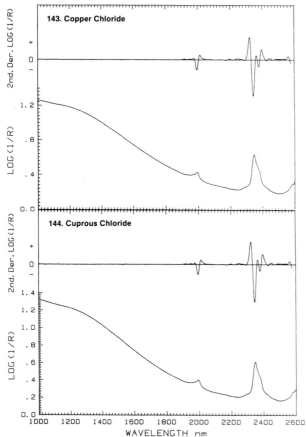

143. Copper Chloride

144. Cuprous Chloride

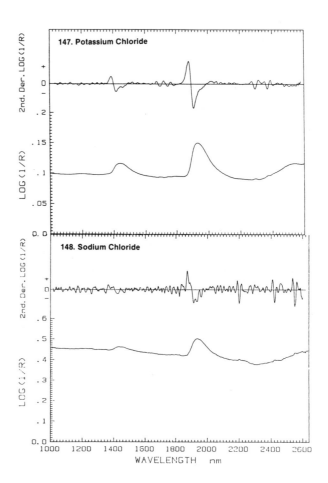

147. Potassium Chloride

148. Sodium Chloride

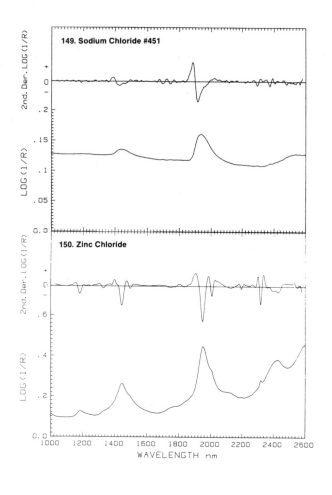

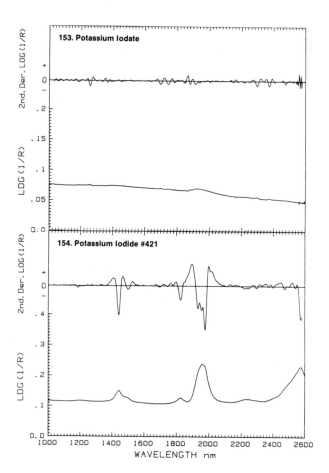

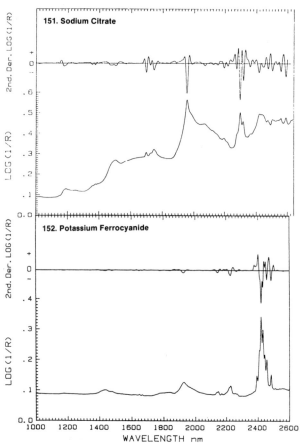

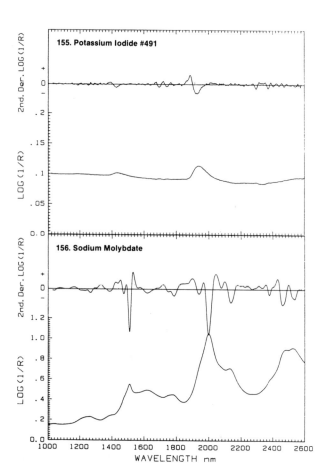

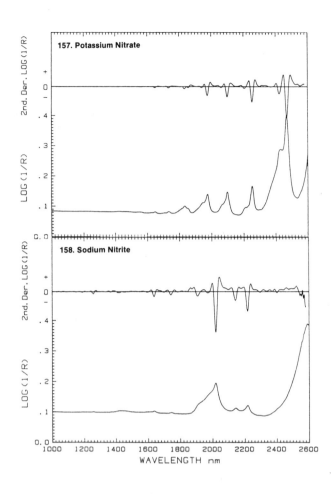

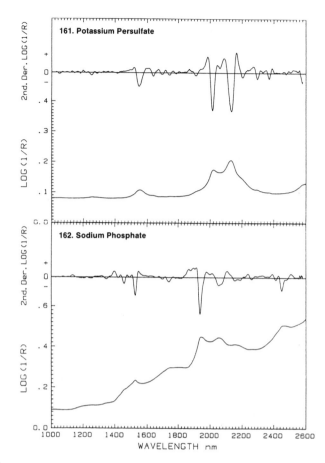

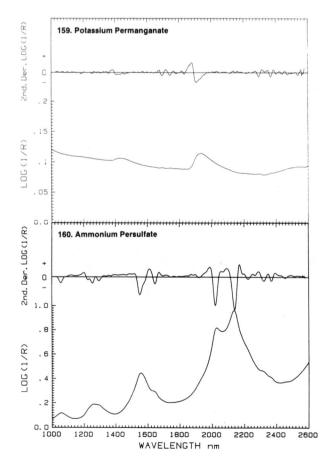

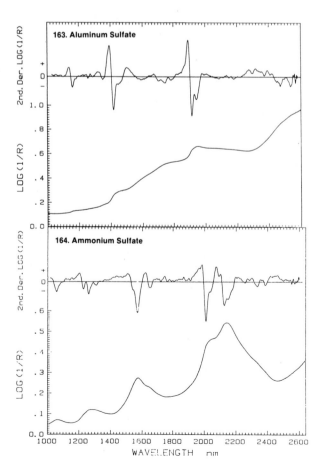

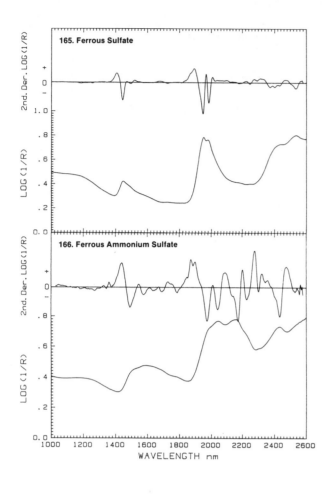

165. Ferrous Sulfate

166. Ferrous Ammonium Sulfate

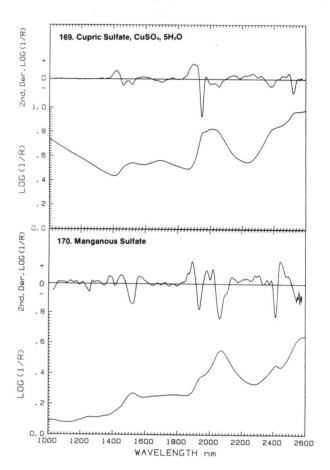

169. Cupric Sulfate, CuSO₄, 5H₂O

170. Manganous Sulfate

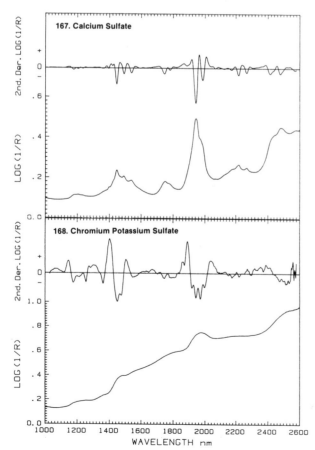

167. Calcium Sulfate

168. Chromium Potassium Sulfate

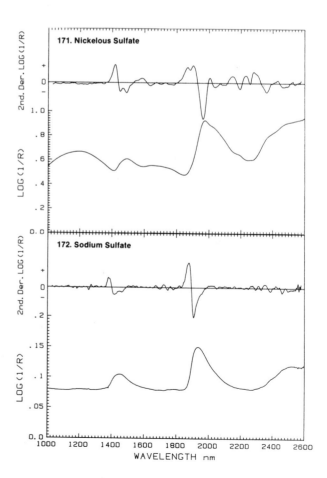

171. Nickelous Sulfate

172. Sodium Sulfate

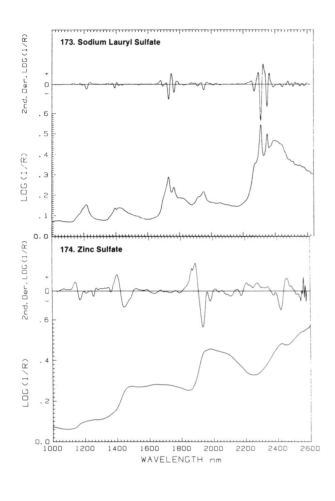

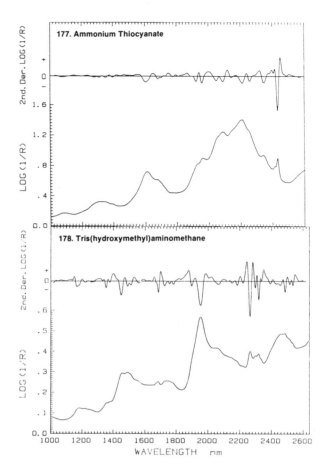

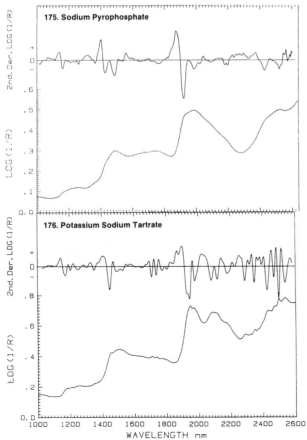

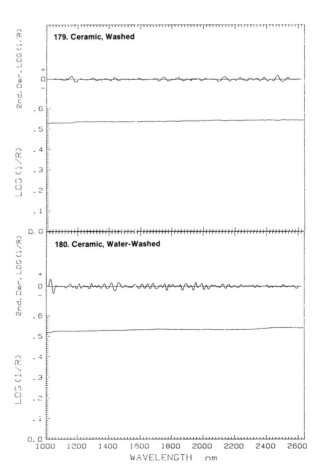

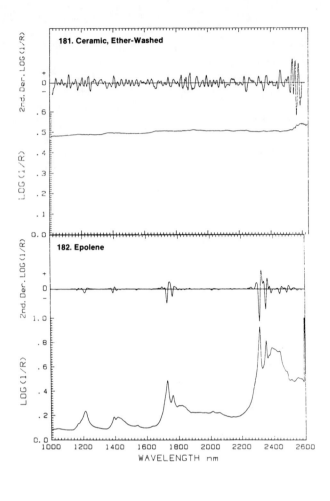

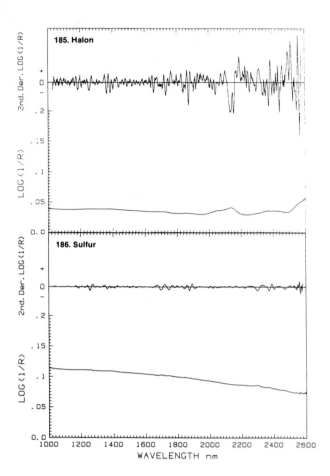

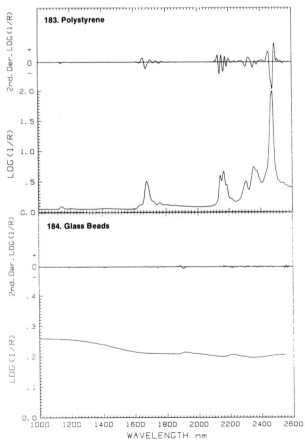

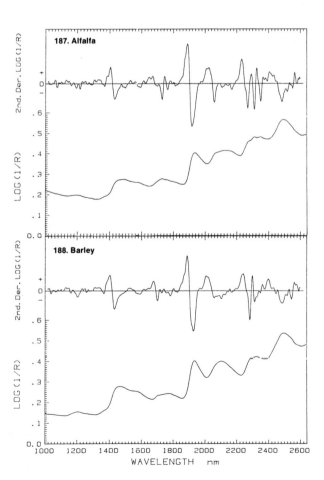

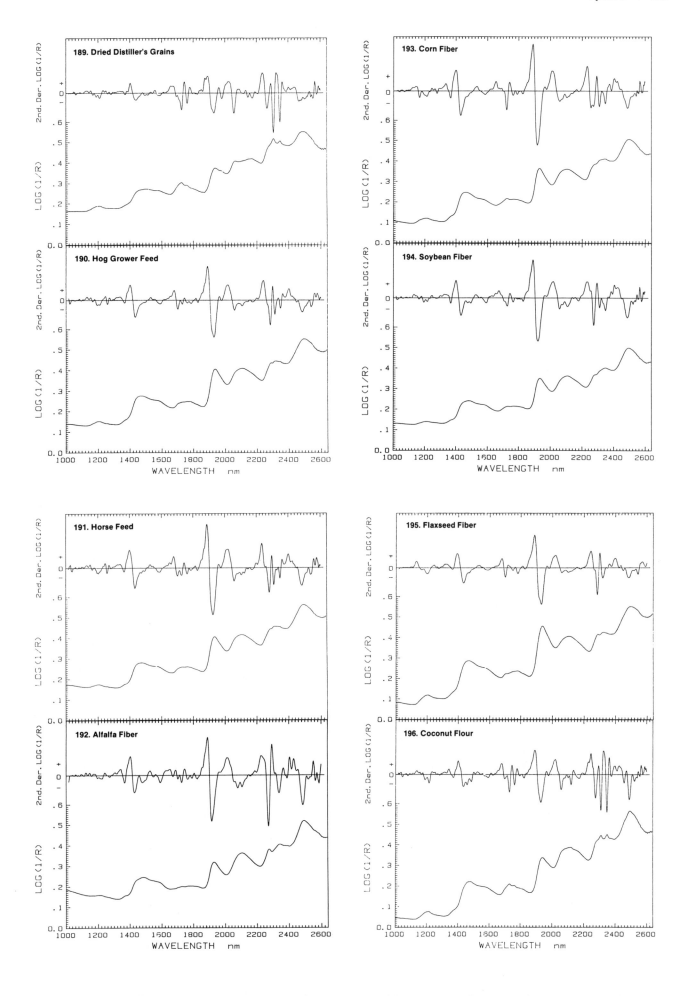

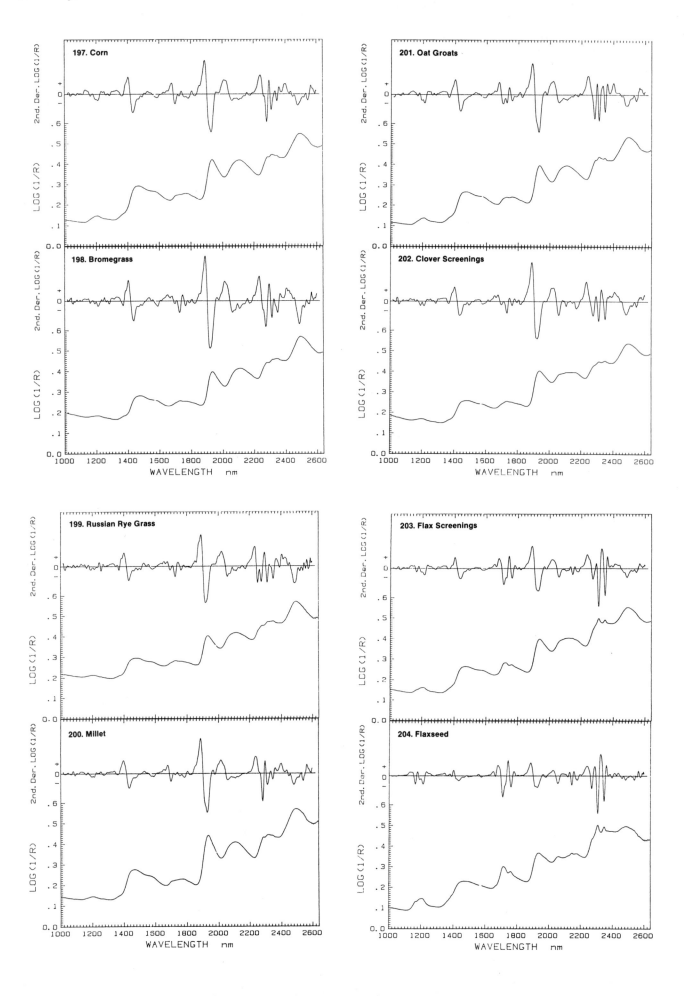

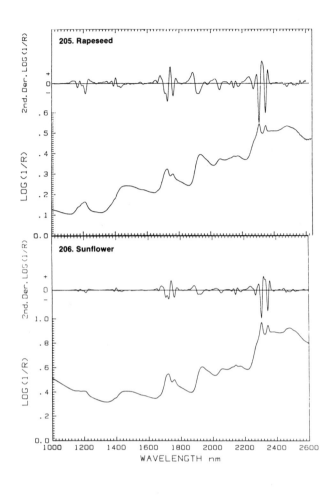

205. Rapeseed

206. Sunflower

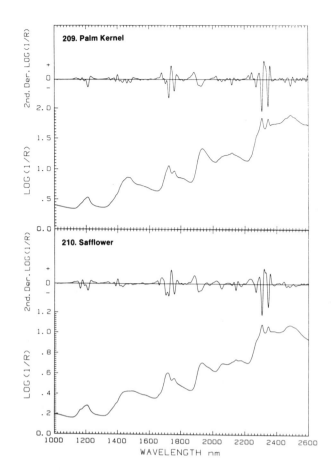

209. Palm Kernel

210. Safflower

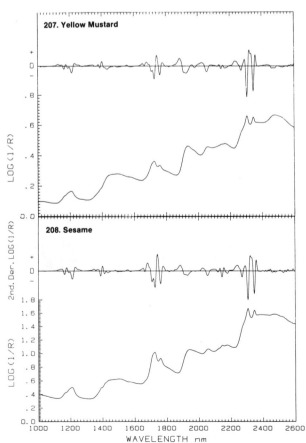

207. Yellow Mustard

208. Sesame

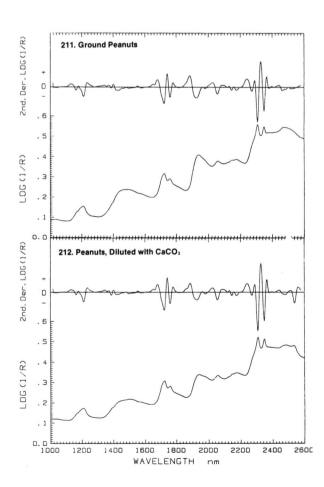

211. Ground Peanuts

212. Peanuts, Diluted with CaCO₃

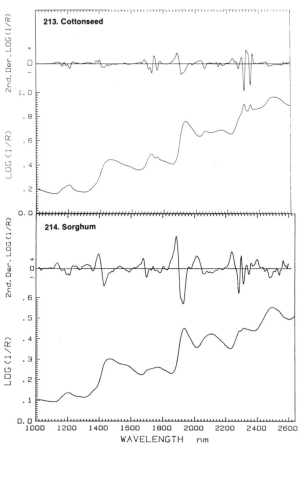

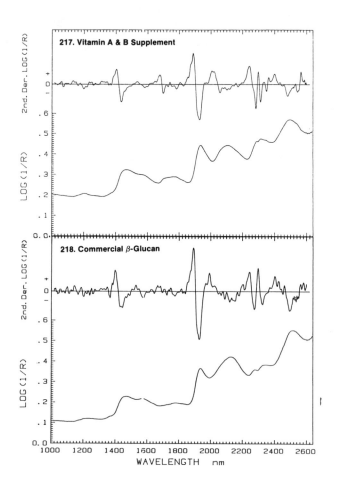

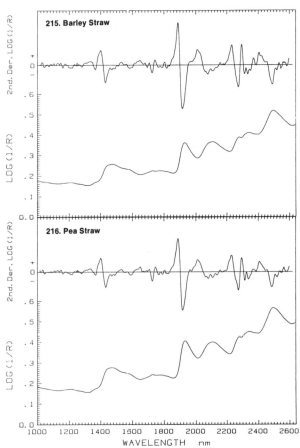

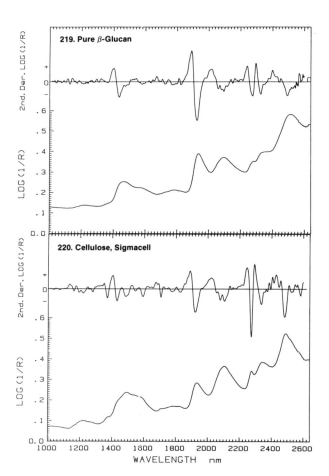

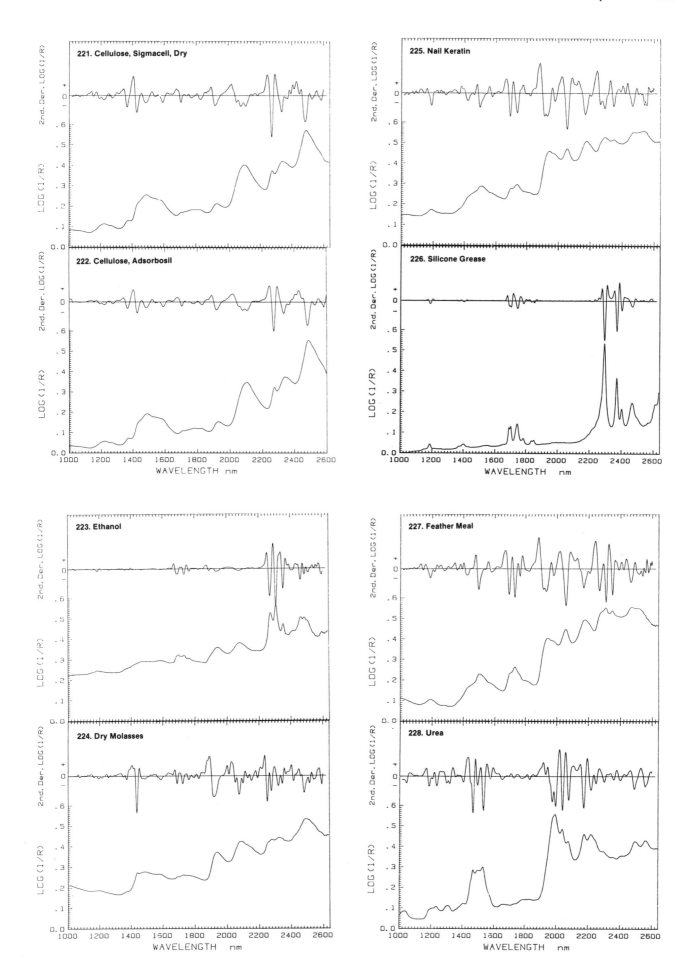

221. Cellulose, Sigmacell, Dry

222. Cellulose, Adsorbosil

223. Ethanol

224. Dry Molasses

225. Nail Keratin

226. Silicone Grease

227. Feather Meal

228. Urea

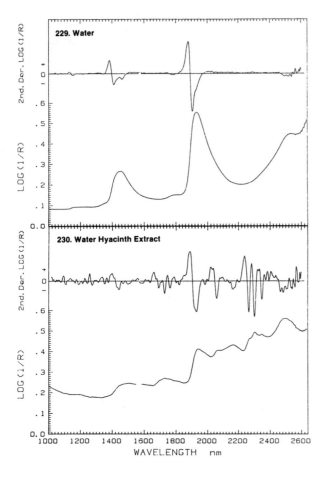

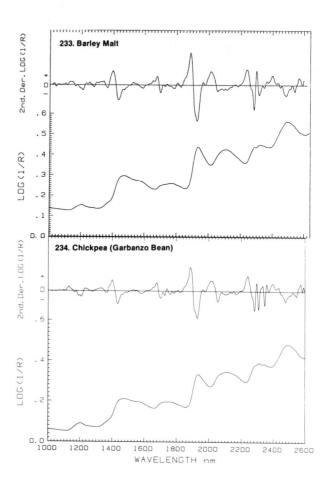

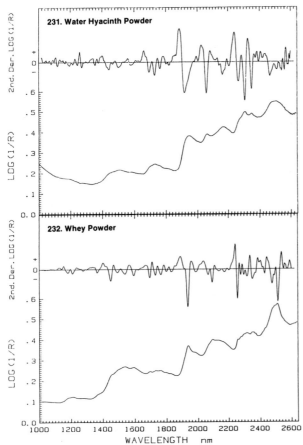

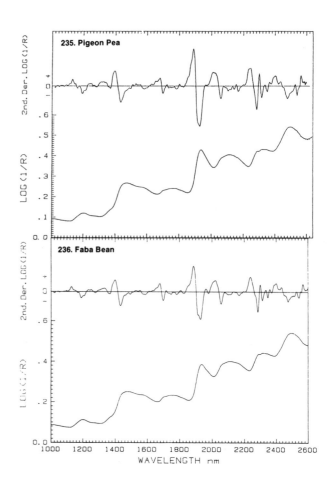

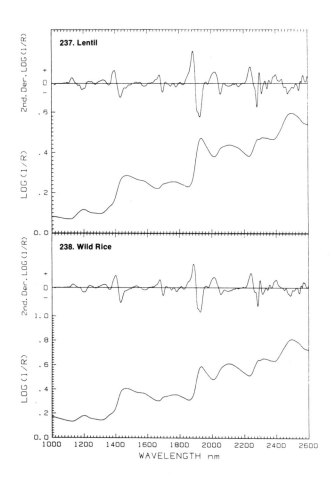

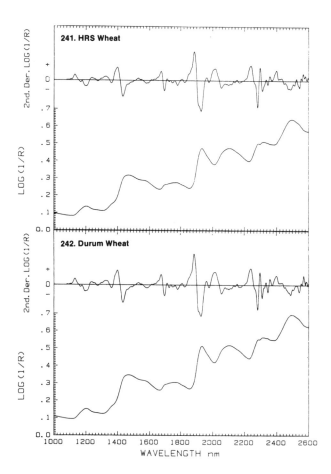

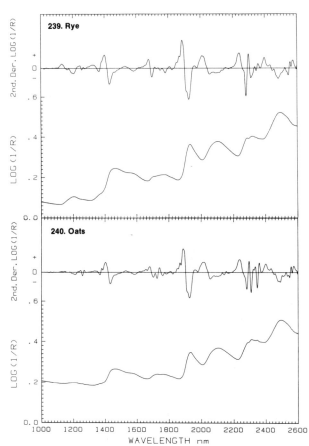

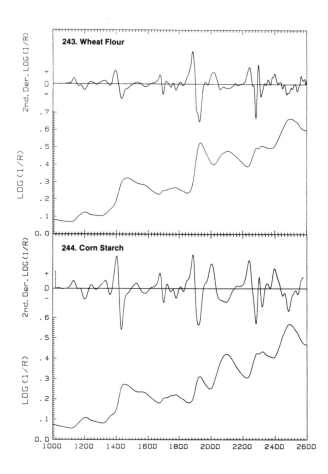

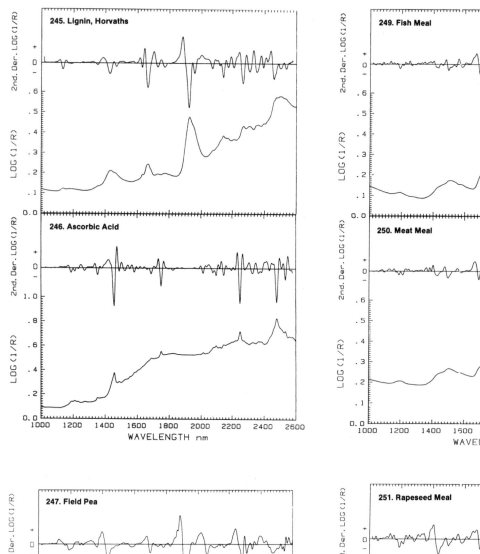

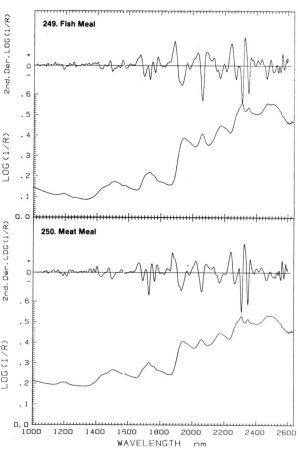

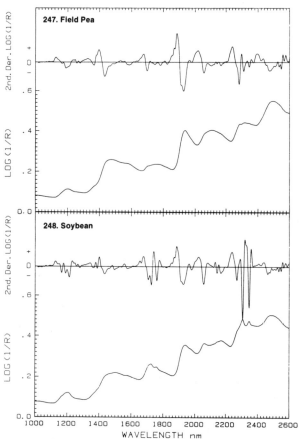

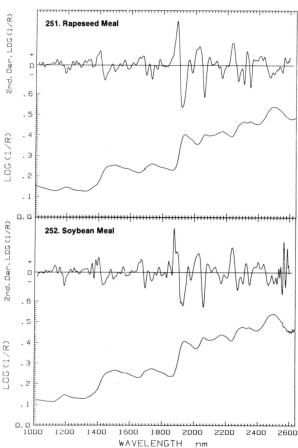

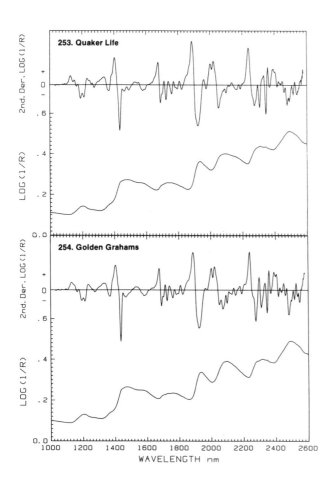

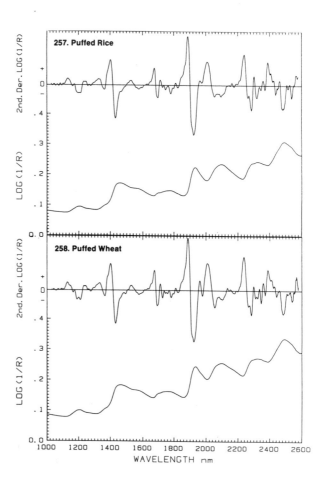

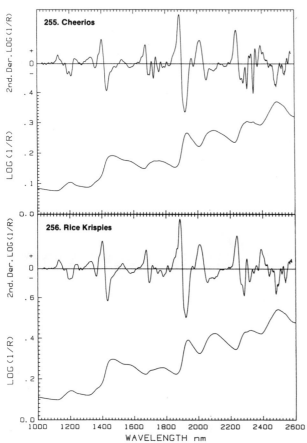

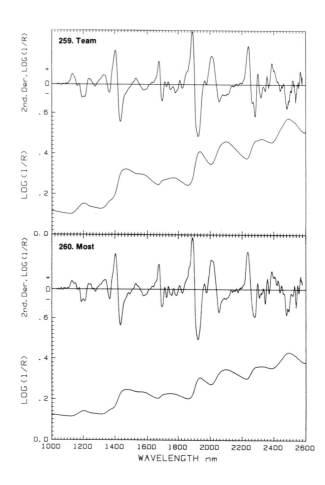

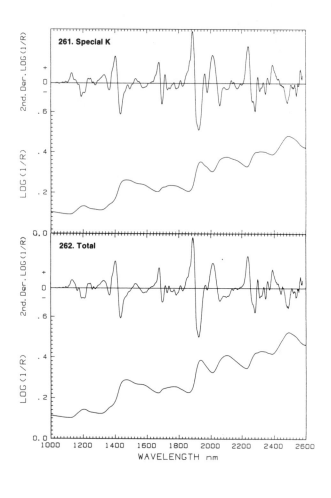

261. Special K

262. Total

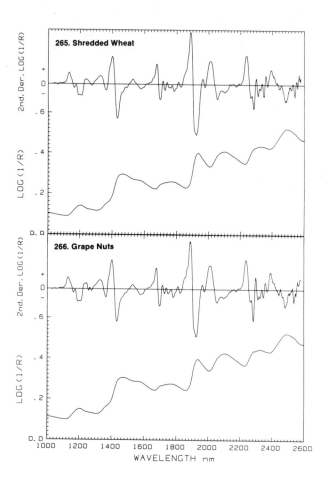

265. Shredded Wheat

266. Grape Nuts

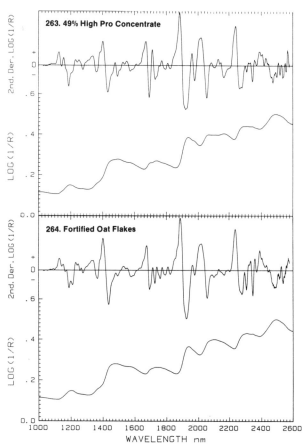

263. 49% High Pro Concentrate

264. Fortified Oat Flakes

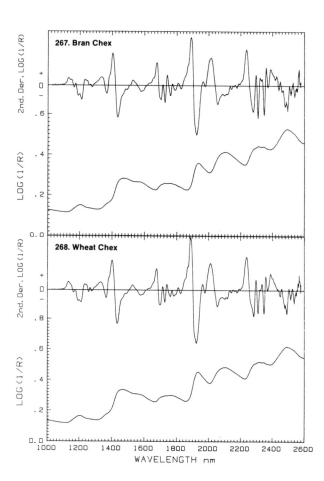

267. Bran Chex

268. Wheat Chex

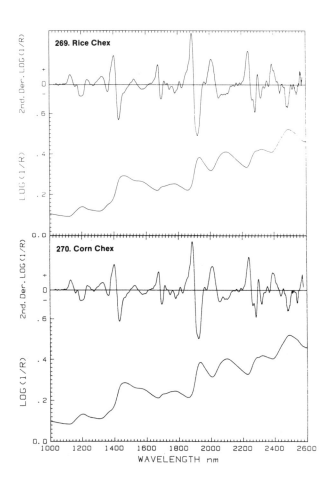

269. Rice Chex

270. Corn Chex

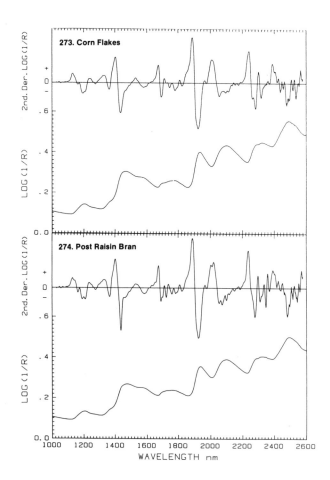

273. Corn Flakes

274. Post Raisin Bran

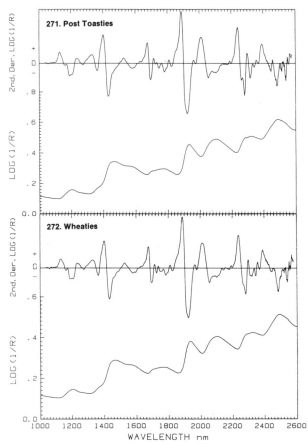

271. Post Toasties

272. Wheaties

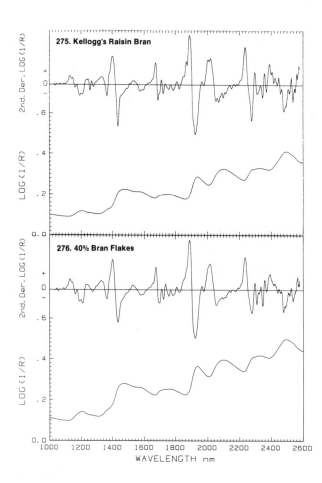

275. Kellogg's Raisin Bran

276. 40% Bran Flakes

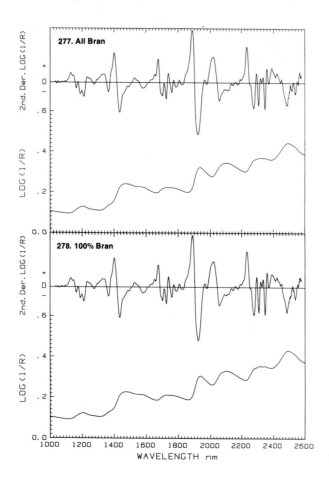

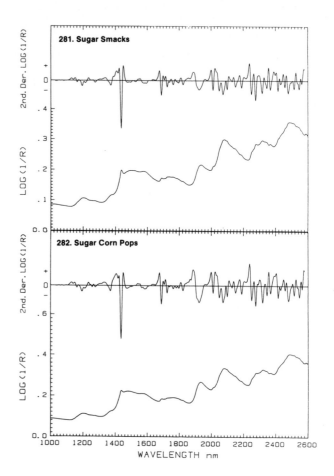

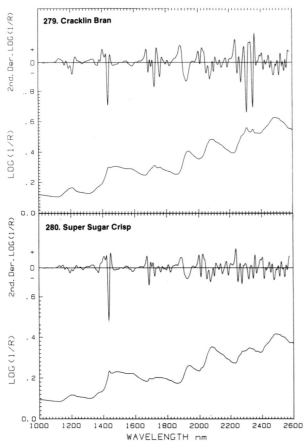

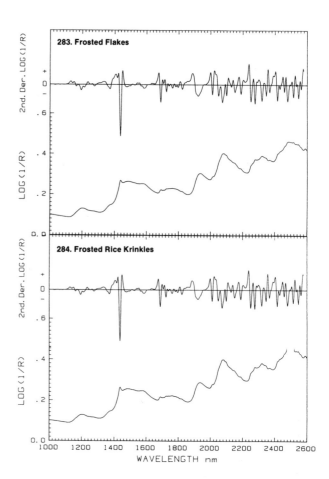

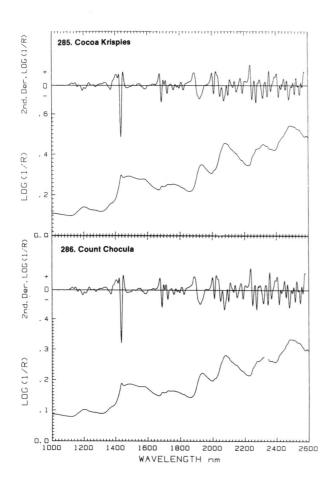

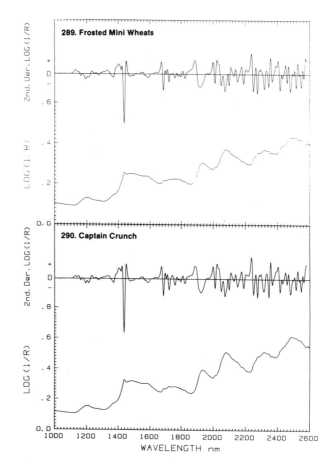

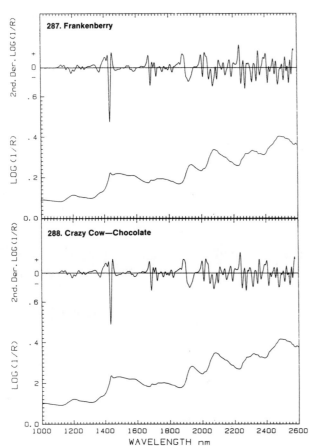

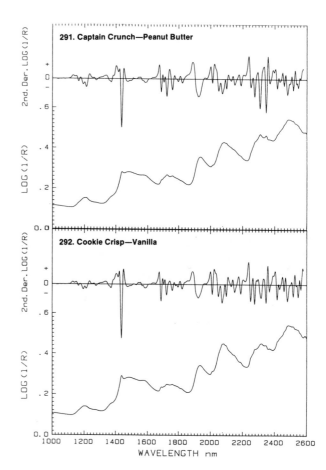

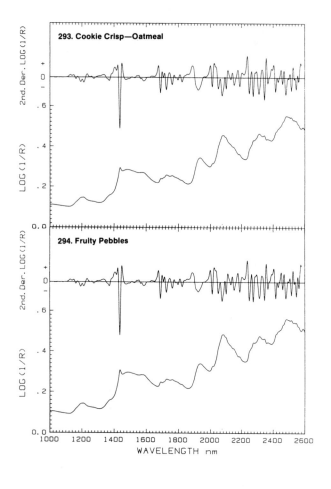

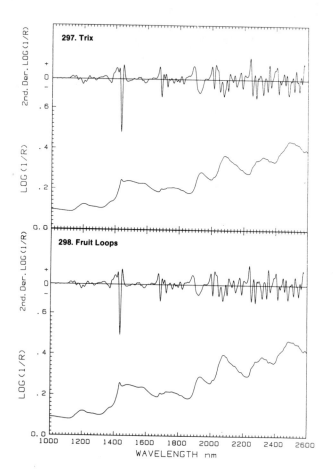

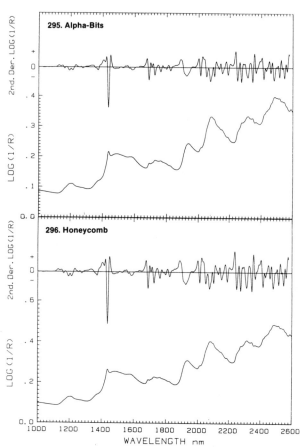

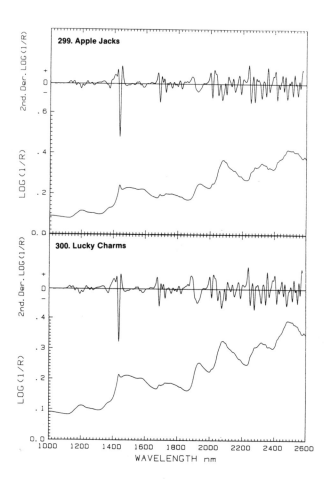

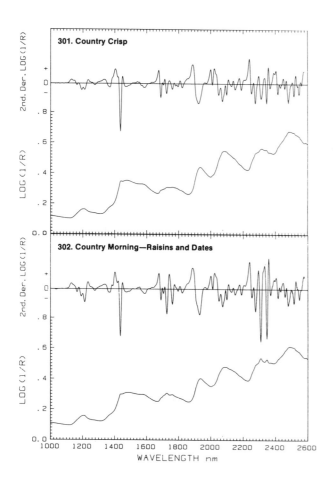

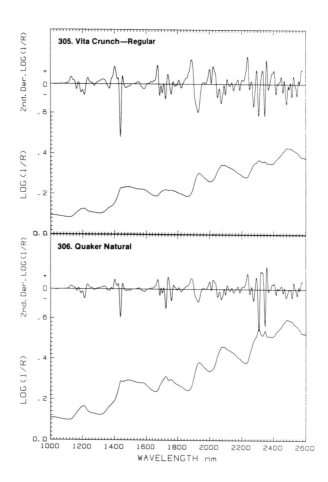

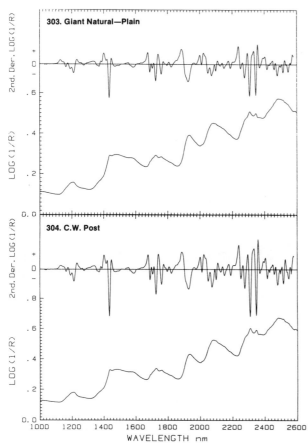

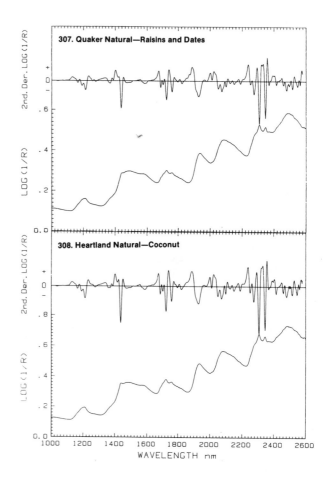

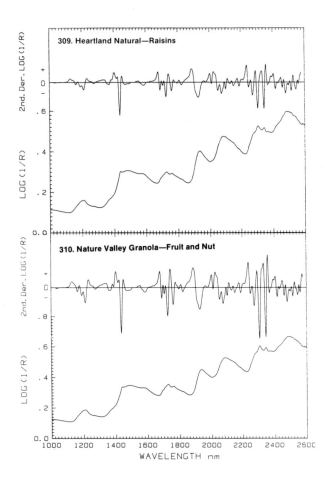

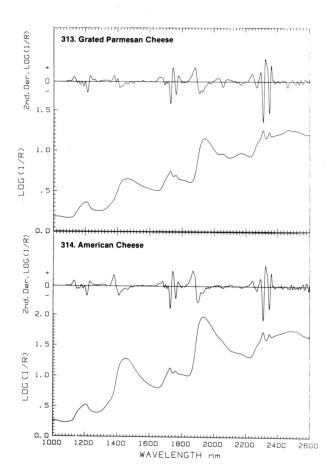

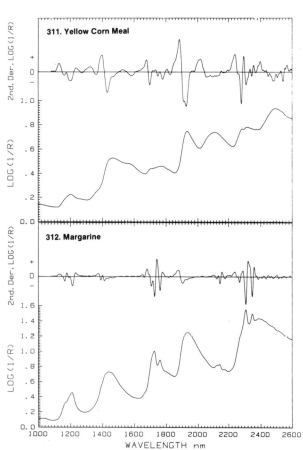

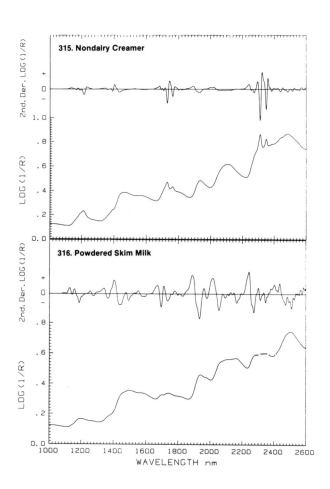

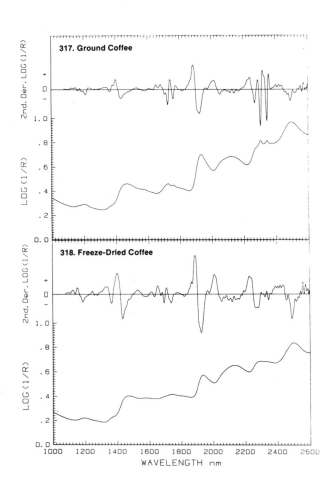

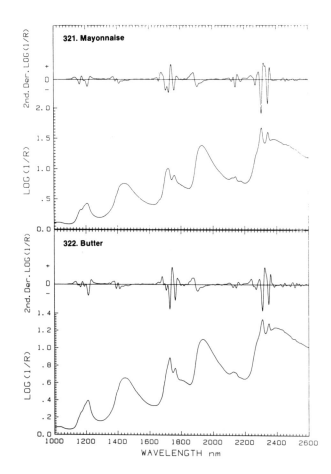

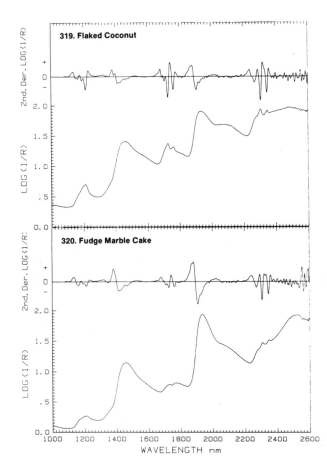

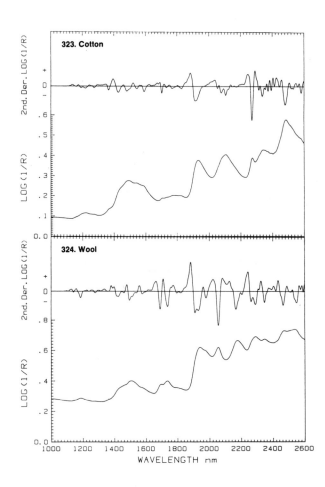

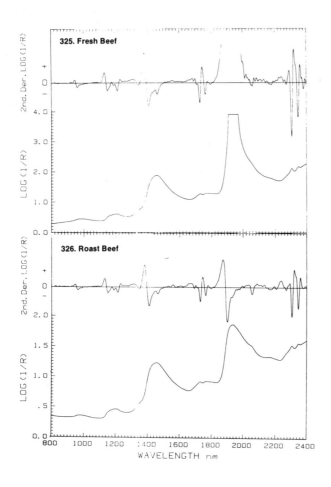

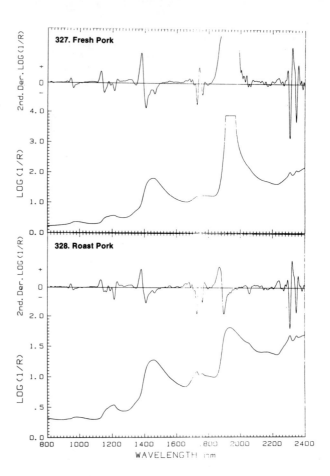

KEY TO NEAR-INFRARED SPECTRA

rs refer to spectrum numbers in the section beginning on page 239.

AMINO ACIDS
D-Alanine, 1
L-Alanine, 2
L-Arginine, 3
L-Asparagine, 4
L-Aspartic acid, 5
L-Cysteine, 6
L-Cystine, 7
L-Glutamic acid, 8
L-Glutamine, 9
Glycine, 10
L-Histidine, 11
L-Hydroxyproline, 25
L-Isoleucine, 12
L-Leucine, 13
D-Lysine, 14
L-Lysine, 15
L-Methionine, 16
DL-Phenylalanine, 18
L-Phenylalanine, 17
L-Proline, 19
L-Serine, 20
L-Threonine, 21
L-Tryptophan, 22
L-Tyrosine, 23
L-Valine, 24

BREAKFAST CEREALS
All Bran, 277
Alpha-Bits, 295
Apple Jacks, 299
Bran, 100%, 278
Bran Chex, 267
Bran Flakes, 40%, 276
Captain Crunch, 290
Captain Crunch-peanut butter, 291
Cheerios, 255
Cocoa Krispies, 285
Cookie Crisp-oatmeal, 293
Cookie Crisp-vanilla, 292
Corn Chex, 270
Corn Flakes, 273
Count Chocula, 286
Country Crisp, 301
Country Morning-raisins and dates, 302
Cracklin Bran, 279
Crazy Cow-chocolate, 288
C. W. Post, 304
Fortified Oat Flakes, 264
Frankenberry, 287
Frosted Flakes, 283
Frosted Mini Wheats, 289
Frosted Rice Krinkles, 284
Fruit Loops, 298
Fruity Pebbles, 294
Giant Natural-plain, 303
Golden Grahams, 254
Grape Nuts, 266
Heartland Natural-coconut, 308
Heartland Natural-raisins, 309
High Pro Concentrate, 40%, 263

Honeycomb, 296
Kellogg's Raisin Bran, 275
Lucky Charms, 300
Most, 260
Nature Valley Granola-fruit and nut, 310
Post Raisin Bran, 274
Post Toasties, 271
Puffed Rice, 257
Puffed Wheat, 258
Quaker Life, 253
Quaker Natural, 306
Quaker Natural—raisins and dates, 307
Rice Chex, 269
Rice Krispies, 256
Shredded Wheat, 265
Special K, 261
Sugar Corn Pops, 282
Sugar Smacks, 281
Super Sugar Crisp, 280
Team, 259
Total, 262
Trix, 297
Vita Crunch-regular, 305
Wheat Chex, 268
Wheaties, 272

CHEMICALS
Alumina, absorption, 125
Aluminum sulfate, 163
Ammonium acetate, 126
Ammonium chloride, 136
Ammonium persulfate, 160
Ammonium sulfate, 164
Ammonium thiocyanate, 177
Barium chloride, 137
Calcium carbonate, 133
Calcium chloride #401, 138
Calcium chloride #401, after drying, 139
Calcium chloride #481, 140
Calcium chloride #481, after drying, 141
Calcium chloride #481, after still further
   drying, 142
Calcium sulfate, 167
Ceramic, ether-washed, 181
Ceramic, washed, 179
Ceramic, water-washed, 180
Chromium potassium sulfate, 168
Copper chloride, 143
Cupric carbonate, 134
Cupric sulfate, $CuSO_4$, $5H_2O$, 169
Cuprous chloride, 144
Epolene, 182
Ferric chloride, 145
Ferrous ammonium sulfate, 166
Ferrous sulfate, 165
Glass beads, 184
Halon, 185
Manganous chloride, 146
Manganous sulfate, 170
Nickelous sulfate, 171
Polystyrene, 183

Potassium borate, 131
Potassium bromide, 132
Potassium chloride, 147
Potassium ferrocyanide, 152
Potassium iodate, 153
Potassium iodide #421, 154
Potassium iodide #491, 155
Potassium nitrate, 157
Potassium permanganate, 159
Potassium persulfate, 161
Potassium sodium tartrate, 176
Sodium acetate, 127
Sodium bicarbonate, 129
Sodium bisulfite, 130
Sodium carbonate, 135
Sodium chloride, 148
Sodium chloride #451, 149
Sodium citrate, 151
Sodium lauryl sulfate, 173
Sodium molybdate, 156
Sodium nitrite, 158
Sodium phosphate, 162
Sodium pyrophosphate, 175
Sodium sulfate, 172
Sulfur, 186
Tris(hydroxymethyl)aminomethane, 178
Zinc acetate, 128
Zinc chloride, 150
Zinc sulfate, 174

MISCELLANEOUS COMMODITIES
Alfalfa, 187
Alfalfa fiber, 192
American cheese, 314
Ascorbic acid, 246
Barley, 188
Barley malt, 233
Barley straw, 215
Beef, fresh, 325
Beef, roast, 326
Bromegrass, 198
Butter, 322
Cake, fudge marble, 320
Cellulose, Adsorbosil, 222
Cellulose, Sigmacell, 220
Cellulose, Sigmacell, dry, 221
Chickpea (garbanzo bean), 234
Clover screenings, 202
Coconut, flaked, 319
Coconut flour, 196
Coffee, freeze-dried, 318
Coffee, ground, 317
Corn, 197
Corn fiber, 193
Corn meal, yellow, 311
Corn starch, 244
Cotton, 323
Cottonseed, 213
Creamer, nondairy, 315
Distiller's grains, dried, 189
Durum wheat, 242

APPENDIX A:
SPECTRA OF AGRICULTURAL PRODUCTS
AND BY-PRODUCTS

The spectra in Appendix A were recorded on a Foss NIRSystems model 6500 scanning spectrophotometer. The second derivative of log(1/R) (D2OD) was determined with a segment of 5 and a gap of 10 wavelength points on NSAS software. A key to the spectra is on page 289.

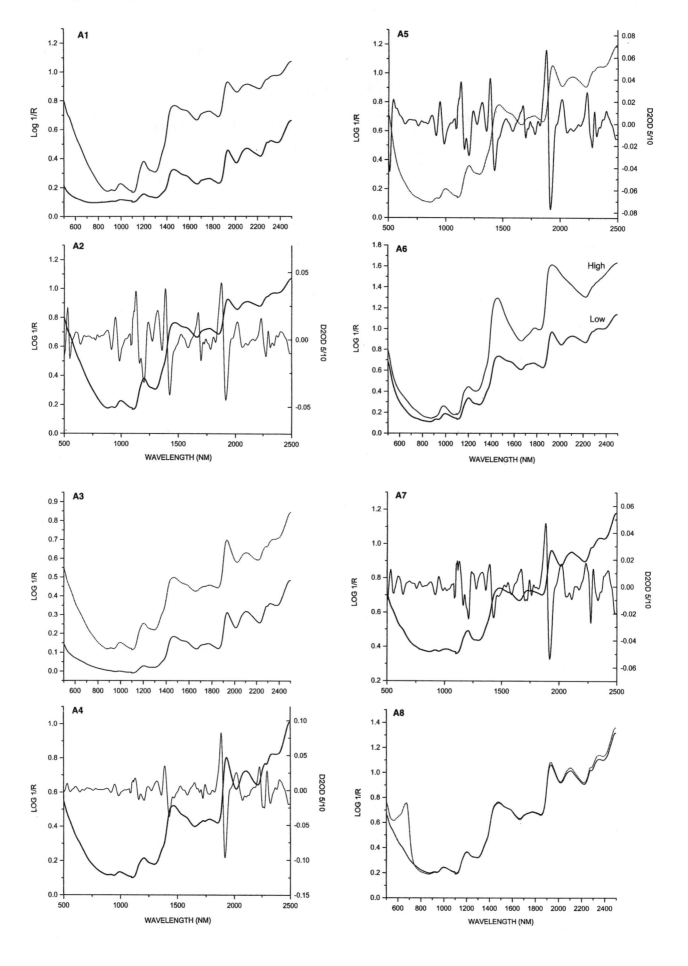

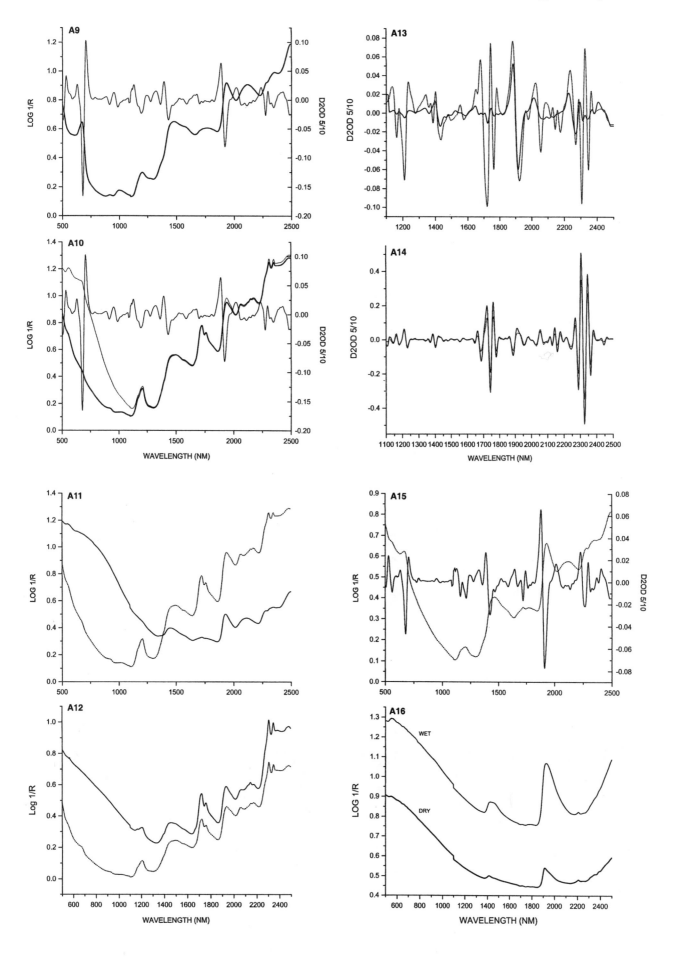

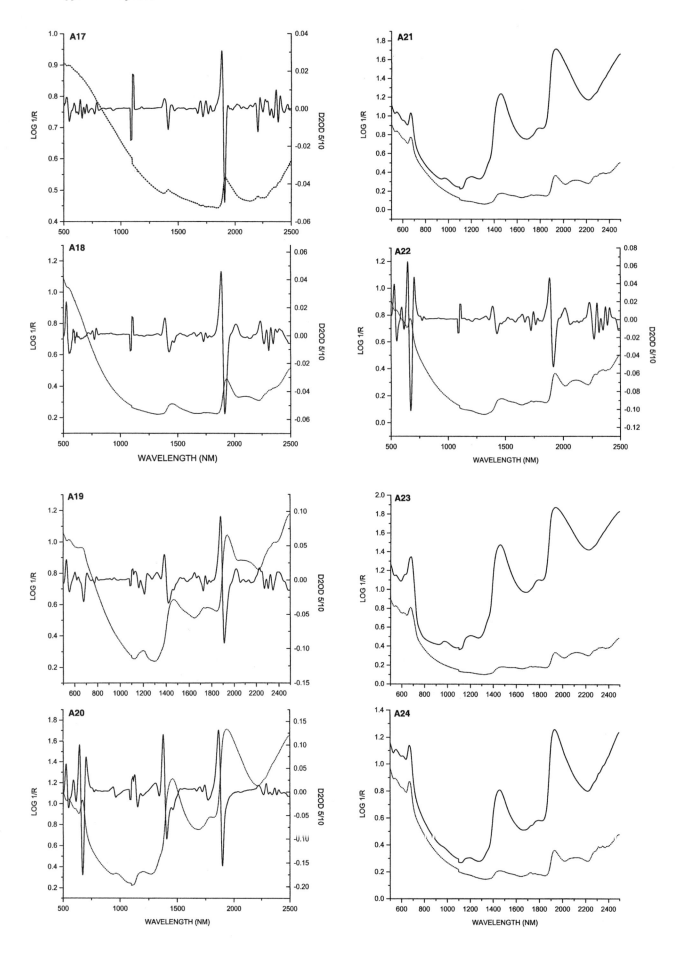

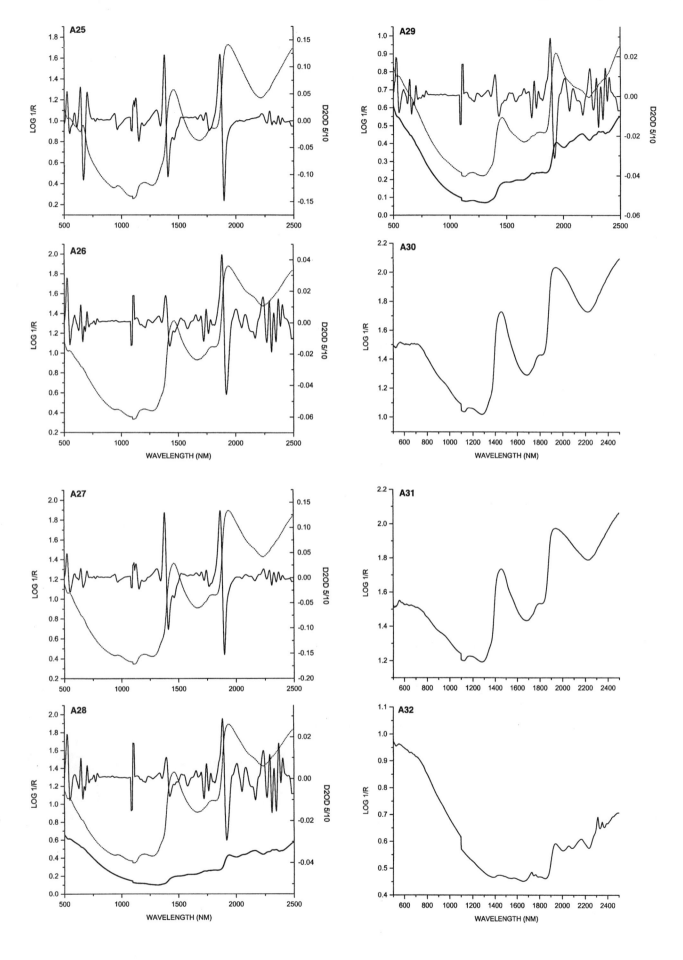

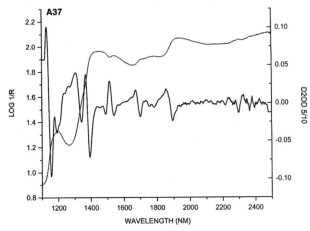

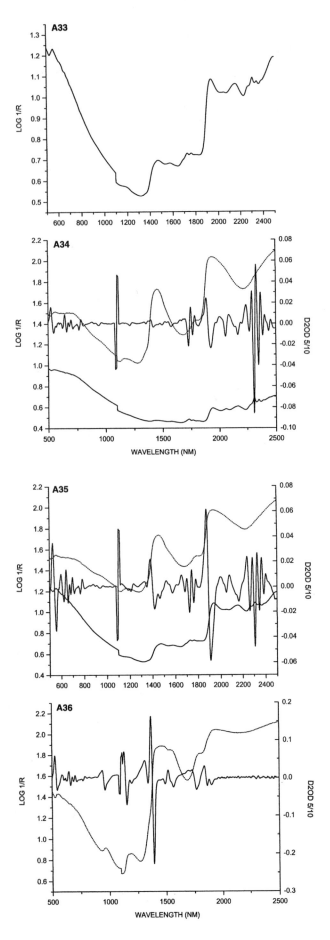

KEY TO SPECTRA IN APPENDIX A

A1. Hard red spring wheat, whole (upper) and ground (lower).

A2. Hard red spring wheat, log(1/*R*) (heavy solid line) and D2OD (solid line).

A3. Barley, whole (dotted line) and ground (solid line).

A4. Oats, whole (heavy solid line) and D2OD (solid line).

A5. Corn, whole (dotted line) and D2OD (solid line).

A6. Corn, high and low moisture.

A7. Soybean, whole (heavy solid line) and D2OD (solid line).

A8. Peas, whole (dotted line) and yellow (solid line).

A9. Lentil (Laird), whole, log(1/*R*) (heavy solid line) and D2OD (solid line), illustrating chlorophyll at 680 nm.

A10. Canola, black (dotted line) and yellow (heavy solid line). The D2OD (solid line) illustrates chlorophyll at 674 nm.

A11. Sunflower (solid line) and canola (dotted line) whole seeds. Note "rabbit ears" on canola that are not present on sunflower.

A12. Sunflower (solid line) and canola (dotted line) ground seeds. Note "rabbit ears" on both.

A13. Sunflower (solid line) and canola (dotted line) whole-seed D2OD. Note prominent oil bands on canola that are not present on sunflower. Both samples contained >50% oil.

A14. Sunflower (solid line) and canola (dotted line) ground-seed D2OD. Note prominent oil bands on both spectra.

A15. Hempseed, whole (dotted line) and D2OD (solid line).

A16. Soil, field, wet and air-dried.

A17. Soil, air-dried log(1/*R*) (dotted line) and D2OD (solid line). Note predominance of water and CH bands.

A18. Peat moss log(1/*R*) (dotted line) and D2OD (solid line). CH bands are different from those in soil.

A19. Borage, log(1/*R*) (dotted line) and D2OD (solid line).

A20. Manure, dairy cow,[a] log(1/*R*) (dotted line) and D2OD (solid line).

A21. Manure, dairy cow,[a] as received (solid line) and air-dried (dotted line).

A22. Manure, dairy cow,[a] air-dried (dotted line) and D2OD (solid line).

A23. Manure, Canada goose,[a] as received and air-dried, log(1/*R*).

A24. Manure, deer,[a] as received and air-dried, log(1/*R*).

A25. Manure, horse,[a] as received, log(1/*R*) (dotted line) and D2OD (solid line).

A26. Manure, turkey, log(1/*R*) (as received, dotted line) and D2OD (air-dried, solid line).

A27. Manure, laying chicken, as received, log(1/*R*) (dotted line) and D2OD (solid line).

A28. Manure, laying chicken, as received (dotted line), air-dried log(1/*R*) (heavy solid line), with air-dried D2OD (solid line).

A29. Manure, broiler chicken, as received (dotted line), air-dried log(1/*R*) (heavy solid line), with air-dried D2OD (solid line).

A30. Manure, hog,[b] as received, log(1/*R*).

A31. Manure, hog, feeder,[b] as received, log(1/*R*).

A32. Manure, hog,[b] air-dried, log(1/*R*).

A33. Manure, hog, feeder,[b] air-dried, log(1/*R*).

A34. Manure, hog,[b] as received (dotted line), air-dried log(1/*R*) (lower solid line), and air-dried D2OD (solid line). Note prominent CH (oil or fat) bands.

A35. Manure, hog, feeder,[b] as received (dotted line), air-dried log(1/*R*) (lower solid line), and air-dried D2OD (solid line).

A36. Manure, hog,[b] from manure pond (1.6%) dry matter, log(1/*R*) (dotted line) and D2OD (solid line).

A37. Fish eggs,[c] log(1/*R*) (dotted line) and D2OD (solid line).

[a] The dairy cow, Canada goose, deer, and horse manures all show chlorophyll at different wavelengths: dairy cow (672 nm), goose, (686 nm), deer (676 nm), and horse (668 nm).
[b] "Hog" manure was from pregnant sows and "hog, feeder" from young, growing hogs.
[c] The fish eggs were scanned from 1,100 to 2,500 nm in a Foss NIRSystems model 5000 instrument.

INDEX